Nonlinear Ordinary Differential Equations: Problems and Solutions

A Sourcebook for Scientists and Engineers

D. W. Jordan and P. Smith

UNIVERSITY PRESS

OXFORD
UNIVERSITY PRESS

Great Clarendon Street, Oxford OX2 6DP

Oxford University Press is a department of the University of Oxford.
It furthers the University's objective of excellence in research, scholarship,
and education by publishing worldwide in

Oxford New York

Auckland Cape Town Dar es Salaam Hong Kong Karachi
Kuala Lumpur Madrid Melbourne Mexico City Nairobi
New Delhi Shanghai Taipei Toronto

With offices in

Argentina Austria Brazil Chile Czech Republic France Greece
Guatemala Hungary Italy Japan Poland Portugal Singapore
South Korea Switzerland Thailand Turkey Ukraine Vietnam

Oxford is a registered trade mark of Oxford University Press
in the UK and in certain other countries

Published in the United States
by Oxford University Press Inc., New York

© D. W. Jordan & P. Smith, 2007

The moral rights of the authors have been asserted
Database right Oxford University Press (maker)

First published 2007

All rights reserved. No part of this publication may be reproduced,
stored in a retrieval system, or transmitted, in any form or by any means,
without the prior permission in writing of Oxford University Press,
or as expressly permitted by law, or under terms agreed with the appropriate
reprographics rights organization. Enquiries concerning reproduction
outside the scope of the above should be sent to the Rights Department,
Oxford University Press, at the address above

You must not circulate this book in any other binding or cover
and you must impose the same condition on any acquirer

British Library Cataloguing in Publication Data

Data available

Library of Congress Cataloging in Publication Data

Data available

Typeset by Newgen Imaging Systems (P) Ltd., Chennai, India
Printed in Great Britain
on acid-free paper by
Biddles Ltd., King's Lynn, Norfolk

ISBN 978–0–19–921203–3

10 9 8 7 6 5 4 3 2

Preface

This handbook contains more than 500 fully solved problems, including 272 diagrams, in qualitative methods for nonlinear differential equations. These comprise all the end-of-chapter problems in the authors' textbook *Nonlinear Ordinary Differential Equations* (4th edition), Oxford University Press (2007), referred to as NODE throughout the text. Some of the questions illustrate significant applications, or extensions of methods, for which room could not be found in NODE.

The solutions are arranged according to the chapter names question-numbering in NODE. Each solution is headed with its associated question. The wording of the problems is the same as in the 4th edition except where occasional clarification has been necessary. Inevitably some questions refer to specific sections, equations and figures in NODE, and, for this reason, the handbook should be viewed as a supplement to NODE. However, many problems can be taken as general freestanding exercises, which can be adapted for coursework, or used for self-tuition.

The development of mathematics computation software in recent years has made the subject more accessible from a numerical and graphical point of view. In NODE and this handbook, Mathematica^{TM} has been used extensively (however the text is not dependent on this software), but there are also available other software and dedicated packages. Such programs are particularly useful for displaying phase diagrams, and for manipulating trigonometric formulae, calculating perturbation series and for handling other complicated algebraic processes.

We can sympathize with readers of earlier editions who worked through the problems, and we are grateful to correspondents who raised queries about questions and answers. We hope that we have dealt with their concerns. We have been receiving requests for the solutions to individual problems and for a solutions manual since the first edition. This handbook attempts to meet this demand (at last!), and also gave us the welcome opportunity to review and refine the problems.

This has been a lengthy and complex operation, and every effort has been made to check the solutions and our LaTeX typesetting. We wish to express our thanks to the School of Computing and Mathematics, Keele University for the use of computing facilities, and to Oxford University Press for the opportunity to make available this supplement to *Nonlinear Ordinary Differential Equations*.

Dominic Jordan and Peter Smith
Keele, 2007

Contents

The chapter headings are those of *Nonlinear Ordinary Differential Equations* but the page numbers refer to this book. The section headings listed below for each chapter are taken from *Nonlinear Ordinary Differential Equations*, and are given for reference and information.

1 Second-order differential equations in the phase plane   1

Phase diagram for the pendulum equation • Autonomous equations in the phase plane • Mechanical analogy for the conservative system $\ddot{x} = f(x)$ • The damped linear oscillator • Nonlinear damping: limit cycles • Some applications • Parameter-dependent conservative systems • Graphical representation of solutions

2 Plane autonomous systems and linearization   63

The general phase plane • Some population models • Linear approximation at equilibrium points • The general solution of linear autonomous plane systems • The phase paths of linear autonomous plane systems • Scaling in the phase diagram for a linear autonomous system • Constructing a phase diagram • Hamiltonian systems

3 Geometrical aspects of plane autonomous systems   133

The index of a point • The index at infinity • The phase diagram at infinity • Limit cycles and other closed paths • Computation of the phase diagram • Homoclinic and heteroclinic paths

4 Periodic solutions; averaging methods   213

An energy-balance method for limit cycles • Amplitude and frequency estimates: polar coordinates • An averaging method for spiral phase paths • Periodic solutions: harmonic balance • The equivalent linear equation by harmonic balance

5 Perturbation methods   251

Nonautonomos systems: forced oscillations • The direct perturbation method for the undamped Duffing equation • Forced oscillations far from resonance • Forced oscillations near resonance with weak excitation • The amplitude equation for the undamped pendulum • The amplitude equation for a damped pendulum • Soft and hard springs • Amplitude-phase perturbation for the pendulum equation • Periodic solutions of autonomous equations (Lindstedt's method) • Forced oscillation of a self-excited equation • The perturbation method and Fourier series • Homoclinic bifurcation: an example

6 Singular perturbation methods   289

Non-uniform approximation to functions on an interval • Coordinate perturbation • Lighthill's method • Time-scaling for series solutions of autonomous equations • The multiple-scale technique applied to saddle points and nodes • Matching approximation on an interval • A matching technique for differential equations

7 Forced oscillations: harmonic and subharmonic response, stability, and entrainment   339

General forced periodic solutions • Harmonic solutions, transients, and stability for Duffing's equation • The jump phenomenon • Harmonic oscillations, stability, and transients for the forced van der Pol equation • Frequency entrainment for the van der Pol equation • Subharmonics of Duffing's equation by perturbation • Stability and transients for subharmonics of Duffing's equation

8 Stability   385

Poincaré stability (stability of paths) • Paths and solution curves for general systems • Stability of time solutions: Liapunov stability • Liapunov stability of plane autonomous linear systems • Structure of the solutions of n-dimensional linear systems • Structure of n-dimensional inhomogeneous linear systems • Stability and boundedness for linear systems • Stability of linear systems with constant coefficients • Linear approximation at equilibrium points for first-order systems in n variables • Stability of a class of nonautonomous linear systems in n dimensions • Stability of the zero solution of nearly linear systems

9 Stabilty by solution perturbation: Mathieu's equation   417

The stability of forced oscillations by a solution perturbation • Equations with periodic coefficients (Floquet theory) • Mathieu's equation arising from a Duffing equation • Transition curves for Mathieu's equation by perturbation • Mathieu's damped equation arising from a Duffing equation

10 Liapunov methods for determining stability of the zero solution   449

Introducing the Liapunov method • Topograhic systems and the Poincaré-Bendixson theorem • Liapunov stability of the zero solution • Asymptotic stability of the zero solution • Extending weak Liapunov functions to asymptotic stability • A more general theory for autonomous systems • A test for instability of the zero solution: n dimensions • Stability and the linear approximation in two dimensions • Exponential function of a matrix • Stability and the linear approximation for nth order autonomous systems • Special systems

11 The existence of periodic solutions   485

The Poincaré-Bendixson theorem and periodic solutions • A theorem on the existence of a centre • A theorem on the existence of a limit cycle • Van der Pol's equation with large parameter

12 Bifurcations and manifolds   497

Examples of simple bifurcations • The fold and the cusp • Further types of bifurcation • Hopf bifrcations • Higher-order systems: manifolds • Linear approximation: centre manifolds

13 Poincaré sequences, homoclinic bifurcation, and chaos   533

Poincaré sequences • Poincaré sections for non-autonomous systems • Subharmonics and period doubling • Homoclinic paths, strange attractors and chaos • The Duffing oscillator • A discrete system: the logistic difference equation • Liapunov exponents and difference equations • Homoclinic bifurcation for forced systems • The horseshoe map • Melnikov's method for detecting homoclinic bifurcation • Liapunov's exponents and differential equations • Power spectra • Some characteristic features of chaotic oscillations

References   585

1 Second-order differential equations in the phase plane

• 1.1 Locate the equilibrium points and sketch the phase diagrams in their neighbourhood for the following equations:

(i) $\ddot{x} - k\dot{x} = 0$.
(ii) $\ddot{x} - 8x\dot{x} = 0$.
(iii) $\ddot{x} = k (|x| > 0)$, $\ddot{x} = 0$ $(|x| < 1)$.
(iv) $\ddot{x} + 3\dot{x} + 2x = 0$.
(v) $\ddot{x} - 4x + 40x = 0$.
(vi) $\ddot{x} + 3|\dot{x}| + 2x = 0$.
(vii) $\ddot{x} + k\,\text{sgn}(\dot{x}) + c\,\text{sgn}(x) = 0$, $(c > k)$. Show that the path starting at $(x_0, 0)$ reaches $((c-k)^2 x_0 / (c+k)^2, 0)$ after one circuit of the origin. Deduce that the origin is a spiral point.
(viii) $\ddot{x} + x\,\text{sgn}(x) = 0$.

1.1. For the general equation $\ddot{x} = f(x, \dot{x})$, (see eqn (1.6)), equilibrium points lie on the x axis, and are given by all solutions of $f(x, 0) = 0$, and the phase paths in the plane (x, y) $(y = \dot{x})$ are given by all solutions of the first-order equation

$$\frac{dy}{dx} = \frac{f(x, y)}{y}.$$

Note that scales on the x and y axes are not always the same. Even though explicit equations for the phase paths can be found for problems (i) to (viii) below, it is often easier to compute and plot phase paths numerically from $\ddot{x} = f(x, \dot{x})$, if a suitable computer program is available. This is usually achieved by solving $\dot{x} = y$, $\dot{y} = f(x, y)$ treated as simultaneous differential equations, so that $(x(t), y(t))$ are obtained parametrically in terms of t. The phase diagrams shown here have been computed using *Mathematica*.

(i) $\ddot{x} - k\dot{x} = 0$. In this problem $f(x, y) = ky$. Since $f(x, 0) = 0$ for all x, the whole x axis consists of equilibrium points. The differential equation for the phase paths is given by

$$\frac{dy}{dx} = k.$$

The general solution is $y = kx + C$, where C is an arbitrary constant. The phase paths for $k > 0$ and $k < 0$ are shown in Figure 1.1.

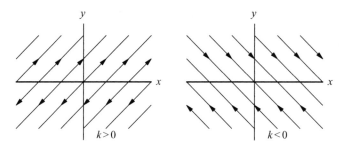

Figure 1.1 Problem 1.1(i): $\ddot{x} - k\dot{x}$.

(ii) $\ddot{x} - 8x\dot{x} = 0$. In this problem $f(x,y) = 8xy$. Since $f(x,0) = 0$, every point on the x axis is an equilibrium point. The differential equation for the phase paths is given by

$$\frac{dy}{dx} = 8x,$$

which has the general solution $y = 4x^2 + C$. The phase paths are shown in Figure 1.2.

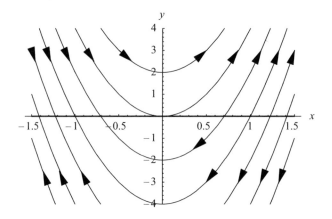

Figure 1.2 Problem 1.1(ii): $\ddot{x} - 8x\dot{x} = 0$.

(iii) $\ddot{x} = k$ ($|x| > 1$); $\ddot{x} = 0$ ($|x| < 1$). In this problem

$$f(x,y) = \begin{cases} k & (|x| > 1) \\ 0 & (|x| < 1). \end{cases}$$

Since $f(x,0) = 0$ for $|x| < 1$, but is non-zero outside this interval, all points in $|x| < 1$ on the x axis are equilibrium points. The differential equations for the phase paths are given by

$$\frac{dy}{dx} = 0, \ (|x| < 1), \quad \frac{dy}{dx} = \frac{k}{y}, \ (|x| > 1).$$

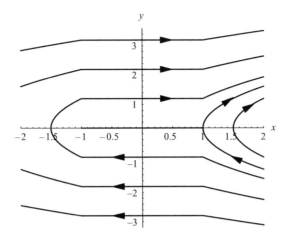

Figure 1.3 Problem 1.1(iii): $\ddot{x} = k \; (|x| > 1); \; \ddot{x} = 0 \; (|x| < 1)$.

Hence the families of paths are

$$y = C, \; (|x| < 1), \quad \tfrac{1}{2}y^2 = kx + C, \; (|x| > 1).$$

Some paths are shown in Figure 1.3 (see also Section 1.4 in NODE).

(iv) $\ddot{x} + 3\dot{x} + 2x = 0$. In this problem $f(x, y) = -2x - 3y$, and there is a single equilibrium point, at the origin. This is a linear differential equation which exhibits strong damping (see Section 1.4) so that the origin is a node. The equation has the characteristic equation

$$m^2 + 3m + 2 = 0, \;\; \text{or} \;\; (m+1)(m+2) = 0.$$

Hence the parametric equations for the phase paths are

$$x = Ae^{-t} + Be^{-2t}, \quad y = \dot{x} = -Ae^{-t} - 2Be^{-2t}.$$

The node is shown in Figure 1.4.

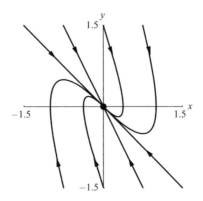

Figure 1.4 Problem 1.1(iv): $\ddot{x} + 3\dot{x} + 2x = 0$, stable node.

4 Nonlinear ordinary differential equations: problems and solutions

(v) $\ddot{x} - 4\dot{x} + 40x = 0$. In this problem $f(x, y) = -40x + 4y$, and there is just one equilibrium point, at the origin. From the results in Section 1.4, this equilibrium point is an unstable spiral. The general solution is

$$x = e^{2t}[A \cos 6t + B \sin 6t],$$

from which y can be found. Spiral paths are shown in Figure 1.5.

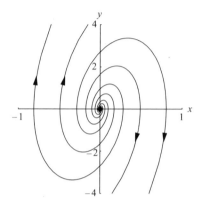

Figure 1.5 Problem 1.1(v): $\ddot{x} - 4\dot{x} + 40x = 0$, unstable spiral.

(vi) $\ddot{x} + 3|\dot{x}| + 2x = 0$, $f(x, y) = -2x - 3|y|$. There is a single equilibrium point, at the origin. The phase diagram is a combination of a stable node for $y > 0$ and an unstable node for $y < 0$ as shown in Figure 1.6. The equilibrium point is unstable.

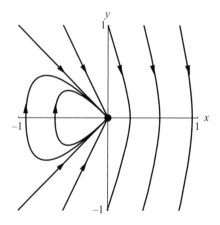

Figure 1.6 Problem 1.1(vi): $\ddot{x} + 3|\dot{x}| + 2x = 0$.

(vii) $\ddot{x} + k\,\text{sgn}\,(\dot{x}) + c\,\text{sgn}\,(x) = 0$, $c > k$. Assume that $k > 0$ and $x_0 > 0$. In this problem $f(x, y) = -k\,\text{sgn}\,(y) - c\,\text{sgn}\,(x)$, and the system has one equilibrium point, at the origin.

By writing
$$y\frac{dy}{dx} = \frac{1}{2}\frac{d}{dx}(y^2)$$
in the equation for the phase paths we obtain
$$y^2 = 2[-k\,\mathrm{sgn}\,(y) - c\,\mathrm{sgn}\,(x)]x + C,$$
where C is a constant. The value of C is assigned separately for each of the four quadrants into which the plane is divided by the coordinate axes, using the requirement that the composite phase paths should be continuous across the axes. For the path starting at $(x_0, 0)$, its equation in $x > 0, y < 0$ is
$$y^2 = 2(k-c)x + C_1.$$
Therefore $C_1 = 2(c-k)x_0$.

Continuity of the path into $x < 0, y < 0$ requires
$$y^2 = 2(k+c)x + 2(c-k)x_0.$$
On the axis $y = 0$, $x = -(c-k)x_0/(c+k)$.

The path in the quadrant $x < 0, y > 0$ is
$$y^2 = 2(c-k)x + C_2.$$
By continuity, $C_2 = (c-k)^2 x_0/(c+k)$.

Finally the path in the quadrant $x > 0, y > 0$ is
$$y^2 = -(c+k)x + C_2.$$
This path cuts the positive x axis at $x = x_1 = (c-k)^2 x_0/(c+k)^2$ as required. Since $c > k$, it follows that $x_1 < x_0$. After n circuits $x_n = \gamma^n x_0$ where $\gamma = (c-k)^2/(c+k)^2$. Since $\gamma < 1$, then $x_n \to 0$. Hence the phase diagram (not shown) is a stable spiral made by matching parabolas on the axes.

(viii) $\ddot{x} + x\,\mathrm{sgn}\,(x) = 0$. The system has a single equilibrium point, at the origin, and $f(x, y) = -x\,\mathrm{sgn}\,(x)$. The phase paths are given by
$$y^2 = -x^2 + C_1, \ (x > 0), \quad y^2 = x^2 + C_2, \ (x < 0).$$
The phase diagram is a centre for $x > 0$ joined to a saddle for $x < 0$ as shown in Figure 1.7.

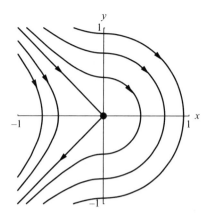

Figure 1.7 Problem 1.1(viii): $\ddot{x} + x\,\text{sgn}(x) = 0$.

• **1.2** Sketch the phase diagram for the equation $\ddot{x} = -x - \alpha x^3$, considering all values of α. Check the stability of the equilibrium points by the method of Section 1.7.

1.2. $\ddot{x} = -x - \alpha x^3$.
Case (i). $\alpha > 0$. The equation has a single equilibrium point, at $x = 1$. The phase paths are given by

$$\frac{dy}{dx} = -\frac{x(1+\alpha x^2)}{y},$$

which is a separable first-order equation. The general solution is given by

$$\int y\,dy = -\int x(1+\alpha x^2)\,dx + C, \tag{i}$$

so that

$$\tfrac{1}{2}y^2 = -\tfrac{1}{2}x^2 - \tfrac{1}{4}x^4 + C.$$

The phase diagram is shown in Figure 1.8 with $\alpha = 1$, and the origin can be seen to be a centre.
Case (ii). $\alpha < 0$. There are now three equilibrium points: at $x = 0$ and at $x = \pm 1/\sqrt{\alpha}$. The phase paths are still given by (i), but computed in this case with $\alpha = -1$ (see Figure 1.9). There is a centre at $(0, 0)$ and saddles at $(\pm 1, 0)$.

This equation is a parameter-dependent system with parameter α as discussed in Section 1.7. As in eqn (1.62), let $f(x,\alpha) = -x - \alpha x^3$. Figure 1.10 shows that in the region above $x = 0$, $f(x,\alpha)$ is positive for all α, which according to Section 1.7 (in NODE) implies that the origin is stable. The other equilibrium points are unstable.

1 : Second-order differential equations in the phase plane   7

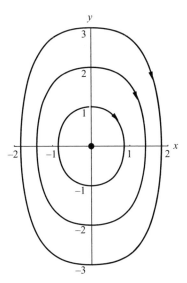

Figure 1.8   Problem 1.2: Phase diagram for $\alpha = 1$.

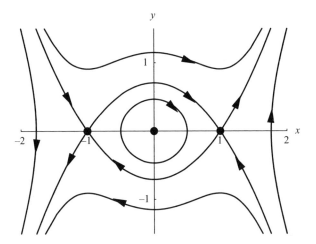

Figure 1.9   Problem 1.2: Phase diagram for $\alpha = -1$.

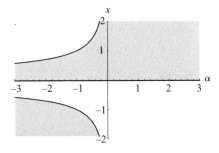

Figure 1.10   Problem 1.2: The diagram shows the boundary $x(1 - \alpha x^2) = 0$; the shaded regions indicate $f(x, \alpha) > 0$.

8 Nonlinear ordinary differential equations: problems and solutions

- 1.3 A certain dynamical system is governed by the equation $\ddot{x}+\dot{x}^2+x=0$. Show that the origin is a centre in the phase plane, and that the open and closed paths are separated by the path $2y^2 = 1 - 2x$.

1.3. $\ddot{x}+\dot{x}^2+x=0$. The phase paths in the (x, y) plane are given by the differential equation

$$\frac{dy}{dx} = \frac{-y^2 - x}{y}.$$

By putting

$$y\frac{dy}{dx} = \frac{1}{2}\frac{d}{dx}(y^2),$$

the equation can be expressed in the form

$$\frac{d(y^2)}{dx} + 2y^2 = -2x,$$

which is a linear equation for y^2. Hence

$$y^2 = Ce^{-2x} - x + \tfrac{1}{2},$$

which is the equation for the phase paths.

The equation has a single equilibrium point, at the origin. Near the origin for y small, $\ddot{x}+x \approx 0$ which is the equation for simple harmonic motion (see Example 1.2 in NODE). This approximation indicates that the origin is a centre.

If the constant $C < 0$, then $Ce^{-2x} \to -\infty$ as $x \to \infty$, which implies that $-x + \tfrac{1}{2} + Ce^{-2x}$ must be zero for a negative value of x. There is also a positive solution for x The paths are closed for $C < 0$ since any path is reflected in the x axis. If $C \geq 0$, then the equation $-x + \tfrac{1}{2} + Ce^{-2x} = 0$ has exactly one solution and this is positive. To see this sketch the line $z = x - \tfrac{1}{2}$ and the exponential curve $z = Ce^{-2x}$ for positive and negative values for C and see where they intersect. The curve bounding the closed paths is the parabola $y^2 = -x + \tfrac{1}{2}$. The phase diagram is shown in Figure 1.11.

- 1.4 Sketch the phase diagrams for the equation $\ddot{x}+e^x=a$, for $a<0$, $a=0$, and $a>0$.

1.4. $\ddot{x}+e^x=a$. The phase paths in the (x, y) plane are given by

$$y\frac{dy}{dx} = a - e^x, \qquad (i)$$

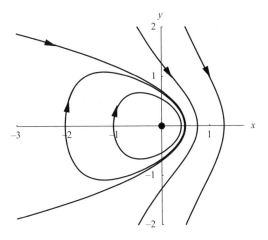

Figure 1.11 Problem 1.3.

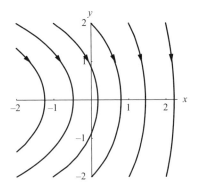

Figure 1.12 Problem 1.4: $a < 0$.

which has the general solution

$$\tfrac{1}{2}y^2 = ax - e^x + C. \tag{ii}$$

Case (a), $a < 0$. The system has no equilibrium points. From (i), dy/dx is never zero, negative for $y > 0$ and positive for $y < 0$. Some phase paths are shown in Figure 1.12.

Case (b), $a = 0$. The system has no equilibrium points. As in (a), dy/dx is never zero. Some phase paths are shown in Figure 1.13.

Case (c), $a > 0$. This equation has one equilibrium point at $x = \ln a$. The potential $V(x)$ (see Section 1.3) of this conservative system is

$$V(x) = \int (-a + e^x) dx = -ax + e^x,$$

which has the expected stationary value at $x = \ln a$. Since $V''(\ln a) = e^{\ln a} = a > 0$, the stationary point is a minimum which implies a centre in the phase diagram. Some phase paths are shown in Figure 1.14 for $a = 2$.

10 Nonlinear ordinary differential equations: problems and solutions

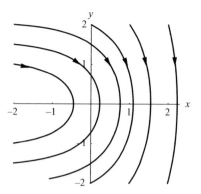

Figure 1.13 Problem 1.4: $a=0$.

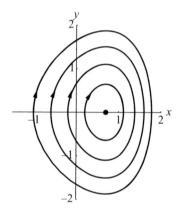

Figure 1.14 Problem 1.4: $a>0$.

- **1.5** Sketch the phase diagrams for the equation $\ddot{x}-e^x=a$, for $a<0$, $a=0$, and $a>0$.

1.5. $\ddot{x}-e^x=a$. The differential equation of the phase paths is given by

$$y\frac{dy}{dx}=a+e^x,$$

which has the general solution

$$\tfrac{1}{2}y^2=ax+e^x+C.$$

Case (a), $a<0$. There is a single equilibrium point, at $x=\ln(-a)$. The potential $\mathcal{V}(x)$ (see Section 1.3 in NODE) of this conservative system is

$$\mathcal{V}(x)=\int(-a-e^x)dx=-ax-e^x,$$

which has the expected stationary value at $x = \ln(-a)$. Since

$$\mathcal{V}''(\ln(-a)) = -e^{\ln(-a)} = a < 0,$$

the stationary point is a maximum, indicating a saddle at $x = \ln(-a)$. Some phase paths are shown in Figure 1.15.

Case (b), $a > 0$. The equation has no equilibrium points. Some typical phase paths are shown in Figure 1.16.

Case (c), $a = 0$. Again the equation has no equilibrium points, and the phase diagram has the main features indicated in Figure 1.16 for the case $a > 0$, that is, phase paths have positive slope for $y > 0$ and negative slope for $y < 0$.

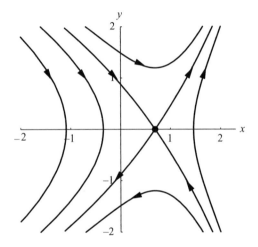

Figure 1.15 Problem 1.5: $a < 0$.

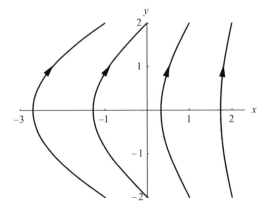

Figure 1.16 Problem 1.5: $a > 0$.

- **1.6** The potential energy $V(x)$ of a conservative system is continuous, and is strictly increasing for $x < -1$, zero for $|x| \leq 1$, and strictly decreasing for $x > 1$. Locate the equilibrium points and sketch the phase diagram for the system.

1.6. From Section 1.3, a system with potential $V(x)$ has the governing equation

$$\ddot{x} = -\frac{dV(x)}{dx}.$$

Equilibrium points occur where $\ddot{x} = 0$, or where $dV(x)/dx = 0$, which means that all points on the x axis such that $|x| \leq 1$ are equilibrium points. Also, the phase paths are given by

$$\tfrac{1}{2}y^2 = V(x) + C.$$

Therefore the paths in the interval $|x| \leq 1$ are the straight lines $y = C$. Since $V(x)$ is strictly increasing for $x < -1$, the paths must resemble the left-hand half of a centre at $x = -1$. In the same way the paths for $x > 1$ must be the right-hand half of a centre. A schematic phase diagram is shown in Figure 1.17.

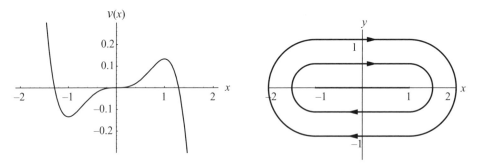

Figure 1.17 Problem 1.6: This diagram shows some phase paths for the equation with $V(x) = x + 1$, $(x < 1)$, $V(x) = -x + 1$, $(x > 1)$.

- **1.7** Figure 1.33 (in NODE) shows a pendulum striking an inclined wall. Sketch the phase diagram of this 'impact oscillator', for α positive and α negative, when (i) there is no loss of energy at impact, (ii) the magnitude of the velocity is halved on impact.

1.7. Assume the approximate pendulum equation (1.1), namely

$$\ddot{\theta} + \omega^2 \theta = 0,$$

and assume that the amplitude of the oscillations does not exceed $\theta = \tfrac{1}{2}\pi$ (thus avoiding any complications arising from impacts above the point of suspension).

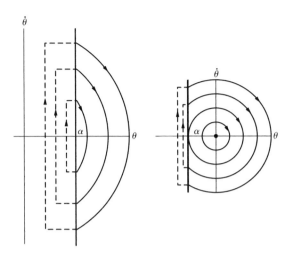

Figure 1.18 Problem 1.7: Perfect rebound for $\alpha > 0$ and $\alpha < 0$.

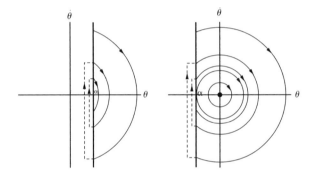

Figure 1.19 Problem 1.7: The rebound speed is half that of the impact speed.

(i) Perfect rebound with no loss of energy. In all cases the phase diagram consists of segments of a centre cut off at $\theta = \alpha$, and the return path after impact will depend on the rebound velocity after impact. The dashed lines in Figure 1.18 indicate the rebound velocity which has the same magnitude as the impact velocity.

(ii) In these phase diagrams (see Figure 1.19) the rebound speed is half that of the impact speed.

- 1.8 Show that the time elapsed, T, along a phase path C of the system $\dot{x} = y$, $\dot{y} = f(x,y)$ is given, in a form alternative to (1.13), by

$$T = \int_C (y^2 + f^2)^{-(1/2)} ds,$$

where ds is an element of distance along C.
By writing $\delta s \approx (y^2 + f^2)^{\frac{1}{2}} \delta t$, indicate, very roughly, equal time intervals along the phase paths of the system $\dot{x} = y$, $\dot{y} = 2x$.

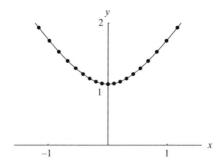

Figure 1.20 Problem 1.8: The phase path $y = (1 + 2x^2)^{1/2}$ is shown with equal time steps $\delta t = 0.1$.

1.8. Let C be a segment of a phase path from A to B, traced out by a representative point P between times T_A and t_B, and $s(t)$ be the arc length along C measured from A to P. Along the path

$$\delta s \approx [(\delta x)^2 + (\delta y)^2]^{1/2},$$

so

$$\frac{\delta s}{\delta t} \approx \left[\left(\frac{\delta x}{\delta t}\right)^2 + \left(\frac{\delta y}{\delta t}\right)^2\right]^{1/2}.$$

In the limit $\delta t \to 0$, the velocity of P is ds/dt given by

$$\frac{ds}{dt} = [\dot{x}^2 + \dot{y}^2]^{1/2}. \tag{i}$$

The transient time T is given by

$$T = t_B - t_A = \int_{t_A}^{t_B} dt = \int_C \frac{ds}{ds/dt} = \int_C \frac{ds}{[\dot{x}^2 + \dot{y}^2]^{1/2}} = \int_C (y^2 + f^2)^{-(1/2)} ds, \tag{ii}$$

since $\dot{x} = y$ and $\dot{y} = f$.

For the case $\dot{x} = y$, $\dot{y} = 2x$ the phase paths consist of the family of hyperbolas $y^2 - 2x^2 = \alpha$, where α is an arbitrary constant. From (i), a small time interval δt corresponds to a step length δs along a phase path given approximately by

$$\delta s \approx [\dot{x}^2 + \dot{y}^2]^{1/2} \delta t = (y^2 + 4x^2)^{1/2} \delta t = (\alpha + 6x^2)^{1/2} \delta t.$$

Given a value of the parameter α, the step lengths for a constant δt are determined by the factor $(\alpha + 6x^2)^{1/2}$, and tend to be comparatively shorter when the phase path is closer to the origin. This is illustrated in Figure 1.20 for the branch $y = (1 - 2x^2)^{1/2}$.

- 1.9 On the phase diagram for the equation $\ddot{x} + x = 0$, the phase paths are circles. Use (1.13) in the form $\delta t \approx \delta x / y$ to indicate, roughly, equal time steps along several phase paths.

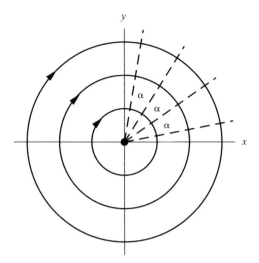

Figure 1.21 Problem 1.9.

1.9. The phase paths of the simple harmonic oscillator $\ddot{x}+x=0$ are given by $dy/dx = -x/y$, which has the general solution $x^2 + y^2 = C^2$, $C > 0$. The paths can be represented parametrically by $x = C\cos\theta$, $y = C\sin\theta$, where θ is the polar angle. By (1.13), an increment in time is given by

$$\delta t \approx \frac{\delta x}{y} = \frac{-C\sin\theta\,\delta\theta}{C\sin\theta} = -\delta\theta.$$

This formula can be integrated to give $t = -\theta + B$. Hence equal time steps are equivalent to equal steps in the polar angle θ. All phase paths are circles centred at the origin and the time taken between radii subtending the same angle, say α, at the origin as shown in Figure 1.21.

• 1.10 Repeat Problem 1.9 for the equation $\ddot{x} + 9x = 0$, in which the phase paths are ellipses.

1.10. The phase paths of $\ddot{x} + 9x = 0$ are given by $dy/dx = -9x/y$, which has the general solution $9x^2 + y^2 = C^2$, $C > 0$. The paths are concentric ellipses. The paths can be represented parametrically by $x = \frac{1}{3}C\cos\theta$, $y = C\sin\theta$, where θ is the polar angle. By (1.13), an increment in time is given by

$$\delta t \approx \frac{\delta x}{y} = \frac{-(1/3)C\sin\theta\,\delta\theta}{C\sin\theta} = -\frac{1}{3}\delta\theta.$$

Hence equal time steps are equivalent on all paths to the lengths of segments cut by equal polar angles α as shown in Figure 1.22.

16 Nonlinear ordinary differential equations: problems and solutions

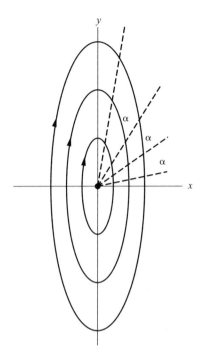

Figure 1.22 Problem 1.10.

- **1.11** The pendulum equation, $\ddot{x} + \omega^2 \sin x = 0$, can be approximated for moderate amplitudes by the equation $\ddot{x} + \omega^2(x - \frac{1}{6}x^3) = 0$. Sketch the phase diagram for the latter equation, and explain the differences between it and Figure 1.2 (in NODE).

1.11. For small $|x|$, the Taylor expansion of $\sin x$ is given by

$$\sin x = x - \tfrac{1}{6}x^3 + O(x^5).$$

Hence for small $|x|$, the pendulum equation $\ddot{x} + \omega^2 \sin x = 0$ can be approximated by

$$\ddot{x} + \omega^2 \left(x - \frac{1}{6}x^3 \right) = 0.$$

If x is unrestricted this equation has three equilibrium points, at $x = 0$ and $x = \pm\sqrt{6} \approx \pm 2.45$. The pendulum equation has equilibrium points at $x = n\pi$, $(n = 0, 1, 2, \ldots)$. Obviously, the approximate equation is not periodic in x, and the equilibrium points at $x = \pm\sqrt{6}$ differ considerably from those of the pendulum equation. We can put $\omega = 1$ without loss since time can always be rescaled by putting $t' = \omega t$. Figure 1.23 shows the phase diagrams for both equations for amplitudes up to 2. The solid curves are phase paths of the approximation and the dashed

1: Second-order differential equations in the phase plane

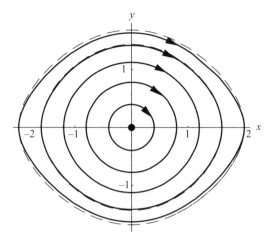

Figure 1.23 Problem 1.11: The solid curves represent the phase paths of the approximate equation $\ddot{x} + x - \frac{1}{6}x^3 = 0$, and the dashed curves show the phase paths of $\ddot{x} + \sin x = 0$.

curves those of the pendulum equation. For $|x| < 1.5$, the phase paths are visually indistinguishable. The closed phase paths indicate periodic solutions, but the periods will increase with increasing amplitude.

- **1.12** The displacement, x, of a spring-mounted mass under the action of Coulomb dry friction is assumed to satisfy
$$m\ddot{x} + cx = -F_0 \text{sgn}(\dot{x}),$$
where m, c and F_0 are positive constants (Section 1.6). The motion starts at $t=0$, with $x = x_0 > 3F_0/c$ and $\dot{x} = 0$. Subsequently, whenever $x = -\alpha$, where $(2F_0/c) - x_0 < -\alpha < 0$ and $\dot{x} > 0$, a trigger operates, to increase suddenly the forward velocity so that the kinetic energy increases by a constant amount E. Show that if $E > 8F_0^2/c$, a periodic motion exists, and show that the largest value of x in the periodic motion is equal to $F_0/c + E/(4F_0)$.

1.12. The equation for Coulomb dry friction is

$$m\ddot{x} + cx = -F_0 \text{sgn}(\dot{x}) = \begin{cases} -F_0 & \dot{x} > 0 \\ F_0 & \dot{x} < 0 \end{cases}.$$

For $\dot{x} = y < 0$, the differential equation for the phase paths is given by

$$m\frac{dy}{dx} = \frac{F_0 - cx}{y}.$$

Integrating this separable equation, we obtain

$$\tfrac{1}{2}my^2 = -\tfrac{1}{2c}(F_0 - cx)^2 + B_1 = \tfrac{1}{2c}(F_0 - cx_0)^2 - \tfrac{1}{2c}(F_0 - cx)^2,$$

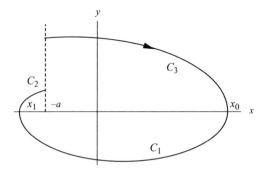

Figure 1.24 Problem 1.12:

using the initial conditions $x(0)=x_0$, $y(0)=0$. This segment of the path is denoted in Figure 1.24 by C_1. It meets the x axis again where

$$F_0 - cx = -F_0 + cx_0, \text{ so that } x = x_1 = \frac{2F_0}{c} - x_0.$$

For $y > 0$, phase paths are given by

$$\frac{1}{2}my^2 = -\frac{1}{2c}(F_0 + cx)^2 + B_2. \qquad (i)$$

Denote this segment by C_2. It is the continuation into $y > 0$ of C_1 from $x = x_1$, $y = 0$. Hence

$$B_2 = \frac{1}{2c}(F_0 + cx_1)^2 = \frac{1}{2c}(3F_0 - x_0)^2,$$

so that

$$x_1 = \frac{2F_0}{c} - x_0.$$

The condition $x_1 = -x_0 + (2F_0/c) < -\alpha$ ensures that the 'trigger' operates within the range of x illustrated.

Denote the segment which meets C_1 at $x = x_0$ by C_3. From (i), its equation is

$$\frac{1}{2}my^2 = -\frac{1}{2c}(F_0+cx)^2 + B_3 = -\frac{1}{2c}(F_0+cx)^2 + \frac{1}{2c}(F_0+cx_0)^2.$$

At $x = -\alpha$, the energy on C_2 is

$$E_2 = \frac{1}{2c}[(3F_0 - cx_0)^2 - (F_0 - c\alpha)^2],$$

whilst on C_3, the energy is

$$E_3 = \frac{1}{2c}[(F_0 + cx_0)^2 - (F_0 - c\alpha)^2].$$

At $x = -\alpha$, the energy increases by E. Therefore

$$E = E_3 - E_2$$
$$= \frac{1}{2c}[(F_0+cx_0)^2 - (F_0-c\alpha)^2] - \frac{1}{2c}[(3F_0-cx_0)^2 - (F_0-c\alpha)^2]$$
$$= \frac{1}{2c}[(F_0+cx_0)^2 - (3F_0-cx_0)^2]$$
$$= \frac{1}{2c}[-8F_0^2 + 8F_0 cx_0]$$

A periodic solution occurs if the initial displacement is

$$x_0 = \frac{E}{4F_0} + \frac{F_0}{c}.$$

Note that the results are independent of α. For a cycle to be possible, we must have $x_0 > 3F_0/c$. Therefore E and F_0 must satisfy the inequality

$$\frac{E}{4F_0} + \frac{F_0}{c} > \frac{3F_0}{c}, \quad \text{or} \quad E > \frac{8F_0^2}{c}.$$

- **1.13** In Problem 1.12, suppose that the energy is increased by E at $x = -\alpha$ for both $\dot{x} < 0$ and $\dot{x} > 0$; that is, there are two injections of energy per cycle. Show that periodic motion is possible if $E > 6F_0^2/c$, and find the amplitude of the oscillation.

1.13. Refer to the previous problem for the equations of the phase paths in $y > 0$ and $y < 0$. The system experiences an increase in kinetic energy for both y positive and y negative. The periodic path consists of four curves whose equations are listed below:

$$C_1: mcy^2 + (F_0 - cx)^2 = (F_0 - cx_0)^2$$
$$C_2: mcy^2 + (F_0 - cx)^2 = (F_0 - cx_1)^2$$
$$C_3: mcy^2 + (F_0 + cx)^2 = (F_0 + cx_1)^2$$
$$C_4: mcy^2 + (F_0 + cx)^2 = (F_0 + cx_0)^2$$

The paths, the point $(x_0, 0)$ where the paths C_1 and C_4 meet, and the point $(x_1, 0)$ where the paths C_2 and C_3 meet are shown in Figure 1.24. At $x = -\alpha$ the energy is increased by E for both positive and negative y. The discontinuities at $x = -\alpha$ are shown in Figure 1.25. Therefore, at $x = -\alpha$,

$$E = \frac{1}{2c}[(F_0-cx_1)^2 - (F_0+c\alpha)^2 - (F_0-cx_0)^2 + (F_0+c\alpha)^2],$$

$$E = \frac{1}{2c}[(F_0+cx_0)^2 - (F_0+c\alpha)^2 - (F_0-cx_1)^2 + (F_0+c\alpha)^2].$$

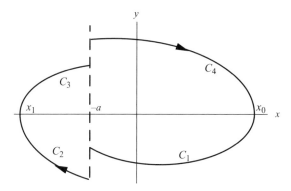

Figure 1.25 Problem 1.13.

Simplifying these results

$$E = \frac{1}{2c}[-2F_0 c x_1 + c^2 x_1^2 + 2F_0 c x_0 - c^2 x_0^2],$$

$$E = \frac{1}{2c}[2F_0 c x_0 + c^2 x_0^2 - 2F_0 c x_1 - c^2 x_1^2],$$

Elimination of E gives $x_1 = -x_0$, and

$$x_0 = -x_1 = \frac{E}{2F_0}.$$

- **1.14** The 'friction pendulum' consists of a pendulum attached to a sleeve, which embraces a close-fitting cylinder (Figure 1.34 in NODE). The cylinder is turned at a constant rate $\Omega > 0$. The sleeve is subject to Coulomb dry friction through the couple $G = -F_0 \text{sgn}(\dot{\theta} - \Omega)$. Write down the equation of motion, the equilibrium states, and sketch the phase diagram.

1.14. Taking moments about the spindle, the equation of motion is

$$mga \sin \theta + F_0 \text{sgn}(\dot{\theta} - \Omega) = -ma^2 \ddot{\theta}.$$

Equilibrium positions of the pendulum occur where $\ddot{\theta} = \dot{\theta} = 0$, that is where

$$mga \sin \theta - F_0 \text{sgn}(-\Omega) = mga \sin \theta + F_0 = 0,$$

assuming that $\Omega > 0$. Assume also that $F_0 > 0$. The differential equation is invariant under the change of variable $\theta' = \theta + 2n\pi$ so all phase diagrams are periodic with period 2π in θ.
If $F_0 < mga$, there are two equilibrium points; at

$$\theta = \sin^{-1}\left(\frac{F_0}{mga}\right) \text{ and } \pi - \sin^{-1}\left(\frac{F_0}{mga}\right):$$

note that in the second case the pendulum bob is above the sleeve.

The phase diagram with the parameters $\Omega = 1$, $g/a = 2$ and $F_0/(ma^2) = 1$ is shown in Figure 1.26. There is a centre at $\theta = \sin^{-1}(\frac{1}{2})$ and a saddle point at $x = \pi - \sin^{-1}(\frac{1}{2})$. Discontinuities in the slope occur on the line $\dot\theta = \Omega = 1$. On this line between $\theta = \sin^{-1}(-\frac{1}{2})$ and $\theta = \sin^{-1}(\frac{1}{2})$, phase paths meet from above and below in the positive direction of θ.

Suppose that a representative point P arrives somewhere on the segment AB in Figure 1.26. The angular velocity at this point is given by $\dot\theta(t) = \Omega$ (i.e. it is in time with the rotation of the spindle at this point). It therefore turns to move along AB, in the direction of increasing θ. It cannot leave AB into the regions $\dot\theta > \Omega$ or $\dot\theta < \Omega$ since it must not oppose the prevailing directions. Therefore the representative point continues along AB with constant velocity Ω, apparently 'sticking' to the spindle, until is arrives at B where it is diverted on to the ellipse. Its subsequent motion is then periodic.

If $F_0 = mga$ there is one equilibrium position at $\theta = \frac{1}{2}\pi$, in which the pendulum is horizontal. In this critical case the centre and the saddle merge at $\theta = \frac{1}{2}\pi$ so that the equilibrium point is a hybrid centre/saddle point.

If $F_0 > mga$, there are no equilibrium positions. The phase diagram for $\Omega = 1$, $g/a = 1$ and $F_0/(ma^2) = 2$ is shown in Figure 1.27. All phase paths approach the line $\dot\theta = \Omega$, which is a

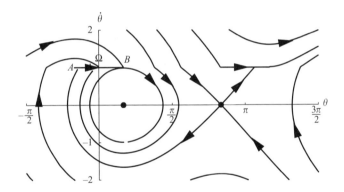

Figure 1.26 Problem 1.14: Typical phase diagram for the friction-driven pendulum for $F_0 < mga$.

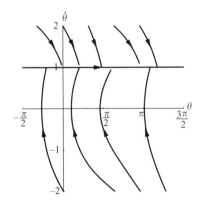

Figure 1.27 Problem 1.14: Typical phase diagram for the friction-driven pendulum for $F_0 > mga$.

- **1.15** By plotting 'potential energy' of the nonlinear conservative system $\ddot{x} = x^4 - x^2$, construct the phase diagram of the system. A particular path has the initial conditions $x = \frac{1}{2}$, $\dot{x} = 0$ at $t = 0$. Is the subsequent motion periodic?

1.15. From NODE, (1.29), the potential function for the conservative system defined by

$$\ddot{x} = x^4 - x^2$$

is given by

$$\mathcal{V}(x) = -\int (x^4 - x^2) dx = \frac{1}{3}x^3 - \frac{1}{5}x^5.$$

Its graph is shown in the upper diagram in Figure 1.28. The system has three equilibrium points: at $x = 0$ and $x = \pm 1$. The equilibrium point at $x = -1$ corresponds to a minimum of the potential function which generates a centre in the phase diagram, and there is a maximum at $x = 1$ which implies a saddle point. The origin is a point of inflection of $\mathcal{V}(x)$. Near the origin $\ddot{x} = -x^2$ has a cusp in the phase plane. The two phase paths from the origin are given

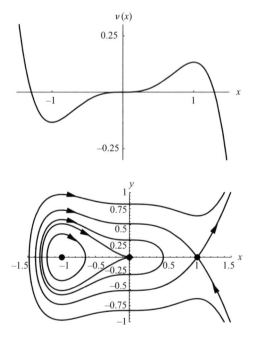

Figure 1.28 Problem 1.15: Potential energy and phase diagram for the conservative system $\ddot{x} = x^4 - x^2$.

by $\frac{1}{2}y^2 = -\frac{2}{3}x^3$ approximately and they only exist for $x \leq 0$. Generally the equations for the phase paths can be found explicitly as

$$\tfrac{1}{2}y^2 + \mathcal{V}(x) = \tfrac{1}{2}y^2 + \tfrac{1}{3}x^3 - \tfrac{1}{5}x^5 = C.$$

A selection of phase paths is shown in the lower diagram of Figure 1.28 including the path which starts at $x(0) = -\frac{1}{2}$, $y(0) = 0$. The closed phase path indicates periodic motion.

- **1.16** The system $\ddot{x} + x = -F_0 \operatorname{sgn}(\dot{x})$, $F_0 > 0$, has the initial conditions $x = x_0 > 0$, $\dot{x} = 0$. Show that the phase path will spiral exactly n times before entering equilibrium (Section 1.6) if $(4n-1)F_0 < x_0 < (4n+1)F_0$.

1.16. The system is governed by the equation

$$\ddot{x} + x = -F_0 \operatorname{sgn}(\dot{x}) = \begin{cases} -F_0 & (\dot{x} > 0) \\ F_0 & (\dot{x} < 0). \end{cases}$$

For $y > 0$, the differential equation of the phase paths is

$$\frac{dy}{dx} = \frac{-x - F_0}{y}.$$

Integrating, the solutions can be expressed as

$$(x + F_0)^2 + y^2 = A. \tag{i}$$

Similarly, for $y < 0$, the phase paths are given by

$$(x - F_0)^2 + y^2 = B. \tag{ii}$$

For $y > 0$ the phase paths are semicircles centred at $(-F_0, 0)$, and for $y < 0$ they are semicircles centred at $(F_0, 0)$. The equation has a line of equilibrium points for which $-1 < x < 1$. The semicircle paths are matched as shown in Figure 1.29 (drawn with $F_0 = 1$), and a path eventually meets the x axis between $x = -1$ and $x = 1$ either from above or below depending on the initial value x_0. We have to insert a path from $(-1, 0)$ to the origin and a path from $(1, 0)$ to the origin to complete the phase diagram.

Let the path which starts at $(x_0, 0)$ next cut the x axis at $(x_1, 0)$. From (ii) the path is

$$(x - F_0)^2 + y^2 = (x_0 - F_0)^2, \quad (y < 0).$$

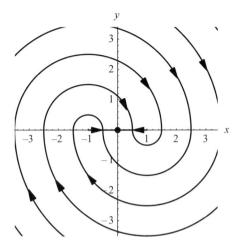

Figure 1.29 Problem 1.16.

from which it follows that $x_1 = 2F_0 - x_0$. Assume that $2F_0 - x_0 < -F_0$, that is, $x_0 > 3F_0$ so that the path continues. The continuation of the path lies on the semicircle

$$(x + F_0)^2 + y^2 = (x_1 + F_0)^2 = (3F_0 - x_0)^2, \quad (y > 0).$$

Assume that it meets the x axis again at $x = x_2$. Hence $x_2 = x_0 - 4F_0$. The spiral will continue if $x_2 > F_0$ or $x_0 > 5F_0$ and terminate if $x_0 < 5F_0$. Hence just one cycle of the spiral occurs if $3F_0 < x_0 < 5F_0$.

If the spiral continues then $x = x_2$ becomes the new initial point and a further spiral occurs if

$$3F_0 < x_2 < 5F_0 \text{ or } 7F_0 < x_0 < 9F_0.$$

Continuing this process, a phase path will spiral just n times if

$$(4n - 1)F_0 < x_0 < (4n + 1)x_0.$$

- **1.17** A pendulum of length a has a bob of mass m which is subject to a horizontal force $m\omega^2 a \sin\theta$, where θ is the inclination to the downward vertical. Show that the equation of motion is $\ddot{\theta} = \omega^2(\cos\theta - \lambda)\sin\theta$, where $\lambda = g/(\omega^2 a)$. Investigate the stability of the equilibrium states by the method of NODE, Section 1.7 for parameter-dependent systems. Sketch the phase diagrams for various λ.

1.17. The forces acting on the bob are shown in Figure 1.30. Taking moments about O

$$m\omega^2 a \sin\theta \cdot a\cos\theta - mga\sin\theta = ma^2\ddot{\theta},$$

1 : Second-order differential equations in the phase plane   25

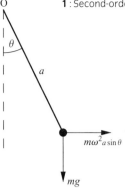

Figure 1.30 Problem 1.17.

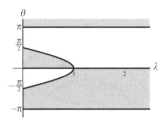

Figure 1.31 Problem 1.17: The graph of $f(\theta,\lambda)=0$ where the regions in which $f(\theta,\lambda)>0$ are shaded.

or

$$\ddot{\theta} = \omega^2(\cos\theta - \lambda)\sin\theta = f(\theta,\lambda),$$

in the notation of NODE, Section 1.7, where $\lambda/(\omega^2 a)$. Let $\omega=1$: time can be scaled to eliminate ω. The curves in the (θ,λ) given by $f(\theta,\lambda)=0$ are shown in Figure 1.31 with the regions where $f(\theta,\lambda)>0$ are shaded. If $\lambda<1$, then the pendulum has four equilibrium points at $\theta=\pm\cos^{-1}\lambda$, $\theta=0$ and $\theta=\pi$. The diagram is periodic with period 2π in θ so that the equilibrium point at $\theta=-\pi$ is the same as that at $\theta=\pi$. Any curves above the shaded regions indicate stable equilibrium points (centres) and any curves below shaded regions indicate unstable equilibrium points (saddles). Hence, for $\lambda<1$, $\theta=\pm\cos^{-1}\lambda$ are stable points, whilst $\theta=0$ and $\theta=\pi$ are both unstable. The equations of the phase paths can be found by integrating

$$\dot{\theta}\frac{d\dot{\theta}}{d\theta} = \sin\theta\cos\theta - \lambda\sin\theta.$$

The general solution is

$$\dot{\theta}^2 = \sin^2\theta + 2\cos\theta + C.$$

The phase diagram is shown in Figure 1.32 for $\lambda=0.4$. As expected from the stability diagram, there are centres at $\theta=\pm\cos^{-1}\lambda$ and saddles at $x=0$ and $x=\pi$.

If $\lambda\geq 1$, there are two equilibrium point at $\theta=\pi$ (or $-\pi$). The phase diagram is shown in Figure 1.33 with $\lambda=2$. The origin now becomes a stable centre but $\theta=\pi$ remains a saddle.

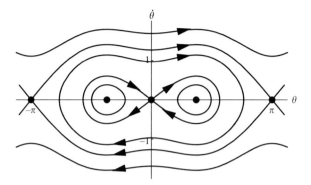

Figure 1.32 Problem 1.17: Phase diagram for $\lambda = 0.4 < 1$.

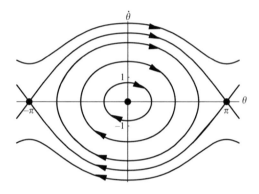

Figure 1.33 Problem 1.17: Phase diagram for $\lambda = 2 > 1$.

- **1.18** Investigate the stability of the equilibrium points of the parameter-dependent system $\ddot{x} = (x - \lambda)(x^2 - \lambda)$.

1.18. The equation is

$$\ddot{x} = (x - \lambda)(x^2 - \lambda) = f(x, \lambda)$$

in the notation of NODE, Section 1.7. The system is in equilibrium on the line $x = \lambda$ and the parabola $x^2 = \lambda$. These boundaries are shown in Figure 1.34 together with the shaded regions in which $f(x, \lambda) > 0$.

- $\lambda \leq 0$. There is one equilibrium point, an unstable saddle at $x = \lambda$.
- $0 < \lambda < 1$. There are three equilibrium points: at $x = -\sqrt{\lambda}$ (saddle), $x = \lambda$ (centre) and $x = \sqrt{\lambda}$ (saddle).
- $\lambda = 1$. This is a critical case in which $f(x, \lambda)$ is positive on both sides of $x = 1$. The equilibrium point is an unstable hybrid centre/saddle.
- $\lambda > 1$. There are three equilibrium points: at $x = -\sqrt{\lambda}$ (saddle), $x = \sqrt{\lambda}$ (centre) and $x = \lambda$ (saddle).

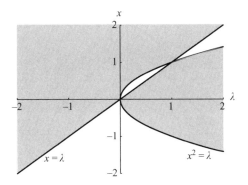

Figure 1.34 Problem 1.18.

• **1.19** If a bead slides on a smooth parabolic wire rotating with constant angular velocity ω about a vertical axis, then the distance x of the particle from the axis of rotation satisfies $(1+x^2)\ddot{x} + (g - \omega^2 + \dot{x}^2)x = 0$. Analyse the motion of the bead in the phase plane.

1.19. The differential equation of the bead is

$$(1+x^2)\ddot{x} + (g - \omega^2 + \dot{x}^2)x = 0.$$

The equation represents the motion of a bead sliding on a rotating parabolic wire with its lowest point at the origin. The variable x represents distance from the axis of rotation. Put $\dot{x} = y$ and $g - \omega^2 = \lambda$; then equilibrium points occur where

$$y = 0 \text{ and } (\lambda + y^2)x = 0.$$

If $\lambda \neq 0$, all points on the x axis of the phase diagram are equilibrium points, and if $\lambda = 0$ there is a single equilibrium point, at the origin.

The differential equation of the phase paths is

$$\frac{dy}{dx} = -\frac{(\lambda + y^2)x}{(1+x^2)y},$$

which is a separable first-order equation. Hence, separating the variables and integrating

$$\int \frac{y\,dy}{\lambda + y^2} = -\int \frac{x\,dx}{1+x^2} + C,$$

or

$$\tfrac{1}{2}\ln|\lambda + y^2| = -\tfrac{1}{2}\ln(1+x^2) + C,$$

or

$$(\lambda + y^2)(1+x^2) = A.$$

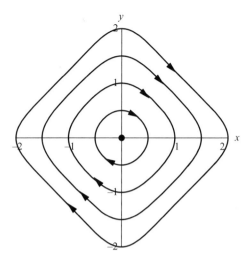

Figure 1.35 Problem 1.19: Phase diagram for $\lambda = 1$.

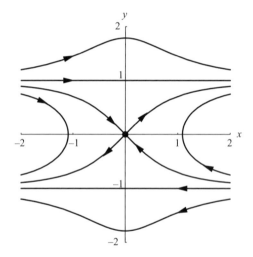

Figure 1.36 Problem 1.19: Phase diagram for $\lambda = -1$.

- $\lambda > 0$. The phase diagram for $\lambda = 1$ is shown in Figure 1.35 which implies that the origin is a centre. In this mode, for low angular rates, the bead oscillates about the lowest point of the parabola.
- $\lambda < 0$. The phase diagram for $\lambda = -1$ is plotted in Figure 1.36 which shows that the origin is a saddle. For higher angular rates the origin becomes unstable and the bead will theoretically go off to infinity. Note that $y = \pm 1$ are phase paths which means, for example, that the bead starting from $x = 0$ with velocity $y = 1$ will move outwards at a constant rate.
- $\lambda = 1$. The phase diagram is shown in Figure 1.37. If the bead is placed at rest at any point on the wire then it will remain in that position subsequently.

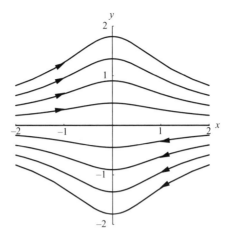

Figure 1.37 Problem 1.19: Phase diagram for $\lambda = 0$.

- **1.20** A particle is attached to a fixed point O on a smooth horizontal plane by an elastic string. When unstretched, the length of the string is $2a$. The equation of motion of the particle, which is constrained to move on a straight line through O, is

$\ddot{x} = -x + a \, \text{sgn}(x)$, $|x| > a$ (when the string is stretched),
$\ddot{x} = 0$, $|x| \leq a$ (when the string is slack),

x being the displacement from O. Find the equilibrium points and the equations of the phase paths, and sketch the phase diagram.

1.20. The equation of motion of the particle is

$$\ddot{x} = -x + a \, \text{sgn}(x), \quad (|x| > a)$$
$$\ddot{x} = 0, \quad (|x| \leq a).$$

All points in the interval $|x| \leq a$, $y = 0$ are equilibrium points. The phase paths as follows.

(i) $x > a$. The differential equation is

$$\frac{dy}{dx} = \frac{-x + a}{y},$$

which has the general solution

$$y^2 + (x - a)^2 = C_1.$$

These phase paths are semicircles centred at $(a, 0)$.

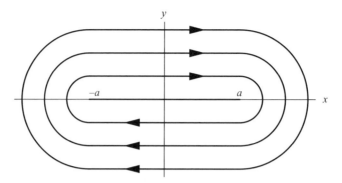

Figure 1.38 Problem 1.20.

(ii) $-a \leq x \leq a$. The phase paths are the straight lines $y = C_2$.
(iii) $x < -a$. The differential equation is

$$\frac{dy}{dx} = \frac{-x-a}{y},$$

which has the general solution

$$y^2 + (x+a)^2 = C_3.$$

These phase paths are semicircles centred at $(-a, 0)$.

A sketch of the phase paths is shown in Figure 1.38. All paths are closed which means that all solutions are periodic.

- **1.21** The equation of motion of a conservative system is $\ddot{x} + g(x) = 0$, where $g(0) = 0$, and $g(x)$ is strictly increasing for all x, and

$$\int_0^x g(u)du \to \infty \text{ as } x \to \pm\infty. \tag{i}$$

Show that the motion is always periodic.
By considering $g(x) = xe^{-x^2}$, show that if (i) does not hold, the motions are not all necessarily periodic.

1.21. The equation for the phase paths is

$$\frac{dy}{dx} = -\frac{g(x)}{y}.$$

The variables separate to give the general solution in the form

$$\frac{1}{2}y^2 = -\int_0^x g(u)du + C. \tag{i}$$

Write
$$\int_0^x g(u)\,du = G(x). \qquad (ii)$$

Then (i) defines two families of paths where $C > 0$;

$$y = \sqrt{2}\{C - G(x)\}^{1/2} \text{ when } G(x) < C; \qquad (iii)$$

and the reflection in the x axis;

$$y = -\sqrt{2}\{C - G(x)\}^{1/2} \text{ when } G(x) < C. \qquad (iv)$$

Since $g(x) < 0$ when $x < 0$, and $g(x) > 0$ when $x > 0$, then $G(x)$ is strictly increasing to $+\infty$ as $x \to -\infty$ and $x \to \infty$. Also $G(x)$ is continuous and $G(0) = 0$. Therefore, given any value of $C > 0$, $G(x)$ takes the value C at exactly two values of x, one negative and the other positive.

Consider the family of positive solutions (iii). Take any positive value of the constant C. At the two points where $G(x) = C$, we have $y(x) = 0$. Between them $y(x) > 0$, and the graph of the path cuts the x axis at right angles (see Section 1.2). When the corresponding reflected curve (iv) ($y < 0$) is joined to this one, we have a smooth closed curve. By varying the parameter C the process generates a family of closed curves nested around the origin (which is therefore a centre), and all motions are periodic.

If $g(x) = xe^{-x^2}$, then $G(x) = \frac{1}{2}(1 - e^{-x^2})$, which does not go to infinity as $x \to \pm\infty$. The solutions (iii) and (iv) become

$$y = \pm\sqrt{2}\{B + \tfrac{1}{2}e^{-x^2}\}^{1/2}, \text{ where } B = C - \tfrac{1}{2}.$$

If $-\tfrac{1}{2} < B < \tfrac{1}{2}$ (i.e. if $0 < C < 1$) the above analysis holds; there is a family of closed curves surrounding the origin. These represent periodic motions. However, if $B > \tfrac{1}{2}$, the corresponding paths do not meet the x axis, but run from $x = -\infty$ to $x = +\infty$ outside the central region. These are not periodic motions.

- 1.22 The wave function $u(x,t)$ satisfies the partial differential equation
$$\frac{\partial^2 u}{\partial x^2} + \alpha \frac{\partial u}{\partial x} + \beta u^3 + \gamma \frac{\partial u}{\partial t} = 0.$$
where α, β and γ are positive constants. Show that there exist travelling wave solutions of the form $u(x,t) = U(x - ct)$ for any c, where $U(\zeta)$ satisfies
$$\frac{d^2 U}{d\zeta^2} + (\alpha - \gamma c)\frac{dU}{d\zeta} + \beta U^3 = 0.$$
Using Problem 1.21, show that when $c = \alpha/\gamma$, all such waves are periodic.

1.22. The wave function $u(x,t)$ satisfies the partial differential equation

$$\frac{\partial^2 u}{\partial x^2} + \alpha \frac{\partial u}{\partial x} + \beta u^3 + \gamma \frac{\partial u}{\partial t} = 0.$$

Let $u(x,t) = U(x-ct)$ and $\zeta = x - ct$. Then

$$\frac{\partial u}{\partial x} = \frac{dU}{d\zeta}, \quad \frac{\partial^2 u}{\partial x^2} = \frac{d^2 U}{d\zeta^2}, \quad \frac{\partial u}{\partial t} = -c\frac{dU}{d\zeta},$$

so that the partial differential equation becomes the ordinary differential equation

$$\frac{d^2 U}{d\zeta^2} + (\alpha - \gamma c)\frac{dU}{d\zeta} + \beta U^3 = 0.$$

If $c = \alpha/\gamma$, the equation becomes

$$\frac{d^2 U}{d\zeta^2} + \beta U^3 = 0,$$

which can be compared with the conservative system in Problem 1.21. In this case $g(U) = \beta U^3$. Obviously $g(U) < 0$ for $U < 0$, $g(U) > 0$ for $U > 0$ and $g(0) = 0$. Also

$$\beta \int_0^U v^3 dv = \frac{\beta}{4} U^4 \to \infty, \text{ as } U \to \pm\infty.$$

Therefore by Problem 1.21 these waves are all periodic.

- **1.23** The linear oscillator $\ddot{x} + \dot{x} + x = 0$ is set in motion with initial conditions $x = 0$, $\dot{x} = v$, at $t = 0$. After the first and each subsequent cycle the kinetic energy is instantaneously increased by a constant, E, in such a manner as to increase $\dot{x}$. Show that if $E = \frac{1}{2}v^2(1 - e^{4\pi/\sqrt{3}})$, a periodic motion occurs. Find the maximum value of x in a cycle.

1.23. The oscillator has the equation $\ddot{x} + \dot{x} + x = 0$, with initial conditions $x(0) = 0$, $\dot{x}(0) = v$. It is easier to solve this equation for x in terms of t rather than to use eqn (1.9) for the phase paths. The characteristic equation is $m^2 + m + 1 = 0$, with roots $m = \frac{1}{2}(-1 \pm \sqrt{3}i)$. The general (real) solution is therefore

$$x(t) = e^{-\frac{1}{2}t}[A\cos(\tfrac{1}{2}\sqrt{3}t) + B\sin(\tfrac{1}{2}\sqrt{3}t)]. \tag{i}$$

Also we shall require $\dot{x}(t)$:

$$\dot{x}(t) = \frac{v}{\sqrt{3}} e^{-\frac{1}{2}t}[\sqrt{3}\cos(\tfrac{1}{2}\sqrt{3}t) - \sin(\tfrac{1}{2}\sqrt{3}t)]. \tag{ii}$$

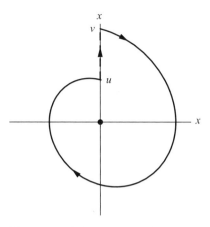

Figure 1.39 Problem 1.23: The limit cycle, with the jump along the y axis.

The first circuit is completed by time $t = 4\pi/\sqrt{3}$. $\dot{x}$ is then equal to u, say, where u is given by

$$u = \dot{x}(4\pi/\sqrt{3}) = ve^{-2\pi/\sqrt{3}}. \tag{iii}$$

At this moment(see Figure 1.39) the oscillator receives an impulsive increment E of kinetic energy, of such magnitude as to return the velocity $\dot{x}$ from its value u to the given initial velocity v. From (iii)

$$E = \tfrac{1}{2}v^2 - \tfrac{1}{2}u^2 = \tfrac{1}{2}v^2(1 - e^{-4\pi/\sqrt{3}}). \tag{iv}$$

The second cycle then duplicates the first, since its initial conditions are *physically* equivalent to those for the first cycle, and similarly for all the subsequent cycles. The motion is therefore periodic, with period $T = 4\pi/\sqrt{3}$.

The turning points of $x(t)$ occur where $\dot{x}(t) = 0$; that is, where $\tan(\sqrt{3}/2) = \sqrt{3}$ (from (ii)). This has two solutions in the range 0 and 2π. These are

$$t = \frac{2\pi}{3\sqrt{3}} \text{ and } t = \frac{4\pi}{3\sqrt{3}}$$

(by noting that $\tan^{-1}\sqrt{3} = \tfrac{1}{3}\pi$). From (i) the corresponding values of x are

$$x = ve^{-\pi/(3\sqrt{3})} \text{ and } x = -ve^{-2\pi/(3\sqrt{3})}.$$

The overall maximum of $x(t)$ is therefore $ve^{-\pi/(3\sqrt{3})}$.

- **1.24** Show how phase paths of Problem 1.23 having arbitrary initial conditions spiral on to a limit cycle. Sketch the phase diagram.

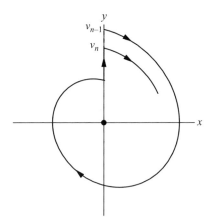

Figure 1.40 Problem 1.24: The limit cycle with the jump along the y axis.

1.24. (Refer to Problem 1.23.) The system is the same as that of Problem 1.23, but with the initial conditions $x(0) = 0$, $\dot{x}(0) = v_0 > 0$, where v_0 is arbitrary. Suppose the impulsive energy increment at the end of every cycle is E, an arbitrary positive constant. v_n will represent the value of $\dot{x}$ at the end of the nth cycle, following the energy increment delivered at the end of that cycle, and it serves as the initial condition for the next cycle (see Figure 1.40).

For the first cycle, starting at $x = 0$, $\dot{x}(0) = v_0$, we have (as in Problem 1.23)

$$\tfrac{1}{2}v_1^2 - \tfrac{1}{2}v_0^2 e^{-4\pi/\sqrt{3}} = E,$$

or

$$v_1^2 = 2E + v_0^2 e^{-4\pi/\sqrt{3}}. \tag{i}$$

For the second cycle (starting at v_1)

$$v_2^2 = 2E + v_1^2 e^{-4\pi/\sqrt{3}}, \tag{ii}$$

and so on. For the nth cycle

$$v_n^2 = 2E + v_{n-1}^2 e^{-4\pi/\sqrt{3}}. \tag{iii}$$

By successive substitution we obtain

$$v_n^2 = 2E(1 + e^{-\rho} + \cdots + e^{-(n-1)\rho} + v_0^2 e^{-n\rho}), \tag{iv}$$

in which we have written for brevity $\rho = 4\pi/\sqrt{3}$.

By using the usual formula for the sum of a geometric series (iv) reduces to

$$v_n^2 - v_0^2 = (1 - e^{-n\rho})\left(\frac{2E}{1 - e^{-\rho}} - v_0^2\right) \quad \text{for all } n \geq 1. \tag{v}$$

(A) In the special case when $E = \frac{1}{2}v_0^2(1-e^{-\rho})$, the right-hand side of (v) is zero, so $v_0^2 = v_1^2 = \cdots = v_n^2$, which corresponds to the periodic solution in Problem 1.23.

(B) If $v_0^2 < 2E/(1-e^{-\rho})$, the sequence $v_0, v_1, \ldots, v_n$ is strictly increasing, and

$$\lim_{n\to\infty} v_n^2 = \frac{2E}{1-e^{-\rho}}.$$

The limit cycle in (A) is approached from inside.

(C) If $v_0^2 = 2E/(1-e^{-\rho})$, the sequence is strictly decreasing and

$$\lim_{n\to\infty} v_n^2 = \frac{2E}{1-e^{P}-\rho};$$

so the limit cycle in (A) is approached from the outside.

The more general initial conditions $x(0) = X$, $\dot{x}(0) = V$, where X and V are both arbitrary, correspond to one of the categories (A), (B) or (C); so the same limit (A) is approached.

- **1.25** The kinetic energy, T, and the potential energy, V, of a system with one degree of freedom are given by
$$T = T_0(x) + \dot{x}T_1(x) + \dot{x}^2 T_2(x), \quad V = V(x).$$
Use Lagrange's equation
$$\frac{d}{dt}\left(\frac{\partial T}{\partial \dot{x}}\right) - \frac{\partial T}{\partial x} = -\frac{\partial V}{\partial x}$$
to obtain the equation of motion of the system. Show that the equilibrium points are stationary points of $T_0(x) - V(x)$, and that the phase paths are given by the energy equation $T_2(x)\dot{x}^2 - T_0(x) + V(x) = \text{constant}$.

1.25. The kinetic and potential energies are given by
$$T = T_0(x) + \dot{x}T_1(x) + \dot{x}^2 T_2(x), \quad V = V(x).$$

Applying Lagrange's equation

$$\frac{d}{dt}\left(\frac{\partial T}{\partial \dot{x}}\right) - \frac{\partial T}{\partial x} = -\frac{\partial V}{\partial x},$$

the equation of motion is

$$\frac{d}{dt}(2T_2\dot{x} + T_1) - (T_2'\dot{x}^2 + T_1'\dot{x} + T_0') = -V',$$

or

$$2T_2\ddot{x} + T_2'\dot{x}^2 - T_0' = -V'. \tag{i}$$

Equilibrium points, where $\ddot{x} = \dot{x} = 0$, occur where $T_0' - \mathcal{V}' = 0$, that is, at the stationary points of the energy function $T_0(x) - \mathcal{V}(x)$. Let $y = \dot{x}$. Equation (i) can be expressed in the form

$$\frac{d}{dx}(T_2(x)y^2) - T_0'(x) + \mathcal{V}'(x) = 0,$$

which can be integrated to give the phase paths, namely

$$T_2(x)y^2 - T_0(x) + \mathcal{V}(x) = C.$$

- **1.26** Sketch the phase diagram for the equation $\ddot{x} = -f(x + \dot{x})$, where

$$f(u) = \begin{cases} f_0 & u \geq c, \\ f_0 u/c & |u| \leq c, \\ -f_0 & u \leq -c \end{cases}$$

where f_0, c are constants, $f_0 > 0$, and $c > 0$. How does the system behave as $c \to 0$?

1.26. The system is governed by the equation $\ddot{x} = -f(x + \dot{x})$, where

$$f(u) = \begin{cases} f_0 & u > c \\ f_0 u/c & |u| \leq c \\ -f_0 & u < -c \end{cases}$$

Let $y = \dot{x}$. The phase paths are as follows.

- $x + y > c$, $\ddot{x} = -f_0$. The equation for the phase paths is

$$\frac{dy}{dx} = -\frac{f_0}{y} \quad \Rightarrow \quad \frac{1}{2}y^2 = -f_0 x + C_1.$$

The phase paths are parabolas with their axes along the x axis.
- $|x + y| \leq c$, $\ddot{x} = -f_0(x + \dot{x})/c$. It is easier to solve the linear equation

$$c\ddot{x} + f_0 \dot{x} + f_0 x = 0$$

parametrically in terms of t. The characteristic equation is

$$cm^2 + f_0 m + f_0 = 0.$$

which has the roots

$$m_1, m_2 = \frac{1}{2c}[-f_0 \pm \sqrt{(f_0^2 - 4cf_0)}].$$

Therefore

$$x = Ae^{m_1 t} + Be^{m_2 t}$$

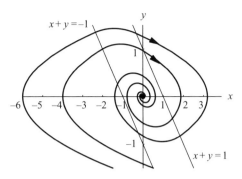

Figure 1.41 Problem 1.26: The spirals are shown for the parameter values $f_0 = 0.25$ and $c = 1$. Note that scales on the axes are not the same in the drawing.

The roots are both real and negative if $f_0 > 4c$, which means that the phase diagram between the lines $x + y = c$ and $x + y = -c$ is a stable node. If $f_0 < 4c$, then the phase diagram is a stable spiral.
- $x + y < -c$, $\ddot{x} = f_0$. The phase paths are given by

$$\tfrac{1}{2} y^2 = f_0 x + C_2,$$

which again are parabolas but pointing in the opposite direction.

Figure 1.41 shows a phase diagram for the spiral case. The spiral between the lines $x = y = 1$ and $x + y = -1$ is linked with the parabolas on either side of the two lines. The total picture is a stable spiral. A similar matching occurs with the stable node.

As $c \to 0$, the lines $x + y = c$ and $x + c = -1$ merge and the spiral disappears leaving a centre created by the joining of the parabolas.

- **1.27** Sketch the phase diagram for the equation $\ddot{x} = u$, where
$u = -\text{sgn}\,(\sqrt{2}|x|^{1/2}\text{sgn}\,(x) + \dot{x})$.
(u is an elementary control variable which can switch between $+1$ and -1. The curve $\sqrt{2}|x|^{1/2}\text{sgn}\,(x) + y = 0$ is called the switching curve.)

1.27. The control equation is

$$\ddot{x} = -\text{sgn}\,[\sqrt{2}|x|^{1/2}\text{sgn}\,(x) + \dot{x}].$$

The equilibrium point satisfies

$$\text{sgn}\,[\sqrt{2}|x|^{1/2}\text{sgn}\,(x)] = 0, \quad \text{or} \quad |x|^{1/2}\text{sgn}\,(x) = 0,$$

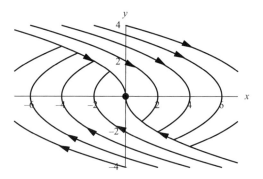

Figure 1.42 Problem 1.27.

of which $x = 0$ is the only solution. In the phase plane the boundary between the two modes of the phase diagram is the switching curve

$$y = - \operatorname{sgn}[\sqrt{2}|x|^{1/2}\operatorname{sgn}(x)],$$

which is two half parabolas which meet at the origin as shown in Figure 1.42. There are distinct families of phase paths on either side of this curve.

- $\sqrt{2}|x|^{1/2}\operatorname{sgn}(x) + y > 0$. The equation is $\ddot{x} = -1$ so that $dy/dx = -1/y$ and the phase paths are given by the parabolas $y^2 = -2x + C_1$
- $\sqrt{2}|x|^{1/2}\operatorname{sgn}(x) + y < 0$. In this case $\ddot{x} = 1$ so that the phase paths are given by $y^2 = 2x + C_2$.

When the parabolic paths reach the switching curve their only exit is along the switching curve into the equilibrium point at the origin.

- **1.28** The relativistic equation for an oscillator is

$$\frac{d}{dt}\left\{\frac{m_0 \dot{x}}{\sqrt{[1 - (\dot{x}/c)^2]}}\right\} + kx = 0, \quad |\dot{x}| < c$$

where m_0, c and k are positive constants. Show that the phase paths are given by

$$\frac{m_0 c^2}{\sqrt{[1 - (y/c)^2]}} + \frac{1}{2}kx^2 = \text{constant}.$$

If $y = 0$ when $x = a$, show that the period, T, of an oscillation is given by

$$T = \frac{4}{c\sqrt{\varepsilon}}\int_0^a \frac{[1 + \varepsilon(a^2 - x^2)]dx}{\sqrt{(a^2 - x^2)}\sqrt{[2 + \varepsilon(a^2 - x^2)]}}, \quad \varepsilon = \frac{k}{2m_0 c^2}.$$

The constant ε is small; by expanding the integrand in powers of ε show that

$$T \approx \frac{\pi\sqrt{2}}{c}\left(\varepsilon^{-(1/2)} + \frac{3}{8}\varepsilon^{1/2}a^2\right).$$

1.28. The equation of the oscillator is

$$\frac{d}{dt}\left\{\frac{m_0\dot{x}}{\sqrt{[1-(\dot{x}/c)^2]}}\right\}+kx=0,$$

which has one equilibrium point at the origin. Also the phase plane is restricted to $|\dot{x}|<c$. Let $y=\dot{x}$ and

$$f(y)=\frac{m_0 y}{\sqrt{[1-(y/c)^2]}}.$$

Then the equation of the oscillator is

$$y\frac{df(y)}{dy}+kx=0, \quad \text{or} \quad yf'(y)\frac{dy}{dx}+kx=0.$$

This is a separable first-order equation with solution

$$\int yf'(y)dy=-k\int dx+C,$$

which after integration by parts leads to

$$yf(y)-\int f(y)dy=-\frac{1}{2}kx^2+C,$$

or

$$\frac{m_0 y^2}{\sqrt{[1-(y/c)^2]}}-\int \frac{m_0 y\,dy}{\sqrt{[1-(y/c)^2]}}=-\frac{1}{2}kx^2+C,$$

or

$$\frac{m_0 y^2}{\sqrt{[1-(y/c)^2]}}+m_0 c^2\sqrt{[1-(y/c)^2]}=-\frac{1}{2}kx^2+C,$$

so that

$$\frac{m_0 c^2}{\sqrt{[1-(y/c)^2]}}=-\frac{1}{2}kx^2+C, \tag{i}$$

as required. A sketch of the phase diagram is shown in Figure 1.43. It can be seen that the origin is a centre. The particular path through $(a,0)$ is, from (i),

$$\frac{m_0 c^2}{\sqrt{[1-(y/c)^2]}}=-\frac{1}{2}kx^2+m_0 c^2+\frac{1}{2}ka^2,$$

or

$$\frac{1}{\sqrt{[1-(y/c)^2]}}=1+\varepsilon(a^2-x^2),$$

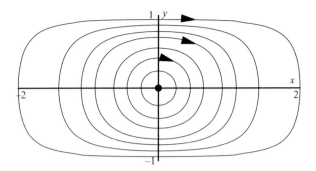

Figure 1.43 Problem 1.28: Phase diagram for $k=1$, $c=1$ and $m_0=1$.

where $\varepsilon = k/(2m_0c^2)$. Solve this equation for y:

$$y = \frac{dx}{dt} = \frac{c\sqrt{\varepsilon}\sqrt{[a^2-x^2]}\sqrt{[2+\varepsilon(a^2-x^2)]}}{1+\varepsilon(a^2-x^2)}.$$

Therefore

$$T = \frac{4}{c\sqrt{\varepsilon}} \int_0^a \frac{1+\varepsilon(a^2-x^2)dx}{\sqrt{(a^2-x^2)}\sqrt{[2+\varepsilon(a^2-x^2)]}}; \qquad \text{(ii)}$$

the integral is multiplied by 4 since integration between 0 and a covers a quarter of the period, and the time over each quarter is the same by symmetry.

Expand the integrand in powers of ε for small ε using a Taylor series. Then

$$\frac{1+\varepsilon(a^2-x^2)}{\sqrt{(a^2-x^2)}\sqrt{[2+\varepsilon(a^2-x^2)]}} \approx 2^{-(1/2)}\left[\frac{1}{\sqrt{(a^2-x^2)}} + \frac{3}{4}\varepsilon\sqrt{(a^2-x^2)}\right].$$

Hence

$$T \approx \frac{2\sqrt{2}}{c\sqrt{\varepsilon}} \int_0^a \left[\frac{1}{\sqrt{(a^2-x^2)}} + \frac{3}{4}\varepsilon\sqrt{(a^2-x^2)}\right] dx$$

$$= \frac{2\sqrt{2}}{c\sqrt{\varepsilon}} \left[\sin^{-1}(x/a) + \frac{3}{8}\varepsilon\{x\sqrt{(a^2-x^2)} + a^2\sin^{-1}(x/a)\}\right]_0^a$$

$$= \frac{2\sqrt{2}}{c\sqrt{\varepsilon}} \left(\frac{1}{2}\pi + \frac{3}{16}\varepsilon\pi a^2\right)$$

$$= \frac{\pi\sqrt{2}}{c} \left(\frac{1}{\varepsilon^{1/2}} + \frac{3}{8}\varepsilon^{1/2}a^2\right)$$

as $\varepsilon \to 0$.

- **1.29** A mass m is attached to the mid-point of an elastic string of length $2a$ and stiffness λ (see Figure 1.35 in NODE or Figure 1.44). There is no gravity acting, and the tension is zero in the equilibrium position. Obtain the equation of motion for transverse oscillations and sketch the phase paths.

1.29. Assume that oscillations occur in the direction of x (see Figure 1.44). By symmetry we can assume that the tensions in the strings on either side of m are both given by T. The equation of motion for m is
$$2T\sin\theta = -m\ddot{x}.$$
Assuming Hooke's law,
$$T = \lambda \times \text{extension} = \lambda[\sqrt{(x^2 + a^2)} - a].$$
Therefore
$$m\ddot{x} = -\frac{2kx[\sqrt{(x^2+a^2)} - a]}{\sqrt{(x^2+a^2)}}. \tag{i}$$
There is one expected equilibrium point at $x = 0$. This is a conservative system with potential (see NODE, Section 1.3)
$$\mathcal{V}(x) = 2k\int\left(x - \frac{ax}{\sqrt{(x^2+a^2)}}\right)dx = k[x^2 - a\sqrt{(x^2+a^2)}]. \tag{ii}$$
The equation of motion (i) can be expressed in the dimensionless form
$$X'' = -\frac{X\sqrt{(X^2+1)} - 1}{\sqrt{(X^2+1)}}$$
after putting $x = aX$ and $t = m\tau/(2k)$. The phase diagram in the plane $(X, Y = X')$ is shown in Figure 1.45. From (ii) the potential energy $\mathcal{V}$ has a minimum at $x = 0$ (or $X = 0$) so that the origin is a centre.

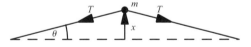

Figure 1.44 Problem 1.29.

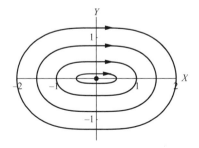

Figure 1.45 Problem 1.29: Phase diagram.

- **1.30** The system
$$\ddot{x} + x = F(v_0 - \dot{x})$$
is subject to the friction law
$$F(u) = \begin{cases} 1 & u > \varepsilon \\ u/\varepsilon & -\varepsilon < u < \varepsilon \\ -1 & u < -\varepsilon \end{cases}$$
where $u = v_0 - \dot{x}$ is the slip velocity and $v_0 > \varepsilon > 0$. Find explicit equations for the phase paths in the $(x, y = \dot{x})$ plane. Compute a phase diagram for $\varepsilon = 0.2$, $v_0 = 1$ (say). Explain using the phase diagram that the equilibrium point at $(1, 0)$ is a centre, and that all paths which start outside the circle $(x-1)^2 + y^2 = (v_0 - \varepsilon)^2$ eventually approach this circle.

1.30. The equation of the friction problem is
$$\ddot{x} + x = F(v_0 - \dot{x}),$$
where
$$F(u) = \begin{cases} 1 & u > \varepsilon \\ u/\varepsilon & -\varepsilon < u < \varepsilon \\ -1 & u < -\varepsilon. \end{cases}$$
The complete phase diagram is a combination of phase diagrams matched along the lines $y = v_0 + \varepsilon$ and $y = v_0 - \varepsilon$.

- $y > v_0 + \varepsilon$. In this region $F = 1$. Hence the phase paths satisfy
$$\frac{dy}{dx} = -\frac{x+1}{y},$$
which has the general solution
$$(x+1)^2 + y^2 = C_1.$$
The phase paths are arcs of circles centred at $(-1, 0)$.
- $v_0 - \varepsilon < y < v_0 + \varepsilon$. The differential equation is
$$\ddot{x} + x = \frac{1}{\varepsilon}(v_0 - \dot{x}), \quad \text{or } \varepsilon\ddot{x} + \dot{x} + \varepsilon x = v_0,$$
which is an equation of linear damping. The characteristic equation is
$$\varepsilon m^2 + m + \varepsilon = 0,$$
which has the solutions
$$m_1, m_2 = \frac{1}{2\varepsilon}[-1 \pm \sqrt{(1 - 4\varepsilon^2)}].$$

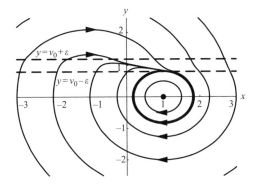

Figure 1.46 Problem 1.30: Phase diagram of $\ddot{x} + x = F(v_0 - \dot{x})$ for $\varepsilon = 0.2$, $v_0 = 1$.

Since ε is small, the solutions are both real and negative. The general solution is

$$x = Ae^{m_1 t} + Be^{m_2 t} + \frac{v_0}{\varepsilon}.$$

which is the solution for a stable node centred at $(v_0/\varepsilon, 0)$.
- $y < v_0 - \varepsilon$. With $F = -1$, the phase paths are given by

$$(x-1)^2 + y^2 = C_2,$$

which are arcs of circles centred at $(1, 0)$.

Figure 1.46 shows a computed phase diagram for the oscillator with the parameters $\varepsilon = 0.2$, $v_0 = 1$. The equilibrium point at $(1, 0)$ is a centre. The phase paths between $y = v_0 + \varepsilon$ and $y = v_0 - \varepsilon$ are parts of those of a stable node centred at $x = v_0/\varepsilon = 5$, $y = 0$. All paths which start outside the circle

$$(x-1)^2 + y^2 = (v_0 - \varepsilon)^2 = 0.8^2,$$

eventually approach this periodic solution.

- **1.31** The system

$$\ddot{x} + x = F(\dot{x}),$$

where

$$F(\dot{x}) = \begin{cases} k\dot{x} + 1 & \dot{x} < v_0 \\ 0 & \dot{x} = v_0 \\ -k\dot{x} - 1 & \dot{x} > v_0 \end{cases},$$

and $k > 0$, is a possible model for Coulomb dry friction with damping. If $k < 2$, show that the equilibrium point is an unstable spiral. Compute the phase paths for, say, $k = 0.5$, $v_0 = 1$. Using the phase diagram discuss the motion of the system, and describe the limit cycle.

1.31. The equation for the Coulomb friction is

$$\ddot{x} + x = F(\dot{x}),$$

where

$$F(\dot{x}) = \begin{cases} k\dot{x} + 1 & \dot{x} < v_0 \\ 0 & \dot{x} = v_0 \\ -k\dot{x} - 1 & \dot{x} > v_0 \end{cases}$$

For $y < v_0$, the equation of motion is

$$\ddot{x} - k\dot{x} + x = 1.$$

The system has one equilibrium point at $x = 1$ which is unstable, a spiral if $k < 2$ and a node if $k > 2$. For $y > v_0$, the equation of motion is

$$\ddot{x} + k\dot{x} + x = -1,$$

which as part of a phase diagram of a stable spiral or node centred at $x = -1$. These families of paths meet at the line $y = v_0$.

Assume that $k < 2$. For $y < v_0$, the phase paths have zero slope on the line $-x + ky + 1 = 0$, which meets the line $y = v_0$ at $x = kv_0 + 1$. Similarly, the phase paths for $y > v_0$ have zero slope along the line $x + ky + 1 = 0$ which meets the line $y = v_0$ at $x = -kv_0 - 1$. On the phase diagram insert a phase path on $y = v_0$ between $x = -kv_0 - 1$ and $x = kv_0 + 1$ along which phase paths meet pointing in the direction of positive x. In this singular situation the only exit is along the line until $x = kv_0 + 1$ is reached where the path continues for $y < v_0$. See Figure 1.47. This particular path continues as the limit cycle. Paths spiral into the limit cycle from external and internal points.

The section of phase path on $y = v_0$ corresponds to dry friction in which two surfaces stick for a time. This occurs in every period of the limit cycle.

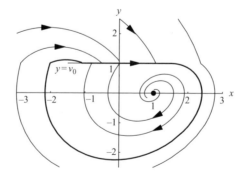

Figure 1.47 Problem 1.31: The phase diagram with $k = 0.5$ and $v = 1$. The thickest curve is the limit cycle.

- **1.32** A pendulum with magnetic bob oscillates in a vertical plane over a magnet, which repels the bob according to the inverse square law, so that the equation of motion is (Figure 1.36 in NODE)
$$ma^2\ddot\theta = -mga\sin\theta + Fh\sin\phi,$$
where $h > a$ and $F = c/(a^2 + h^2 - 2ah\cos\theta)$ and c is a constant. Find the equilibrium positions of the bob, and classify them as centres and saddle points according to the parameters of the problem. Describe the motion of the pendulum.

1.32. Take moments about the point of suspension of the pendulum. Then

$$Fh\sin\phi - mga\sin\theta = ma^2\ddot\theta. \tag{i}$$

where, by the inverse square law,

$$F = \frac{c}{a^2 + h^2 - 2ah\cos\theta}, \quad \tan\phi = \frac{a\sin\theta}{h - a\cos\theta}. \tag{ii}$$

Elimination of f and ϕ in (i) using (ii) leads to an equation in θ

$$ma\ddot\theta = \frac{ch\sin\theta}{(a^2 + h^2 - 2ah\cos\theta)^{3/2}} - mg\sin\theta.$$

There are equilibrium points at $\theta = n\pi$, $(n = 0, \pm 1, \pm 2, \ldots)$ and where

$$a^2 + h^2 - 2ah\cos\theta = \left(\frac{ch}{mg}\right)^{2/3},$$

that is where

$$\cos\theta = \frac{a^2 + h^2 - (ch/mg)^{2/3}}{2ah}.$$

This equation has solutions if

$$-1 \leq \frac{a^2 + h^2 - (ch/mg)^{2/3}}{2ah} \leq 1,$$

or

$$\frac{mg(a-h)^3}{h} \leq c \leq \frac{mg(a+h)^3}{h}. \tag{iii}$$

If c lies outside this interval then the pendulum does not have an inclined equilibrium position. If it exists let the angle of the inclined equilibrium be $\theta = \theta_1$ for $0 < \theta_1 < \pi$. Obviously $\theta = -\theta_1$ and $2n\pi \pm \theta_1$ will also be solutions.

This is a conservative system with potential $\mathcal{V}(\theta)$ (see Section 1.3) such that

$$\mathcal{V}'(\theta) = -\frac{ch\sin\theta}{ma(a^2+h^2-2ah\cos\theta)^{3/2}} + \frac{g\sin\theta}{a}.$$

The nature of the stationary points can be determined by the sign of the second derivative at each point. Thus

$$\mathcal{V}''(\theta) = -\frac{ch\cos\theta}{ma(a^2+h^2-2ah\cos\theta)^{3/2}}$$

$$+\frac{3ch^2\sin^2\theta}{m(a^2+h^2-2ah\cos\theta)^{5/2}} + \frac{g\cos\theta}{a}.$$

- $\theta = n\pi$ (n even).

$$\mathcal{V}''(n\pi) = -\frac{ch}{ma(h-a)^3} + \frac{g}{a}.$$

It follows that $\theta = n\pi$ is a centre if $c < mg(h-a)^3/h$ and a saddle if $c > mg(h-a)^3/h$.
- $\theta = n\pi$ (n odd).

$$\mathcal{V}''(n\pi) = \frac{ch}{ma(h+a)^3} - \frac{g}{a}.$$

Therefore $\theta = n\pi$ is a centre if $c > mg(h+a)^3/h$ and a saddle if $c < mg(h+a)^3/h$.
- $\theta = \theta_1$ subject to $mg(a-h)^3 \leq ch \leq mg(a+h)^3$.

$$\mathcal{V}''(\theta_1) = -\frac{ch\cos\theta_1}{ma(a^2+h^2-2ah\cos\theta_1)^{3/2}}$$

$$+\frac{3ch^2\sin^2\theta_1}{m(a^2+h^2-2ah\cos\theta_1)^{5/2}} + \frac{g\cos\theta_1}{a}$$

$$= \left(\frac{mg}{ch}\right)^{-(5/3)}\frac{ch^2\sin^2\theta_1}{m}$$

$$> 0.$$

Note that $\mathcal{V}(-\theta_1)$ is also positive Therefore, if they exist, all inclined equilibrium points are centres.

Suppose that the parameters a, h and m are fixed, and that c can be increased from zero. The behaviour of the bob is as follows:

- $0 < c < mg(a-h)^3/h$. There are two equilibrium positions: the bob vertically below the suspension point which is a stable centre, or the bob above which is an unstable saddle,
- c takes the intermediate values defined by (iii). Both the highest and lowest points becomes a saddles. The inclined equilibrium points are centres.
- $c > mg(a+h)^3/h$. The lowest point remains a saddle but the highest point switches back to a saddle.

- **1.33** A pendulum with equation $\ddot{x} + \sin x = 0$ oscillates with amplitude a. Show that its period, T, is equal to $4K(\beta)$, where $\beta = \sin^2 \tfrac{1}{2}a$ and

$$K(\beta) = \int_0^{\frac{1}{2}\pi} \frac{d\phi}{\sqrt{(1 - \beta \sin^2 \phi)}}.$$

The function $K(\beta)$ has the power series representation

$$K(\beta) = \tfrac{1}{2}\pi \left[1 + \left(\tfrac{1}{2}\right)^2 \beta + \left(\tfrac{1.3}{2.4}\right)^2 \beta^2 + \cdots \right], \quad |\beta| < 1.$$

Deduce that, for small amplitudes,

$$T = 2\pi \left(1 + \tfrac{1}{16}a^2 + \tfrac{11}{3072}a^4\right) + O(a^6).$$

1.33. The pendulum equation $\ddot{x} + \sin x = 0$ can be integrated once to give the equation of the phase paths in the form

$$\tfrac{1}{2}\dot{x}^2 - \cos x = C = -\cos a, \tag{i}$$

using the condition that $x = a$ when $\dot{x} = 0$. The origin is a centre about which the paths are symmetric in both the x and $y = \dot{x}$ axes. Without loss of generality assume that $t = 0$ initially. The pendulum completes the first cycle when $x = 2\pi$. From (i) the quarter period is

$$K = \int_0^K dt = \frac{1}{\sqrt{2}} \int_0^a \frac{dx}{\sqrt{(\cos x - \cos a)}} = \frac{1}{2} \int_0^a \frac{dx}{\sqrt{(\sin^2(1/2)a - \sin^2(1/2)x)}}.$$

Now apply the substitution $\sin \tfrac{1}{2}x = \sin \tfrac{1}{2}a \sin \phi$ so that the limits are replaced by $\phi = 0$ and $\phi = \tfrac{1}{2}\pi$. Since

$$\tfrac{1}{2} \cos \tfrac{1}{2}x \frac{dx}{d\phi} = \sin \tfrac{1}{2}a \cos \phi,$$

then

$$K(\beta) = \int_0^{\frac{1}{2}\pi} \frac{d\phi}{\sqrt{(1 - \beta \sin^2 \phi)}}.$$

For small β, expand the integrand in powers of β using the binomial expansion so that

$$K(\beta) = \int_0^{(\frac{1}{2})\pi} \left(1 + \tfrac{1}{2}\beta \sin^2 \phi + \tfrac{3}{8}\beta^2 \sin^4 \phi + \cdots \right) d\phi$$

$$= \tfrac{1}{2}\pi + \tfrac{1}{8}\pi\beta + \tfrac{9}{128}\beta^2 \pi + \cdots$$

Now expand β in powers of a:

$$\beta = \sin^2 \frac{1}{2}a = \frac{1}{2}a - \frac{1}{48}a^3 + O(a^5).$$

Finally

$$T = 4K(\beta)$$

$$= 2\pi \left[1 + \frac{1}{4}\left(\frac{1}{4}a^2 - \frac{1}{48}a^4\right) + \frac{9}{1024}a^4 + O(a^6)\right]$$

$$= 2\pi \left(1 + \frac{1}{16}a^2 + \frac{11}{3072}a^4 \right) + O(a^6).$$

as $a \to 0$.

- **1.34** Repeat Problem 1.33 with the equation $\ddot{x} + x - \varepsilon x^3 = 0$ ($\varepsilon > 0$), and show that

$$T = \frac{4\sqrt{2}}{\sqrt{(2 - \varepsilon a^2)}} K(\beta), \quad \beta = \frac{\varepsilon a^2}{2 - \varepsilon a^2},$$

and that

$$T = 2\pi \left(1 + \frac{3}{8}\varepsilon a^2 + \frac{57}{256}\varepsilon^2 a^4 \right) + O(\varepsilon^3 a^6)$$

as $\varepsilon a^2 \to 0$.

1.34. The damped equation $\ddot{x} + x - \epsilon x^3 = 0$ has phase paths given by

$$\tfrac{1}{2}\dot{x}^2 = \tfrac{1}{4}\epsilon x^4 - \tfrac{1}{2}x^2 + C = \tfrac{1}{4}\epsilon x^4 - \tfrac{1}{2}x^2 - \tfrac{1}{4}\epsilon a^4 + \tfrac{1}{2}a^2,$$
$$= \tfrac{1}{2}(x^2 - a^2)(\epsilon x^2 + \epsilon a^2 - 2)$$

assuming that $x = a$ when $\dot{x} = 0$. The equation has equilibrium points at $x = 0$ and $x = \pm 1/\sqrt{\epsilon}$. Oscillations about the origin (which is a centre) occur if $a\sqrt{\epsilon} < 1$. In this case the period T is given by

$$T = 4\sqrt{2} \int_0^a \frac{dx}{\sqrt{(a^2 x^2)}\sqrt{(2 - \epsilon a^2 - \epsilon x^2)}}$$

$$= 4\sqrt{2} \int_0^{(1/2)\pi} \frac{d\phi}{\sqrt{(2 - \epsilon a^2 - \epsilon a^2 \sin^2 \phi)}} \quad \text{(substituting } x = \sin\phi\text{)}$$

$$= \frac{4\sqrt{2}}{\sqrt{(2 - \epsilon a^2)}} K(\beta)$$

where $\beta = \epsilon a^2/(2 - \epsilon a^2)$ and

$$K(\beta) = \int_0^{(1/2)\pi} \frac{d\phi}{\sqrt{(1 - \beta \sin^2 \phi)}}.$$

From the previous problem, with $\mu = \epsilon a^2$,

$$T = \frac{2\pi\sqrt{2}}{\sqrt{(2-\mu)}} \left[1 + \frac{\mu}{4(2-\mu)} + \frac{9\mu^2}{64(2-\mu)^2} + O(\mu^3) \right]$$

$$= \frac{2\pi\sqrt{2}}{\sqrt{(2-\mu)}} \left[1 + \frac{1}{8}\mu\left(1 + \frac{1}{2}\mu\right) + \frac{9\mu^2}{256} + O(\mu^3) \right]$$

$$= 2\pi \left(1 + \frac{\mu}{4} + \frac{3\mu^2}{32} \right) \left(1 + \frac{\mu}{8} + \frac{25\mu^2}{256} \right) + O(\mu^3)$$

$$= 2\pi \left(1 + \frac{3\mu}{8} + \frac{57\mu^2}{256} \right) + O(\mu^3)$$

as $\mu \to 0$.

- **1.35** Show that the equations of the form $\ddot{x} + g(x)\dot{x}^2 + h(x) = 0$ are effectively conservative. (Find a transformation of x which puts the equations into the usual conservative form. Compare with NODE, eqn (1.59).)

1.35. The significant feature of the equation

$$\ddot{x} + g(x)\dot{x}^2 + h(x) = 0 \tag{i}$$

is the $\dot{x}^2$ term. Let $z = f(x)$, where $f(x)$ is twice differentiable and it is assumed that $z = f(x)$ can be uniquely inverted into $x = f^{-1}(z)$. Differentiating

$$\dot{z} = f'(x)\dot{x}, \quad \ddot{z} = f'(x)\ddot{x} + f''(x)\dot{x}^2.$$

Therefore

$$\dot{x} = \frac{\dot{z}}{f'(x)}, \quad \ddot{x} = \frac{\ddot{z}}{f'(x)} - \frac{f''(x)\dot{x}^2}{f'(x)} = \frac{\ddot{z}}{f'(x)} - \frac{f''(x)\dot{z}^2}{f'(x)^3}.$$

Substitution of these derivatives into (i) results in

$$\ddot{z} - \frac{f''(x)}{f'(x)^2}\dot{z}^2 + \frac{g(x)}{f'(x)}\dot{z}^2 + f'(x)h(x) = 0.$$

The $\dot{z}^2$ can be eliminated by choosing $f'(x)$ so that

$$\frac{f''(x)}{f'(x)} = g(x).$$

Aside from a constant we can put

$$f'(x) = \exp\left[\int^x g(u)du\right],$$

and a further integration leads to

$$f(x) = \int \exp\left[\int^x g(u)du\right]dx.$$

In terms of z the equation becomes

$$\ddot{z} + p(z) = 0,$$

where $p(z) = f'(f^{-1}(z))h(f^{-1}(z))$. Obviously this equation is conservative of the form (1.23).

- **1.36** Sketch the phase diagrams for the following.
  (i) $\dot{x} = y$, $\dot{y} = 0$, (ii) $\dot{x} = y$, $\dot{y} = 1$, (iii) $\dot{x} = y$, $\dot{y} = y$.

1.36. (i) $\dot{x} = y$, $\dot{y} = 0$. All points on the x axis are equilibrium points. The solutions are $x = t + A$ and $y = B$. The phase paths are lines parallel to the x axis (see Figure 1.48).
(ii) $\dot{x} = y$, $\dot{y} = 1$. There are no equilibrium points. The equation for phase paths is

$$\frac{dy}{dx} = \frac{1}{y},$$

whose general solution is given by $y^2 = 2x + C$. The phase paths are congruent parabolas with the x axis as the common axis (see Figure 1.48).
(iii) $\dot{x} = y$, $\dot{y} = y$. All points on the x axis are equilibrium points. The phase paths are given by

$$\frac{dy}{dx} = 1 \quad \Rightarrow \quad y = x + C,$$

which are parallel inclined straight lines (see Figure 1.49). All equilibrium points are unstable.

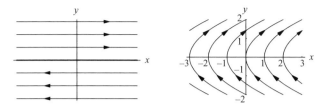

Figure 1.48 Problem 1.36: The phase diagrams for (i) and (ii).

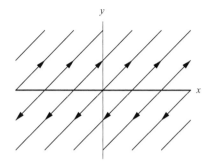

Figure 1.49 Problem 1.36: The phase diagram (iii).

- **1.37** Show that the phase plane for the equation
$$\ddot{x} - \varepsilon x \dot{x} + x = 0, \quad \varepsilon > 0$$
has a centre at the origin, by finding the equation of the phase paths.

1.37. The differential equation for the phase paths of
$$\ddot{x} - \varepsilon x \dot{x} + x = 0,$$
is
$$y \frac{dy}{dx} - \varepsilon xy + x = 0.$$
This is a separable equation having the general solution
$$\varepsilon \int x \, dx = \int \frac{y \, dy}{y - \varepsilon^{-1}} + C = \int \left(1 + \frac{\varepsilon^{-1}}{y - \varepsilon^{-1}}\right) dy + C,$$

or

$$\tfrac{1}{2}\varepsilon x^2 = y + \varepsilon^{-1} \ln |y - \varepsilon^{-1}| + C, \tag{i}$$

where C is a constant. Note that there is a singular solution $y = \varepsilon^{-1}$. The system has a single equilibrium point, at the origin.

To establish a centre it is sufficient to show that all paths in some neighbourhood of the point are closed, so we may restrict consideration to the region $y < \varepsilon^{-1}$. On this range put

$$F(y) = -y - \varepsilon^{-1} \ln|y - \varepsilon^{-1}| = -y - \varepsilon^{-1} \ln(\varepsilon^{-1} - y). \tag{ii}$$

Then from (i) and (ii) we can express the paths as the union of two families of curves:

$$x = \sqrt{2}\varepsilon^{-(1/2)}\{C - F(y)\}^{1/2} \geq 0, \tag{iii}$$

and

$$x = -\sqrt{2}\varepsilon^{-(1/2)}\{C - F(y)\}^{1/2}, \tag{iv}$$

(wherever $C - F(y)$ is non-negative). The curves in (iv) are the reflections in the y axis of those in (iii), and the families join up smoothly across this axis.

Evidently, for $y < \varepsilon^{-1}$,

$$F(0) = -\varepsilon^{-1} \ln(\varepsilon^{-1}) \tag{v}$$

and

$$F'(y) = \frac{y}{\varepsilon^{-1} - y} = \begin{cases} < 0 & \text{if } y < 0 \\ \text{zero} & \text{if } y = 0 \\ > 0 & \text{if } y > 0 \end{cases}. \tag{vi}$$

Therefore $F(y)$ has a minimum at $y = 0$. Also $F(y)$ is strictly increasing in both directions away from $y = 0$ and (from (ii)) $F(y) \to +\infty$ as $y \to -\infty$ and as $y \to \varepsilon^{-1}$ from below.

Consider eqn (iii), using (v) and (vi). If

$$-\varepsilon^{-1} \ln(\varepsilon^{-1}) < C < \infty \tag{vii}$$

there are exactly two values in the range $-\infty < y < \varepsilon^{-1}$ at which the factor $C - F(y)$, and hence x, becomes zero, and between these values $x > 0$. The corresponding reflected path segment given by (iv) completes a closed path, having parameter C. A representative phase diagram is given in Figure 1.50. The unclosed paths correspond to values of $y > \varepsilon^{-1}$: their boundary is the singular solution mentioned above.

- **1.38** Show that the equation $\ddot{x} + x + \varepsilon x^3 = 0$ ($\varepsilon > 0$) with $x(0) = a$, $\dot{x}(0) = 0$ has phase paths given by

$$\dot{x}^2 + x^2 + \tfrac{1}{2}\varepsilon x^4 = (1 + \tfrac{1}{2}\varepsilon a^2)a^2.$$

Show that the origin is a centre. Are all phase paths closed, and hence all solutions periodic?

1.38. The differential equation of the phase paths of

$$\ddot{x} + x + \varepsilon x^3 = 0, \quad (\varepsilon > 0)$$

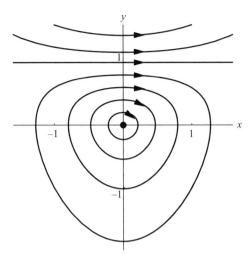

Figure 1.50 Problem 1.37: The phase diagram for $\ddot{x} - \varepsilon x \dot{x} + x = 0$ with $\varepsilon = 1$.

is given by

$$y\frac{dy}{dx} = -x - \varepsilon x^3.$$

Given the conditions $x(0) = a$ and $\dot{x}(0) = 0$, integration of the differential equation gives the phase paths

$$\dot{x}^2 + x^2 + \tfrac{1}{2}\varepsilon x^4 = \text{constant} = (1 + \tfrac{1}{2}\varepsilon a^2)a^2.$$

The equation has a single equilibrium point, at the origin. This is a conservative system (see NODE, Section 1.3) with potential function

$$\mathcal{V}(x) = \int (x + \varepsilon x^3) dx = \frac{1}{2}x^2 + \frac{1}{4}\varepsilon x^4.$$

Differentiating $\mathcal{V}(x)$ twice we obtain

$$\mathcal{V}'(x) = x + \tfrac{1}{2}\varepsilon x^3, \quad \mathcal{V}''(x) = 1 + \tfrac{3}{2}\varepsilon x^2.$$

Therefore $\mathcal{V}'(0) = 0$ and $\mathcal{V}''(0) = 1 > 0$ which means that $\mathcal{V}(x)$ has a minimum at the origin. Locally the origin is a centre. However, $\mathcal{V}'(x) < 0$ for $x < 0$ and $\mathcal{V}'(x) > 0$ for $x < 0$, and also $\mathcal{V}(x) \to \infty$ as $x \to \pm\infty$. These conditions imply that, for every a, $\dot{x}$ is also zero at $x = -a$ and y is continuous between zero at $x = -a$ and $x = a$. Hence every path is closed and all solutions periodic.

- **1.39** Locate the equilibrium points of the equation
$$\ddot{x}+\lambda+x^3-x=0.$$
in the x, λ plane. Show that the phase paths are given by
$$\tfrac{1}{2}\dot{x}^2+\lambda x+\tfrac{1}{4}\lambda x^4-\tfrac{1}{2}x^2=\text{constant}.$$
Investigate the stability of the equilibrium points.

1.39. Consider the parameter-dependent system

$$\ddot{x}+\lambda+x^3-x=0.$$

Using the notation of Section 1.7 (in NODE), let $f(x, \lambda) = x - x^3 - \lambda$. Equilibrium points occur where $f(x, \lambda) = 0$ which is shown as the curve in Figure 1.51. The function $f(x, \lambda)$ is positive in the shaded region. Points on the curve $\lambda = x - x^3$ above the shaded areas are stable and all other points are unstable. Treating λ as a function of x, $x - x^3$ has stationary points at $x = \pm 1/\sqrt{3}$ where $\lambda = \pm 2/\sqrt{3}$ as indicated in Figure 1.51. Therefore if $-2/\sqrt{3} < \lambda < 2/\sqrt{3}$ the equation has three equilibrium points; if $\lambda = \pm 2/\sqrt{3}$ the equation has two; for all other values of λ the equation has one equilibrium point. The phase paths satisfy the differential equation

$$y\frac{dy}{dx} = -x^3 + x - \lambda,$$

where $y = \dot{x}$. Integrating, the phase paths are given by

$$\tfrac{1}{2}y^2 + \lambda x + \tfrac{1}{4}x^4 - \tfrac{1}{2}x^2 = \text{constant}.$$

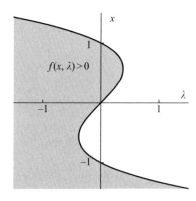

Figure 1.51 Problem 1.39: Graph showing equilibrium points on $\lambda = x - x^3$.

- **1.40** Burgers' equation
$$\frac{\partial \phi}{\partial t} + \phi \frac{\partial \phi}{\partial x} = c \frac{\partial^2 \phi}{\partial x^2}$$
shows diffusion and nonlinear effects in fluid mechanics (see Logan (1994)). Find the equation for permanent waves by putting $\phi(x,t) = U(x-ct)$, where c is the constant wave speed. Find the equilibrium points and the phase paths for the resulting equation and interpret the phase diagram.

1.40. Let $\phi(x,t) = U(x - ct)$ in Burgers' equation
$$\frac{\partial \phi}{\partial t} + \phi \frac{\partial \phi}{\partial x} = c \frac{\partial^2 \phi}{\partial x^2},$$
so that $U(x - ct)$ satisfies the ordinary differential equation
$$-cU'(w) + U(w)U'(w) = cU''(w),$$
where $w = x - ct$. All values of w are equilibrium points. Let $V = U'$. Then the phase paths in the (U, V) plane are given by
$$c \frac{dV}{dU} = U - c,$$
which has the general solution
$$cV = \tfrac{1}{2}(U - c)^2 + A. \tag{i}$$
The phase paths are congruent parabolas all with the axis $U = c$ as shown in Figure 1.52. Phase paths are bounded for $V < 0$ and unbounded for $V > 0$: the latter do not have an obvious physical interpretation.

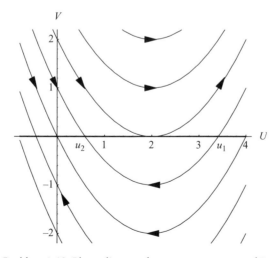

Figure 1.52 Problem 1.40: Phase diagram for permanent waves of Burger's equation.

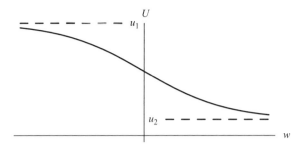

Figure 1.53 Problem 1.40: A permanent waveform of Burgers' equation.

Burgers' equation describes a convection-diffusion process, and a solution $U(x - ct)$ is the shape of a wavefront. For $U' < 0$, the wavefront starts at $U = u_1$, say, and terminates at $U = u_2$. We can obtain an explicit form for the wave if we assume that $U \to u_1$ as $w \to \infty$, and $U \to u_2$ as $w \to -\infty$, and $U' \to 0$ in both cases, with $u_2 < u_1$. Hence from (i),

$$A = -\tfrac{1}{2}(u_1 - c)^2 = -\tfrac{1}{2}(u_2 - c)^2,$$

so that $c = \tfrac{1}{2}(u_1 + u_2)$ and $A = -\tfrac{1}{8}(u_1 - u_2)^2$. Hence (i) becomes

$$-c\dot{U} = (U - u_1)(U - u_2).$$

This is a separable equation with solution

$$w = \frac{u_1 + u_2}{u_2 - u_1} \int \left(\frac{dU}{U - u_1} + \frac{dU}{u_2 - U} \right) = \frac{u_1 + u_2}{u_2 - u_1} \ln \left[\frac{u_2 - U}{U - u_1} \right].$$

Solving for U we obtain

$$U(x - ct) = u_1 + \frac{u_2 - u_1}{1 + \exp[(u_2 - u_1)(x - ct)/(u_1 + u_2)]}.$$

The shape of the waveform is indicated in Figure 1.53.

- **1.41** A uniform rod of mass m and length L is smoothly pivoted at one end and held in a vertical position of equilibrium by two unstretched horizontal springs, each of stiffness k, attached to the other end as shown in Figure 1.37 (in NODE) or Figure 1.54. The rod is free to oscillate in a vertical plane through the springs and the rod. Find the potential energy $\mathcal{V}(\theta)$ of the system when the rod is inclined at an angle θ to the upward vertical. For small θ confirm that

$$\mathcal{V}(\theta) \approx (kL - \tfrac{1}{4}mg)L\theta^2,$$

Sketch the phase diagram for small $|\theta|$, and discuss the stability of this inverted pendulum.

1: Second-order differential equations in the phase plane

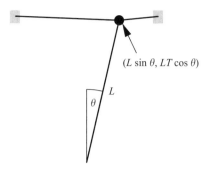

Figure 1.54 Problem 1.41: Inverted pendulum inclined at angle θ.

1.41. Let the distance between the supports be $2a$, and let the origin be at the hinge with horizontal and vertical axes. The supports have coordinates (a, L) and $(-a, L)$. The potential energy includes contributions from the height of the bob and the springs. We assume that the springs obey Hooke's law which states that

$$\text{potential energy} = \tfrac{1}{2}(\text{stiffness}) \times (\text{extension})^2.$$

Therefore the potential energy is

$$V(\theta) = \tfrac{1}{2}mgL\cos\theta - \tfrac{1}{2}mgL + \tfrac{1}{2}k\{\sqrt{[(a - L\sin\theta)^2 + (L - L\cos\theta)^2]} - a\}^2$$

$$+ \tfrac{1}{2}k\{\sqrt{[(a + L\sin\theta)^2 + (L - L\cos\theta)^2]} - a\}^2$$

defined so that $V(0) = 0$.

For $|\theta|$ small, use the approximations $\sin\theta \approx \theta$ and $\cos\theta \approx 1 - \tfrac{1}{2}\theta^2$. Then

$$V(\theta) \approx -\tfrac{1}{4}mgL\theta^2 + \tfrac{1}{2}k\{\sqrt{[(a - L\theta)^2 + L\theta^2]} - a\}^2$$

$$+ \tfrac{1}{2}k\{\sqrt{[(a + L\theta)^2 + L\theta^2]} - a\}^2,$$

$$= -\tfrac{1}{4}mgL\theta^2 + \tfrac{1}{2}k\left[a\left(1 - \frac{2L\theta}{a} + \frac{2L^2\theta^2}{a^2}\right)^{1/2} - a\right]^2$$

$$+ \tfrac{1}{2}k\left[a\left(1 + \frac{2L\theta}{a} + \frac{2L^2\theta^2}{a^2}\right)^{1/2} - a\right]^2$$

$$\approx -\tfrac{1}{4}mgL\theta^2 + kL^2\theta^2$$

as required.

The potential energy is a minimum if $Lk > \frac{1}{4}mg$ (springs with strong stiffness) which means that the bob oscillates about equilibrium in a centre. If $Lk < \frac{1}{4}mg$ (weak stiffness) then the potential energy has a maximum (saddle point in the phase diagram) and the bob is in unstable equilibrium. The phase diagrams are typically those shown in Figure 1.12 in NODE.

- **1.42** Two stars, each with gravitational mass μ, are orbiting each other under their mutual gravitational forces in such a way that their orbits are circles of radius a. A satellite of relatively negligible mass is moving on a straight line through the mass centre G such that the line is perpendicular to the plane of the mutual orbits of this binary system. Explain why the satellite will continue to move on this line. If z is the displacement of the satellite from G, show that
$$\ddot{z} = -\frac{2\mu z}{(a^2 + z^2)^{3/2}}.$$
Obtain the equations of the phase paths. What type of equilibrium point is $z=0$?

1.42. Let F be the gravitational force on the satellite S (mass m) due to one of the stars as shown in Figure 1.55. By the inverse square law
$$F = \frac{m\mu}{z^2 + a^2}.$$

Resolution in the direction GS gives
$$-\frac{2\mu \cos\theta}{z^2 + a^2} = m\ddot{z},$$

or
$$\frac{2\mu z}{(z^2 + a^2)^{\frac{3}{2}}} = \ddot{z}. \qquad (i)$$

Transverse forces balance which means that the satellite will continue to move along the z axis.

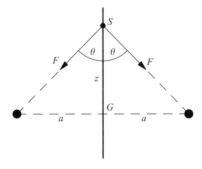

Figure 1.55 Problem 1.42.

The satellite has a single equilibrium point, at G. The phase paths are given by

$$\frac{1}{2}\dot{z}^2 = -2\mu \int \frac{z\,dz}{(z^2+a^2)^{3/2}} = \frac{2\mu}{(z^2+a^2)^{1/2}} + C.$$

From (i), near the origin

$$\ddot{z} \approx -\frac{3\mu z}{a^3},$$

which indicates that the equilibrium point is a centre.

- 1.43 A long wire is bent into the shape of a smooth curve with equation $z = f(x)$ in a fixed vertical (x, z) plane (assume that $f'(x)$ and $f''(x)$ are continuous). A bead of mass m can slide on the wire: assume friction is negligible. Find the kinetic and potential energies of the bead, and write down the equation of the phase paths. Explain why the method of Section 1.3 concerning the phase diagrams for stationary values of the potential energy still holds.

1.43. The bead and wire are shown in Figure 1.56. The components of the velocity of the bead are given by $(\dot{x}, \dot{z} = f'(x)\dot{x})$. The kinetic and potential energies are

$$T = \frac{1}{2}m(\dot{x}^2 + \dot{z}^2) = \frac{1}{2}m(1 + f'(x)^2)y^2, \quad V = mgy = mgf(x),$$

where $y = \dot{x}$ in the phase plane.

Equilibrium points occur where $\mathcal{V}'(x) = 0$, that is, at the stationary points of the curve $z = f(x)$. Suppose that a stationary value occurs at $x = x_1$, so that $f'(x_1) = 0$. At the equilibrium point let $C_1 = mgf(x_1)$. Suppose that $x = x_1$ is a minimum. Then for $C > C_1$ but sufficiently close to C_1, the phase path will be

$$y^2 = \frac{2[C - mgf(x)]}{m[1 + f'(x)^2]},$$

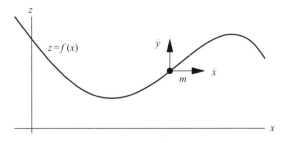

Figure 1.56 Problem 1.43.

and y will be zero where x satisfies $C_1 = mgf(x)$. As in Figure 1.12 (in NODE), the equilibrium point will be a centre. Similar arguments apply to the other types of stationary values.

- **1.44** In the previous problem suppose that friction between the bead and the wire is included. Assume linear damping in which motion is opposed by a frictional force proportional (factor k) to the velocity. Show that the equation of motion of the bead is given by
$$m(1+f'(x)^2)\ddot{x} + mf''(x)\dot{x}^2 + k\dot{x}(1+f'(x)^2) + mgf'(x) = 0,$$
where m is its mass.

Suppose that the wire has the parabolic shape given by $z = x^2$ and that dimensions are chosen so that $k = m$ and $g = 1$. Compute the phase diagram in the neighbourhood of the origin, and explain general features of the diagram near and further away from the origin. (Further theory and experimental work on motion on tracks can be found in the book by Virgin (2000).)

1.44. Let R be the normal reaction of the bead on the wire and let F be the frictional force opposing the motion as shown in Figure 1.57. The horizontal and vertical equations of motion are
$$-R\sin\theta - F\cos\theta = m\ddot{x}, \tag{i}$$
$$R\cos\theta - mg - F\sin\theta = m\ddot{z}, \tag{ii}$$
where θ is the inclination of the tangent of the curve at the bead. Since $z = f(x)$, then $\dot{z} = f'(x)\dot{x}$ and $\ddot{z} = f''(x)\dot{x}^2 + f'(x)\ddot{x}$. Eliminate R between (i) and (ii):
$$-[mg + k\dot{x} f'(x)]\sin\theta - F(\sin^2\theta + \cos^2\theta) = m[f''(x)\dot{x}^2 + f''(x)\ddot{x}] + m\ddot{x}\cos\theta, \tag{iii}$$
where $\sin\theta = f'(x)/\sqrt{[1+f'(x)^2]}$ and $\cos\theta = 1/\sqrt{[1+f'(x)^2]}$. The frictional force F is proportional to the velocity and opposes the motion: therefore
$$F = -k(\dot{x}\cos\theta + \dot{z}\sin\theta) = -\dot{x}\sqrt{[1+f'(x)^2]}.$$

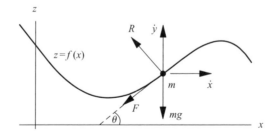

Figure 1.57 Problem 1.44.

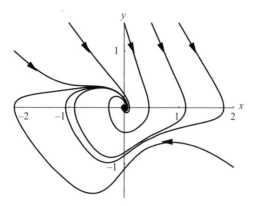

Figure 1.58 Problem 1.44: Phase diagram for $(1+4x^2)\ddot{x} + 2\dot{x}^2 + \dot{x}(1+4x^2) = -2x$.

Finally, the equation of motion is

$$m(1+f'(x)^2)\ddot{x} + mf''(x)\dot{x}^2 + k\dot{x}(1+f'(x)^2) + mgf'(x) = 0.$$

If $f(x) = x^2$, $m = k$ and $g = 1$, then the equation of motion becomes

$$(1+4x^2)\ddot{x} + 2\dot{x}^2 + \dot{x}(1+4x^2) + 2x = 0.$$

The phase diagram is shown in Figure 1.58 in the region $-2 \le x \le 2$, $-1.5 \le y \le 1.5$. For small x and y,

$$\ddot{x} + \dot{x} + 2x \approx 0,$$

neglecting the x^2 and y^2 terms. Locally the phase diagram is a stable spiral, although further away from the origin the spiral shape is distorted by the nonlinear terms.

2 Plane autonomous systems and linearization

• 2.1 Sketch phase diagrams for the following linear systems and classify the equilibrium point:

(i) $\dot{x} = x - 5y$, $\dot{y} = x - y$;
(ii) $\dot{x} = x + y$, $\dot{y} = x - 2y$;
(iii) $\dot{x} = -4x + 2y$, $\dot{y} = 3x - 2y$;
(iv) $\dot{x} = x + 2y$, $\dot{y} = 2x + 2y$;
(v) $\dot{x} = 4x - 2y$, $\dot{y} = 3x - y$;
(vi) $\dot{x} = 2x + y$, $\dot{y} = -x + y$.

2.1. A classification table for equilibrium points of the general linear system

$$\dot{x} = ax + by, \quad \dot{y} = cx + dy$$

is given in Section 2.5 (see also Figure 2.10 in NODE). The key parameters are $p = a + d$, $q = ad - bc$, $\Delta = p^2 - 4q$. All the systems below have an isolated equilibrium point at the origin. The scales on the axes are the same for each phase diagram but actual scales are unnecessary since the equations are homogeneous in x and y. Directions are determined by continuity from directions of $\dot{x}$ and $\dot{y}$ at convenient points in the plane.

Alternatively, classification can be decided by finding the eigenvalues of the matrix of coefficients:

$$A = \begin{bmatrix} a & b \\ c & d \end{bmatrix}.$$

(i) $\dot{x} = x - 5y$, $\dot{y} = x - y$. The parameters are

$$p = 1 - 1 = 0, \quad q = -1 + 5 = 4 > 0, \quad \Delta = 0 - 16 = -16 < 0.$$

Therefore the origin is a centre as shown in Figure 2.1(i).
The eigenvalues of

$$A = \begin{bmatrix} 1 & -5 \\ 1 & -1 \end{bmatrix}$$

are given by

$$\begin{vmatrix} 1-\lambda & -5 \\ 1 & -1-\lambda \end{vmatrix} = \lambda^2 + 4 = 0.$$

The eigenvalues take the imaginary values $\pm 2i$, which is to be expected for a centre.

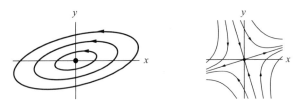

Figure 2.1 Problem 2.1(i): $\dot{x} = x - 5y$, $\dot{y} = x - y$, centre; (ii) $\dot{x} = x + y$, $\dot{y} = x - 2y$, saddle.

(ii) $\dot{x} = x + y$, $\dot{y} = x - 2y$. The parameters are

$$p = 1 - 2 = -1 < 0, \quad q = -2 - 1 = -3 < 0, \quad \Delta = 1 + 12 = 13 > 0.$$

Therefore the origin is a saddle. Its asymptotes can be found by putting $y = mx$ into the equation for the phase paths which is

$$\frac{dy}{dx} = \frac{x - 2y}{x + y}.$$

The result is

$$m = \frac{1 - 2m}{1 + m}, \quad \text{so that } m^2 + 3m - 1 = 0.$$

Therefore the slopes of the asymptotes are

$$m_1, m_2 = \tfrac{1}{2}(-3 \pm \sqrt{13}).$$

The asymptotes and some phase paths are shown in Figure 2.1(ii).

(iii) $\dot{x} = -4x + 2y$, $\dot{y} = 3x - 2y$. The parameters are

$$p = -4 - 2 = -6 < 0, \quad q = 8 - 6 = 2 > 0, \quad \Delta = 36 - 8 = 28 > 0.$$

Therefore the origin is a stable node. The radial straight paths are given by $y = mx$ where

$$m = \frac{3 - 2m}{-4 + 2m} \quad \text{or } 2m^2 - 2m - 3 = 0.$$

Hence the radial paths are

$$y = m_1 x, \quad y = m_2 x, \quad \text{where } m_1, m_2 = \frac{1}{2}(1 \pm \sqrt{7}).$$

The radial paths and some phase paths are shown in Figure 2.2(iii).

(iv) $\dot{x} = x + 2y$, $\dot{y} = 2x + 2y$. The parameters are

$$p = 1 + 2 = 3 > 0, \quad q = 2 - 4 = -2 < 0, \quad \Delta = 9 + 8 = 17 > 0.$$

The origin is a saddle. The slopes of the asymptotes are

$$m_1, m_2 = \tfrac{1}{2}(1 \pm \sqrt{17}).$$

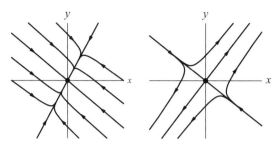

Figure 2.2 Problem 2.1(iii) :$\dot{x} = -4x + 2y$, $\dot{y} = 3x - 2y$, stable node; (iv) $\dot{x} = x + 2y$, $\dot{y} = 2x + 2y$, saddle.

See Figure 2.2(iv)

(v) $\dot{x} = 4x - 2y$, $\dot{y} = 3x - y$. The parameters are

$$p = 4 - 1 = 3 > 0, \quad q = -4 + 6 = 2 > 0, \quad \Delta = 9 - 8 = 1 > 0.$$

Therefore the origin is an unstable node. The radial paths have slopes $m_1 = \frac{1}{2}$ and $m_2 = 3$ and equations

$$y = \tfrac{1}{2}x, \quad y = 3x.$$

The phase diagram is shown in Figure 2.3(v).

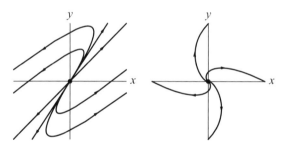

Figure 2.3 Problem 2.1(v): $\dot{x} = 4x - 2y$, $\dot{y} = 3x - y$, unstable node; (vi) $\dot{x} = 2x + y$, $\dot{y} = -x + y$, unstable spiral.

(vi) $\dot{x} = 2x + y$, $\dot{y} = -x + y$. The parameters are

$$p = 2 + 1 = 3 > 0, \quad q = 2 + 1 = 3 > 0, \quad \Delta = 9 - 12 = -3 < 0.$$

The origin is an unstable spiral. Some phase paths are shown in Figure 2.3(vi).

- **2.2** Some of the following systems either generate a single eigenvalue, or a zero eigenvalue, or in other ways vary the types illustrated in Section 2.5. Sketch their phase diagrams
  (i) $\dot{x} = 3x - y$, $\dot{y} = x + y$;
  (ii) $\dot{x} = x - y$, $\dot{y} = 2x - 2y$;
  (iii) $\dot{x} = x$, $\dot{y} = 2x - 3y$;

(iv) $\dot{x} = x, \dot{y} = x + 3y$;
(v) $\dot{x} = -y, \dot{y} = 2x - 4y$;
(vi) $\dot{x} = x, \dot{y} = y$;
(vii) $\dot{x} = 0, \dot{y} = x$.

2.2. Note that the scales on both axes are the same.

(i) $\dot{x} = 3x - y, \dot{y} = x + y$. Using the classification table (See Section 2.5)

$$p = 3 + 1 = 4 > 0, \quad q = 3 + 1 = 4 > 0, \quad \Delta = 16 - 16 = 0.$$

Hence the origin is an unstable degenerate node with a repeated eigenvalue of $m = 1$. The straight line $y = x$ contains radial paths (Figure 2.4).

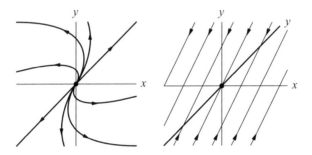

Figure 2.4 Problem 2.2(i): $\dot{x} = 3x - y, \dot{y} = x + y$, unstable degenerate node; (ii) $\dot{x} = x - y, \dot{y} = 2x - 2y$, parallel paths.

(ii) $\dot{x} = x - y, \dot{y} = 2x - 2y$. All points on the line $y = x$ are equilibrium points. The phase paths are given by

$$\frac{dy}{dx} = 2 \quad \Rightarrow \quad y = 2x + C,$$

which is a family of parallel straight lines (Figure 2.4(ii)).

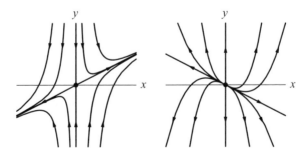

Figure 2.5 Problem 2.2(iii): $\dot{x} = x, \dot{y} = 2x - 3y$, saddle; (iv) $\dot{x} = x, \dot{y} = x + 3y$, unstable node with the y axis as radial paths.

(iii) $\dot{x} = x$, $\dot{y} = 2x - 3y$. The parameters are

$$p = 1 - 3 = -2 < 0, \quad q = -3 < 0, \quad \Delta = 4 + 12 = 16 > 0,$$

which implies that the equilibrium point is a saddle. From the first equation, the axis $x = 0$ is a solution, as also is $y = \frac{1}{2}x$. These lines are the asymptotes of the saddle point (Figure 2.5).

(iv) $\dot{x} = x$, $\dot{y} = x + 3y$. The parameters are

$$p = 1 + 3 = 4 > 0, \quad q = 3 > 0, \quad \Delta = 16 - 12 = 4 > 0,$$

which is an unstable node with radial paths along $x = 0$ and $y = -\frac{1}{2}x$ (Figure 2.5).

(v) $\dot{x} = -y$, $\dot{y} = 2x - 4y$. The parameters are

$$p = -4 < 0, \quad q = 2 > 0, \quad \Delta = 16 - 8 = 8 > 0,$$

This is a stable node but with eigenvalues $2 \pm 2\sqrt{2}$ of differing signs. This produces a similar phase diagram to that for Problem 1(iii).

(vi) $\dot{x} = x$, $\dot{y} = y$. The parameters are

$$p = 2, \quad q = 1, \quad \Delta = 4 - 4 = 0,$$

which makes it a degenerate case between an unstable node and an unstable spiral. The phase paths are given by

$$\frac{dy}{dx} = \frac{y}{x} \Rightarrow y = Cx,$$

which is a family of radial straight lines as shown in Figure 2.6(vi): it is a it star-shaped phase diagram.

(vii) $\dot{x} = 0$, $\dot{y} = x$. All points on the y axis are equilibrium points. The parameter values are $p = q = \Delta = 0$ which makes this a degenerate case. The equations can be solved directly to give

$$x = C, \quad y = \int x\,dt + D = \int C\,dt + D = Ct + D.$$

Hence the phase diagram (shown in Figure 2.6(vii)) consists of all lines $x = C$ parallel to the y axis.

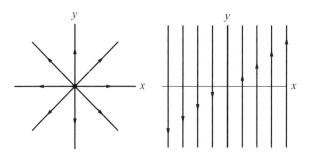

Figure 2.6 Problem 2.2(vi): $\dot{x} = x$, $\dot{y} = y$, saddle; (vii) $\dot{x} = x$, $\dot{y} = x$, unstable node with the y axis as radial paths.

- 2.3 Locate and classify the equilibrium points of the following systems. Sketch the phase diagrams: it will often be helpful to obtain isoclines and path directions at other points in the plane.

(i) $\dot{x} = x - y$, $\dot{y} = x + y - 2xy$;
(ii) $\dot{x} = ye^y$, $\dot{y} = 1 - x^2$;
(iii) $\dot{x} = 1 - xy$, $\dot{y} = (x-1)y$;
(iv) $\dot{x} = (1 + x - 2y)x$, $\dot{y} = (x-1)y$;
(v) $\dot{x} = x - y$, $\dot{y} = x^2 - 1$;
(vi) $\dot{x} = -6y + 2xy - 8$, $\dot{y} = y^2 - x^2$;
(vii) $\dot{x} = 4 - 4x^2 - y^2$, $\dot{y} = 3xy$;
(viii) $\dot{x} = -y\sqrt{(1-x^2)}$, $\dot{y} = x\sqrt{(1-x^2)}$ for $|x| \leq 1$;
(ix) $\dot{x} = \sin y$, $\dot{y} = -\sin x$;
(x) $\dot{x} = \sin x \cos y$, $\dot{y} = \sin y \cos x$.

2.3. For the system $\dot{x} = X(x, y)$, $\dot{y} = Y(x, y)$, the equilibrium points are given by solutions of $X(x, y) = 0$, $Y(x, y) = 0$. The linear approximations (Section 2.3) near each equilibrium point are classified using the table in Section 2.5, or Figure 2.10 (both in NODE). Curve sketching can be helped by plotting the isoclines $Y(x, y) = 0$ (phase paths locally parallel to the x axis) and $X(x, y) = 0$ (phase paths locally parallel to the y axis). Since these problems are nonlinear, scales along the axes are now significant.

(i) $\dot{x} = x - y$, $\dot{y} = x + y - 2xy$. The equilibrium points are given by

$$x - y = 0, \quad x + y - 2xy = 0.$$

There are two equilibrium points, at $(0, 0)$ and $(1, 1)$.

(a) $(0, 0)$. The linear approximation is

$$\dot{x} = x - y, \quad \dot{y} \approx x + y.$$

Hence the parameters are

$$p = 2 > 0, \quad q = 1 + 1 = 2 > 0, \quad \Delta = 4 - 8 = -4 < 0,$$

which means that the origin is locally an unstable spiral.

(b) $(1, 1)$. Put $x = 1 + \xi$ and $y = 1 + \eta$. The linear approximation is

$$\dot{\xi} = \xi - \eta, \quad \dot{\eta} \approx -\xi - \eta.$$

For this linear approximation the parameters are

$$p = 0, \quad q = -2 < 0, \quad \Delta = 0 + 8 = 4 > 0,$$

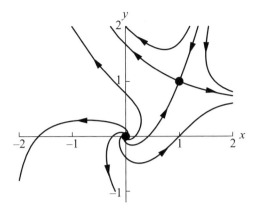

Figure 2.7 Problem 2.3(i): $\dot{x} = x - y$, $\dot{y} = x + y - 2xy$.

which means that $(1, 1)$ is locally a saddle point with asymptotes in the directions of the slopes $1 \pm \sqrt{2}$.

The zero-slope isocline is the curve $x + y - 2xy = 0$ and the infinite-slope isocline is the line $y = x$. A computed phase diagram is shown in Figure 2.7.

(ii) $\dot{x} = ye^y$, $\dot{y} = 1 - x^2$. The equilibrium points are given by

$$ye^y = 0, \quad 1 - x^2 = 0.$$

Therefore there are two equilibrium points, at $(1, 0)$ and $(-1, 0)$.

(a) $(1, 0)$. Put $x = 1 + \xi$. The linear approximation is

$$\dot{\xi} \approx y, \quad \dot{y} \approx -2\xi.$$

The parameters are

$$p = 0, \quad q = 2 > 0, \quad \Delta = -8 < 0,$$

from which we infer that the $(1, 0)$ is a centre.

(b) $(-1, 0)$. Put $x = -1 + \xi$. The linear approximation is

$$\dot{\xi} \approx y, \quad \dot{y} \approx 2\xi.$$

The parameters are

$$p = 0, \quad q = -2 < 0, \quad \Delta = 8 > 0,$$

which implies that $(-1, 0)$ is a saddle. The phase diagram is shown in Figure 2.8. Note that the isoclines of zero slope are the straight lines $x = \pm 1$.

(iii) $\dot{x} = 1 - xy$, $\dot{y} = (x - 1)y$. The equilibrium points are given by

$$1 - xy = 0, \quad (x - 1)y = 0,$$

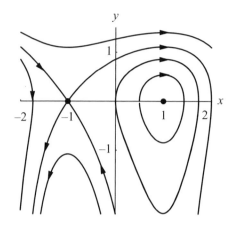

Figure 2.8 Problem 2.3(ii) : $\dot{x} = ye^y$, $\dot{y} = 1 - x^2$.

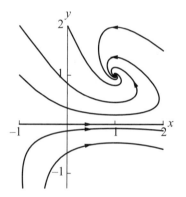

Figure 2.9 Problem 2.3(iii): $\dot{x} = 1 - xy$, $\dot{y} = (x - 1)y$.

which has the single solution $(1, 1)$. Let $x = 1 + \xi$ and $y = 1 + \eta$. Then the linear approximation is

$$\dot{\xi} \approx -\xi - \eta, \quad \dot{\eta} \approx \xi.$$

The parameters are

$$p = -1 < 0, \quad q = 1 > 0, \quad \Delta = 1 - 4 = -3 < 0,$$

which means that $(1, 1)$ is a stable spiral. Note that $y = 0$ is a phase path. The phase diagram is shown in Figure 2.9.

(iv) $\dot{x} = (1 + x - 2y)x$, $\dot{y} = (x - 1)y$. The equilibrium points are given by

$$(1 + x - 2y)x = 0, \quad (x - 1)y = 0.$$

There are three equilibrium points: at $(0, 0)$, $(1, 1)$ and $(-1, 0)$. Note that the axes $x = 0$ and $y = 0$ are phase paths. The straight line $x = 1$ is an isocline of zero slope.

(a) $(0, 0)$. The linear approximation is

$$\dot{x} \approx x, \quad \dot{y} = -y.$$

The parameters are

$$p = 0, \quad q = -1 < 0, \quad \Delta = -4 < 0,$$

which implies that $(0, 0)$ is a saddle.

(b) $(1, 1)$. Let $x = 1 + \xi$ and $y = 1 + \eta$. The linear approximation is

$$\dot{\xi} = (2 + \xi - 2 - 2\eta)(1 + \xi) \approx \xi - 2\eta, \quad \dot{\eta} = \xi.$$

The parameters are

$$p = 1 > 0, \quad q = 2 > 0, \quad \Delta = 1 - 4 = -3 < 0.$$

Hence $(1, 1)$ is an unstable spiral

(c) $(-1, 0)$. Let $x = -1 + \xi$. Then the linear approximation is

$$\dot{\xi} \approx -\xi + 2y, \quad \dot{y} \approx -2y.$$

Hence the parameters are

$$p = -3 < 0, \quad q = 2 > 0, \quad \Delta = 9 - 8 = 1 > 0,$$

which means that $(-1, 0)$ is a stable node.

The phase diagram is shown in Figure 2.10.

(v) $\dot{x} = x - y$, $\dot{y} = x^2 - 1$. The equilibrium points are given by

$$x - y = 0, \quad x^2 - 1 = 0.$$

Therefore the equilibrium points occur at $(1, 1)$ and $(-1, -1)$. The isoclines of zero slope are the lines $x = \pm 1$, and the isocline of infinite slope is the line $y = x$.

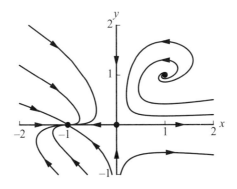

Figure 2.10 Problem 2.3(iv): $\dot{x} = (1 + x - 2y)x$, $\dot{y} = (x - 1)y$.

(a) $(1, 1)$. Let $x = 1 + \xi$ and $y = 1 + \eta$. The linear approximation is

$$\dot\xi = \xi - \eta, \quad \dot\eta \approx 2\xi,$$

which has the parameters

$$p = 1 > 0, \quad q = 2 > 0, \quad \Delta = 1 - 8 = -7 < 0.$$

Hence $(1, 1)$ is an unstable spiral.

(b) $(-1, -1)$. Let $x = -1 + \xi$ and $y = -1 + \eta$. The linear approximation is

$$\dot\xi = \xi - \eta, \quad \dot\eta \approx -2\xi,$$

which has the parameters

$$p = 1 > 0, \quad q = -2 < 0, \quad \Delta = 1 + 8 = 9 > 0.$$

Therefore $(-1, -1)$ is a saddle point. The phase diagram is shown in Figure 2.11.

(vi) $\dot x = -6y + 2xy - 8$, $\dot y = y^2 - x^2$. The equilibrium points are given by

$$-3y + xy - 4 = 0, \quad y^2 - x^2 = (y - x)(y + x) = 0.$$

If $y = -x$, the first equation has no real solutions, whilst for $y = x$, there are two solutions, leading to equilibrium points at $(-1, -1)$ and $(4, 4)$.

(a) $(-1, -1)$. Let $x = -1 + \xi$ and $y = -1 + \eta$. The linear approximation is

$$\dot\xi \approx -2\xi - 8\eta, \quad \dot\eta \approx 2\xi - 2\eta,$$

which has the parameters

$$p = -4 < 0, \quad q = 20 > 0, \quad \Delta = 16 - 80 = -64 < 0.$$

Hence $(-1, -1)$ is a stable spiral.

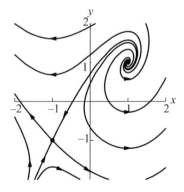

Figure 2.11 Problem 2.3(v): $\dot x = x - y$, $\dot y = x^2 - 1$.

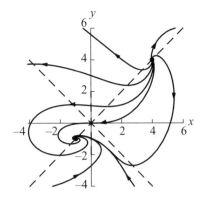

Figure 2.12 Problem 2.3(vi): $\dot{x} = -6x + 2xy - 8$, $\dot{y} = y^2 - x^2$: the dashed lines are the isoclines of zero slope.

(b) (4, 4). Let $x = 4 + \xi$ and $y = 4 + \eta$. The linear approximation is

$$\dot{\xi} \approx 8\xi + 2\eta, \quad \dot{\eta} \approx -8\xi + 8\eta,$$

which has the parameters

$$p = 16 > 0, \quad q = 80 > 0, \quad \Delta = 256 - 320 = -64 < 0.$$

Hence (4, 4) is an unstable spiral.
    The phase diagram is shown in Figure 2.12.

(vii) $\dot{x} = 4 - 4x^2 - y^2$, $\dot{y} = 3xy$. The equilibrium points are solutions of

$$4 - 4x^2 - y^2 = 0, \quad 3xy = 0.$$

The complete set of solutions is (0, 2), (0, −2), (1, 0) and (−1, 0). The x axis is a phase path, and the y axis is a zero-slope isocline.
    (a) (0, 2). Let $y = 2 + \eta$. The linear approximation is

$$\dot{x} \approx -4\eta, \quad \dot{\eta} \approx 6x,$$

which has the parameters

$$p = 0, \quad q = 24 > 0, \quad \Delta = -96 < 0.$$

Hence (0, 2) is a centre.
    (b) (0, −2). Let $y = -2 + \eta$. The linear approximation is

$$\dot{x} \approx 4\eta, \quad \dot{\eta} \approx -6x,$$

which has the parameter values

$$p = 0, \quad q = 24 > 0, \quad \Delta = -96 < 0.$$

Therefore (0, −2) is also a centre.

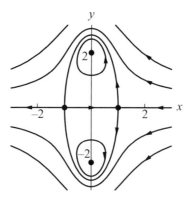

Figure 2.13 Problem 2.3(vii): $\dot{x} = 4 - 4x^2 - y^2$, $\dot{y} = 3xy$.

(c) $(1, 0)$. Let $x = 1 + \xi$. Then the linear approximation is

$$\dot{\xi} \approx -8\xi, \quad \dot{y} \approx 3y,$$

which has the parameter values

$$p = -8 + 3 = -5 < 0, \quad q = -24 < 0, \Delta = 25 + 96 = 121 > 0.$$

This equilibrium point is a saddle.

(d) $(-1, 0)$. Let $x = -1 + \xi$. The linear approximation is

$$\dot{\xi} \approx 8\xi, \quad \dot{y} \approx -3y,$$

which has the parameter values

$$p = 8 - 3 = 5 > 0, \quad q = -24 < 0, \quad \Delta = 25 + 96 = 121 > 0.$$

The equilibrium point is also a saddle.

The phase diagram is shown in Figure 2.13.

(viii) $\dot{x} = -y\sqrt{(1 - x^2)}$, $\dot{y} = x\sqrt{(1 - x^2)}$, for $|x| \leq 1$. The equilibrium points include the origin $(0, 0)$ and all points on the lines $x = \pm 1$. The equations are real only in the strip $|x| \leq 1$. The phase paths are given by

$$\frac{dy}{dx} = -\frac{x}{y},$$

which has the general solution $x^2 + y^2 = C$. All phase paths in the strip $|x| < 1$ are circles which means that the origin is a centre. The phase diagram is shown in Figure 2.14.

(ix) $\dot{x} = \sin y$, $\dot{y} = -\sin x$. Equilibrium points occur where both $\sin y = 0$ and $\sin x = 0$. Hence there is an infinite set of such points at $(m\pi, n\pi)$ where $m = 0, \pm 1, \pm 2, \ldots$ and $n = 0 \pm 1, \pm 2, \ldots$. Since the equations are unchanged by the transformations $x \to x + 2m\pi$,

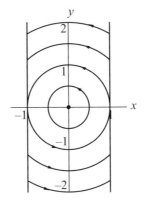

Figure 2.14 Problem 2.3(viii): $\dot{x} = -y\sqrt{(1-x^2)}$, $\dot{y} = x\sqrt{(1-x^2)}$, for $|x| \leq 1$.

$y \to y + 2n\pi$, the phase diagram is periodic with period 2π in both the x and y directions. The equations of the phase paths can be found from

$$\frac{dy}{dx} = -\frac{\sin x}{\sin y}.$$

This is a separable equation with general solution $\cos x + \cos y = C$. Note that this system is Hamiltonian (Section 2.8) from which we infer that any simple equilibrium points will be centres or saddle points. Near the origin

$$\dot{x} \approx y, \quad \dot{y} \approx -x,$$

which indicates a centre. Near $(\pi, 0)$, let $x = \pi + \xi$. Then the linear approximation is

$$\dot{\xi} \approx y, \quad \dot{y} \approx \xi,$$

which indicates a saddle. In fact the centres and saddles alternate in both the x and y directions. The phase diagram is shown in Figure 2.15.

(x) $\dot{x} = \sin x \cos y$, $\dot{y} = \sin y \cos x$. The consistent pairings of $\dot{x} = 0$ and $\dot{y} = 0$ are

$$\sin x = 0, \quad \sin y = 0, \quad \text{and} \quad \cos y = 0, \quad \cos x = 0.$$

Therefore there are equilibrium points at

$$x = m\pi, \quad y = n\pi, \quad \text{and at } x = \tfrac{1}{2}(2p+1)\pi, \quad y = \tfrac{1}{2}(2q+1)\pi,$$

where $m, n, p, q = 0, \pm 1, \pm 2, \ldots$. There are the obvious singular solutions given by the straight lines $x = r\pi$ and $y = s\pi$, where $r, s = 0, \pm 1, \pm 2, \ldots$. Near the origin the linear approximation is

$$\dot{x} \approx x, \quad \dot{y} \approx y$$

Locally the phase paths are given by

$$\frac{dy}{dx} = \frac{y}{x} \Rightarrow y = Cx.$$

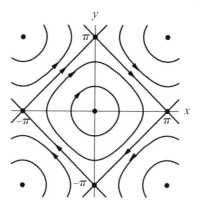

Figure 2.15 Problem 2.3(ix): $\dot{x} = \sin y$, $\dot{y} = -\sin x$.

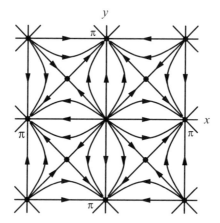

Figure 2.16 Problem 2.3(x): $\dot{x} = \sin x \cos y$, $\dot{y} = \sin y \cos x$.

Hence the origin (and similarly all other grid points) have locally star-shaped phase diagrams. It can also be verified that the lines $y = x + p\pi$ and $y = -x + p\pi$ for $p = 0, \pm 1, \pm 2, \ldots$ are also phase paths (separatrices) and that these equilibrium points are saddle points. The phase diagram, which is periodic with period 2π in both the x and y directions, is shown in Figure 2.16.

- **2.4** Construct phase diagrams for the following differential equations, using the phase plane in which $y = \dot{x}$.
  (i) $\ddot{x} + x - x^3 = 0$;
  (ii) $\ddot{x} + x + x^3 = 0$;
  (iii) $\ddot{x} + \dot{x} + x - x^3 = 0$;
  (iv) $\ddot{x} + \dot{x} + x + x^3 = 0$;
  (v) $\ddot{x} = (2\cos x - 1)\sin x$.

2.4. (i) $\ddot{x} + x - x^3 = 0$, $\dot{x} = y$. The system has three equilibrium points: at $(-1,0)$, $(0,0)$ and $(1,0)$. This is a conservative system with potential $V(x) = \frac{1}{2}x^2 - \frac{1}{4}x^4$, which has a local minimum at $x = 0$ and local maxima at $x = \pm 1$ (see NODE, Example 1.6). Hence $(0,0)$ is a centre, and $(\pm 1, 0)$ are saddles. The phase paths are given by

$$\frac{dy}{dx} = \frac{-x + x^3}{y} \Rightarrow 2y^2 = -x^4 + 2x^2 + C.$$

The phase diagram is shown in Figure 1.13 (in NODE).

(ii) $\ddot{x} + x + x^3 = 0$, $\dot{x} = y$. This is a conservative system (see NODE, Section 1.3) with one equilibrium point at the origin. The potential $V(x) = \frac{1}{2}x^2 + \frac{1}{4}x^4$ has a local minimum at $x = 0$. The origin is therefore a centre. The equation for the phase paths is given by

$$\frac{dy}{dx} = \frac{-x - x^3}{y} \Rightarrow 2y^2 = -x^4 - 2x^2 + C.$$

The phase diagram is shown in Figure 2.17.

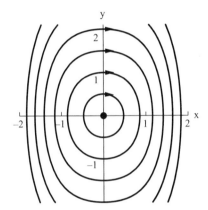

Figure 2.17 Problem 2.4(ii): $\dot{x} = y$, $\dot{y} = -x - x^3$.

(iii) $\ddot{x} + \dot{x} + x - x^3 = 0$, $\dot{x} = y$. This is (i) with damping. The system still has three equilibrium points at $(-1,0)$, $(0,0)$ and $(1,0)$.

(a) $(-1, 0)$. Let $x = -1 + \xi$. Then the linear approximation is

$$\dot{\xi} = y, \quad \dot{y} \approx 2\xi - y,$$

which has the parameter values

$$p = -1 < 0, \quad q = -2 < 0.$$

Therefore $(-1, 0)$ is a saddle point.

(b) $(0,0)$. Then $\dot{x} = y$ and $\dot{y} \approx -x - y$, which has the parameter values

$$p = -1 < 0, \quad q = 1 > 0, \quad \Delta = 1 - 4 = -3 < 0.$$

Therefore $(0,0)$ is a stable spiral.

(c) $(1,0)$. As in (a) this equilibrium point is a saddle. The phase diagram is shown in Figure 2.18.

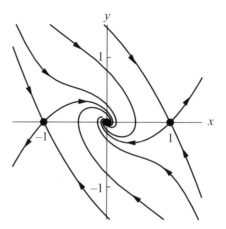

Figure 2.18 Problem 2.4(iii): $\dot{x} = y$, $\dot{y} = -y - x + x^3$.

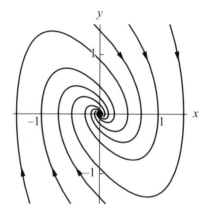

Figure 2.19 Problem 2.4(iv): $\dot{x} = y$, $\dot{y} = -y - x - x^3$.

(iv) $\ddot{x} + \dot{x} + x + x^3 = 0$, $\dot{x} = y$. This is (ii) with damping. The system has one equilibrium point at the origin, where the linear approximation is $\dot{x} = y$, $\dot{y} = -x - y$. This implies a stable spiral as shown in Figure 2.19.

(v) $\ddot{x} = (2\cos x - 1)\sin x$, $\dot{x} = y$. Equilibrium points occur where $\sin x = 0$, and where $\cos x = \frac{1}{2}$, that is, respectively, at

$$x = n\pi, \quad (n = 0, \pm 1, \pm 2, \ldots), \quad \text{and} \quad x = \pm\tfrac{1}{3}\pi + 2m\pi, \quad (m = 0, \pm 1, \pm 2, \ldots).$$

This is a conservative system (see Section 1.3) with potential

$$\mathcal{V}(x) = -\int (2\cos x - 1)\sin x\, dx = -\sin^2 x - \cos x.$$

Its second derivative is given by

$$\mathcal{V}''(x) = -2 + 4\sin^2 x + \cos x.$$

2 : Plane autonomous systems and linearization   79

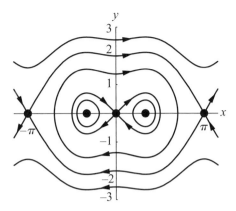

Figure 2.20 Problem 2.4(v): $\dot{x} = y$, $\dot{y} = (2\cos x - 1)\sin x$.

For the equilibrium points $x = n\pi$,

$$V''(x) = -2 + \cos n\pi < 0,$$

which means that $V(x)$ has maximum values there, giving saddles, whilst at $x = (\pm\tfrac{1}{3}\pi + 2m\pi)$,

$$V''(x) = -2 + 4\sin^2(\pm\tfrac{1}{3}\pi + 2m\pi) + \cos(\pm\tfrac{1}{3}\pi + 2m\pi)$$

$$= -2 + 6 + 1 = 5 > 0,$$

giving centres at these points. Note that the phase diagram shown in Figure 2.20 is periodic with period 2π in the x direction.

• **2.5** Confirm that the system $\dot{x} = x - 5y$, $\dot{y} = x - y$ consists of a centre. By substituting into the equation for the paths or otherwise show that the family of ellipses given by

$$x^2 - 2xy + 5y^2 = \text{constant}$$

describes the paths. Show that the axes are inclined at about $13.3°$ (the major axis) and $-76.7°$ (the minor axis) to the x direction, and that the ratio of major to minor axis length is about 2.62.

2.5. The system $\dot{x} = x - 5y$, $\dot{y} = x - y$ has also been investigated in Problem 2(i) and the answer includes the phase diagram which has a centre at the origin. The phase paths are given by

$$\frac{dy}{dx} = \frac{x - y}{x - 5y},$$

which is a standard homogeneous equation. The substitution $y = zx$ is required. Then in terms of z and x, the equation becomes

$$x\frac{dz}{dx} + z = \frac{1 - z}{1 - 5z},$$

or
$$x\frac{dz}{dx} = \frac{5z^2 - 2z + 1}{1 - 5z}.$$

This is a separable equation with solution
$$\int \frac{(1 - 5z)dz}{5z^2 - 2z + 1} = \int \frac{dx}{x} + B.$$

Hence
$$-\ln|5z^2 - 2z + 1| = 2\ln|x| + C.$$

which can be simplified to
$$x^2 - 2xy + 5y^2 = D. \tag{i}$$

This quadratic form defines all the phase paths.

Consider orthogonal axes (x', y') which are a rotation of (x, y) through an angle α counterclockwise. Then
$$x = x'\cos\alpha - y'\sin\alpha, \quad y = x'\sin\alpha + y'\cos\alpha.$$

Substitute x and y into (i) so that
$$(x'\cos\alpha - y'\sin\alpha)^2 - 2(x'\cos\alpha - y'\sin\alpha)(x'\sin\alpha + y'\cos\alpha) + 5(x'\sin\alpha + y'\cos\alpha)^2 = D,$$

or
$$x'^2(3 - \sin 2\alpha - 2\cos 2\alpha) + 2x'y'(2\sin 2\alpha - \cos 2\alpha)$$
$$+ y'^2(3 + \sin 2\alpha + 2\cos 2\alpha) = D.$$

The new axes are in the directions of the major and minor axes of the elliptic paths if the coefficient of $x'y'$ is zero. This is so if $\tan 2\alpha = \frac{1}{2}$. Hence the directions of the axes are approximately $13.3°$ and $-76.7°$. In terms of the new coordinates a typical ellipse is
$$x'^2(3 - \sqrt{5}) + y'^2(3 + \sqrt{5}) = \text{constant}.$$

Hence the ratio of major and minor axes is
$$\sqrt{\left(\frac{3 + \sqrt{5}}{3 - \sqrt{5}}\right)} = 2.62.$$

- **2.6** The family of curves which are orthogonal to the family described by the equation $(dy/dx) = f(x, y)$ is given by the solution of $(dy/dx) = -[1/f(x, y)]$. (These are called orthogonal trajectories of the first family.) Prove that the family which is orthogonal to a centre that is associated with a linear system is a node.

2.6. Let the linear system be $\dot{x} = ax + by$, $\dot{y} = cx + dy$ with an equilibrium point at $(0,0)$. The origin is a centre if
$$p = a + d = 0, \quad q = ad - bc > 0.,$$
and it follows that $\Delta = p^2 - 4q = -4q < 0$. The phase paths are given by the differential equation
$$\frac{dy}{dx} = \frac{cx + dy}{ax + by}.$$
The orthogonal phase paths are given by
$$\frac{dy}{dx} = -\frac{ax + by}{cx + dy}.$$
This equivalent to either of the following linear systems:
$$\dot{x} = cx + dy, \ \dot{y} = -ax - by, \ \text{or} \ \dot{x} = -cx - dy, \ \dot{y} = ax + by.$$
In both cases $q = ad - bc$ which is positive, and
$$\Delta = p^2 - 4q = (c - b)^2 - 4(-a^2 - bc) = (c - b)^2 + 4a^2 > 0.$$
Therefore from the table in Section 2.5, the orthogonal phase diagram is a node which can be either stable or unstable.

- 2.7 Show that the origin is a spiral point of the system $\dot{x} = -y - x\sqrt{(x^2 + y^2)}$, $\dot{y} = x - y\sqrt{(x^2 + y^2)}$, but a centre for its linear approximation.

2.7. The system is
$$\dot{x} = -y - x\sqrt{(x^2 + y^2)}, \quad \dot{y} = x - y\sqrt{(x^2 + y^2)}, \tag{i}$$
which has one equilibrium point, at the origin. Exact solutions can be found if we switch to polar coordinates (r, θ) given by $x = r\cos\theta$, $y = r\sin\theta$. In terms of r and θ, the equations become
$$\dot{r}\cos\theta - r\sin\theta \dot{\theta} = -r\sin\theta - r^2\cos\theta,$$
$$\dot{r}\sin\theta + r\cos\theta \dot{\theta} = r\cos\theta - r^2\sin\theta.$$
Solving for $\dot{r}$ and $\dot{\theta}$, we obtain
$$\dot{r} = -r^2, \quad \dot{\theta} = 1.$$
The phase paths are given by
$$\frac{dr}{d\theta} = -r^2,$$
which can be integrated to give the spiral curves $r = 1/(\theta + C)$. As $\theta \to \infty$, $r \to 0$ which implies that the origin is a stable spiral.

82 Nonlinear ordinary differential equations: problems and solutions

The linear approximation to (i) near the origin is, however,

$$\dot{x} \approx -y, \quad \dot{y} = x,$$

which is the linear system for a centre.

This problem is a counter-example to the conjecture that a centre for a linear approximation implies that the full system also has a centre.

• **2.8** Show that the systems $\dot{x} = y$, $\dot{y} = -x - y^2$, and $\dot{x} = x + y_1$, $\dot{y}_1 = -2x - y_1 - (x + y_1)^2$, both represent the equation $\ddot{x} + \dot{x}^2 + x = 0$ in different (x, y) and (x, y_1) phase planes. Obtain the equation of the phase planes in each case.

2.8. Eliminate y between $\dot{x} = y$ and $\dot{x} = -x - y^2$. Then

$$\ddot{x} = -x - \dot{x}^2. \tag{i}$$

The elimination of y_1 between $\dot{x} = x + y_1$ and $\dot{y}_1 = -2x - y_1 - (x + y_1)^2$ gives

$$\ddot{x} - \dot{x} = -2x - \dot{x} + x - \dot{x}^2, \text{ or } \ddot{x} = -x - \dot{x}^2,$$

which agrees with (i).

Phase paths for $\dot{x} = y$, $\dot{x} = -x - y^2$.

The differential equation of the phase paths in the (x, y) plane is given by

$$\frac{dy}{dx} = \frac{-x - y^2}{y},$$

or

$$\frac{d(y^2)}{dx} + 2y^2 = -2x.$$

This first-order equation has the general solution

$$y^2 = Ae^{-2x} - x + \tfrac{1}{2}. \tag{ii}$$

Phase paths for $\dot{x} = x + y_1$, $\dot{y}_1 = -2x - y_1 - (x + y_1)^2$.

The phase paths in the (x, y_1) plane will be given by (ii) with y replaced by $x + y_1$, that is,

$$(x + y_1)^2 = Ae^{-2x} - x + \tfrac{1}{2}.$$

• **2.9** Use eqn (2.9) in the form $\delta s \approx \delta t \sqrt{(X^2 + Y^2)}$ to mark off approximately equal time steps on some of the phase paths of $\dot{x} = xy$, $\dot{y} = xy - y^2$.

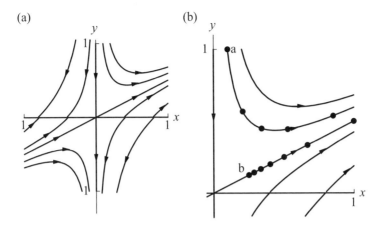

Figure 2.21 Problem 2.9: $\dot{x} = xy$, $\dot{y} = xy - y^2$. (a) Shows the general features of the phase diagram, whilst (b) shows a smaller section of the phase diagram in the first quadrant marked at equal time intervals.

2.9. All points on the x axis are equilibrium points of the system $\dot{x} = xy$, $\dot{y} = xy - y^2$. The differential equation for the phase paths is given by

$$\frac{dy}{dx} = \frac{xy - y^2}{xy} = \frac{x - y}{x},$$

which is the same equation as that for the linear system $\dot{x} = x$, $\dot{y} = x - y$. The parameters for this linear system, which has an equilibrium point at the origin, are $p = 0$, $q = -1 < 0$ which signifies a saddle point. Therefore the phase paths of this saddle point are the same as those of the nonlinear equation but the sense of the paths are different as shown in Figure 2.21(a), since the x axis is a line of equilibrium points.

In Figure 2.21(b), the formula for an element of arc of length $\delta s \approx \delta t \sqrt{(X^2 + Y^2)}$, where $X = xy$ and $Y = xy - y^2$, has been used. Two sets of equal time steps starting at a : $(0.25, 0.125)$ and b : $(0.1, 1)$ are shown by the succession of dots in the direction of the paths. For these time steps $\delta t = 1$ was chosen.

- **2.10** Obtain approximations to the phase paths described by eqn (2.12) in the neighbourhood of the equilibrium point $x = b/d$, $y = a/c$ for the predator–prey problem $\dot{x} = ax - cxy$, $\dot{y} = -by + dxy$, $(a, b, c, d) > 0$ (see NODE, Example 2.3). (Write $x = b/d + \xi$, $y = a/c + \eta$, and expand the logarithms to second-order terms in ξ and η.)

2.10. The phase paths for the predator–prey problem

$$\dot{x} = ax - cxy, \quad \dot{y} = -by + dxy,$$

are given by (see eqn (2.12))

$$a \ln y + b \ln x - cy - xd = C.$$

84 Nonlinear ordinary differential equations: problems and solutions

The equations have an equilibrium point at $(b/d, a/c)$. Close to this equilibrium point let

$$x = \frac{b}{d} + \xi, \quad y = \frac{a}{c} + \eta.$$

Then

$$a \ln y + b \ln x - cy - xd$$

$$= a \ln \left(\frac{a}{c} + \eta\right) + b \ln \left(\frac{b}{d} + \xi\right) - c \left(\frac{a}{c} + \eta\right) - d \left(\frac{b}{d} + \xi\right)$$

$$= \ln \left(\frac{a}{c}\right) + a \ln \left(1 + \frac{c\eta}{a}\right) + b \ln \left(\frac{b}{d}\right) + b \ln \left(1 + \frac{d\xi}{b}\right) - a - c\eta - b - d\xi$$

$$\approx -(a+b) + a \left(\frac{c\eta}{a} - \frac{c^2\eta^2}{2a^2}\right) + b \left(\frac{d\xi}{b} - \frac{d^2\xi^2}{2b^2}\right) - c\eta - d\xi$$

$$= -(a+b) - \frac{d^2\xi^2}{2b} - \frac{c^2\eta^2}{2a},$$

using standard Taylor expansions for the logarithms. Therefore close to the equilibrium point the phase paths are ellipses with equation

$$\frac{d^2\xi^2}{b} + \frac{c^2\eta^2}{a} = \text{constant}.$$

- **2.11** For the system $\dot{x} = ax + by$, $\dot{y} = cx + dy$, where $ad - bc = 0$, show that all points on the line $cx + dy = 0$ are equilibrium points. Sketch the phase diagram for the system $\dot{x} = x - 2y$, $\dot{y} = 2x - 4y$.

2.11. For the linear system $\dot{x} = ax + by$, $\dot{y} = cx + dy$, the parameters are $p = a + d$, $q = ad - bc$ and $\Delta = p^2 - 4q = (a+d)^2 > 0$. Equilibrium points are given by

$$ax + by = 0, \quad cx + dy = 0.$$

Since $ad - bc = 0$, there are solutions other than $x = 0$, $y = 0$ which means that x and y can satisfy both equations. Hence all points on $cx + dy = 0$ (or, equivalently, $ax + by = 0$) are equilibrium points.

For the particular problem, $\dot{x} = x - 2y$, $\dot{y} = 2x - 4y$, $ad - bc = -4 + 4 = 0$, so that all points on the line $x - 2y = 0$ are equilibrium points. The phase paths are given by

$$\frac{dy}{dx} = \frac{2x - 4y}{x - 2y} = 2,$$

which has the general solution $y = 2x + C$. The phase paths are parallel straight lines with the sense of the paths as shown in Figure 2.22.

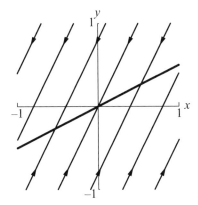

Figure 2.22 Problem 2.11: $\dot{x} = x - 2y$, $\dot{y} = 2x - 4y$: equilibrium points lie on the line $x = 2y$.

• **2.12** The interaction between two species is governed by the deterministic model $\dot{H} = (a_1 - b_1 H - c_1)H$, $\dot{P} = (-a_2 + c_2 H)P$, where H is the population of the host (prey), and P is that of the parasite (or predator), all constants being positive. (Compare NODE, Example 2.3: the term $-b_1 H^2$ represents interference with the host population when it gets too large.) Assuming that $a_1 c_2 - b_1 a_2 > 0$, find the equilibrium states for the populations, and find how they vary with time from various initial populations.

2.12. The host(H)–parasite(P) problem is governed by the model

$$\dot{H} = (a_1 - b_1 H - c_1 P)H, \quad \dot{P} = (-a_2 + c_2 H)P.$$

The system is in equilibrium at the points (order (H, P))

$$(0,0), \quad \left(\frac{a_1}{b_1}, 0\right), \quad \left(\frac{a_2}{c_2}, \frac{a_1 c_2 - b_1 a_2}{c_1 c_2}\right) = \left(\frac{a_2}{c_2}, \frac{D}{c_1 c_2}\right),$$

say, where $D = a_1 c_2 - b_1 a_2$.

I. $(0, 0)$. Near the origin

$$\dot{H} \approx a_1 H, \quad \dot{P} \approx -a_2 P.$$

This is a saddle point with separatrices $P = 0$ and $H = 0$.

II. $(a_1/b_1, 0)$. Let $H = (a_1/b_1) + \xi$. Then near the equilibrium point,

$$\dot{\xi} \approx -a_1 \xi - \frac{c_1 a_1}{b_1} P, \quad \dot{P} \approx \frac{a_1 c_2 - a_2 b_1}{b_1}.$$

The parameters for the linear approximation are

$$p = -a_1 + \frac{D}{b_1}, \quad q = -\frac{a_1 D}{b_1} < 0.$$

Hence this equilibrium point is also a saddle point.

III. $(a_2/c_2, D/(c_1 c_2))$. Let
$$H = \frac{a_2}{c_2} + \xi, \quad P = \frac{D}{c_1 c_2} + \eta.$$

Then
$$\dot{\xi} = \left[a_1 - b_1 \left(\frac{a_2}{c_2} + \xi \right) - c_1 \left(\frac{D}{c_1 c_2} + \eta \right) \right] \left(\frac{a_2}{c_2} + \xi \right)$$
$$\approx -\frac{a_2 b_1}{c_2} \xi - \frac{a_2 c_1}{c_2} \eta,$$

and
$$\dot{\eta} \approx \frac{D}{c_1} \xi.$$

The parameters associated with this linear approximation are
$$p = -\frac{a_2 b_1}{c_2} > 0, \quad q = \frac{a_2 D}{c_2} > 0, \quad \Delta = \frac{a_2^2 b_1^2}{c_2^2} - \frac{4 a_2 D}{c_2}.$$

Therefore the equilibrium point is

a stable spiral if $\Delta > \dfrac{a_2 b_1^2}{4 c_2}$, or a stable node if $\Delta < \dfrac{a_2 b_1^2}{4 c_2}$.

Figure 2.23 shows the phase diagram for the system
$$\dot{H} = (2 - H - P)H, \quad \dot{P} = (-1 + H)P,$$

for which $p = -1$, $q = 1 > 0$ and $\Delta = -3 < 0$ at $(1, 1)$. Therefore the equilibrium point is locally a stable spiral.

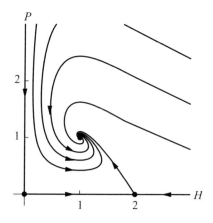

Figure 2.23 Problem 2.12: $\dot{H} = (2 - H - P)H$, $\dot{P} = (-1 + xH)P$.

- **2.13** With the same terminology as in Problem 2.12, analyze the system $\dot{H} = (a_1 - b_1 H - c_1 P)H$, $\dot{P} = (a_2 - b_2 P + c_2 H)P$, all the constants being positive. (In this model the parasite can survive on alternative food supplies, although the prevalence of the host encourages growth in population.) Find the equilibrium states. Confirm that the parasite population can persist even if the host dies out.

2.13. In this model of host–parasite (H, P) populations

$$\dot{H} = (a_1 - b_1 H - c_1 P)H, \quad \dot{P} = (a_2 - b_2 P + c_2 H)P.$$

There are equilibrium points at $(0,0)$, $(a_1/b_1, 0)$, $(0, a_2/b_2)$ and

$$\left(\frac{a_1 b_2 - c_1 a_2}{b_1 b_2 + c_1 c_2}, \frac{a_2 b_1 + a_1 c_2}{b_1 b_2 + c_1 c_2} \right),$$

provided $a_1 b_2 \geq c_1 a_2$.

The parasite population can persist if the parameters satisfy $a_1 b_2 = c_1 a_2$, which is consistent with the existence of an alternative food supply.

- **2.14** Consider the host–parasite population model $\dot{H} = (a_1 - c_1 P)H$, $\dot{P} = (a_2 - c_2(P/H))P$, where the constants are positive. Analyse the system in the H, P plane.

2.14. In this model of host–parasite (H, P) populations

$$\dot{H} = (a_1 - c_1 P)H, \quad \dot{P} = \left(a_2 - \frac{c_2 P}{H} \right) P,$$

where $H > 0$. The populations have one equilibrium state, at

$$(H, P) = \left(\frac{a_1 c_2}{a_2 c_1}, \frac{a_1}{c_1} \right).$$

Note that the P axis is a *singular* line since $\dot{P}$ is unbounded there.
Let

$$H = \frac{a_1 c_2}{a_2 c_1} + \xi, \quad P = \frac{a_1}{c_1} + \eta.$$

Then

$$\dot{\xi} = (a_1 - a_1 - c_1 \eta) \left(\frac{a_1 c_2}{a_2 c_1} + \xi \right) \approx -\frac{a_1 c_2 \eta}{a_2},$$

and
$$\dot{\eta} = \left[a_2 - c_2\left(\frac{a_1}{c_1} + \eta\right)\left(\frac{a_1 c_2}{a_2 c_1} + \xi\right)^{-1}\right]\left(\frac{a_1}{c_1} + \eta\right) \approx \frac{a_2^2}{c_2}\xi - a_2\eta.$$

The parameters associated with this linear approximation are
$$p = -a_2 < 0, \quad q = a_1 a_2 > 0, \quad \Delta = a_2^2 - 4a_1 a_2.$$

Therefore by the table in Section 2.5, the equilibrium point is a stable node if $a_2 > 4a_1$ and a stable spiral if $a_2 < 4a_1$.

- **2.15** In the population model $\dot{F} = -\alpha F + \beta\mu(M)F$, $\dot{M} = -\alpha M + \gamma\mu(M)F$, where $\alpha > 0$, $\beta > 0$, $\gamma > 0$, F and M are the female and male populations. In both cases the death rates are α. The birth rate is governed by the coefficient $\mu(M) = 1 - e^{-kM}$, $k > 0$, so that for large M the birth rate for females is βF and that for males is γF, the rates being unequal in general. Show that if $\beta > \alpha$ then there are two equilibrium points, at $(0,0)$ and at
$$\left(-\frac{\beta}{\gamma k}\ln\left[\frac{\beta-\alpha}{\beta}\right], -\frac{1}{k}\ln\left[\frac{\beta-\alpha}{\beta}\right]\right).$$
Show that the origin is stable and that the other equilibrium point is a saddle point, according to their linear approximations. Verify that $M = \gamma F/\beta$ is a particular solution. Sketch the phase diagram and discuss the stability of the populations.

2.15. A male–female population is modelled by the birth and death equations
$$\dot{F} = -\alpha F + \beta\mu(M)F, \quad \dot{M} = -\alpha M + \gamma\mu(M)F, \tag{i}$$
where $\mu(M) = 1 - e^{-kM}$. Equilibrium occurs where
$$-F[\alpha + \beta(1 - e^{-kM})] = 0, \quad -\alpha M + \gamma(1 - e^{-kM})F = 0.$$
The equations have two equilibrium points; at
$$(0,0) \quad \text{and} \quad \left(-\frac{\beta}{\gamma k}\ln\left[\frac{\beta-\alpha}{\beta}\right], -\frac{1}{k}\ln\left[\frac{\beta-\alpha}{\beta}\right]\right)$$
in the (F, M) plane.

I. $(0,0)$. Near the origin
$$\dot{F} \approx -\alpha F, \quad \dot{M} \approx -\alpha M.$$
The phase paths are given by
$$\frac{dM}{dF} = \frac{M}{F},$$

so that $M = CF$, where C is an arbitrary constant. The paths are are straight lines into the origin, which implies that the origin is stable.

II. For the other equilibrium point, let

$$(F_0, M_0) = \left(-\frac{\beta}{\gamma k} \ln\left[\frac{\beta - \alpha}{\beta}\right], -\frac{1}{k} \ln\left[\frac{\beta - \alpha}{\beta}\right]\right).$$

Let $F = F_0 + \xi$ and $M = M_0 + \eta$ Then, from (i),

$$\dot\xi = -\alpha(F_0 + \xi) + \beta(1 - e^{-k(M_0+\eta)})(F_0 + \xi)$$
$$\approx -\alpha F_0 - \alpha\xi + \beta[1 - e^{-kM_0}(1 - k\eta)](F_0 + \xi)$$
$$= \beta k F_0 e^{-kM_0} \eta$$
$$= -\frac{\beta(\beta - \alpha)}{\gamma} \ln\left[\frac{\beta - \alpha}{\beta}\right] \eta.$$

$$\dot\eta = -\alpha(M_0 + \eta) + \gamma(1 - e^{-k(M_0+\eta)})(F_0 + \xi)$$
$$\approx -\alpha M_0 - \alpha\eta + \gamma[1 - e^{-kM_0}(1 - k\eta)](F_0 + \xi)$$
$$= \frac{\alpha\gamma}{\beta}\xi + (-\alpha + \gamma k e^{-kM_0} F_0)\eta$$
$$= \frac{\alpha\gamma}{\beta}\xi - \left(\alpha + (\beta - \alpha)\ln\left[\frac{\beta - \alpha}{\beta}\right]\right)\eta.$$

The parameters associated with this linear approximation are

$$p = -\left(\alpha + (\beta - \alpha)\ln\left[\frac{\beta - \alpha}{\beta}\right]\right), \quad q = \alpha(\beta - \alpha)\ln\left[\frac{\beta - \alpha}{\beta}\right] < 0.$$

Hence (F_0, M_0) is a saddle point.

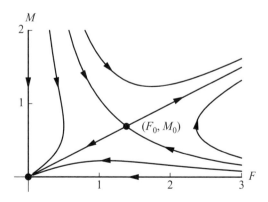

Figure 2.24 Problem 2.15: Population model with $\alpha = 0.5$, $\beta = 1$, $\gamma = 0.5$ and $k = 1$.

It can be verified by direct substitution in the differential equations that $M = \gamma F/\beta$ satisfies both equations, and therefore defines two phase paths which are separatrices through the saddle point. The phase diagram with parameters $\alpha = 0.5$, $\beta = 1$, $\gamma = 0.5$ and $k = 1$ is shown in Figure 2.24.

- 2.16 A rumour spreads through a closed population of constant size $N+1$. At time t the total population can be classified into three categories:
x persons who are ignorant of the rumour;
y persons who are actively spreading the rumour;
z persons who have heard the rumour but have stopped spreading it: if two persons who are spreading the rumour meet then they stop spreading it.
The contact rate between any two categories is a constant, μ.
Show that the equations
$$\dot{x} = -\mu xy, \quad \dot{y} = \mu[xy - y(y-1) - yz]$$
give a deterministic model of the problem. Find the equations of the phase paths and sketch the phase diagram.
Show that, when initially $y = 1$ and $x = N$, the number of people who ultimately never hear the rumour is x_1, where
$$2N + 1 - 2x_1 + N\ln(x_1/N) = 0.$$

2.16. In incremental form the equations are the limits as $\delta t \to 0$ of
$$\delta x = -\mu xy \delta t, \quad \delta y = \mu[xy - y(y-1) - yz]\delta t.$$

Contact frequencies between any two groups are assumed to be proportional to the product of the population sizes. Thus, the decrease in the number of those who do not know the rumour $-\delta x$ must be proportional to $xy\delta t$, and the number δy of those who are actively spreading the rumour must increase at a rate proportional to contacts between x and y, and decrease at a rate proportional to meetings between spreaders, $y(y-1)$, and between y and those who already know, z. Hence
$$\delta y = \mu[xy - y(y-1) - yz]\delta t,$$
and the differential equation follows in the limit $\delta t \to 0$. Since the population has constant size $N+1$, the third equation is $x + y + z = N = 1$.
Substitute for z in the differential equations in the question. Then x and y satisfy
$$\dot{x} = -\mu xy, \quad \dot{y} = \mu[xy - y(y-1) - y(N+1-x-y)] = \mu(2xy - Ny). \tag{i}$$

Therefore the differential equation for the phase paths is
$$\frac{dy}{dx} = \frac{\dot{y}}{\dot{x}} = \frac{N-2x}{x}.$$

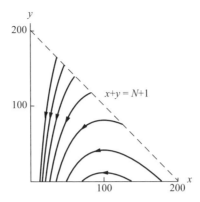

Figure 2.25 Problem 2.16: Epidemic model with $N = 200$.

This is a separable first-order equation with general solution

$$y = \int \left(\frac{N}{x} - 2\right) dx = N \ln x - 2x + C, \quad (x > 0). \tag{ii}$$

The second equation in (i) has the solution $y = 0$, which must correspondence to a line of equilibrium points since $\dot{x} = 0$ also. In the model this means that no one is spreading the rumour, but there may be a (constant) number of individuals who do not know the rumour. Linearization is not really helpful since the points on the x axis will be non-standard equilibrium points. The phase paths can be plotted using the curves given by (ii): some phase paths are shown in Figure 2.25. Note that $x + y \leq N$, so that the phase diagram is bounded by this line.

If the initial conditions are $y = 1$ and $x = N$, then, from (ii)

$$1 = N \ln N - 2N + C, \text{ so that } C = 2N + 1 - N \ln N.$$

Hence on this path

$$y = N \ln x - N \ln N - 2x + 2N + 1.$$

The number of individuals x_1 who never hear the rumour occurs where $y = 0$. Therefore x_1 satisfies

$$N \ln x_1 - N \ln N - 2x_1 + 2N + 1 = 0.$$

- **2.17** The one-dimensional steady flow of a gas with viscosity and heat conduction satisfies the equations

$$\frac{\mu_0}{\rho c_1} \frac{dv}{dx} = \sqrt{(2v)}[2v - \sqrt{(2v)} + \theta],$$

$$\frac{k}{gR\rho c_1} \frac{d\theta}{dx} = \sqrt{(2v)}\left[\frac{\theta}{\gamma - 1} - v + \sqrt{(2v)} - c\right],$$

where $v = u^2/(2c_1)^2$, $c = c_2^2/c_1^2$ and $\theta = gRT/c_1^2 = p/(\rho c_1^2)$. In this notation, x is measured in the direction of flow, u is the velocity, T is the temperature, ρ is the density, p the pressure,

R the gas constant, k the coefficient of thermal conductivity, μ_0 the coefficient of viscosity, γ the ratio of the specific heats, and c_1, c_2 are arbitrary constants. Find the equilibrium states of the system.

2.17. The one-dimensional steady flow of a gas satisfies

$$\frac{\mu_0}{\rho c_1}\frac{dv}{dx} = \sqrt{(2v)}[2v - \sqrt{(2v)} + \theta],$$

$$\frac{k}{gR\rho c_1}\frac{d\theta}{dx} = \sqrt{(2v)}\left[\frac{\theta}{\gamma - 1} - v + \sqrt{(2v)} - c\right].$$

where $v \geq 0$. Equilibrium points in the (θ, v) plane occur where

$$\sqrt{(2v)}[2v - \sqrt{(2v)} + \theta] = 0, \quad \sqrt{(2v)}\left[\frac{\theta}{\gamma - 1} - v + \sqrt{(2v)} - c\right] = 0.$$

Since both equations are satisfied by $v = 0$ for all θ, all points on the θ are in equilibrium. Equilibrium also occurs where

$$2v - \sqrt{(2v)} + \theta = 0, \quad \frac{\theta}{\gamma - 1} - v + \sqrt{(2v)} - c = 0. \tag{i}$$

Elimination of θ leads to the quadratic equation

$$(\gamma + 1)v - \gamma\sqrt{(2v)} + c(\gamma - 1) = 0.$$

in $\sqrt{v}$. This equation has two solutions

$$\sqrt{(2v)} = \frac{1}{\gamma + 1}[\gamma \pm \sqrt{\{\gamma^2 - 2c(\gamma^2 - 1)\}}].$$

The ratio γ usually satisfies $\gamma > 1$. There will be two stationary values for $\sqrt{(2v)}$ if

$$\sqrt{\{\gamma^2 - 2c(\gamma^2 - 1)\}} < \gamma, \quad \text{or} \quad -2(\gamma^2 - 1)c < 0,$$

which is not possible since $c > 0$. Therefore there is one equilibrium value for v. The corresponding value for θ can be found from either of the equations in (i).

• 2.18 A particle moves under a central attractive force γ/r^α per unit mass, where r, θ are the polar coordinates of the particle in its plane of motion. Show that

$$\frac{d^2u}{d\theta^2} + u = \frac{\gamma}{h^2}u^{\alpha-2}.$$

where $u = r^{-1}$, h is the angular momentum about the origin per unit mass of the particle, and γ is a constant. Find the non-trivial equilibrium point in the u, $du/d\theta$ plane and classify it according to its linear approximation. What can you say about the stability of the circular orbit under this central force?

2.18. In Figure 2.26, the position of the particle of mass m is (r, θ) in polar coordinates. The radial and transverse equations of motion are, under the influence of the central force $m\gamma/r^\alpha$ are

$$-\frac{m\gamma}{r^\alpha} = m(\ddot{r} - r\dot{\theta}^2), \tag{i}$$

$$m\frac{d}{dt}(r^2\dot{\theta}) = 0. \tag{ii}$$

From (ii) it follows that

$$mr^2\dot{\theta} = \text{constant} = mh, \tag{iii}$$

say, where mh is the (constant) angular momentum of the particle. Now eliminate $\dot{\theta}$ between (i) and (ii) so that

$$\ddot{r} - \frac{h^2}{r^3} = -\frac{\gamma}{r^\alpha}. \tag{iv}$$

Using the identity

$$\ddot{r} = \dot{\theta}\frac{d}{d\theta}\left(\dot{\theta}\frac{dr}{d\theta}\right),$$

and the change of variable $u = 1/r$, eqn (iv) can be expressed in the form

$$\frac{d^2 u}{d\theta^2} + u = \frac{\gamma}{h^2}u^{\alpha-2}.$$

Let $v = du/d\theta$. Then the differential equation of the phase paths in the (θ, p) plane is

$$\frac{dv}{d\theta} = \frac{k}{h^2}u^{\alpha-2} - u.$$

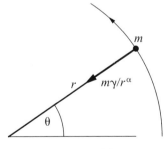

Figure 2.26 Problem 2.18.

There are equilibrium points; at $u = 0$ (physically of no interest) and at $u = u_0 = (h^2/k)^{1/(\alpha-3)}$ provided $\alpha \neq 3$, in which case there is just one equilibrium point at $u = 0$.

Let $u = u_0 + u'$, where $|u'|$ is small. Then

$$\frac{dv}{d\theta} = \frac{k}{h^2}(u_0 + u')^{\alpha-2} - u_0 - u'$$

$$\approx \frac{k}{h^2}u_0^{\alpha-2}\left[1 + (\alpha - 2)\left(\frac{u'}{u_0}\right)\right] - u_0 - u'$$

$$= (\alpha - 3)u'$$

Hence the linear approximation of the equilibrium point indicates that it is a centre if $\alpha < 3$, and a saddle point $\alpha > 3$.

The equilibrium point $u = u_0$, $v = 0$ corresponds to a circular orbit of the particle, which is stable if $\alpha < 3$ and unstable if $\alpha > 3$. The gravitational inverse-square law gives a stable orbit.

- **2.19** The relativistic equation for the central orbit of a planet is

$$\frac{d^2u}{d\theta^2} + u = k + \varepsilon u^2,$$

where $u = 1/r$, and r, θ are the polar coordinates of the planet in the plane of its motion. The term εu^2 is the 'Einstein correction', and k and ε are positive constants, with ε very small. Find the equilibrium point which corresponds to a perturbation of the Newtonian orbit. Show that the equilibrium point is a centre in the $u, du/d\theta$ plane according to the linear approximation. Confirm this by using the potential energy method of NODE, Section 1.3.

2.19. The relativistic equation is

$$\frac{d^2u}{d\theta^2} + u = k + \varepsilon u^2 \tag{i}$$

(see Problem 2.19). Aside from the correction term k, the polar equation can be derived as in Problem 2.18 assuming the inverse-square law. The equilibrium points are given by

$$\varepsilon u^2 - u + k = 0,$$

that is,

$$u = \begin{cases} u_1 \\ u_2 \end{cases} = \frac{1}{2\varepsilon}[1 \pm \sqrt{(1 - 4k\varepsilon)}].$$

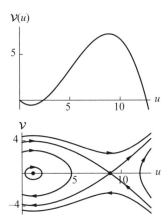

Figure 2.27 Problem 2.19: The phase diagram for the relativistic equation has been computed for $k = 1$ and $\varepsilon = 0.1$: the equilibrium points are at $u = 1.127$ and $u = 8.873$.

Let $v = du/d\theta$, and $u = u_i + u'_i$, $(i = 1, 2)$, where $|u'_i|$ is small. Substituting into (i), we have

$$\frac{d^2 u'_i}{d\theta^2} = -u_i - u'_i + k + \varepsilon(u_i + u'_i)^2$$

$$\approx -u_i - u'_i + k + u_i^2 + 2u_i u'_i$$

$$= \pm\sqrt{(1 - 4k\varepsilon)} u'_i$$

Hence u_1 is a saddle, and u_2 is a centre, assuming that ε is sufficiently small to make $4k\varepsilon < 1$. In the notation of Section 1.3, the potential associated with eqn (i) is

$$\mathcal{V}(u) = -\int (k + \varepsilon u^2 - u)du = -ku - \frac{1}{3}\varepsilon u^3 + \frac{1}{2}u^2.$$

A graph of the potential $\mathcal{V}(u)$ and the corresponding phase diagram are shown in Figure 2.27. The equilibrium points correspond to circular orbits. In terms of $r = 1/u$, the inner orbit is unstable whilst the outer orbit is stable.

- **2.20** A top is set spinning at an axial rate n about its pivotal point, which is fixed in space. The equations of its motion, in terms of the angles θ and μ are (see Figure 2.14 in NODE)

$$A\ddot\theta - A(\Omega + \dot\mu)^2 \sin\theta \cos\theta + Cn(\Omega + \dot\mu) \sin\theta - Mgh\sin\theta = 0,$$
$$A\dot\theta^2 + A(\Omega + \dot\mu)^2 \sin^2\theta + 2Mgh\cos\theta = E;$$

where (A, A, C) are the principal moments of inertia about O, M is the mass of the top, h is the distance between the mass centre and the pivot, and E is a constant. Show that an equilibrium state is given by $\theta = \alpha$, after elimination of Ω between

$$A\Omega^2 \cos\alpha - Cn\Omega + Mgh = 0, \quad \text{and} \quad A\Omega^2 \sin^2\alpha + 2Mgh\cos\alpha = E.$$

Suppose that $E = 2Mgh$, so that $\theta = 0$ is an equilibrium state. Show that, close to this state, θ satisfies

$$A\ddot{\theta} + [(C - A)\Omega^2 - Mgh]\theta = 0.$$

For what condition on Ω is the motion stable?

2.20. The equations of the motion of the top in terms of the angles θ and μ are

$$A\ddot{\theta} - A(\Omega + \dot{\mu})^2 \sin\theta \cos\theta + Cn(\Omega + \dot{\mu}) \sin\theta - Mgh \sin\theta = 0, \qquad \text{(i)}$$

$$A\dot{\theta}^2 + A(\Omega + \dot{\mu})^2 \sin^2\theta + 2Mgh\cos\theta = E. \qquad \text{(ii)}$$

$\dot{\mu}$ can be eliminated between these equations to obtain a second-order differential equation in θ. The equilibrium states of the top are then given by putting $\ddot{\theta} = 0$ and $\dot{\theta} = 0$ in this equation in θ, but this is equivalent to the elimination of Ω between (i) and (ii) with both $\dot{\theta}$ and $\dot{\mu}$ zero, that is, between

$$(-A\Omega^2 \cos\theta + Cn\Omega - Mgh) \sin\theta = 0, \qquad \text{(iii)}$$

and

$$A\Omega^2 \sin^2\theta + 2Mgh \cos\theta = E. \qquad \text{(iv)}$$

If $E = 2Mgh$, then eqn (iv) becomes

$$A\Omega^2 \sin^2\theta = 2Mgh(1 - \cos\theta).$$

Hence $\theta = 0$ is a solution of this equation and (iii), and must be an equilibrium point in which the top spins about its axis; which is vertical. Also in this state $\Omega = n$ and $\mu = 0$. For small $|\theta|$, eqn (i) becomes

$$\ddot{\theta} + [(C - A)\Omega^2 - Mgh]\theta \approx 0.$$

The vertical spin is stable if $(C - A)\Omega^2 > Mgh$.

• 2.21 Three gravitating particles with gravitational masses μ_1, μ_2, μ_3, move in a plane so that they always remain at the vertices of an equilateral triangle $P_1 P_2 P_3$ with varying side-length $a(t)$ as shown in Figure 2.15 (in NODE). The triangle rotates in the plane with spin $\Omega(t)$ about the combined mass-centre G. If the position vectors of the particles are $\mathbf{r}_1$, $\mathbf{r}_2$, $\mathbf{r}_3$, relative to G, show that the equations of motion are

$$\ddot{\mathbf{r}}_i = -\frac{\mu_1 + \mu_2 + \mu_3}{a^3}\mathbf{r}_i, \quad (i = 1, 2, 3).$$

If $|\mathbf{r}_i| = r_i$, deduce the polar equations

$$\ddot{r}_i - r_i\Omega^2 = -\frac{\mu_1 + \mu_2 + \mu_3}{a^3}r_i, \quad r_i^2\Omega = \text{constant}, \quad (i = 1, 2, 3).$$

Explain why a satisfies

$$\ddot{a} - a\Omega^2 = -\frac{\mu_1 + \mu_2 + \mu_3}{a^2}, \quad a^2\Omega = \text{constant} = K,$$

say, and that solutions of these equations completely determine the position vectors. Express the equation in non-dimensionless form by the substitutions $a = K^2/(\mu_1 + \mu_2 + \mu_3)$, $t = K^3\tau/(\mu_1 + \mu_2 + \mu_3)^2$, sketch the phase diagram for the equation in u obtained by eliminating Ω, and discuss possible motions of this Lagrange configuration.

2.21. The configuration is shown in Figure 2.28. Since G is the mass-centre,

$$\mu_1\mathbf{r}_1 + \mu_2\mathbf{r}_2 + \mu_3\mathbf{r}_3 = 0. \tag{i}$$

The equation of motion fo P_1 is

$$\mu_1\ddot{\mathbf{r}}_1 = -\frac{\mu_1\mu_2(\mathbf{r}_2 - \mathbf{r}_1)}{a^3} - \frac{\mu_1\mu_3(\mathbf{r}_3 - \mathbf{r}_1)}{a^3},$$

or, using (i),

$$\ddot{\mathbf{r}}_1 = -\frac{(\mu_1 + \mu_2 + \mu_3)}{a^3}\mathbf{r}_1,$$

with similar equations for $\mathbf{r}_2$ and $\mathbf{r}_3$.

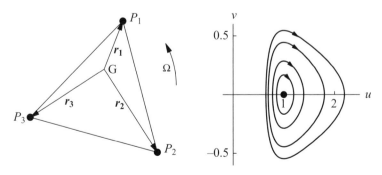

Figure 2.28 Problem 2.21: Lagrange equilateral configuration for a three-body problem with $P_1P_2 = P_2P_3 = P_3P_1 = a(t)$: phase diagram in (u, v) plane.

In fixed axes let the polar equations of P_1 be $r_1 = |\mathbf{r}_1|$ and θ_1. Since the triangle rotates with spin Ω, it follows that $\dot\theta_1 = \Omega$. Similarly for P_2 and P_3, the other polar angles also satisfy $\dot\theta_2 = \dot\theta_3 = \Omega$. The radial and transverse polar equations are therefore

$$\ddot r_i - r_i \dot\theta_i^2 = \ddot r_i - r_i \Omega^2 = -\frac{\mu_1 + \mu_2 + \mu_3}{a^3} r_i, \tag{ii}$$

and

$$\frac{\mathrm{d}}{\mathrm{d}t}(r_i^2 \dot\theta_i) = 0, \text{ which implies } r_i^2 \dot\theta_i = r_i^2 \Omega = \text{constant}. \tag{iii}$$

for $i = 1, 2, 3$.

Throughout the motion, the equilateral triangle varies in size as it rotates, but the ratio r_1/a remains constant with time, and similarly for r_2/a and r_3/a. From (ii) and (iii), it follows that

$$\ddot a - a\Omega^2 = -\frac{\mu_1 + \mu_2 + \mu_3}{a^2}, \tag{iv}$$

and

$$a^2 \Omega = \text{constant} = K, \text{ (say)}. \tag{v}$$

Elimination of Ω between (iv) and (v) leads to

$$\ddot a - \frac{K^2}{a^3} = -\frac{\mu_1 + \mu_2 + \mu_3}{a^2}, \quad (a > 0).$$

The equation can be expressed in the dimensionless form

$$\frac{\mathrm{d}^2 u}{\mathrm{d}\tau^2} = u'' = \frac{1-u}{u^3}, \quad (u > 0),$$

using the substitutions $a = K^2 u/(\mu_1 + \mu_2 + \mu_3)$, $t = K^3 \tau/(\mu_1 + \mu_2 + \mu_3)^2$.
If $v = u'$, then the equation for the phase paths is

$$v \frac{\mathrm{d}v}{\mathrm{d}u} = \frac{1-u}{u^3},$$

which has the general solution

$$v^2 = -\frac{1}{u^2} + \frac{2}{u} + C.$$

In the (u, v) phase plane, the system has a single equilibrium point at $u = 1$, which is a centre (see Figure 2.28). This implies that the lengths of the sides of the equilateral triangle oscillate with time. The fixed point corresponds to a Lagrange motion in which the three masses remain at the vertices of a fixed equilateral triangle as they rotate in circular orbits about G. (For information, in general, the orbits are similar ellipses although this is not proved here.)

- 2.22 A disc of radius a is freely pivoted at its centre A so that it can turn in a vertical plane. A spring, natural length $2a$ and stiffness λ connects a point B on the circumference of the disc to a fixed pont O, distance $2a$ above A. Show that θ satisfies
$$I\ddot{\theta} = -Ta\sin\phi, \quad T = \lambda a[(5 - 4\cos\theta)^{1/2} - 2],$$
where T is the tension in the spring, I is the moment of inertia of the disc about A, $\widehat{OAB} = \theta$, and $\widehat{ABO} = \phi$. Find the equilibrium states of the disc, and their stability.

2.22. The configuration of the system is shown in Figure 2.29. Taking moments about A,
$$-Ta\sin(\pi - \phi) = I\ddot{\theta}, \quad \text{or} \quad -Ta\sin\phi = I\ddot{\theta}, \tag{i}$$
where, from triangle ABO, using the sine and cosine rules,
$$\frac{OB}{\sin\theta} = \frac{2a}{\sin\phi}, \quad OB^2 = 4a^2 + a^2 - 4a^2\cos\theta.$$
Elimination of OB between these equations leads to
$$\sin\phi = \frac{2a\sin\theta}{OB} = \frac{2\sin\theta}{\sqrt{(5 - 4\cos\theta)}}. \tag{ii}$$

By Hooke's law the tension,
$$T = \lambda(OB - 2a) = \lambda a[\sqrt{(5 - 4\cos\theta)} - 2]. \tag{iii}$$
where λ is a constant. Elimination of T and ϕ between eqns (i), (ii) and (iii) leads to the differential equation for θ:
$$I\ddot{\theta} + 2\lambda a^2[1 - 2\{5 - 4\cos\theta)\}^{-(1/2)}]\sin\theta = 0.$$

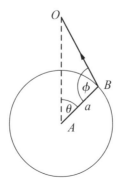

Figure 2.29 Problem 2.22.

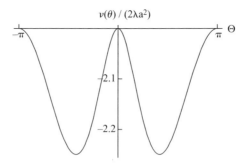

Figure 2.30 Problem 2.22:

Equilibrium occurs where $\ddot\theta = 0$, namely where

$$[1 - 2\{5 - 4\cos\theta\}^{-(1/2)}]\sin\theta = 0.$$

The solutions are $\theta = 0$, $\theta = \pi$ and $\theta = \pm\cos^{-1}(\tfrac{1}{4})$.

This is a conservative system so that we can investigate stability of the equilibrium points by using the energy method of Section 1.3. The potential energy $V(\theta)$ satisfies

$$\frac{dV}{d\theta} = 2\lambda a^2 [1 - 2(5 - 4\cos\theta)^{-(1/2)}]\sin\theta.$$

Hence, we can choose

$$V(\theta) = 2\lambda a^2 \int [1 - 2(5 - 4\cos\theta)^{-(1/2)}]\sin\theta\, d\theta$$

$$= 2\lambda a^2 [-\sqrt{(5 - 4\cos\theta)} - \cos\theta].$$

The graph of $V(\theta)$ versus θ is shown in Figure 2.30. It can be seen that $V(\theta)$ has maxima at $\theta = 0$ and $\theta = \pi$. These correspond to unstable positions of equilibrium. Minima occur at $\pm\cos^{-1}(\tfrac{1}{4})$ which indicates stable equilibrium.

- **2.23** A man rows a boat across a river of width a occupying the strip $0 \le x \le a$ in the x, y plane, always rowing towards a fixed point on one bank, say $(0, 0)$. He rows at a constant speed u relative to the water, and the river flows at a constant speed v. Show that

$$\dot x = -ux/\sqrt{(x^2 + y^2)}, \quad \dot y = v - uy/\sqrt{(x^2 + y^2)},$$

where (x, y) are the coordinates of the boat. Show that the phase paths are given by $y + \sqrt{(x^2 + y^2)} = Ax^{1-\alpha}$, where $\alpha = v/u$. Sketch the phase diagram for $\alpha < 1$ and interpret it. What kind of point is the origin? What happens to the boat if $\alpha > 1$?

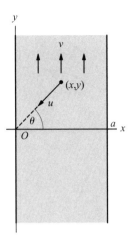

Figure 2.31 Problem 2.23.

2.23. A plan view of the river and the boat is shown in Figure 2.31. If (x, y) are the coordinates of the boat and the θ is the angle between the x axis and the radius to the boat, then the velocity components $(\dot{x}, \dot{y})$ of the boat are given by

$$\dot{x} = -u \cos \theta = -\frac{ux}{\sqrt{(x^2 + y^2)}},$$

$$\dot{y} = v - u \sin \theta = v - \frac{uy}{\sqrt{(x^2 + y^2)}},$$

since the boat always points towards the origin. The phase paths are given by

$$\frac{dy}{dx} = \frac{\dot{y}}{\dot{x}} = \frac{uy - v\sqrt{(x^2 + y^2)}}{ux}.$$

This is a first-order homogeneous equation for which we use the substitution $y = wx$. Therefore the equation becomes, in terms of w and x,

$$x \frac{dw}{dx} + w = \frac{uw - v\sqrt{(1 + w^2)}}{u},$$

or

$$x \frac{dw}{dx} = -\frac{v}{u}\sqrt{(1 + w^2)}.$$

This is a separable equation with solution

$$\int \frac{dw}{\sqrt{(1 + w^2)}} = -\frac{v}{u} \int \frac{dx}{x} = -\frac{v}{u} \ln x + C, \quad (x > 0),$$

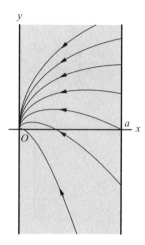

Figure 2.32 Problem 2.23: Phase diagram with $\alpha = \tfrac{1}{2}$ and $a = 1$.

where C is an arbitrary constant. For the left-hand side use the substitution $w = \tan\theta$ so that

$$\int \frac{dw}{\sqrt{(1+w^2)}} = \int \sec\theta \, d\theta = \ln(\sec\theta + \tan\theta)$$

$$= \ln[\sqrt{(1+w^2)} + w] = [\sqrt{(x^2+y^2)} + y]/x.$$

Therefore the solution can be expressed in the implicit form

$$y = -\sqrt{(x^2+y^2)} + Ax^{1-\alpha}. \tag{i}$$

where A is a positive constant and $\alpha = v/u$. The origin is a *singular* point (i.e. not an equilibrium point) where solutions of the differential equations cross. A phase diagram for $\alpha = 0.5$ and $a = 1$ is shown in Figure 2.32.

If $\alpha > 1$, then the river flow speed is greater than the speed of the boat. From (i) it can be seen that the paths no longer pass through the origin. The rower cannot reach the origin from *any* point on the opposite bank.

- 2.24 In a simple model of a national economy, $\dot{I} = I - \alpha C$, $\dot{C} = \beta(I - C - G)$, where I is the national income, C is the rate of consumer spending and G the rate of government expenditure; the constants α and β satisfy $1 < \alpha < \infty$, $1 \leq \beta < \infty$. Show that if the rate of government expenditure G_0 is constant there is an equilibrium state. Classify the equilibrium state and show that the economy oscillates when $\beta = 1$.

Consider the situation when the government expenditure is related to the national income by the rule $G = G_0 + kI$, where $k > 0$. Show that there is no equilibrium state if $k \leq (\alpha-1)/\alpha$. How does the economy then behave?

Discuss an economy in which $G = G_0 + kI^2$, and show that there are two equilibrium states if $G_0 < (\alpha - 1)^2/(4k\alpha^2)$.

2.24. The economy model is governed by the equations

$$\dot{I} = I - \alpha C, \quad \dot{C} = \beta(I - C - G). \tag{i}$$

(i) $G = G_0$. Equilibrium occurs where

$$I - \alpha C = 0, \text{ and } I - C - G_0 = 0,$$

which has the solution

$$C = \frac{G_0}{\alpha - 1}, \quad I = \frac{\alpha G_0}{\alpha - 1}.$$

The equations are linear so that we can read off the usual parameters from (i):

$$p = 1 - \beta < 0, \quad q = -\beta + \alpha\beta > 0,$$

$$\Delta = (1 - \beta)^2 - 4\beta(\alpha - 1) = (1 + \beta)^2 - 4\beta\alpha.$$

From the table in Section 2.5, the equilibrium point is a stablenode if $\beta > 1$ and $(1+\beta)^2 > 4\beta\alpha$, and a stable spiral if $\beta > 1$ and $(1+\beta)^2 < 4\beta\alpha$. If $\beta = 1$, then the equilibrium point is a centre. In the latter case the economy will oscillate with amplitude which is dependent on the initial conditions.

(ii) $G = G_0 + kI$. The model now becomes

$$\dot{I} = I - \alpha C, \quad \dot{C} = \beta[(1 - k)I - C - G_0]. \tag{ii}$$

Equilibrium occurs where

$$C = \frac{G_0}{\alpha - 1 - k\alpha}, \quad I = \frac{\alpha G_0}{\alpha - 1 - k\alpha}.$$

If $k < (\alpha - 1)/\alpha$ the equilibrium point is in the first quadrant: otherwise the model has no equilibrium there (since C and I must both be positive). At the equilibrium point the parameter values are

$$p = 1 - \beta < 0, \quad q = \beta(\alpha - 1 - \alpha k), \quad \Delta = (1 - \beta)^2 - 4\beta(\alpha - 1 - \alpha k).$$

The point is a saddle point if $k > (\alpha - 1)/\alpha$ (in the third quadrant). If $k < (\alpha - 1)/\alpha$, then the equilibrium point, in the first quadrant, is a stable node if $(1 - \beta)^2 > 4\beta(\alpha - 1 - k)$ or a stable spiral if $(1 - \beta)^2 < 4\beta(\alpha - 1 - k)$. Two phase diagrams for the case $k > (\alpha - 1)/\alpha$ are shown in Figure 2.33.

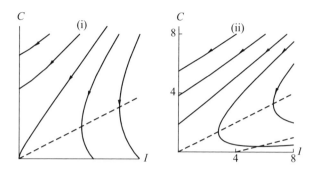

Figure 2.33 Problem 2.24: The phase diagrams are shown for the following parameters: (i) $\alpha = 2$, $\beta = 1$, $k = 2$, $G_0 = 1$; (ii) $\alpha = 2$, $\beta = 1$, $k = \frac{3}{4}$, $G_0 = 1$. The dashed lines show the isoclines of zero and infinite slopes.

(iii) $G = G_0 + kI^2$. The equations become

$$\dot{I} = I - \alpha C, \quad \dot{C} = \beta(I - C - G_0 - kI^2).$$

Equilibrium states are given by

$$I - \alpha C = 0 \quad \beta(I - C - G_0 - kI^2) = 0.$$

Eliminating I, C satisfies

$$C = \frac{1}{2k\alpha^2}\{(\alpha - 1) \pm \sqrt{[(\alpha - 1)^2 - 4G_0 k\alpha^2]}\}$$

and I can be found from $I = \alpha C$. There are two real positive solutions if $G_0 < (\alpha - 1)^2/(4k\alpha^2)$.

- **2.25** Let $f(x)$ and $g(y)$ have local minima at $x = a$ and $y = b$ respectively. Show that $f(x) + g(y)$ has a minimum at (a, b). Deduce that there exists a neighbourhood (a, b) in which all solutions of the family of equations $f(x) + g(y) = $ constant represent closed curves surrounding (a, b).
  Show that $(0, 0)$ is a centre for the system $\dot{x} = y^5$, $\dot{y} = -x^3$, and that all paths are closed curves.

2.25. Define the open intervals $I_1 : \varepsilon > |x - a| > 0$ and $I_2 : \varepsilon > |y - b| > 0$. Since $f(x)$ has a minimum at $x = a$, and $g(y)$ has a minimum at $y = b$, there exists an ε such that $f(x) > f(a)$ for all $x \in I_1$, and $g(y) > g(b)$ for all $y \in I_2$.
For the function of two variables $f(x) + g(y)$,

$$f(x) + g(y) > f(a) + g(b) \text{ for all } (x, y) \in I_1 \times I_2.$$

Therefore (a, b) is a local minimum of $f(x) + g(y)$, and for some neighbourhood of (a, b), there exists a constant $c_1 > f(a, b)$ such that the curves $f(x) + g(b) = c$ are closed curves about (a, b) for all C such that $f(a, b) < c < c_1$.

For the system

$$\dot{x} = y^5, \quad \dot{y} = -x^3,$$

which has a single equilibrium point, at $(0,0)$, the phase paths are given by

$$\frac{dy}{dx} = -\frac{x^3}{y^5}.$$

The solutions are given by

$$\tfrac{1}{6}y^6 + \tfrac{1}{4}x^4 = \text{constant}.$$

In the notation above, let $f(x) = \tfrac{1}{4}x^4$, which has a minimum at $x = 0$, and let $g(y) = \tfrac{1}{6}y^6$, which has a minimum at $y = 0$. By the result above $\tfrac{1}{6}y^6 + \tfrac{1}{4}x^4$ has minimum at $(0,0)$. Hence, the origin is surrounded by a nest of closed paths, and is therefore a centre.

- **2.26** For the predator–prey problem in NODE, Section 2.2, show, by using Problem 2.25, that all solutions in $x > 0$, $y > 0$ are periodic.

2.26. The predator–prey equations are, from Section 2.2,

$$\dot{x} = ax - cxy, \quad \dot{y} = -by + dxy,$$

and the equation of the phase paths is

$$b \ln x - dx + a \ln y - cy = C,$$

a constant. Equilibrium occurs at $(b/d, a/c)$. In the notation of Problem 2.25, let $f(x) = b \ln x - dx$ and $g(y) = a \ln y - cy$. Since

$$f'(x) = \frac{b}{x} - d, \quad f''(x) = -\frac{b}{x^2}, \quad g'(y) = \frac{a}{y} - c, \quad g''(y) = -\frac{a}{y^2},$$

it can be verified that $f(x)$ has a minimum at $x = b/d$, and that $g(y)$ has a minimum at $y = a/c$. The level curves of the surface $f(x) + g(y)$ cover the whole of the first quadrant about a minimum at $(b/d, a/c)$, which is a centre. Hence all solutions are periodic about the centre.

- **2.27** Show that the phase paths of the Hamiltonian system $\dot{x} = -\partial H/\partial y$, $\dot{y} = \partial H/\partial x$ are given by $H(x, y) = $ constant. Equilibrium points occur at the stationary points of $H(x, y)$. If (x_0, y_0) is an equilibrium point, show that (x_0, y_0) is stable according to the linear approximation if $H(x, y)$ has a maximum or a minimum at the point. (Assume that all the second derivatives of H are non-zero at x_0, y_0.)

2.27. A first-order system is said to be Hamiltonian if there exists a differentiable function $H(x, y)$ such that

$$\dot{x} = -\frac{\partial H}{\partial y}, \quad \dot{y} = \frac{\partial H}{\partial x}. \qquad (i)$$

The phase paths of the system are given by

$$\frac{dy}{dx} = -\frac{\partial H/\partial x}{\partial H/\partial y},$$

which, using the chain rule, can be expressed in the form

$$\frac{\partial H}{\partial x} + \frac{\partial H}{\partial y}\frac{dy}{dx} = \frac{dH(x, y)}{dx} = 0,$$

treating y as a function x. Therefore the general solution is $H(x, y) = $ constant. Equilibrium points occur at the stationary points of $H(x, y)$.

Let (x_0, y_0) be a stationary point of $H(x, y)$. Consider the perturbation $x = x_0 + x'$ and $y = y_0 + y'$. Then eqns (i) become

$$\dot{x} = -H_y(x_0 + x', y_0 + y')$$
$$\approx -[H_y(x_0, y_0) + H_{yx}(x_0, y_0)x' + H_{yy}(x_0, y_0)y']$$
$$= -H_{yx}(x_0, y_0)x' - H_{yy}(x_0, y_0)y' = -Bx' - Cy',$$

say, and

$$\dot{y} = H_x(x_0 + x', y_0 + y')$$
$$\approx H_x(x_0, y_0) + H_{xx}(x_0, y_0)x' + H_{xy}(x_0, y_0)y'$$
$$= H_{xx}(x_0, y_0)x' + H_{xy}(x_0, y_0)y' = Ax' + By'$$

say, using the first two terms of the Taylor series in both cases. The second derivative test for functions of two variables says that (x_0, y_0) is a maximum or minimum if the second derivatives satisfy $AC - B^2 > 0$. For the linear approximation above, the parameters are $p = -B + B = 0$ and $q = -B^2 + AC$. For stability we require $q > 0$ which is the same condition as for a stationary maximum or minimum of $H(x, y)$.

• 2.28 The equilibrium points of the nonlinear parameter-dependent system $\dot{x} = y$, $\dot{y} = f(x, y, \lambda)$ lie on the curve $f(x, 0, \lambda) = 0$ in the x, λ plane. Show that an equilibrium point (x_1, λ_1) is stable and that all neighbouring solutions tend to this point (according to the linear approximation) if $f_x(x_1, 0, \lambda_1) < 0$ and $f_y(x_1, 0, \lambda_1) < 0$.
Investigate the stability if $\dot{x} = y$, $\dot{y} = -y + x^2 - \lambda x$.

2.28. Consider the system $\dot{x} = y$, $\dot{y} = f(x, y, \lambda)$. Equilibrium points occur where $y = 0$ and $f(x, 0, \lambda)$. Let $x = x_1 + x'$ and $y = y'$. The linear approximation is

$$\dot{x}' = y', \quad \dot{y}' = f(x_1 + x', y', \lambda) \approx f_x(x_1, 0, \lambda_1) x' + f_y(x_1, 0, \lambda) y'.$$

The parameters of the linear approximation are

$$p = f_y(x_1, 0, \lambda_1), \quad q = -f_x(x_1, 0, \lambda_1).$$

The equilibrium point is stable if $p < 0$ and $q > 0$, which is equivalent to

$$f_y(x_1, 0, \lambda_1) < 0, \quad f_x(x_1, 0, \lambda_1) < 0.$$

In the example

$$\dot{x} = y, \quad \dot{y} = -y + x^2 - \lambda x,$$

$f(x, y, \lambda) = -y + x^2 - \lambda x$. The equilibrium points are at $(0, 0)$ and $(\lambda, 0)$. The first derivatives are

$$f_x(x, 0, \lambda) = 2x - \lambda, \quad f_y(x, 0, \lambda) = -1.$$

At $(0, 0)$,

$$f_x(0, 0, \lambda) = -\lambda < 0$$

if $\lambda > 0$, and

$$f_y(0, 0, \lambda) = -1 < 0,$$

for all λ. Hence $(0, 0)$ is stable if $\lambda > 0$.
At $(\lambda, 0)$,

$$f_x\left(\tfrac{1}{2}\lambda, 0, \lambda\right) = \lambda < 0,$$

if $\lambda < 0$, and

$$f_y\left(\tfrac{1}{2}\lambda, 0, \lambda\right) = -1.$$

Therefore the point $(\lambda, 0)$ is stable if $\lambda < 0$.

- 2.29 Find the equations for the phase paths for the general epidemic described (Section 2.2) by the system
$$\dot{x} = -\beta xy, \quad \dot{y} = \beta xy - \gamma y, \quad \dot{z} = \gamma y.$$
Sketch the phase diagram in the (x, y) plane. Confirm that the number of infectives reaches its maximum when $x = \gamma/\beta$.

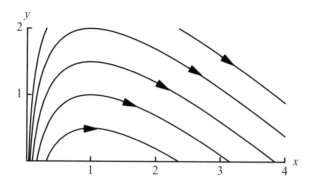

Figure 2.34 Problem 2.29: The phase diagram is drawn for $\gamma/\beta = 1$.

2.29. The general epidemic equations are

$$\dot{x} = -\beta x y, \quad \dot{y} = \beta x y - \gamma y, \quad \dot{z} = \gamma y$$

(see Example 2.4). From the first two equations equilibrium occurs at $y = 0$ for all $x \geq 0$. The phase paths in the x, y plane are given by

$$\frac{dy}{dx} = \frac{\dot{y}}{\dot{x}} = -\frac{\beta x - \gamma}{\beta x}.$$

This is a separable equation with general solution

$$y = -\int \frac{\beta x - \gamma}{\beta x} dx = -x + \frac{\gamma}{\beta} \ln x + C,$$

noting that both x and y must be positive. A phase diagram is shown in Figure 2.34. From the equation $\dot{y} = \beta x y - \gamma y$, $\dot{y} = 0$ where $x = \gamma/\beta$. The maxima lie on the line $x = 1$ in Figure 2.34 where $dy/dx = 0$.

- 2.30 Two species x and y are competing for a common food supply. Their growth equations are

$$\dot{x} = x(1 - x - y), \quad \dot{y} = y(3 - x - \tfrac{3}{2}y), \quad (x, y > 0).$$

Classify the equilibrium points using linear approximations. Draw a sketch indicating the slopes of the phase paths in $x \geq 0$, $y \geq 0$. If $x = x_0 > 0$, $y = y_0 > 0$ initially, what do you expect the long-term outcome of the species to be? Confirm your conclusions numerically by computing phase paths.

2.30. The system

$$\dot{x} = x(1 - x - y), \quad \dot{y} = y\left(3 - x - \tfrac{3}{2}y\right)$$

is in equilibrium at $(0,0)$, $(0,2)$ and $(1,0)$, the solutions of

$$x(1 - x - y) = 0, \text{ and } y\left(3 - x - \tfrac{3}{2}y\right) = 0.$$

- $(0,0)$. The linear approximation is

$$\dot{x} \approx x, \quad \dot{y} \approx 3y,$$

which is an unstable star-shaped equilibrium point;
- $(0,2)$. Let $x = \xi$, $y = 2 + \eta$. Then

$$\dot{\xi} = \xi[1 - \xi - (2 + \eta)] \approx -\xi, \quad \dot{\eta} = (2 + \eta)\left[3 - \xi - \tfrac{3}{2}(2 + \eta)\right] \approx -2\xi - 3\eta,$$

which implies a stable node.
- $(1,0)$. Let $x = 1 + \xi$, $y = \eta$. The

$$\dot{\xi} = (1 + \xi)[1 - (1 + \xi) - \eta] \approx -\xi - \eta, \quad \dot{\eta} = \eta\left[3 - (1 + \xi) - \tfrac{3}{2}\eta\right] \approx 2\eta,$$

which implies a saddle.

The isoclines on which $dy/dx = 0$ are the straight lines $y = 0$ and $3 - x - \tfrac{3}{2}y = 0$, and the isoclines on which $dy/dx = \infty$ are $x = 0$ and $1 - x - y = 0$. These are shown in Figure 2.35 together with the signs of the slopes of the paths between these lines. The computed phase diagram is also shown in this figure. For any initial point $x = x_0 > 0$, $y = y_0 > 0$, all phase paths approach $(0, 2)$ asymptotically, which means that species x ultimately dies out.

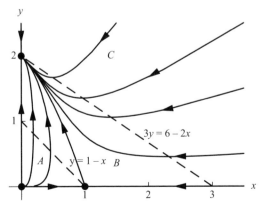

Figure 2.35 Problem 2.30: The phase diagram for $\dot{x} = x(1 - x - y)$, $\dot{y} = y(3 - x - \tfrac{3}{2}y)$. In the phase diagram, A is the region in which $\dot{x} > 0$, $\dot{y} > 0$, B is the region in which $\dot{x} < 0$, $\dot{y} > 0$ and C is the region in which $\dot{x} < 0$, $\dot{y} < 0$.

- **2.31** Sketch the phase diagram for the competing species x and y for which
$$\dot{x} = (1 - x^2 - y^2)x, \quad \dot{y} = (\tfrac{5}{4} - x - y)y.$$

2.31. The competing species equations are
$$\dot{x} = (1 - x^2 - y^2)x, \quad \dot{y} = (1.25 - x - y)y.$$

In the the quadrant $x \geq 0$, $y \geq 0$, the equilibrium points are given by
$$(0,0), \quad (0, 1.250), \quad (1, 0), \quad (0.294, 0.956), \quad (0.956, 0.294).$$
A computed phase diagram is shown in Figure 2.36. The dashed line and arc are respectively the isoclines of zero and infinite slopes. From the figure we can see that $(0,0)$ is a unstable star-shaped point. The points $(1,0)$ and $(0.294, 0.956)$ are saddle points, and $(0,1)$ and $(0.956, 0.294)$ are stable nodes. These interpretations from the phase diagram can be confirmed by finding the linear approximations at each equilibrium point.

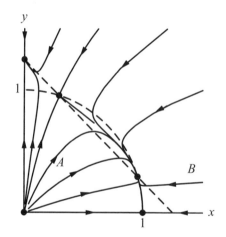

Figure 2.36 Problem 2.31: The phase diagram for $\dot{x} = x(1 - x^2 - y^2)$, $\dot{y} = y(1.25 - x - y)$.

- **2.32** A space satellite is in free flight on the line joining, and between, a planet (mass m_1) and its moon (mass m_2), which are at a fixed distance a apart. Show that
$$-\frac{\gamma m_1}{x^2} + \frac{\gamma m_2}{(a - x)^2} = \ddot{x},$$
where x is the distance of the satellite from the planet and γ the gravitational constant. Show that the equilibrium point is unstable according to the linear approximation.

2.32. If G_1 and G_2 are the gravitational forces on the satellite due, respectively, to the planet and the moon, then the equation of motion is

$$-G_1 + G_2 = m\ddot{x},$$

where m is the mass of the satellite. By the law of gravitation,

$$G_1 = \frac{\gamma m m_1}{x^2}, \quad G_2 = \frac{\gamma m m_2}{(a-x)^2}.$$

Hence x satisfies the equation

$$-\frac{\gamma m_1}{x^2} + \frac{\gamma m_2}{(a-x)^2} = \ddot{x}.$$

Put the equation in dimensionless form by the changes of variable $x = az$ and $t = a\tau/\sqrt{(\gamma m_2)}$, so that z satisfies

$$-\frac{\lambda}{z^2} + \frac{1}{(1-z)^2} = \frac{d^2 z}{d\tau^2} = z'', \tag{i}$$

say. Equilibrium occurs where

$$-\frac{\lambda}{z^2} + \frac{1}{(1-z)^2} = 0, \text{ or } (\lambda - 1)z^2 - 2\lambda z + \lambda = 0.$$

Therefore

$$z = \frac{\lambda \pm \sqrt{\lambda}}{\lambda - 1} = z_1,$$

say, provided $\lambda \neq 1$. If $\lambda = 1$ (planet and moon have the same masses) then $z = \frac{1}{2}$. If $\lambda > 1$, then for equilibrium between the bodies choose the minus sign, so that the solution being considered is $z = (\lambda - \sqrt{\lambda})/(\lambda - 1)$. The case $\lambda < 1$ can be deduced using the transformations $z \to 1 - z$ and $\lambda \to 1/\lambda$.

Let $z = z_1 + \zeta$. Then (i) becomes

$$\zeta'' = -\frac{\lambda}{(z_1 + \zeta)^2} + \frac{1}{[1 - (z_1 + \zeta)]^2}$$

$$\approx -\frac{\lambda}{z_1^2}\left(1 - \frac{2\zeta}{z_1}\right) + \frac{1}{(1-z_1)^2}\left(1 + \frac{2z_1\zeta}{1-z_1}\right)$$

$$= 2\left(\frac{\lambda}{z_1^3} + \frac{z_1}{1-z_1^3}\right)\zeta$$

Since $0 < z_1 < \frac{1}{2}$, the coefficient of ζ on the right-hand side of this equation is positive. Hence the equilibrium is unstable since the solution can have local exponential growth.

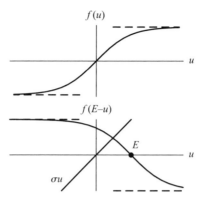

Figure 2.37 Problem 2.33.

• 2.33 The system
$$\dot V_1 = -\sigma V_1 + f(E - V_2), \quad \dot V_2 = -\sigma V_2 + f(E - V_1), \quad \sigma > 0, E > 0$$
represents (Andronov and Chaikin 1949) a model of a triggered sweeping circuit for an oscilloscope. The conditions on $f(u)$ are: $f(u)$ continuous on $-\infty < u < \infty$, $f(-u) = -f(u)$, $f(u)$ tends to a limit as $u \to \infty$, and is monotonic decreasing (see Figure 3.20 in NODE).

Show by a geometrical argument that there is always at least one equilibrium point (v_0, v_0) say, and that when $f'(E - v_0) < \sigma$ it is the only one; and deduce by taking the linear approximation that it is a stable node. (Note that $f'(E - v) = -df(E - v)/dv$.)

Show that when $f'(E - v_0) > \sigma$ there are two others, at $(V', (1/\sigma)f(E - V'))$ and $((1/\sigma)f(E - V'), V)$ respectively for some V'. Show that these are stable nodes, and that the one at (v_0, v_0) is a saddle point.

2.33. The triggered sweeping circuit is governed by the equation
$$\dot V_1 = -\sigma V_1 + f(E - V_2), \quad \dot V_2 = -\sigma V_2 + f(E - V_1).$$

Typical graphs of $f(u)$ and $f(E-u)$ versus u are shown in Figure 2.37. The point of intersection of σu with $F(E-u)$ is shown in the second figure, but there may be further points of intersection. It is convenient to make changes of variable to standard notation at this point. Let $x = V_1$, $y = V_2$, $g(y) = f(E - y)$ and $g(x) = f(E - x)$. The equations can then be expressed in the form
$$\dot x = -\sigma x + g(y), \quad \dot y = g(x) - \sigma y. \qquad (i)$$

Equilibrium points occur where the curve $\sigma x = g(y)$ intersects its *inverse* curve $\sigma y = g(x)$ as shown in Figure 2.38(a). There always exists one point of intersection at $x = v_0$ where v_0 satisfies $\sigma v_0 = g(v_0)$, but there may be two further points as shown in the enlarged section of

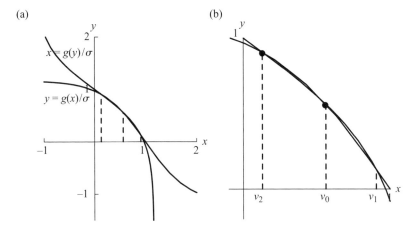

Figure 2.38 Problem 2.33: The figures has been drawn with the value $\sigma = 1$ and the function $g(x) = 1.163 \tanh[1.2(1-x)]$. The intersections occur at $v_0 = 0.561$, $v_1 = 0.906$ and $v_2 = 0.130$, approximately, as indicated in the enlarged inset shown in (b).

the figure shown in Figure 2.38(b). The latter will occur if the slope of $y = g(x)/\sigma$ at (v_0, v_0) is less than -1, since the curves are mutual inverses. Therefore there are three equilibrium points if and only if $g'(v_0) < -\sigma$, and one at $x = v_0$ if $g'(v_0) > -\sigma$. When there are three such points they are labelled as (v_0, v_0), (v_1, v_2) and (v_2, v_1) as shown in the figure. Since the curves are mutual inverse functions, then $\sigma v_1 = g(v_2)$ and $\sigma v_2 = g(v_1)$.

(i) $0 > g'(v_0) > -\sigma$. For the only equilibrium point $x = v_0$, let $x = v_0 + \xi$ and $y = v_0 + \eta$. Then the linear approximation of (i) is given by

$$\dot{\xi} = -\sigma(v_0 + \xi) + g(v_0 + \eta) \approx -\sigma\xi + g'(v_0)\eta,$$

$$\dot{\eta} = g(v_0) + \xi) - \sigma(v_0 + \eta) \approx g'(v_0)\xi - \sigma\eta.$$

The parameters are

$$p = -2\sigma < 0, \quad q = \sigma^2 - g'(v_0)^2 > 0, \quad \Delta = 4g'(v_0)^2 > 0.$$

Therefore (v_0, v_0) is a stable node.

(ii) $g'(v_0) < -\sigma$. From (i) the equilibrium point (v_0, v_0) has the parameter values

$$p = -2\sigma < 0, \quad q = \sigma^2 - g'(v_0)^2 < 0.$$

Therefore (v_0, v_0) is a saddle point.

For the point (v_1, v_2), let $x = v_1 + \xi$ and $y = v_2 + \eta$. Then substitution in (i) leads to

$$\dot{\xi} = -\sigma(v_1 + \xi) + g(v_2 + \eta) \approx -\sigma\xi + g'(v_2)\eta,$$

$$\dot{\eta} = g(v_1) + \xi) - \sigma(v_2 + \eta) \approx g'(v_1)\xi - \sigma\eta.$$

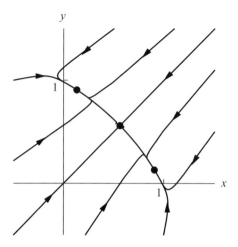

Figure 2.39 Problem 2.33: Phase diagram for $\dot{x} = -\sigma x + 1.163\tanh[1.2(1-y)]$, $\dot{y} = 1.163\tanh[1.2(1-x)] - \sigma y$ with $\sigma = 1$, showing a saddle and two stable nodes.

The parameters of this linear approximation are

$$p = -2\sigma < 0, \quad q = \sigma^2 - g'(v_1)g'(v_2), \quad \Delta = 4g'(v_1)g'(v_2) > 0.$$

Consider the intersection of the curves $C_1 : y = g(x)/\sigma$ and $C_2 : x = g(y)/\sigma$ at $x = v_1$. If the slope of C_1 is m_1, and of C_2 is m_2 at $x = v_1$, then $m_1 > m_2$ (see Figure 2.38). Therefore, since

$$m_1 = \frac{g'(v_1)}{\sigma} \quad \text{and} \quad m_2 = \frac{\sigma}{g'(v_2)},$$

it follows that $g'(v_1)g'(v_2) < \sigma^2$ (remember that $g'(v_1)$ and $g'(v_2)$ are negative which affects manipulation of the inequalities). Hence for the equilibrium point (v_1, v_2), $q > 0$ which means that the equilibrium point is a stable node.

A similar argument can be used to show that (v_2, v_1) is also a node: simply interchange v_1 and v_2 in the analysis above. A phase diagram for case (ii) (three equilibrium points) is shown in Figure 2.39 for the same parameter values as in Figure 2.38.

- **2.34** Investigate the equilibrium points of $\dot{x} = a - x^2$, $\dot{y} = x - y$. Show that the system has a saddle and a stable node for $a > 0$ but no equilibrium points if $a < 0$. The system is said to undergo a **bifurcation** as a increases through $a = 0$. This bifurcation is an example of a **saddle-node bifurcation**. Draw phase diagrams for $a = 1$ and $a = -1$.

2.34. The parameter-dependent system is

$$\dot{x} = a - x^2, \quad \dot{y} = x - y.$$

Equilibrium points occur where $a - x^2 = 0$ and $x - y = 0$. If $a < 0$, then there are no equilibrium points. If $a > 0$, there are equilibrium points at $(\sqrt{a}, \sqrt{a})$ and $(-\sqrt{a}, -\sqrt{a})$.

(a) $(\sqrt{a}, \sqrt{a})$. Let $x = \sqrt{a} + \xi$, $y = \sqrt{a} + \eta$. Then the linear approximation is

$$\dot{\xi} \approx -2\sqrt{a}\xi, \quad \dot{\eta} = \xi - \eta.$$

The associated parameters are

$$p = -2\sqrt{a} - 1 < 0, \quad q = 2\sqrt{a} > 0, \quad \Delta = 4\sqrt{a} + 1 > 0.$$

Hence $(\sqrt{a}, \sqrt{a})$ is a stable node.

(b) $(-\sqrt{a}, -\sqrt{a})$. Let $x = -\sqrt{a} + \xi$, $y = -\sqrt{a} + \eta$. Then the linear approximation is

$$\dot{\xi} \approx 2\sqrt{a}\xi, \quad \dot{\eta} = \xi - \eta.$$

The parameter q is given by $q = -2\sqrt{a} < 0$, so that $(-\sqrt{a}, -\sqrt{a})$ is a saddle point. Some phase paths for the case $a = 1$ are shown in Figure 2.40.

- **2.35** Figure 2.16 (in NODE) represents a circuit for activating an electric arc A which has the voltage-current characteristic shown. Show that $L\dot{I} = V - V_a(I)$, $RC\dot{V} = -RI - V + E$ where $V_a(I)$ has the general shape shown in Figure 2.16 (in NODE). By forming the linear approximating equations near the equilibrium points find the conditions on E, L, C, R and V'_a for stable working, assuming that $V = E - RI$ meets the curve $V = V_a(I)$ in three points of intersection.

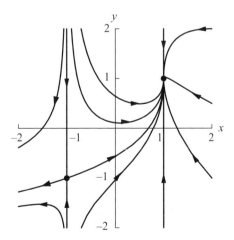

Figure 2.40 Problem 2.34: Phase diagram for $\dot{x} = a - x^2$, $\dot{y} = x - y$ for $a = 1$, showing a stable node at $(1, 1)$ and a saddle point at $(-1, -1)$.

2.35. Consult Figure 2.16 in NODE. For the voltage V, we have on the right- and left-hand sides,
$$V - L\frac{dI}{dt} - V_a(I) = 0, \tag{i}$$
and
$$E - Ri - v = 0. \tag{ii}$$
For the circuit including the capacitance C and the inductance L,
$$-\frac{1}{C}\int (i - I)dt + V = 0, \text{ or } i = I + C\frac{dV}{dt}. \tag{iii}$$
Eliminate i between (ii) and (iii) giving
$$\dot{V} + \frac{1}{RC}V = \frac{E - RI}{RC}. \tag{iv}$$
The required first-order equations are given by (i) and (iv), namely
$$\dot{I} = \frac{V}{L} - \frac{V_a(I)}{L}, \quad \dot{V} = -\frac{1}{C}I - \frac{1}{RC}V + \frac{E}{RC}. \tag{v}$$

Let (I_n, V_n), $(n = 1, 2, 3)$ be the equilibrium points, and consider the perturbations $I = I_n + \xi$, $V = V_n + \eta$. Then, from (iv),
$$\dot{\xi} = \frac{1}{L}[V_n - V_a(I_n + \xi)] \approx -\frac{1}{L}V_a'(I_n)\xi + \frac{1}{L}\eta,$$
$$\dot{\eta} = -\frac{1}{C}(I_n + \xi) - \frac{1}{RC}(V_n + \eta) + \frac{E}{RC} = -\frac{1}{C}\xi - \frac{1}{RC}\eta,$$
where V_n and I_n satisfy $V_n - V_a(I_n) = 0$ and $-RI_n - V_n + E = 0$ for each value of n. In general in the linear approximations the parameters are
$$p = -\frac{1}{L}V_a'(I_n) - \frac{1}{RC}, \quad q = \frac{V_a'(I_n)}{RCL} + \frac{1}{LC}.$$

- (I_1, V_1). From Figure 2.41, $V_a'(I_1) > 0$ so that $p < 0$ and $q > 0$, which means that the equilibrium is either a stable node or spiral depending on the sign of $\Delta = p^2 - 4q$.
- (I_2, V_2). From Figure 2.41, $V_a'(I_2) < 0$. Stability will depend on the signs of p and q. If $V_a'(I_2) < -L$, then the equilibrium point is a saddle (and unstable). If $V_a(I_2) < -L/(RC)$, then $p < 0$ so that the equilibrium point is an unstable node or spiral. We can combine these instability conditions into the single condition $V_a'(I_2) < \max(-L, -L/(RC))$. If $V_a'(I_2) > \max(-L, -L/(RC))$, then $p < 0$ and $q > 0$ which means that (I_2, V_2) is stable.
- (I_3, V_3). From Figure 2.41, it can be seen that $V_a'(I_3) < 0$ also. Therefore, the results are the same as the previous case with I_2, V_2 replaced by I_3, V_3.

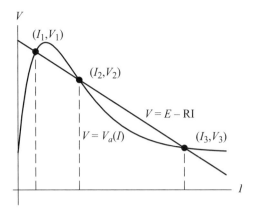

Figure 2.41 Problem 2.35: This figure was plotted using $V_a(I) = 0.05 + xe^{-2.5x}$ and the line $V = 0.2 - 0.06I$ to create three points of intersection.

- 2.36 The equation for the current x in the circuit of Figure 2.17(a) (in NODE) is
$$LC\ddot{x} + RC\dot{x} + x = I.$$
Neglect the grid current, and assume that I depends only on the relative grid potential $e_g : I = I_S$ (saturation current) for $e_g > 0$ and $I = 0$ for $e_g < 0$ (see Figure 2.17(b) in NODE). Assume also that the mutual indictance $M > 0$, so that $e_g <,> 0$ according as $\dot{x} >, < 0$. Find the nature of the phase paths. By considering their successive intersections with the x axis show that a limit cycle is approached from all initial conditions (assume $R^2C < 4L$).

2.36. The equation for the current x is
$$LC\ddot{x} + RC\dot{x} + x = I,$$
where
$$I = \begin{cases} I_s & \dot{x} > 0 \\ 0 & \dot{x} < 0 \end{cases}.$$
The (x, y) phase plane equations with $y = \dot{x}$ are
$$\dot{x} = y, \quad \dot{y} = -\omega^2 x - 2ky + \begin{cases} \omega^2 I_s (> 0) & y > 0 \\ 0 & y < 0 \end{cases},$$
where $\omega^2 = 1/(LC)$ and $k = R/(2L)$.

The solutions are (for $\omega^2 > k^2$, that is, $R^2C > 4L$)
$$x = Ae^{-kt}\cos(\Omega t + \alpha) \text{ for } y < 0,$$
$$x = I_s + Ae^{-kt}\cos(\Omega t + \alpha) \text{ for } y > 0.$$

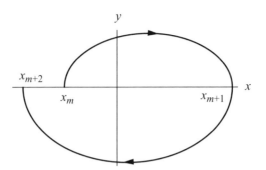

Figure 2.42 Problem 2.36: Three successive intersections x_m, x_{m+1}, x_{m+2} with the x axis.

Consider the sequence of intersections x_m, x_{m+1}, x_{m+2} with the y axis as shown in Figure 2.42. Let $t = t_0$ at $x = x_m$ and $t = t_0 + T$, where $T = \pi/\Omega$, at $x = x_{m+1}$. Therefore

$$x_{m+1} = -e^{-kT} x_m.$$

Similarly

$$x_{m+2} = I_s - e^{-kT} x_{m+1} = I_s + \lambda x_m, \qquad \text{(i)}$$

putting $\lambda = e^{-kT}$. Let m be an even number, say, $2n - 2$. Then iteration of (i) gives

$$x_{2n} = I_s + \lambda(I_s + \lambda(I_s + \lambda(\cdots + \lambda(I_s + \lambda x_0))\cdots))$$

$$= I_s + \lambda I_s + \cdots + \lambda^{n-1} I_s + \lambda^n x_0$$

$$= \frac{1 - \lambda^n}{1 - \lambda} I_s + \lambda^n x_0$$

$$\to \frac{I_s}{1 - \lambda}$$

as $n \to \infty$ for all x_0. Since this limit of x_{2n} exists there must be a limit cycle through $x = I_s/(1-\lambda)$
It cuts the x axis again at $x = -\lambda I_s/(1 - \lambda)$.

- **2.37** For the circuit in Figure 2.17(a) (in NODE) assume that the relation between I and e_g is as in Figure 2.18 (in NODE); that is $I = F(e_g + ke_p)$, where e_g and e_p are relative grid and plate potentials, $k > 0$ is a constant, and in the neighbourhood of the point of inflection, $f(u) = I_0 + au - bu^3$, where $a.0$, $b > 0$. Deduce the equation for x when the DC source E is set so that the operating point is the point of inflection. (A form of Rayleigh's equation is obtained, implying an unstable or a stable limit cycle respectively.)

2.37 As in the previous problem,
$$LC\ddot{x} + RC\dot{x} + x = I,$$
where
$$I = I_0 + a(e_g + ke_p - E_0) - b(e_g + ke_p - E_0)^3,$$
$$e_g = M\frac{dx}{dt}, \quad E - L\frac{dx}{dt} - Rx - e_p = 0.$$
Therefore
$$e_g + ke_p - E_0 = (M - kL)\frac{dx}{dt} - kRx + kE - E_0 = (M - kL)\dot{x} - kRx,$$
when the working point $E_0 = kE$. Finally x satisfies
$$\ddot{x} + \frac{R}{L}\dot{x} + \frac{1}{LC}x = \frac{1}{LC}[I_0 + a[(M - kL)\dot{x} - kRx] - b[(M - kL)\dot{x} - kRx]^3].$$
This can be rearranged in the form
$$\ddot{x} + A\dot{x} + B\dot{x}^3 + \omega^2 x = D,$$
where
$$A = \frac{1}{LC}\left[\frac{R}{L} - a(M - kL)\right], \quad B = \frac{b}{LC}(M - kL)^3, \quad D = \frac{I_0}{LC}.$$
If the right-hand side is reduced to zero by a suitable translation of x, then the result is Rayleigh's equation (see Example 4.6). It has a stable limit cycle if $A < 0$ and $B > 0$, that is, if
$$R < aL(M - kL) \text{ and } M > kL.$$

- 2.38 Figure 2.19(a) (in NODE) represents two identical DC generators connected in parallel, with inductance and resistance L, r. Here R is the resistance of the load. Show that the equations for the currents are
$$L\frac{di_1}{dt} = -(r + R)i_1 - Ri_2 + E(i_1), \quad L\frac{di_2}{dt} = -Ri_1 - (r + R)i_2 + E(i_2).$$
Assuming that $E(i)$ has the characteristics indicated by Figure 2.19(b) (in NODE) show that

(i) when $E'(0) < r$ the state $i_1 = i_2 = 0$ is stable and is otherwise unstable;
(ii) when $E'(0) < r$ there is a stable state $i_1 = -i_2$ (no current flows to R);
(iii) when $E'(0) > r + 2R$ there is a state with $i_1 = i_2$, which is unstable.

2.38. For the upper circuit, through generator $E(i_1)$ and resistance R,

$$E(i_1) - L\frac{di_1}{dt} - ri_1 - R(i_1 + i_2) = 0,$$

and for the lower circuit through the generator $E(i_2)$ and resistance R,

$$E(i_2) - L\frac{di_2}{dt} - ri_2 - R(i_1 + i_2) = 0.$$

These equations can be rearranged into

$$L\frac{di_1}{dt} = -(r+R)i_1 - Ri_2 + E(i_1),$$

$$L\frac{di_2}{dt} = -Ri_1 - (r+R)i_2 + E(i_2).$$

An equilibrium or steady state occurs when the right-hand sides are zero, or when

$$E(i_1) = (r+R)i_1 + Ri_2,$$

$$E(i_2) = Ri_1 + (r+R)i_2.$$

Let $(i_1, i_2) = (a, b)$ be an equilibrium point, and let $i_1 = a + x$, $i_2 = b + y$. Then for small $|x|$ and $|y|$,

$$L\dot{x} \approx [-(r+R) + E'(a)]x - Ry,$$

$$L\dot{y} \approx -Rx + [-(r+R) + E'(b)]y.$$

The parameters associated with this linear approximation are

$$p = E'(a) + E'(b) - 2(r+R), \quad q = [E'(a) - (r+R)][E'(b) - (r+R)] - R^2.$$

(i) The point $(0,0)$ is always an equilibrium point. At this point

$$p = 2E'(0) - 2r - 2R, \quad q = (E'(0) - r - R)^2 - R^2 = (E'(0) - r - 2R)(E'(0) - r).$$

For $0 < E'(0) < r$, $p < 0$ and $q > 0$ which means that the equilibrium point is stable. If $r + 2R > E'(0) > r$, then $q < 0$ which implies that the equilibrium point is a saddle point (unstable). If $E'(0) > r + 2R$, then $p > 0$ and $q > 0$ which imply that the equilibrium point is either an unstable node or unstable spiral.

(ii) Since $E(i)$ is an odd function, the equations for equilibrium are unchanged if i_1 is replaced by $-i_2$, and i_2 by $-i_1$. Let $i_1 = -i_2 = i_0$. It follows that $(i_0, -i_0)$ is an equilibrium state where $E(i_0) = ri_0$. The parameters for the linear approximation are

$$p = -2(r+R) < 0, \quad q = r^2 + 2rR - E'(i_0)^2 > 0$$

if $E'(0) < r$. Hence the equilibrium is a stable node or spiral.

(iii) If $i_1 = i_2 = i_0$, say, then both equilibrium equations are the same so that i_0 satisfies $E(i_0) = (r + 2R)i_0$. Whether non-zero solutions of this equation exist will depend on the slope $E'(0)$, which must be less than $r + 2R$ for existence. The parameters in this case are

$$p = 2[E'(i_0) - r - R], \quad q = [E'(i_0) - (r + R)]^2 - R^2 = [E'(i_0) - r][E'(i_0) - r - 2R].$$

If $r < E'(0) < r + 2R$, then $p < 0$ and $q < 0$ which implies that the equilibrium point is a saddle point.

- **2.39** Show that the Emden–Fowler equation of astrophysics
$$(\xi^2 \eta')' + \xi^\lambda \eta^n = 0$$
is equivalent to the predator–prey model
$$\dot{x} = -x(1 + x + y), \quad \dot{y} = y(\lambda + 1 + nx + y)$$
after the change of variable $x = \xi \eta'/\eta$, $y = \xi^{\lambda-1} \eta^n/\eta'$, $t = \ln|\xi|$

2.39. The Emden–Fowler equation is

$$(\xi^2 \eta')' + \xi^\lambda \eta^n = 0,$$

or

$$2\xi \eta' + \xi^2 \eta'' + \xi^\lambda \eta^n = 0. \qquad (i)$$

Let

$$x = \frac{\xi \eta'}{\eta}, \quad y = \frac{\xi^{\lambda-1} \eta^n}{\eta'}, \quad t = \ln|\xi|.$$

Then, since $\dot{\xi} = \xi$,

$$\dot{x} = \frac{\dot{\xi} \eta'}{\eta} + \frac{\eta \eta'' \dot{\xi}}{\eta} - \frac{\xi \eta'^2 \dot{\xi}}{\eta^2}$$

$$= \frac{\xi \eta'}{\eta} + \frac{\xi^2 \eta''}{\eta} - \frac{\xi^2 \eta'^2}{\eta^2}$$

$$= \frac{\xi \eta'}{\eta} + \frac{1}{\eta}[-\xi^\lambda \eta^n - 2\xi \eta'] - \frac{\xi^2 \eta'^2}{\eta^2} \quad \text{(using (i))}$$

$$= -x - x^2 - (\xi^\lambda \eta^{n-1})$$

$$= -x - x^2 - xy.$$

Similarly

$$\dot{y} = (\lambda - 1)\frac{\xi^{\lambda-1}\dot{\xi}\eta^n}{\eta'} + n\xi^{\lambda-1}\eta^{n-1}\dot{\xi} - \frac{\xi^{\lambda-1}\eta^n\eta''\dot{\xi}}{\eta'^2}$$

$$= (\lambda - 1)\frac{\xi^{\lambda-1}\eta^n}{\eta'} + n\xi^\lambda\eta^{n-1} + \frac{\xi^{\lambda-1}\eta^n}{\eta'}(2\xi\eta' + \xi^\lambda\eta'')$$

$$= (\lambda + 1)y + nxy + y^2,$$

as required.

- **2.40** Show that Blasius' equation
$$\eta''' + \eta\eta'' = 0$$
is transformed by $x = \eta\eta'/\eta''$, $y = \eta'^2/(\eta\eta'')$, $t = \ln|\eta'|$ into
$$\dot{x} = x(1 + x + y), \quad \dot{y} = y(2 + x - y).$$

2.40. The Blasius equation is
$$\eta''' + \eta\eta'' = 0, \tag{i}$$
where $\eta' = d\eta/d\xi$. Let

$$x = \frac{\eta\eta'}{\eta''}, \quad y = \frac{\eta'^2}{\eta\eta''}, \quad t = \ln|\eta'|, \quad \text{or } \eta' = e^t.$$

Differentiating $\eta' = e^t$ with respect to t, we have

$$\eta''\dot{\xi} = e^t = \eta', \text{ so that } \dot{\xi} = \frac{\eta'}{\eta''}. \tag{ii}$$

Then

$$\dot{x} = \frac{\eta'^2\dot{\xi}}{\eta''} + n\dot{\xi} - \frac{\eta\eta'\eta'''\dot{\xi}}{\eta''^2}$$

$$= \frac{\eta'^3}{\eta''^2} + \frac{\eta\eta'}{\eta''} - \frac{\eta\eta'^2\eta'''}{\eta''^2} \quad \text{(using (ii))}$$

$$= \frac{\eta'^3}{\eta''^2} + \frac{\eta\eta'}{\eta''} + \frac{\eta^2\eta'^2}{\eta''^2} \quad \text{(using (i))}$$

$$= xy + x + x^2 = x(1 + x + y)$$

as required. Similarly

$$\dot{y} = \frac{2\eta'\dot{\xi}}{\eta} - \frac{\eta'^3\dot{\xi}}{\eta^2\eta''} - \frac{\eta'^2\eta'''\dot{\xi}}{\eta\eta''^2}$$

$$= \frac{2\eta'^2}{\eta\eta''} - \frac{\eta'^4}{\eta^2\eta''^2} - \frac{\eta'^3\eta'''}{\eta\eta''^3} \quad \text{(using (ii))}$$

$$= \frac{2\eta'^2}{\eta\eta''} - \frac{\eta'^4}{\eta^2\eta''^2} + \frac{\eta'^3}{\eta''^2} \quad \text{(using (i))}$$

$$= 2y - y^2 + xy = y(2 + x - y).$$

- 2.41 Consider the family of linear systems
$$\dot{x} = X\cos\alpha - Y\sin\alpha, \quad \dot{y} = X\sin\alpha + Y\cos\alpha$$
where $X = ax + by$, $Y = cx + dy$, and a, b, c, d are constants and α is a parameter. Show that the parameters are
$$p = (a + d)\cos\alpha + (b - c)\sin\alpha, \quad q = ad - bc.$$
Deduce that the origin is a saddle point for all α if $ad < bc$. If $a = 2$, $b = c = d = 1$, show that the equilibrium point at the origin passes through the sequence stable node, stable spiral, centre, unstable spiral, unstable node as α varies over range π.

2.41. The linear system is

$$\dot{x} = (a\cos\alpha - c\sin\alpha)x + (b\cos\alpha - d\sin\alpha)y,$$
$$\dot{y} = (a\sin\alpha + c\cos\alpha)x + (b\sin\alpha + d\cos\alpha)y.$$

The parameters associated with the origin of this linear system are

$$p = (a + d)\cos\alpha + (b - c)\sin\alpha,$$
$$q = (a\cos\alpha - c\sin\alpha)(b\sin\alpha + d\cos\alpha) - (b\cos\alpha - d\sin\alpha)(a\sin\alpha + c\cos\alpha)$$
$$= ad - bc.$$

If $ad - bc < 0$, then $q < 0$, which means that the origin is a saddle point for all α.
If $a = 2$, $b = c = d = 1$, then $p = 3\cos\alpha$, $q = 1 > 0$ and $\Delta = p^2 - 4q = 9\cos^2\alpha - 4$. The different stabilities can be seen most easily by sketching the graphs of p and Δ against α as shown in Figure 2.43. The nature of the equilibrium at the origin changes where either p or Δ changes sign at A, B and C ($q > 0$ so that a saddle is not possible). The classification is as follows:

- Interval OA: $p > 0$, $\Delta > 0$, unstable node.
- Point A: $p > 0$, $\Delta = 0$, degenerate unstable node.
- Interval AB: $p > 0$, $\Delta < 0$, unstable spiral.

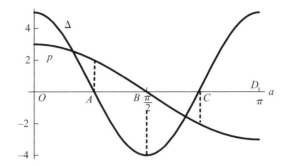

Figure 2.43 Problem 2.41: The graphs shows $p = 3\cos\alpha$ and $\Delta = 9\cos^2\alpha - 4$ for $0 \le \alpha \le \pi$.

- Point B: $p = 0$, centre.
- Interval BC: $p < 0$, $\Delta < 0$, stable spiral.
- Point C: $p < 0$, $\Delta = 0$, degenerate stable node.
- Interval CD, $p < 0$, $\Delta > 0$, stable node.

- 2.42 Show that, given $X(x, y)$, a system equivalent to the equation $\ddot{x} + h(x, \dot{x}) = 0$ is

$$\dot{x} = X(x, y), \quad \dot{y} = -\left\{h(x, X) + X\frac{\partial X}{\partial x}\right\} \bigg/ \frac{\partial X}{\partial y}.$$

2.42. Consider the equation $\ddot{x} + h(x, \dot{x}) = 0$. Let $\dot{x} = X(x, y)$. Then

$$\ddot{x} = \frac{\partial X}{\partial x}\dot{x} + \frac{\partial X}{\partial y}\dot{y} = -h(x, X).$$

by applying a chain rule. Therefore

$$\dot{y} = -\left\{h(x, X) + X\frac{\partial X}{\partial x}\right\} \bigg/ \frac{\partial X}{\partial y}.$$

This problem illustrates how a given differential equation can be represented in different phase planes (x, y) where $\dot{x}$ and $\dot{y}$ are defined through the function $X(x, y)$.

- 2.43 The following system models two species with populations N_1 and N_2 competing for a common food supply:

$$\dot{N}_1 = \{a_1 - d_1(bN_1 + cN_2)\}N_1, \quad \dot{N}_2 = \{a_2 - d_2(bN_1 + cN_2)\}N_2.$$

Classify the equilibrium points of the system assuming that all coefficients are positive. Show that if $a_1 d_2 > a_2 d_1$ then the species N_2 dies out and the species N_1 approaches a limiting size (Volterra's Exclusion Principle).

2.43. The two-species model is governed by the equations
$$\dot{N}_1 = \{a_1 - d_1(bN_1 + cN_2)\}N_1,$$
$$\dot{N}_2 = \{a_2 - d_2(bN_1 + cN_2)\}N_2.$$

Equilibrium occurs where
$$\{a_1 - d_1(bN_1 + cN_2)\}N_1 = 0, \quad \{a_2 - d_2(bN_1 + cN_2)\}N_2 = 0.$$

Assume that $a_1 d_2 \neq a_2 d_1$ (otherwise there is a line of equilibrium points along $bN_1 + cN_2 = a_1/d_1$). In the (N_1, N_2) plane, there are three equilibrium points, at
$$(0,0), \quad (0, v_2), \quad (v_1, 0),$$
where $v_2 = a_2/(cd_2)$ and $v_1 = a_1/(bd_1)$.

- $(0,0)$. The linear approximation is
$$\dot{N}_1 \approx a_1 N_1, \quad \dot{N}_2 \approx a_2 N_2.$$

The parameters are (NODE, Section 2.5)
$$p = a_1 + a_2 > 0, \quad q = a_1 a_2 > 0, \quad \Delta = (a_1 + a_2)^2 - 4a_1 a_2$$
$$= (a_1 - a_2)^2 > 0.$$

Therefore the origin is an unstable node.

- $(0, v_2)$. Let $N_1 = \xi$ and $N_2 = v_2 + \eta$. Then
$$\dot{\xi} \approx \frac{a_1 d_2 - a_2 d_1}{d_2} \xi,$$

$$\dot{\eta} \approx -\frac{a_2 b}{c} \xi - a_2 \eta.$$

The parameters are
$$p = \frac{a_1 d_2 - a_2 d_1}{d_2} - a_2, \quad q = -\frac{(a_1 d_2 - a_2 d_1)a_2}{d_2},$$

where classification depends on the signs of p and q using the table in Figure 2.10 in Section 2.5 of NODE.

- $(v_1, 0)$. Let $N_1 = v_1 + \xi$ and $N_2 = \eta$. Then
$$\dot{\xi} \approx -a_1 \xi - \frac{ca_1}{b} \eta.$$

$$\dot{\eta} \approx \frac{a_2 d_1 - a_1 d_2}{d_1} \eta.$$

The parameters are

$$p = \frac{a_2 d_1 - a_1 d_2}{d_1} - a_1, \quad q = -\frac{(a_2 d_1 - a_1 d_2)a_1}{d_1},$$

where classification depends on the signs of p and q using the table.

If $a_1 d_2 > a_2 d_1$, then, for $(0, v_2)$, $p > 0$ and $q < 0$ making the point a saddle. Also, for $(v_1, 0)$, $p < 0$ and $q > 0$ implying that the equilibrium point is stable, either a node or spiral. This is the only stable point so that eventually $N_2 \to 0$ and $N_1 \to v_1$.

- **2.44** Show that the system

$$\dot{x} = X(x, y) = -x + y, \quad \dot{y} = Y(x, y) = \frac{4x^2}{1 + 3x^2} - y$$

has three equilibrium points, at $(0,0)$, $(\frac{1}{3}, \frac{1}{3})$ and $(1, 1)$. Classify each equilibrium point. Sketch the isoclines $X(x, y) = 0$ and $Y(x, y) = 0$, and indicate the regions where dy/dx is positive, and where dy/dx is negative. Sketch the phase diagram of the system.

2.44. The system is

$$\dot{x} = X(x, y) = -x + y, \quad \dot{y} = Y(x, y) = \frac{4x^2}{1 + 3x^2} - y.$$

Equilibrium occurs where

$$x = y, \quad \frac{4x^2}{1 + 3x^2} = y.$$

Eliminate y leaving

$$x = \frac{4x^2}{1 + 3x^2}, \quad \text{or } 3x^3 - 4x^2 + x = 0.$$

Therefore $x = 0, 1, \frac{1}{3}$ and the equilibrium points are $(0, 0)$, $(\frac{1}{3}, \frac{1}{3})$ and $(1, 1)$.

- Equilibrium point $(0, 0)$. The linear approximation is

$$\dot{x} = -x + y, \quad \dot{y} \approx -y.$$

The associated parameters are

$$p = -2 < 0, \quad q = 1 > 0, \quad \Delta = p^2 - 4q = 0.$$

Hence $(0, 0)$ is a degenerate stable node.

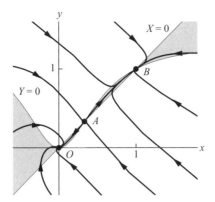

Figure 2.44 Problem 2.44: The phase diagram for $\dot{x} = -x + y$, $\dot{y} = [4x^2/(1+3x^2)] - y$. Equilibrium points lie at $O : (0,0)$, $A : (\frac{1}{3}, \frac{1}{3})$ and $B : (1, 1)$. The shaded regions, bounded by $X(x, y) = 0$ and $Y(x, y) = 0$ indicate where $dy/dx > 0$.

- Equilibrium point $(\frac{1}{3}, \frac{1}{3})$. Let $x = \frac{1}{3} + \xi$ and $y = \frac{1}{3} + \eta$. Then for small $|\xi|$ and $|\eta|$, the linear approximation is

$$\dot{\xi} = -\xi + \eta, \quad \dot{\eta} = \frac{4((1/3) + \xi)^2}{1 + 3((1/3) + \xi)^2} - \frac{1}{3} - \eta \approx \frac{3}{2}\xi - \eta.$$

The associated parameters are

$$p = -2, \quad q = 1 - \tfrac{3}{2} = -\tfrac{1}{2}.$$

Therefore $(\frac{1}{3}, \frac{1}{3})$ is a saddle point.

- Equilibrium point $(1, 1)$. Let $x = 1 + \xi$ and $y = 1 + \eta$. Then the linear approximation is

$$\dot{\xi} = -\xi + \eta, \quad \dot{\eta} = \frac{4(1 + \xi)^2}{1 + 3(1 + \xi)^2} \approx \tfrac{1}{2}\xi - \eta.$$

The parameters are

$$p = -2 < 0, \quad q = 1 - \tfrac{1}{2} = \tfrac{1}{2} > 0, \quad \Delta = 4 - \tfrac{1}{4} = \tfrac{15}{16} > 0.$$

Therefore $(1, 1)$ is a stable node.

The phase diagram is shown in Figure 2.44 together with the isoclines $X(x, y) = 0$ and $Y(x, y) = 0$.

- **2.45** Show that the systems (A) $\dot{x} = P(x, y)$, $\dot{y} = Q(x, y)$ and (B) $\dot{x} = Q(x, y)$, $\dot{y} = P(x, y)$ have the same equilibrium points. Suppose that system (A) has three equilibrium points which, according to their linear approximations are, (a) a stable spiral, (b) an unstable node, (c) a saddle point. To what extent can the equilibrium points in (B) be classified from this information?

2.45. Comparison is to be made between the systems

(A) $$\dot{x} = P(x, y), \quad \dot{y} = Q(x, y)$$

and

(B) $$\dot{y} = Q(x, y), \quad \dot{y} = P(x, y).$$

Assume that in a neighbourhood of an equilibrium point

$$P(x, y) \approx ax + by, \quad Q(x, y) \approx cx + dy.$$

Then the linear approximation for system (A) is

$$\dot{x} = ax + by, \quad \dot{y} = cx + dy,$$

and for system (B),

$$\dot{x} = cx + dy, \quad \dot{y} = ax + by.$$

Hence the parameters classifying equilibrium points for (A) are

$$p_A = a + d, \quad q_A = ad - bc, \quad \Delta_A = (a + d)^2 - 4(ad - bc) = (a - d)^2 + 4bc,$$

and for (B) are

$$p_B = b + c, \quad q_B = bc - ad, \quad \Delta_B = (b + c)^2 - 4(bc - ad) = (b - c)^2 + 4ad.$$

(a) *A* has a stable spiral, which means that

$$p_A = a + d < 0, \quad q_A = ad - bc > 0, \quad \Delta_A = (a - d)^2 + 4bc < 0.$$

It can be seen that $q_B = bc - ad < 0$. Therefore the corresponding equilibrium point for (B) is a saddle.

(b) *A* has a an unstable node, which requires

$$p_A = a + d > 0, \quad q_A = bc - ad > 0 \text{ and } \Delta_A = (a - d)^2 + 4bc > 0.$$

As in (a), $q_B = bc - ad < 0$. Hence the equilibrium point in (B) is also a saddle.

(c) (A) has a saddle. Hence $q_A = ad - bc < 0$. It follows that

$$p_B = b + c, \quad q_B = bc - ad > 0, \quad \Delta_B = (b + c)^2 - 4(bc - ad) = (b - c)^2 + 4ad.$$

All we can say generally is that the equilibrium point is a node or a spiral, where stability will depend on the sign of $p_A = b + c$.

- **2.46** The system defined by the equations
$$\dot{x} = a + x^2 y - (1+b)x, \quad \dot{y} = bx - yx^2, \quad (a \neq 0, b \neq 0)$$
is known as the **Brusselator** and arises in a mathematical model of a chemical reaction (see Jackson (1990)). Show that the system has one equilibrium point at $(a, b/a)$. Classify the equilibrium point in each of the following cases:

(a) $a = 1, b = 2$;
(b) $a = \frac{1}{2}, b = \frac{1}{4}$.

In case (b) draw the isoclines of zero and infinite slope in the phase diagram. Hence sketch the phase diagram.

2.46. The system is
$$\dot{x} = a + x^2 y - (1+b)x, \quad \dot{y} = bx - yx^2.$$
Equilibrium occurs where
$$a + x^2 y - (1+b)x = 0, \quad x(b - xy) = 0,$$
which have just one solution $x = a$, $y = b/a$.
Let $x = a + \xi$, $y = (b/a) + \eta$. Then
$$\dot{\xi} = a + (a+\xi)^2 \left(\frac{b}{a} + \eta\right) - (1+b)(a+\xi) \approx (b-1)\xi + a^2 \eta,$$
and
$$\dot{\eta} = (a+\xi)\left[b - (a+\xi)\left(\frac{b}{a} + \eta\right)\right] \approx b\xi - a^2 \eta$$

The parameters associated with the linear approximation are
$$p = b - 1 - a^2, \quad q = a^2(1 - 2b), \quad \Delta = (b-1)^2 + a^2(6b - 2 + a^2).$$

(a) Case $a = 1$, $b = 2$. The equilibrium point is at $(1, 2)$. The parameters are $p = 0$, $q = -3 < 0$. Therefore $(1, 2)$ is a saddle point.

(b) Case $a = \frac{1}{2}$, $b = \frac{1}{4}$. The equilibrium point is at $(\frac{1}{2}, \frac{1}{2})$. The parameters are $p = -1 < 0$, $q = \frac{1}{8} > 0$, $\Delta = \frac{1}{2} > 0$. Therefore $(\frac{1}{2}, \frac{1}{2})$ is also a stable node. The phase diagram for case (b) is shown in Figure 2.45.

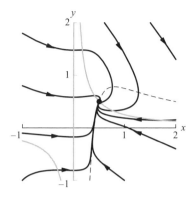

Figure 2.45 Problem 2.46: Phase diagram for $a = \frac{1}{2}$, $b = \frac{1}{4}$ showing a stable node at $(\frac{1}{2}, \frac{1}{2})$. The dashed curve indicates the isocline with infinite slope and the grey curves the isoclines with zero slope (including the y axis).

- **2.47** A Volterra model for the population size $p(t)$ of a species is, in reduced form,

$$\kappa \frac{dp}{dt} = p - p^2 - p \int_0^t p(s)ds, \quad p(0) = p_0,$$

where the integral term represents a toxicity accumulation term (see Small (1989)). Let $x = \ln p$, and show that x satisfies

$$\kappa \ddot{x} + e^x \dot{x} + e^x = 0.$$

Put $y = \dot{x}$, and show that the system is also equivalent to

$$\dot{y} = -(y+1)p/\kappa, \quad \dot{p} = yp.$$

Sketch the phase diagram in the (y, p) plane. Also find the exact equation of the phase paths.

2.47. The Volterra model for the population $p(t)$ satisfies

$$\kappa \frac{dp}{dt} = p - p^2 - p \int_0^t p(s)ds, \quad p(0) = p_0. \tag{i}$$

Differentiate equation (i) with respect to t:

$$\kappa \ddot{p} = \dot{p} - 2p\dot{p} - \dot{p} \int_0^t p(s)ds - p^2$$

$$= \dot{p} - 2p\dot{p} - \frac{\dot{p}}{p}[-\kappa \dot{p} + p - p^2] - p^2, \text{ (using (i))}$$

$$= -p\dot{p} + \kappa \frac{\dot{p}^2}{p} - p^2.$$

Let $p = e^x$ so that $\dot{p} = e^x \dot{x}$ and $\ddot{p} = e^x \ddot{x} + e^x \dot{x}^2$. Then the population equation becomes

$$\kappa(e^x \ddot{x} + e^x \dot{x}^2) = -e^{2x} \dot{x} + \kappa e^x \dot{x}^2 - e^{2x},$$

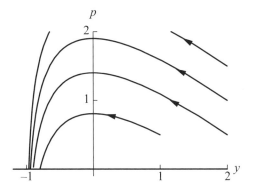

Figure 2.46 Problem 2.47: Phase diagram for the Volterra model with $\kappa = 1$.

or

$$\kappa \ddot{x} + e^x \dot{x} + e^x = 0, \qquad \text{(ii)}$$

as required.

Let $y = \dot{x}$. Then $\dot{p} = e^x \dot{x} = py$. Also

$$\dot{y} = \ddot{x} = -\frac{1}{\kappa}(e^x \dot{x} + e^x) = -\frac{1}{\kappa}(y+1)p,$$

using (ii). Equilibrium occurs for all points on the axis $p = 0$ (for the population model $p \geq 0$). The differential equation for the phase paths in the (y, p) plane is

$$\frac{dy}{dp} = -\frac{y+1}{\kappa y}.$$

For $t = 0$, $\dot{p}(0) = p(0) - p(0)^2 < 0$ assuming that $p(0) > 1$. This is a separable equation with general solution

$$\kappa \int \frac{y}{y+1} dy = -\int dp = -p + C,$$

or

$$\kappa(y - \ln|y+1|) = -p + C.$$

A phase diagram for the model is shown in Figure 2.46 with $\kappa = 1$. In fact this includes all non-zero parameter values for κ, since κ can be eliminated from the from the first-order equations by the transformation $p \to p\kappa$.

3   Geometrical aspects of plane autonomous systems

• **3.1** By considering the variation of path direction on closed curves round the equilibrium points, find the index in each case of Figure 3.29 (in NODE)

3.1. (a) Surround the equilibrium point by a closed curve Γ (see Figure 3.1). Take any point P on Γ, and draw a vector $\mathbf{S}$ at P tangential to the phase path at P. Let ϕ be the angle between a fixed direction and $\mathbf{S}$ measured counterclockwise as shown. As P makes one counterclockwise circuit of Γ, determine how the angle ϕ changes as the direction of $\mathbf{S}$ changes. Any change will be a multiple of 2π. In this example ϕ returns to its original value. Hence the index $I = 0$.

Apply the same method to each of the remaining figures. The indices are

(b) $I = 0$.
(c) $I = 1$.
(d) $I = 1$.
(e) $I = -2$.

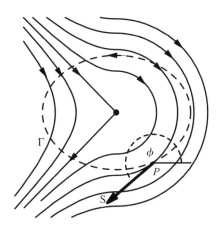

Figure 3.1   Problem 3.1(a)

- 3.2 The motion of a damped pendulum is described by the equations
$$\dot\theta = \omega, \quad \dot\omega = -k\omega - v^2\sin\theta,$$
where $k(> 0)$ and v are constants. Find the indices of all equilibrium states.

3.2. The damped pendulum has the equation
$$\dot\theta = \omega, \quad \dot\omega = -k\omega - v^2\sin\theta.$$

Equilibrium occurs where $\omega = 0$ and $\sin\theta = 0$, that is, at $(n\pi, 0)$, $(n = 0, \pm 1, \pm 2, \ldots)$ in the (θ, ω) phase plane.

Near the origin,
$$\dot\theta = \omega, \quad \dot\omega \approx -v^2\theta - k\omega.$$

The parameters associated with the linear approximation (see Section 2.5) are
$$p = -k < 0, \quad q = v^2 > 0, \quad \Delta = k^2 - 4v^2.$$

Hence the equilibrium point is either a stable node or a stable spiral depending on the sign of Δ. In both cases, however, the index $I = 1$. By the 2π periodicity in θ of the differential equation, the indices of all the equilibrium points $(2\pi n, 0)$, $(n = 0, \pm 1, \pm 2, \ldots)$ also have index $I = 1$.

Near $\theta = \pi$, let $\theta = \pi + \xi$. Then
$$\dot\xi = \omega, \quad \dot\omega = -k\omega - v^2\sin(\pi + \xi) \approx -k\omega + v^2\xi.$$

Therefore $q = -v^2$, that is, the equilibrium point is a saddle point with index $I = -1$. Similarly all the points $((2n + 1)\pi, 0)$, $(n = 0, \pm 1, \pm 2, \ldots)$ are saddle points each with index $I = -1$.

- 3.3 Find the index of the equilibrium points of the following systems: (i) $\dot x = 2xy$, $\dot y = 3x^2 - y^2$; (ii) $\dot x = y^2 - x^4$, $\dot y = x^3 y$; (iii) $\dot x = x - y$, $\dot y = x - y^2$.

3.3. (i) The system
$$\dot x = 2xy, \quad \dot y = 3x^2 - y^2$$
has one equilibrium point, at the origin. Let the curve Γ surrounding the origin be the ellipse $x = \cos\theta$, $y = \sqrt{3}\sin\theta$ for $0 \leq \theta < 2\pi$. Then
$$X(x, y) = 2xy = 2\sqrt{3}\cos\theta\sin\theta = \sqrt{3}\sin 2\theta,$$
$$Y(x, y) = 3x^2 - y^2 = 3(\cos^2\theta - \sin^2\theta) = 3\cos 2\theta.$$

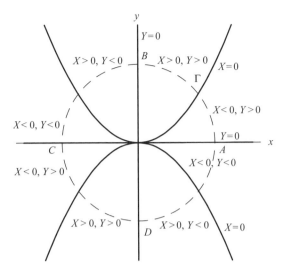

Figure 3.2 Problem 3.3(b)

Then, on Γ,

$$\tan\phi = \frac{Y(x,y)}{X(x,y)} = \frac{3\cos 2\theta}{\sqrt{3}\sin 2\theta} = \sqrt{3}\cot 2\theta = \sqrt{3}\tan[\tfrac{1}{2}\pi - 2\theta].$$

As θ increases from 0 to 2π, ϕ decreases 0 to -4π. Hence $I = -2$.
(ii) The system

$$\dot{x} = y^2 - x^4, \quad \dot{y} = x^3 y$$

has one equilibrium point, at the origin. Let Γ be a circle centred at the origin. Using the method of Theorem 3.3, draw the isoclines $X(x,y) = 0$, that is, $y = \pm x^2$, and $Y(x,y) = 0$, that is, the axes $x = 0$ and $y = 0$, in the phase plane, and mark the regions where X and Y are positive and negative as shown in Figure 3.2. The circle Γ cuts the lines $Y(x,y) = 0$ at the points A, B, C and D. According to Theorem 3.3 we list the sign changes of $\tan\phi = Y/X$ at these points on a counterclockwise circuit of Γ. The signs of X and Y shown in the figure can be used to determine the sign changes:

zero of $Y(x,y)$	A	B	C	D
sign change in $\tan\phi$	+/−	+/−	+/−	+/−

Hence there are $P = 4$ changes from + to −, and $Q = 0$ changes from − to +. The index is given by $I = \tfrac{1}{2}(P - Q) = 2$.
(iii) The system

$$\dot{x} = x - y, \quad \dot{y} = x - y^2$$

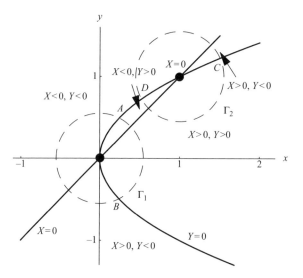

Figure 3.3 Problem 3.3(c)

has two equilibrium points, at $(0,0)$ and $(1,1)$. The isoclines $X(x,y) = x - y = 0$ and $Y(x,y) = x - y^2 = 0$ are shown in Figure 3.3. Surround $(0,0)$ and $(1,1)$ by circles Γ_1 and Γ_2 as shown: $(1,1)$ should be outside Γ_1 and $(0,0)$ outside Γ_2. Let Γ_1 cut $x = y^2$ at A and B, and Γ_2 cut $x = y^2$ at C and D. The sign changes in $\tan\phi = Y/X$ on a counterclockwise circuit of Γ_1 are

zero of $Y(x,y)$	A	B
sign change in $\tan\phi$	$-/+$	$-/+$

Hence $P = 2$ and $Q = 0$ so that at $(0,0)$ the index is $I = \frac{1}{2}(P - Q) = 1$.

For a counterclockwise circuit of Γ_2,

zero of $Y(x,y)$	C	D
sign change in $\tan\phi$	$+/-$	$+/-$

Hence $P = 0$ and $Q = 2$ so that at $(1,1)$ the index is $I = \frac{1}{2}(P - Q) = -1$.

For the origin use result (3.8) with Γ the curve given parametrically by $x = r\cos t$, $y = r\sin t$ $(0 \leq t < 2\pi)$ in (iii). Then, on Γ,

$$X(x,y) = x - y = r(\cos t - \sin t), \quad Y(x,y) = r\cos t - r^2 \sin^2 t.$$

Therefore

$$I_\Gamma = \frac{1}{2\pi} \oint_\Gamma \frac{X dY - Y dX}{X^2 + Y^2}$$

$$= \frac{1}{2\pi} \int_0^{2\pi} \frac{X(dY/dt) - Y(dX/dt)}{X^2 + Y^2} dt$$

$$= \frac{1}{2\pi} \int_0^{2\pi} \frac{(\cos t - \sin t)(-\sin t - 2r \sin t \cos t) + (\cos t - r \sin^2 t)(\sin t + \cos t)}{(\cos t - \sin t)^2 + (\cos t - r \sin^{2t})^2} dt$$

$$= \frac{1}{2\pi} \int_0^{2\pi} \frac{1 - r \sin t(\cos^2 t + 1 - \sin t \cos t)}{(\cos t - \sin t)^2 + (\cos t - r \sin^2 t)^2} dt.$$

Therefore, if $0 < r < 1$, Γ surrounds only the origin for which $I_\Gamma = 1$, whilst for $r > 1$, Γ surrounds both equilibrium points for which $I_\Gamma = 0$; this is the sum of the indices at $(0, 0)$ and $(1, 1)$.

- 3.4 For the linear system $\dot{x} = ax + by$, $\dot{y} = cx + dy$, where $ad - bc \neq 0$, obtain the index at the origin by evaluating

$$I_\Gamma = \int_{s_0}^{s_1} \frac{XY' - YX'}{X^2 + Y^2} ds,$$

showing that it is equal to sgn $(ad - bc)$. (Hint: choose Γ to be the ellipse $(ax+by)^2 + (cx+dy)^2 = 1$.)

3.4. The linear system is

$$\dot{x} = ax + by, \quad \dot{y} = cx + dy, \quad (ad - bc \neq 0).$$

Let Γ be the ellipse

$$(ax + by)^2 + (cx + dy)^2 = 1,$$

which can be represented parametrically by

$$X(x, y) = ax + by = \cos \theta, \quad Y(x, y) = cx + dy = \sin \theta,$$

where $0 \leq \theta < 2\pi$. By solving these equations we obtain

$$x = \frac{d \cos \theta - b \sin \theta}{ad - bc}, \quad y = \frac{-c \cos \theta + a \sin \theta}{ad - bc},$$

since $ad - bc \neq 0$. As θ increases, the ellipse is tracked in a counterclockwise sense if $ad - bc > 0$ and clockwise if $ad - bc < 0$. From (3.7), since $X = ax + by$ and $Y = cx + dy$, but taking account that Γ must be traced anticlockwise

$$I_\Gamma = \frac{1}{2\pi} \int_{s_0}^{s_1} \frac{XY' - YX'}{X^2 + Y^2} ds$$

$$= \frac{\text{sgn}(ad - bc)}{2\pi} \int_0^{2\pi} \frac{X(dY/d\theta) - Y(dX/d\theta)}{X^2 + Y^2} d\theta$$

$$= \frac{\text{sgn}(ad - bc)}{2\pi} \int_0^{2\pi} \frac{\cos\theta(\cos\theta) - \sin\theta(-\cos\theta)}{\cos^2\theta + \sin^2\theta} d\theta$$

$$= \frac{\text{sgn}(ad - bc)}{2\pi} \int_0^{2\pi} d\theta = \text{sgn}(ad - bc).$$

• **3.5** The equation of motion of a bar restrained by springs (see Figure 3.30 in NODE) and attracted by a parallel current-carrying conductor is

$$\ddot{x} + c\left(x - \frac{\lambda}{a - x}\right) = 0,$$

where c (the stiffness of the spring), a and λ are positive constants. Sketch the phase paths for $-x_0 < x < a$, where x_0 is the unstretched length of each spring, and find the indices of the equilibrium points for all $\lambda > 0$.

3.5. The equation of motion of the bar is

$$\ddot{x} + c\left[x - \frac{\lambda}{a - x}\right] = 0. \tag{i}$$

Equilibrium occurs where

$$x(a - x) - \lambda = 0, \text{ or } x^2 - ax + \lambda = 0. \tag{ii}$$

If $\lambda > \frac{1}{4}a^2$, there is no equilibrium state. A typical phase diagram is shown in Figure 3.4.
If $\lambda < \frac{1}{4}a^2$, then the solutions of (ii) are given by

$$x = \tfrac{1}{2}[a \pm \sqrt{(a^2 - 4\lambda)}].$$

Both these solutions are positive. Denote them by x_1 and x_2. Let $x = x_1 + \xi$. Then eqn (i) becomes

$$\ddot{\xi} + c\left[x_1 + \xi - \frac{\lambda}{a - x_1 - \xi}\right] = 0,$$

with linearized approximation

$$\ddot{\xi} + c\left[1 - \frac{\lambda}{(a - x_1)^2}\right]\xi = 0.$$

Therefore the equilibrium point $(x_1, 0)$ is a centre if $\lambda < (a - x_1)^2$, making the coefficient of ξ positive, and a saddle point if $\lambda > (a - x_1)^2$, making it negative. Substituting for x_1 these inequalities become

$$\lambda < \tfrac{1}{4}[a - \sqrt{(a^2 - 4\lambda)}]^2 \quad \text{for a centre,}$$

and

$$\lambda > \tfrac{1}{4}[a - \sqrt{(a^2 - 4\lambda)}]^2 \quad \text{for a saddle,}$$

However, it can be shown that the first inequality (the centre) is not consistent with $\lambda < \tfrac{1}{4}a^2$, by considering the sign of

$$\lambda - \tfrac{1}{4}[a - \sqrt{(a^2 - 4\lambda)}]^2.$$

Hence $(x_1, 0)$ is a saddle.

By a similar argument the linearization near $x = x_2$ leads to the equation

$$\ddot{\xi} + c\left[1 - \frac{\lambda}{(a - x_2)^2}\right]\xi = 0.$$

In this case the critical relation between λ and a is

$$\lambda = (a - x_2)^2, \quad \text{or} \quad \lambda = \tfrac{1}{4}(a + \sqrt{(a^2 - 4\lambda)}),$$

but since $\lambda < \tfrac{1}{4}a^2$ this equilibrium point is a centre. Typical phase diagrams are shown in Figures 3.4 and 3.5.

The index of x_1 (the saddle) is -1, and the index of x_2 (the centre) is $+1$.

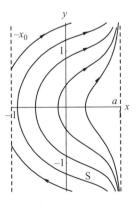

Figure 3.4 Problem 3.5: The phase diagram for the bar with the parameter values $\lambda = 0.5$, $a = 1$, $x_0 = 1$, $c = 1$.

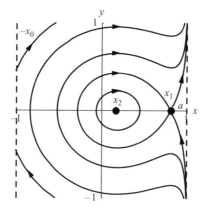

Figure 3.5 Problem 3.5: The phase diagram for the bar with the parameter values $\lambda = 0.15, a = 1, x_0 = 1, c = 1$.

- 3.6 Show that the equation
$$\ddot{x} - \varepsilon(1 - x^2 - \dot{x}^2)\dot{x} + x = 0$$
has an equilibrium point of index 1 at the origin of the phase plane x, y with $\dot{x} = y$. (It also has a limit cycle, $x = \cos t$.) Use NODE, eqn (3.7)(see Problem 3.4), with Γ a circle of radius a to show that, for all a,
$$\int_0^{2\pi} \frac{d\theta}{1 - 2\varepsilon(1 - a^2)\sin\theta\cos\theta + \varepsilon^2(1 - a^2)^2\sin^2\theta} = 2\pi.$$

3.6. The equation
$$\ddot{x} - \varepsilon(1 - x^2 - \dot{x}^2)\dot{x} + x = 0$$
has a single equilibrium point, at the origin. Consider first the special case where Γ is given parametrically by $x = \cos\theta, y = \sin\theta$. Then
$$X(x, y) = y = \sin\theta, \quad Y(x, y) = -x + \varepsilon(1 - x^2 - y^2)y = -\cos\theta,$$
on Γ. By NODE, eqn (3.8) in the text, the index of the origin is given by
$$I_\Gamma = [\phi]_\Gamma = \frac{1}{2\pi} \int_0^{2\pi} \frac{X(dY/d\theta) - Y(dX/d\theta)}{X^2 + Y^2} d\theta$$

$$= \frac{1}{2\pi} \int_0^{2\pi} \frac{\sin\theta(\sin\theta) + \cos\theta(\cos\theta)}{\sin^2\theta + \cos^2\theta} d\theta$$

$$= \frac{1}{2\pi} \int_0^{2\pi} d\theta = 1.$$

The system has only one equilibrium point, at the origin. Hence for *any* simple closed curve Γ surrounding the origin, the change in ϕ in a counterclockwise circuit must be 2π. In particular for the curve $x = a\cos\theta$, $y = a\sin\theta$ $(a > 0)$,

$$2\pi = \int_0^{2\pi} \frac{a\sin\theta[a\sin\theta + \varepsilon(1-a^2)a\cos\theta] - [-a\cos\theta + \varepsilon(1-a^2)a\sin\theta]a\cos\theta}{a^2\sin^2\theta + [-a\cos\theta + \varepsilon(1-a^2)a\sin\theta]^2} d\theta,$$

$$= \int_0^{2\pi} \frac{d\theta}{1 - 2\varepsilon(1-a^2)\sin\theta\cos\theta + \varepsilon^2(1-a^2)^2\sin^2\theta}$$

- **3.7** A limit cycle encloses N nodes, F spirals, C centres, and S saddle points only, all of linear type. Show that $N + F + C - S = 1$.

3.7. We can let the chosen curve Γ be the limit cycle. Taken in either sense the index $I_\Gamma = 1$. By Theorem 3.2, the sum of the indices of the equilibrium points within Γ must be 1. The indices of a linear node, spiral and centre are all 1, whilst the index of the linear saddle is -1. If there are N nodes, F spirals and C centres, then their contribution to the index is $(N + F + C) \times 1 = N + F + C$. If there are S saddles then their contribution is $-S$. Hence $N + F + C - S = 1$.

- **3.8** Given the system
$$\dot{x} = X(x,y)\cos\alpha - Y(x,y)\sin\alpha, \quad \dot{y} = X(x,y)\sin\alpha + Y(x,y)\cos\alpha,$$
where α is a parameter, prove that the index of a simple closed curve which does not meet an equilibrium point is independent of α (see Problem 2.41).

3.8. Compare the two systems

$$\dot{x} = X(x,y), \quad \dot{y} = Y(x,y), \qquad (A)$$

and

$$\left.\begin{array}{l}\dot{x} = P(x,y) = X(x,y)\cos\alpha - Y(x,y)\sin\alpha \\ \dot{y} = Q(x,y) = X(x,y)\sin\alpha + Y(x,y)\cos\alpha\end{array}\right\} \qquad (B)$$

The systems (A) and (B) have the same equilibrium points, which satisfy $X(x,y) = 0$ and $Y(x,y) = 0$. Let Γ be the simple closed curve for both systems. In complex terms

$$X + iY = (P + iQ)e^{i\alpha},$$

so that

$$|X + iY| = |P + iQ|, \quad \arg(X + iY) = \arg(P + iQ) + \alpha.$$

Hence the vector (P, Q) is the vector (X, Y) rotated through $-\alpha$ at each point in the phase plane. Hence the index for Γ with (P, Q) is the same as that of (X, Y) and this is independent of α.

- **3.9** Suppose that the system $\dot{x} = X(x)$ has a finite number of equilibrium points, each of which is either a node, a centre, a spiral or a saddle point, of the elementary types discussed in NODE, Section 2.5, and assume that $I_\infty = 0$. Show that the total number of nodes, centres and spirals is equal to the total number of saddle points plus two.

3.9. According to the Corollary to Theorem 3.4, the sum of all the indices including the point at infinity I_∞ is 2, that is, if the indices of the equilibrium points are I_i, ($i = 1, 2, \ldots, n$), then

$$I_\infty + \sum_{i=1}^{n} I_i = 2.$$

Let there be a total of r centres, nodes and spirals each of which will have index 1, and s saddles each of which will have index -1, where $r + s = n$. Since we are given that $I_\infty = 0$, then $r - s = 2$ as required.

- **3.10** Obtain differential equations describing the behaviour of the linear system, $\dot{x} = ax + by$, $\dot{y} = cx + dy$, at infinity. Sketch the phase diagram, and analyse the system $\dot{x} = 2x - y$, $\dot{y} = 3x - 2y$ near the horizon.

3.10. In the equations

$$\dot{x} = ax + by, \quad \dot{y} = cx + dy$$

let (see NODE, Section 3.2)

$$x_1 = \frac{x}{x^2 + y^2}, \quad y_1 = -\frac{y}{x^2 + y^2}.$$

With $x_1 = x/(x^2 + y^2)$, $y_1 = -y/(x^2 + y^2)$ and $z_1 = x_1 + iy_1$, the transformed system is given by NODE, (3.11)

$$\frac{dz_1}{dt} = -z_1^2(X + iY)$$
$$= -(x_1 + iy_1)^2[ax + by + i(cx + dy)]$$
$$= -(x_1^2 - y_1^2 + 2ix_1y_1)[ax_1 - by_1 + i(cx_1 - dy_1)]/r_1^2$$

where $r_1^2 = x_1^2 + y_1^2$. Therefore, by equating real and imaginary parts of this equation,

$$\dot{x}_1 = [(y_1^2 - x_1^2)(ax_1 - by_1) + 2x_1 y_1(cx_1 - dy_1)]/r_1^2, \qquad \text{(i)}$$

$$\dot{y}_1 = [-2x_1 y_1(ax_1 - by_1) + (y_1^2 - x_1^2)(cx_1 - dy_1)]/r_1^2. \qquad \text{(ii)}$$

For the system $\dot{x} = 2x - y$, $\dot{y} = 3x - 2y$, the coefficients are $a = 2, b = -1, c = 3, d = -2$. The origin is a centre with index 1. Hence (i) and (ii) become

$$\dot{x}_1 = [(y_1^2 - x_1^2)(2x_1 + y_1) + 2x_1 y_1(3x_1 + 2y_1)]/r_1^2$$
$$= (-2x_1^3 + 5x_1^2 y_1 + 6x_1 y_1^2 + y_1^3)/r_1^2,$$
$$\dot{y}_1 = [-2x_1 y_1(2x_1 + y_1) + (y_1^2 - x_1^2)(3x_1 + 2y_1)]/r_1^2$$
$$= (-3x_1^3 - 6x_1^2 y_1 + x_1 y_1^2 + 2y_1^3)/r_1^2$$

The origin in this system is a *singular* point since $\dot{x}_1$ and $\dot{y}_1$ are not defined at $(0,0)$. However, we can define a phase diagram through the equation

$$\frac{dy_1}{dx_1} = \frac{\dot{y}_1}{\dot{x}_1} = \frac{-3x_1^3 - 6x_1^2 y_1 + x_1 y_1^2 + 2y_1^3}{-2x_1^3 + 5x_1^2 y_1 + 6x_1 y_1^2 + y_1^3}.$$

The origin is a higher-order equilibrium point of the equivalent equations

$$\dot{u} = -2u^3 + 5u^2 v + 6uv^2 + v^3,$$
$$\dot{v} = -3u^3 - 6u^2 v + uv^2 + 2v^3,$$

in the (u, v) plane: the phase paths will be identical in the (x_1, y_1) and (u, v) planes. We shall try to find any separatrices by putting $v = ku$. Then

$$\frac{dv}{du} = k = \frac{-3 - 6k + k^2 + 2k^3}{-2 + 5k + 6k^2 + k^3},$$

or

$$k^4 + 4k^3 + 4k^2 + 4k + 3 = 0, \text{ or } (k+1)(k+3)(k^2+1) = 0.$$

The quartic has two real solutions $k = -1$ and $k = -3$. Computed phase paths in the neighbourhood of the origin are shown in Figure 3.6. A counterclockwise circuit of Γ in the figure indicates the the point at infinity has an index of 3. Since the saddle at the origin has index -1, the sum of the indices is 2 which confirms the Corollary to Theorem 3.4.

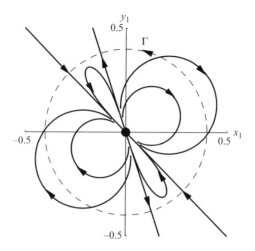

Figure 3.6 Problem 3.10:

- **3.11** A certain system is known to have exactly two equilibrium points, both saddle points. Sketch phase diagrams in which

(i) a separatrix connects the saddle points,
(ii) no separatrix connects them.

For example, the system $\dot{x} = 1 - x^2$, $\dot{y} = xy$ has a saddle connection joining saddle points at $(\pm 1, 0)$. The perturbed system $\dot{x} = 1 - x^2$, $\dot{y} = xy - \varepsilon x^2$ for $0 < \varepsilon \ll 1$ breaks the saddle connection (heteroclinic bifurcation).

3.11. Figure 3.28 (in NODE) shows examples of separatrices which connect two saddle points and separatrices which do not. A possible system to illustrate two saddles is

$$\dot{x} = 1 - x^2, \quad \dot{y} = xy.$$

Some phase paths for the system are shown in Figure 3.7: the saddles are at $(\pm 1, 0)$ and these are the only equilibrium points.

A perturbation of this system can break the link between the saddle points. Consider the equations

$$\dot{x} = 1 - x^2, \quad \dot{y} = xy + \varepsilon x^2,$$

in which ε is a small parameter. With $\varepsilon = 0.2$, the system has saddle equilibrium points at $(1, -0.2)$ and $(-1, 0.2)$. The separatrices only are shown in Figure 3.8.

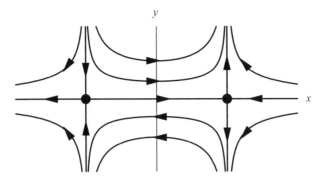

Figure 3.7 Problem 3.11: Phase paths for the system $\dot{x} = 1 - x^2$, $\dot{y} = xy$.

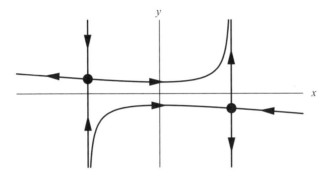

Figure 3.8 Problem 3.11: $\dot{x} = 1 - x^2$, $\dot{y} = xy + \varepsilon x^2$ showing only the separatrices, in the case $\varepsilon = 0.2$.

- **3.12** Deduce the index at infinity for the system $\dot{x} = x - y$, $\dot{y} = x - y^2$ by calculating the indices of the equilibrium points.

3.12. The system $\dot{x} = x - y$, $\dot{y} = x - y^2$ has equilibrium points where $x - y = 0$ and $x - y^2 = 0$. There are two such points, at $(0, 0)$ and $(1, 1)$. Their types are as follows:

- $(0, 0)$. The linear approximations are $\dot{x} = x - y$, $\dot{y} \approx x$. The associated parameters are $p = 1 > 0$, $q = 1 > 0$, $\Delta = p^2 - 4q = 1 - 4 = -3 < 0$. Hence from Section 2.5, Hence $(0, 0)$ is an unstable spiral with index $I_1 = 1$.
- $(1, 1)$. Let $x = 1 + \xi$ and $y = 1 + \eta$. Then the linear approximations are $\dot{\xi} = \xi - \eta$, $\dot{\eta} = 1 + \xi - (1 + \eta)^2 \approx \xi - 2\eta$. The parameters are $p = 1 - 2 = -1 < 0$, $q = -2 + 1 = -1 < 0$ which imply that $(1, 1)$ is a saddle point with index $I_2 = -1$.

By Theorem 3.4, $I_\infty = 2 - I_1 - I_2 = 2 - 1 + 1 = 2$.

- **3.13** Use the geometrical picture of the field (X, Y), in the neighbourhood of an ordinary point (i.e. not an equilibrium point) to confirm Theorem 3.1.

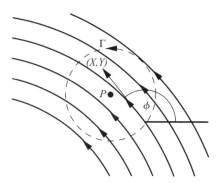

Figure 3.9 Problem 3.13:

3.13. Figure 3.9 shows on ordinary point P of a phase diagram. Surround P by a closed curve Γ (usually a circle centred at P) so that there are no equilibrium points in or on Γ. Since $\dot{x}$ and $\dot{y}$ are continuous and $\dot{x}_P \neq 0$, $\dot{y}_P$ (their values at P), there exists a neighbourhood of P in which $\dot{x}$ and $\dot{y}$ retain the same signs as $\dot{x}_P$ and $\dot{y}_P$ respectively. Assume that Γ is in this neighbourhood. Then the direction of the vector (X, Y) points into the same quadrant at every point of Γ, which establishes the result that $I_\Gamma = 0$.

- 3.14 Suppose that, for two plane systems $\dot{x}_1 = X_1(x_1)$, $\dot{x}_2 = X_2(x_2)$, and for a given closed curve Γ, there is no point on Γ at which X_1 and X_2 are opposite in direction. Show that the index of Γ is the same for both systems.
  The system $\dot{x} = y$, $\dot{y} = x$ has a saddle point at the origin. Show that the index of the origin for the system $\dot{x} = y + cx^2 y$, $\dot{y} = x - cy^2 x$ is likewise -1.

3.14. Consider the two plane systems $\dot{x}_1 = X_1(x_1)$, $\dot{x}_2 = X_2(x_2)$. Let $\phi(s)$ be the angle between X_2 and a fixed direction, and let $\theta(s)$ be the angle between X_1 and X_2 (see Figure 3.10), where s ($\alpha \leq s < \beta$), is the curve parameter for one circuit of Γ. Since X_1 and X_2 are never opposite to each other, that is, $-\pi < \theta < \pi$, then $\theta(\alpha) = \theta(\beta)$. Then

$$I_\Gamma(X_1) = \frac{1}{2\pi}[\theta(s) + \phi(s)]_\alpha^\beta$$

$$= \frac{1}{2\pi}\{[\theta(s)]_\alpha^\beta + [\phi(s)]_\alpha^\beta\}$$

$$= \frac{1}{2\pi}\{\theta(\beta) - \theta(\alpha) + \phi(\beta) - \phi(\alpha)\}$$

$$= \frac{1}{2\pi}[\phi(s)]_\alpha^\beta = I_\Gamma(X_2),$$

as required.

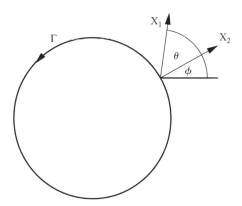

Figure 3.10 Problem 3.14:

The system $\dot{x} = y$, $\dot{y} = x$ has a saddle point at the origin with index -1. The perturbed system is $\dot{x} = y + cx^2y$, $\dot{y} = x - cxy^2$ which also has just one equilibrium point, at the origin. Let

$$X_1 = (y, x), \quad X_2 = (y + cx^2y, x - cxy^2).$$

Let Γ be a circle centre the origin with radius r. On Γ the angle θ between the vectors X_1 and X_2 is given by

$$\cos\theta = \frac{X_1 \cdot X_2}{|X_1||X_2|}$$

$$= \frac{y(y + cx^2y) + x(x - cxy^2)}{\sqrt{(x^2 + y^2)}\sqrt{[(x - cxy^2)^2 + (y + cx^2y)^2]}}$$

$$= \frac{r^2}{r\sqrt{[r^2 + c^2r^2x^2y^2]}}$$

$$= \frac{1}{\sqrt{[1 + c^2x^2y^2]}},$$

where $r = \sqrt{(x^2 + y^2)}$. It follows that $\cos\theta > 0$ for all c so that $\cos\theta$ can never equal -1, that is, θ can never be $-\pi$ on Γ. By the first part of the problem, the index of the perturbed system must also be -1.

- **3.15** Use Problem 3.14 to show that the index of the equilibrium point $x = 0$, $\dot{x} = 0$ for the equation $\ddot{x} + \sin x = 0$ on the usual phase plane has index 1, by comparing the equation $\ddot{x} + x = 0$.

3.15. The systems $\ddot{x} + x = 0$ and $\ddot{x} + \sin x = 0$ both have an equilibrium point at the origin, but the second equation will have further equilibrium points at $x = n\pi$, $(n = \pm 1, \pm 2, \ldots)$. Let Γ be a circle of radius $a < \pi$, centre the origin, so that the only equilibrium point inside Γ is the origin. Let Γ be described parametrically by $x = a\cos\theta$, $y = a\sin\theta$, $(0 < \theta \leq 2\pi)$. In the notation of Problem 3.14, let

$$\mathbf{X}_1 = (y, -x), \quad \mathbf{X}_2 = (y, -\sin x).$$

Then, on Γ,

$$\mathbf{X}_1 = (a\sin\theta, -a\cos\theta), \quad \mathbf{X}_2 = (a\sin\theta, -\sin(a\cos\theta)).$$

Since the vectors $\mathbf{X}_1$ and $\mathbf{X}_2$ have the same first component for all θ, they can never be in opposition. Hence, by Problem 3.14, the origin for both systems must have the same index. With $\dot{x} = y$, the origin of the system $\ddot{x} + x = 0$ is a centre with index 1. Hence the other system must have the same index.

- **3.16** The system
$$\dot{x} = ax + by + P(x, y), \quad \dot{y} = cx + dy + Q(x, y)$$
has an isolated equilibrium point at $(0, 0)$, and $P(x, y) = O(r^2)$, $Q(x, y) = O(r^2)$ as $r \to 0$, where $r^2 = x^2 + y^2$. Assuming that $ad - bc \neq 0$, show that the origin has the same index as its linear approximation.

3.16. The system

$$\dot{x} = ax + by + P(x, y), \quad \dot{y} = cx + dy + Q(x, y)$$

has an isolated equilibrium point at the origin. Let

$$\mathbf{X}_1 = (ax + by, cx + dy), \quad \mathbf{X}_2 = \mathbf{X}_1 + \mathbf{P},$$

where $\mathbf{P} = (P(x, y), Q(x, y))$. Also let Γ be the circle of radius ρ centred at the origin. Let θ be the smaller angle between $\mathbf{X}_1$ and $\mathbf{X}_2$, then

$$\cos\theta = \frac{\mathbf{X}_1 \cdot \mathbf{X}_2}{|\mathbf{X}_1||\mathbf{X}_2|} = \frac{|\mathbf{X}_1|^2 + \mathbf{P} \cdot \mathbf{X}_1}{|\mathbf{X}_1||\mathbf{X}_2|}.$$

Since $|\mathbf{X}_1| = O(\rho)$ and $|\mathbf{X}_2| = O(\rho)$, then

$$\mathbf{P} \cdot \mathbf{X}_1 = O(\rho^3).$$

Therefore if ρ is sufficiently small,

$$|X_1|^2 + P \cdot X_1 = O(\rho^2) + O(\rho^3) > 0.$$

Hence $\cos \theta > 0$ for all θ. The vectors can never point in opposite directions for any point on Γ, so that the nonlinear equations have the same index as its linear approximation.

- **3.17** Show that, on the phase plane with $\dot{x} = y$, $\dot{y} = Y(x, y)$, Y continuous, the index I_Γ of any simple closed curve Γ that encloses all equilibrium points can only be 1, -1, or zero.
  Let $\dot{x} = y$, $\dot{y} = f(x, \lambda)$, with f, $\partial f / \partial x$ and $\partial f / \partial \lambda$ continuous, represent a parameter-dependent system with parameter λ. Show that, at a bifurcation point (NODE, Section 1.7), where an equilibrium point divides as λ varies, the sum of the indices of the equlibrium points resulting from the splitting is unchanged. (Hint: the integrand in eqn (3.7) is continuous.)
  Deduce that the equilibrium points for the system $\dot{x} = y$, $\dot{y} = -\lambda x + x^3$ consist of a saddle point for $\lambda < 0$, and a centre and two saddles for $\lambda > 0$.

3.17. We use Theorem 3.3 in NODE. The system is

$$\dot{x} = X(x, y) = y, \quad \dot{y} = Y(x, y).$$

Let the closed curve Γ be chosen to enclose all the equilibrium points (which must be on the x axis) and to cut the x axis in just two points (see Figure 3.11). In moving from A to B, $X(x, y) = y$ changes from negative to positive. Whether $\tan \phi$ changes from $-\infty$ to ∞ or from ∞ to $-\infty$ depends on the sign of $Y(x, y)$ in AB. Hence at this transit either $P = 1$ or $Q = 1$. Similarly at the transit of the x axis between C and D either $P = 1$ or $Q = 1$. Any combination

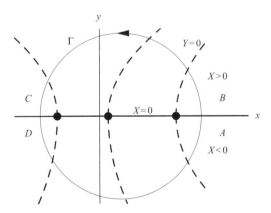

Figure 3.11 Problem 3.17: The figure shows a typical case with three equilibrium points surrounded by a closed curve Γ. The dashed lines represent the isoclines $Y = 0$.

of P and Q at the two intersections are possible. Hence, the index of all the equilibrium points is

$$I = \tfrac{1}{2}(P - Q) = \tfrac{1}{2}(\pm P \pm Q) = 1, 0, \text{ or } -1.$$

This result is also true for any set of adjacent equilibrium points which can be enclosed by a curve Γ.

Consider the system $\dot{x} = y$, $\dot{y} = f(x, \lambda)$. Equilibrium points occur where $f(x, \lambda) = 0$, $y = 0$. Suppose that a bifurcation occurs at $\lambda = \lambda_0$ where the number of equilibrium points changes as λ increases through λ_0. Consider a closed curve Γ which surrounds the bifurcation equilibrium points for all values $|\lambda - \lambda_0| < \varepsilon$ for some $\varepsilon > 0$, and cuts the x axis in just two points. Let Γ be described by $x = x(s)$, $y = y(s)$ for $s_0 \leq s < s_1$. The index (which will be a function λ) is given by eqn (3.7), namely

$$I_\Gamma(\lambda) = \frac{1}{2\pi} \int_{s_0}^{s_1} \frac{XY' - YX'}{X^2 + Y^2} \, ds$$

$$= \frac{1}{2\pi} \int_{s_0}^{s_1} \frac{y(s) df(x(s), \lambda)/ds - f(x(s), \lambda) dy(s)/ds}{y(s)^2 + f(x(s), \lambda)^2} \, ds$$

The integrand is a continuous function of λ (the denominator has no zeros) and the value of the integral, being an index, must be a positive or negative integer or zero for any given value of λ. By continuity it cannot have any jumps. It must therefore retain the same value over the interval $|\lambda - \lambda_0| < \varepsilon$.

Consider the system $\dot{x} = y$, $\dot{y} = -\lambda x + x^3$. Equilibrium occurs where $y = 0$, $-\lambda x + x^3 = 0$. If $\lambda \leq 0$, the equations have one equilibrium point, at $(0, 0)$. If $\lambda > 0$, then the equations have three equilibrium points, at $(-\sqrt{\lambda}, 0)$, $(0, 0)$, $(\sqrt{\lambda}, 0)$. If $\lambda < 0$, the equilibrium point is a centre with index 1. Since this is a conservative system the equilibrium points must be centres (index 1) or saddles (index -1). Hence by the previous theory, the three equilibrium points for $\lambda > 0$ must have a combined index of 1. Therefore, the three equilibrium points must consist of one centre and two saddle points.

- 3.18 Prove a similar result to that of Problem 3.17 for the system $\dot{x} = y$, $\dot{y} = f(x, y, \lambda)$. Deduce that the system $\dot{x} = y$, $\dot{y} = -\lambda x - ky - x^3$, $(k > 0)$, has a saddle point at $(0, 0)$ when $\lambda < 0$ which bifurcates into a stable spiral or node and two saddle points as λ becomes positive.

3.18. Consider the system $\dot{x} = y$, $\dot{y} = f(x, y, \lambda)$. Equilibrium points occur where $f(x, 0, \lambda) = 0$, $y = 0$. Suppose that a bifurcation occurs at $\lambda = \lambda_0$ where the number of equilibrium points changes as λ increases through λ_0. Consider a closed curve Γ which surrounds the bifurcation equilibrium points for all values $|\lambda - \lambda_0| < \varepsilon$ for some $\varepsilon > 0$, and cuts the x axis in just two points. Let Γ be described by $x = x(s)$, $y = y(s)$ for $s_0 \leq s < s_1$. The index (which will be a function λ) is given by eqn (3.7), namely

$$I_\Gamma(\lambda) = \frac{1}{2\pi} \int_{s_0}^{s_1} \frac{XY' - YX'}{X^2 + Y^2} ds$$

$$= \frac{1}{2\pi} \int_{s_0}^{s_1} \frac{y(s) df(x(s), y(s), \lambda)/ds - f(x(s), y(s), \lambda) dy(s)/ds}{y(s)^2 + f(x(s), y(s), \lambda)^2} ds$$

The integrand is a continuous function of λ (the denominator has no zeros) and the value of the integral, being an index, must be a positive or negative integer or zero for any given value of λ. By continuity it cannot have any jumps. It must therefore retain the same value over the interval $|\lambda - \lambda_0| < \varepsilon$.

Consider the example $\dot{x} = y$, $\dot{y} = -\lambda x - ky - x^3$, ($k > 0$). For $\lambda < 0$, the equations have one equilibrium point at the origin. Since $\dot{x} = y$, $\dot{y} \approx -\lambda x$, the origin is a saddle point with index -1. For $\lambda > 0$, the system has three equilibrium points, at $(-\sqrt{\lambda}, 0)$, $(0, 0)$, $(\sqrt{\lambda}, 0)$. As λ increases through 0, the system bifurcates producing three equilibrium points which, by the earlier result must still have a combined index of -1. Since the points have non-zero linear appoximations, we can say that the three equilibrium points must have two points with indices -1 (saddle points) and one with index 1 (a node or a spiral, or a centre). However, the stability can only be checked using linear approximations. For $(0, 0)$, the linear approximation is

$$\dot{x} = y, \quad \dot{y} = -\lambda x - ky - x^3 \approx -\lambda x - ky.$$

The parameters are

$$p = -k < 0, \quad q = \lambda > 0, \quad \Delta = k^2 - 4\lambda.$$

Therefore the origin is a stable node if $k^2 > 4\lambda$ or a stable spiral if $k^2 < 4\lambda$.
The other two equilibrium points are saddles.

- **3.19** A system is known to have three closed paths, C_1, C_2 and C_3, such that C_2 and C_3 are interior to C_1 and such that C_2 and C_3 have no interior points in common. Show that there must be at least one equilibrium point in the region bounded by C_1, C_2 and C_3.

3.19. Figure 3.12 shows a closed phase path C_1 with two closed phase paths C_2 and C_3 within it. All closed paths may have either sense. Individually each closed path has the index 1. Since $I_{C_2} + I_{C_3} = 2$ and $I_{C_3} = 1$, it follows that $I_{C_3} \neq I_{C_2} + I_{C_3}$. Hence there must be at least one equilibrium point in $\mathcal{D}$. The sum of the indices of the equilibrium points in $\mathcal{D}$ is -1.

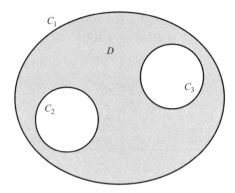

Figure 3.12 Problem 3.19: The shaded region between the closed paths is denoted by $\mathcal{D}$.

- **3.20** For each of the following systems yo are given some information about phase paths and equilibrium points. Sketch phase diagrams consistent with these requirements.

  (i) $x^2 + y^2 = 1$ is a phase path, $(0,0)$ a saddle point, $(\pm\frac{1}{2}, 0)$ centres.

  (ii) $x^2 + y^2 = 1$ is a phase path, $(-\frac{1}{2}, 0)$ a saddle point, $(0,0)$ and $(\frac{1}{2}, 0)$ centres.

  (iii) $x^2 + y^2 = 1$, $x^2 + y^2 = 2$ are phase paths, $(0, \pm\frac{3}{2})$ stabel spirals, $(\pm\frac{3}{2}, 0)$ saddle points, $(0,0)$ a stable spiral.

3.20. (i) Figure 3.13 shows a phase diagram with a closed path $x^2 + y^2 = 1$, a saddle point at $(0,0)$ and centres at $(\pm\frac{1}{2}, 0)$.

(ii) Figure 3.14 shows a phase diagram with a closed path $x^2 + y^2 = 1$, a saddle point at $(-\frac{1}{2}, 0)$, and centres at $(0,0)$ and $(\frac{1}{2}, 0)$.

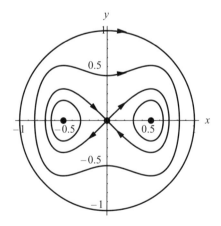

Figure 3.13 Problem 3.20(i):

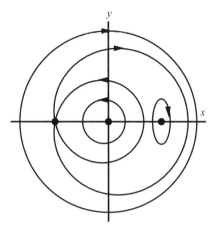

Figure 3.14 Problem 3.20(ii):

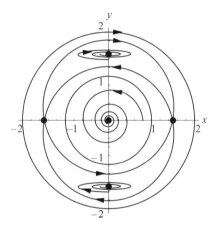

Figure 3.15 Problem 3.20(iii):

(iii) Figure 3.14 shows a phase diagram with a closed path $x^2 + y^2 = 1$, a saddle point at $(-\frac{1}{2}, 0)$, and centres at $(0, 0)$ and $(\frac{1}{2}, 0)$. Figure 3.15 shows a possible phase path configuration which includes two closed paths $x^2 + y^2 = 1$ and $x^2 + y^2 = 2$, stable spirals at $(0, 0)$ and $(0, \pm\frac{3}{2})$ and saddle points at $(\pm\frac{3}{2}, 0)$. Note that the outer closed path has an index 1 which equals the sum of the indices of the five equilibrium points inside the path.

- 3.21 Consider the system
$$\dot{x} = y(z-2), \quad \dot{y} = x(2-z) + 1, \quad x^2 + y^2 + z^2 = 1,$$
which has exactly two equilibrium points, both of which lie on the unit sphere. Project the phase diagram on to the plane $z = -1$ through the point $(0, 0, 1)$. Deduce that $I_\infty = 0$ on this plane (consider the projection of a small circle on the sphere with its centre on the

z axis). Explain, in general terms, why the sum of the indices of the equilibrium points on a sphere is two.

For a certain problem, the phase diagram on the sphere has centres and saddle points only, and it has exactly two saddle points. How many centres has the phase diagram?

3.21. The system is

$$\dot{x} = y(z-2), \quad \dot{y} = x(2-z)+1, \quad x^2+y^2+z^2 = 1.$$

These equations represent a phase diagram on a sphere. Equilibrium occurs where

$$y(z-2) = 0, \quad x(2-z)+1 = 0, \quad x^2+y^2+z^2 = 1.$$

Hence $y = 0$ from the first equation resulting in

$$x(z-2) = 1, \quad x^2+z^2 = 1.$$

Elimination of z leads to

$$x^4 + 3x^2 + 4x + 1 = 0.$$

Solving this equation numerically indicates that there are two real solutions, $x = -0.748$ and $x = -0.340$. The equilibrium points on the sphere have the coordinates $P_1 : (-0.748, 0, 0.663)$ and $P_2 : (-0.340, 0, -0.940)$. Figure 3.16 shows the projection of the phase diagram on the surface of the sphere on to the plane $z = -1$ with $(0, 0, 1)$ as the centre of projection. The point $P : (x, y, z)$, $(x^2 + y^2 + z^2 = 1)$ is projected into the point $Q : (X, Y)$. Figure 3.17 shows the section through the z axis and the points P and Q. If r and R are, respectively, the distances of

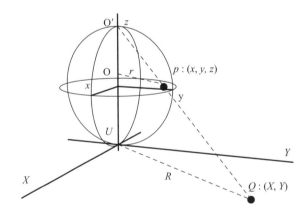

Figure 3.16 Problem 3.21:

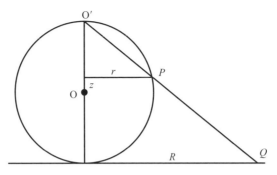

Figure 3.17 Problem 3.21:

the points P and Q from the z axis then, by similar triangles,

$$\frac{r}{1-z} = \frac{R}{2}, \quad \text{or } R = \frac{2r}{1-z}, \quad (-1 \leq z < 1).$$

Using this proportionality rule, it follows that

$$X = \frac{2x}{1-z}, \quad Y = \frac{2y}{1-z}.$$

Hence the equilibrium points P_1 and P_2 map into $Q_1:(-4.439, 0)$ and $Q_2:(-0.351, 0)$ in the (X, Y) plane.

Points at infinity in the (X, Y) plane map into the point $(0, 0, 1)$ on the sphere. Surround this point by a circle $\mathcal{C}$ on the sphere with its centre on the z axis and of sufficiently small radius so that it does not include the equilibrium points P_1 and P_2. Since $\mathcal{C}$ includes no equilibrium points its index is zero. Hence I_∞ for the (X, Y) plane is also zero. Therefore, by Theorem 3.4, the sum of the indices of the equilibrium points on the plane is 2. Since the mapping between the sphere and the plane does not affect the index of any equilibrium point the sum of the indices on the sphere must also be 2.

Since the index of a saddle is -1 and of a centre is 1, there must be four centres on a sphere with two saddles. You can imagine possible configurations of the saddles on the sphere. In one the separatrices of the saddles are connected and divide the surface into four segments with a centre in each. In another the saddles form two non-intersecting figures-of-eight on the surface. In a third one figure-of-eight is within a loop of the other figure-of-eight. In each case there are just four centres.

- **3.22** Show that the following systems have no periodic solutions:

(i) $\dot{x} = y + x^3$, $\dot{y} = x + y + y^3$;
(ii) $\dot{x} = y$, $\dot{y} = -(1 + x^2 + x^4)y - x$.

3.22. Use Bendixson's negative criterion which states that, for the system $\dot{x} = X(x,y)$, $\dot{y} = Y(x,y)$, there are no closed paths in any simply connected region of the phase plane on which $\partial X/\partial x + \partial Y/\partial y$ is of one sign.

(i) $\dot{x} = y + x^3$, $\dot{y} = x + y + y^3$. In this case

$$\frac{\partial X}{\partial x} + \frac{\partial Y}{\partial y} = 3x^2 + 1 + 3y^2 \geq 1 > 0, \quad \text{for all } x, y.$$

Hence this system has no closed paths.

(ii) $\dot{x} = y$, $\dot{y} = -(1 + x^2 + x^4)y - x$. In this case

$$\frac{\partial X}{\partial x} + \frac{\partial Y}{\partial y} = 0 - (1 + x^2 + x^4) \leq -1 < 0, \quad \text{for all } x, y.$$

Hence this system has no closed paths.

- 3.23 (Dulac's test) For the system $\dot{x} = X(x,y)$, $\dot{y} = Y(x,y)$, show that there are no closed paths in a simply connected region in which $\partial(\rho X)/\partial x + \partial(\rho Y)/\partial y$ is of one sign, where $\rho(x,y)$ is any function having continuous first partial derivatives.

3.23. (Dulac's test) Let $\mathcal{D}$ be a simply connected region and $\dot{x} = X(x,y)$, $\dot{y} = Y(x,y)$ a regular system in $\mathcal{D}$. Suppose there exists a continuously differentiable function $\rho(x,y)$ such that

$$\frac{\partial X}{\partial x} + \frac{\partial Y}{\partial y}$$

is of one sign in $\mathcal{D}$. Then the system has no closed phase path in $\mathcal{D}$.

Suppose such a closed path does exist. Let $\mathcal{R}$ denote its interior. Then (see Section 3.4) the divergence theorem applied to the vector field $(\rho X, \rho Y)$ on $\mathcal{R}$ takes the form

$$\oint_\mathcal{C} (\rho X, \rho Y) \cdot \hat{\mathbf{n}} ds = \iint_\mathcal{R} \left(\frac{\partial(\rho X)}{\partial x} + \frac{\partial(\rho Y)}{\partial y} \right) dx dy, \qquad \text{(i)}$$

where ds is an undirected length element on $\mathcal{C}$ and $\hat{\mathbf{n}}$ is the outwardly pointing normal. The vector $(\rho X, \rho Y)$ points along $\mathcal{C}$, and is therefore perpendicular to $\hat{\mathbf{n}}$. The integral on the left in (i) is therefore zero. However, the integrand of the double integral is of one sign in $\mathcal{R}$ so the integral on the right of (i) is non-zero, which is a contradiction. Therefore there can be no closed path in $\mathcal{D}$.

- **3.24** Explain in general terms how Dulac's test (Problem 3.23) and Bendixson's negative criterion may be extended to cover the cases when $\partial(\rho X)/\partial x + \partial(\rho Y)/\partial y$ is of one sign except on isolated points or curves within a simply connected region.

3.24. Suppose that $\partial(\rho X)/\partial x + \partial(\rho Y)/\partial y$ has one sign in a closed curve C, except possibly at a finite number of points or along a finite number of curves, at which it may be zero. These make zero contribution to the integral over $\mathcal{R}$, the interior of C, so

$$\iint_{\mathcal{R}} \left[\frac{\partial}{\partial x}(\rho X) + \frac{\partial}{\partial y}(\rho Y) \right] dx\, dy \neq 0.$$

The proof of Dulac's theorem, Problem 3.23, then follows without change.

- **3.25** For a second-order system $\dot{\mathbf{x}} = \mathbf{X}(\mathbf{x})$, **curl** $(\mathbf{X}) = 0$ and $\mathbf{X} \neq 0$, in a simply connected region $\mathcal{D}$. Show that the system has no closed paths in $\mathcal{D}$. Deduce that
$\dot{x} = y + 2xy, \quad \dot{y} = x + x^2 - y^2$
has no periodic solutions.

3.25. We are given that **curl** $\mathbf{X} = 0$ for the system $\dot{\mathbf{x}} = \mathbf{X}(\mathbf{x})$ in a simply connected region $\mathcal{D}$. Suppose that there exists a closed phase path C in $\mathcal{D}$, whose interior is denoted by $\mathcal{R}$. Then, in vector form, where $\mathbf{X} = (X, Y)$, Green's theorem (or Stokes's theorem in two dimensions) may be written

$$\int_C \mathbf{X} \cdot d\mathbf{r} = \iint_{\mathcal{R}} \mathbf{curl}\, \mathbf{X} \cdot \hat{\mathbf{k}}\, dx\, dy,$$

where $\hat{\mathbf{k}}$ is a unit vector in the positive z direction. By hypothesis, the integral on the right is zero. Therefore

$$\int_C \mathbf{X} \cdot d\mathbf{r} = \int_C (X\, dx + Y\, dy) = 0.$$

Let t represent time. Then over one cycle of the path $t_1 \leq t \leq t_2$,

$$\int_{t_1}^{t_2} (X\, dx + Y\, dy) = \int_{t_1}^{t_2} (X\dot{x} + Y\dot{y})\, dt = \int_{t_1}^{t_2} (X^2 + Y^2)\, dt,$$

so

$$\int_C \mathbf{X} \cdot d\mathbf{r} > 0,$$

since X and Y are not simultaneously zero except possibly at equilibrium points. This contradicts (i), so there exists no closed path C in $\mathcal{D}$.

In the problem
$$\dot{x} = y + 2xy, \quad \dot{y} = x + x^2 + y^2.$$

Hence $\mathbf{X}(\mathbf{x}) = (y + 2xy, x + x^2 - y^2)$, and

$$\operatorname{curl} \mathbf{X} = \left(\frac{\partial (x + x^2 - y^2)}{\partial x} - \frac{\partial (y + 2xy)}{\partial y} \right) \hat{\mathbf{k}} = (1 + 2x - 1 - 2x)\hat{\mathbf{k}} = 0$$

for all x, y, Therefore the system can have no closed paths.

- 3.26 In Problem 3.25 show that $\operatorname{curl} \mathbf{X} = 0$ may be replaced by $\operatorname{curl}(\psi \mathbf{X}) = 0$, where $\psi(x, y)$ is of one sign in $\mathcal{D}$.

3.26. The proof follows as in Problem 3.25. Using the same notation by Green's theorem, supposing that $\operatorname{curl} \mathbf{X} = 0$,

$$\oint_C \psi \mathbf{X} \cdot d\mathbf{r} = \int_\mathcal{R} \operatorname{curl}(\psi \mathbf{X}) \cdot \hat{\mathbf{k}} \, dx dy = 0. \tag{i}$$

But for any closed phase path C bounding a region $\mathcal{D}$,

$$\oint_C \psi \mathbf{X} \cdot d\mathbf{r} = \int_C (\psi X dx + \psi Y dy)$$

$$= \int_{t_1}^{t_2} \left(\psi X \frac{dx}{dt} + \psi Y \frac{dy}{dt} \right) dt$$

$$= \int_{t_1}^{t_2} \psi (X^2 + Y^2) dt \neq 0,$$

since ψ is of one sign in $\mathcal{R}$. This contradicts (i): therefore there are no closed paths in $\mathcal{D}$.

- 3.27 By using Dulac's test (Problem 3.23) with $\rho = e^{-2x}$, show that
$\dot{x} = y, \quad \dot{y} = -x - y + x^2 + y^2$
has no periodic solutions.

3.27. Dulac's test is given in Problem 3.23. For the system

$$\dot{x} = X(x, y) = y, \quad \dot{y} = Y(x, y) = -x - y + x^2 + y^2,$$

let $\rho(x, y) = e^{-2x}$. Then

$$\frac{\partial(\rho X)}{\partial x} + \frac{\partial(\rho Y)}{\partial y} = \frac{\partial}{\partial x}(e^{-2x}y) + \frac{\partial}{\partial y}[e^{-2x}(-x - y + x^2 + y^2)]$$

$$= -2e^{-2x}y + e^{-2x}(-1 + 2y) = -e^{-2x} < 0$$

for all x, y. Therefore the system has no closed phase paths.

Note that the system has two equilibrium points, at $(0, 0)$ (stable spiral) and at $(1, 0)$ (saddle point).

- **3.28** Use Dulac's test (Problem 3.23) to show that
$\dot{x} = x(y - 1), \quad \dot{y} = x + y - 2y^2$,
has no periodic solutions.

3.28. Dulac's test is given in Problem 3.23. The system is

$$\dot{x} = X(x, y) = x(y - 1), \quad \dot{y} = Y(x, y) = x + y - 2y^2.$$

Equilibrium points occur at $(0, 0)$, $(0, \frac{1}{2})$ and $(1, 1)$. The axis $x = 0$ is a solution of the equations. We conclude that paths cannot cross $x = 0$, and there can be no phase paths in the half-plane $x < 0$ since it contains no equilibrium points.

Using the function $\rho(x, y)$, consider

$$\frac{\partial(\rho X)}{\partial x} + \frac{\partial(\rho Y)}{\partial y} = \frac{\partial \rho}{\partial x}x(y - 1) + \rho y + \frac{\partial \rho}{\partial y}(x + y - 2y^2) + \rho(x - 4y)$$

$$= \frac{\partial \rho}{\partial x}x(y - 1) - 3\rho y + \frac{\partial \rho}{\partial y}(x + y - 2y^2) + \rho x.$$

We can eliminate y in this expression by choosing $\rho = x^3$. Then

$$\frac{\partial(\rho X)}{\partial x} + \frac{\partial(\rho Y)}{\partial y} = -3x^3 < 0$$

in the half-plane $x > 0$. Therefore there can be no closed paths in $x > 0$.

- **3.29** Show that the following systems have no periodic solutions:
  (i) $\dot{x} = y, \dot{y} = 1 + x^2 - (1 - x)y$;
  (ii) $\dot{x} = -(1 - x)^3 + xy^2, \dot{y} = y + y^3$;
  (iii) $\dot{x} = 2xy + x^3, \dot{y} = -x^2 + y - y^2 + y^3$;

(iv) $\dot{x} = x$, $\dot{y} = 1 + x + y^2$;
(v) $\dot{x} = y$, $\dot{y} = -1 - x^2$;
(vi) $\dot{x} = 1 - x^3 + y^2$, $\dot{y} = 2xy$;
(vii) $\dot{x} = y$, $\dot{y} = (1 + x^2)y + x^3$.

3.29. (i) $\dot{x} = y$, $\dot{y} = 1 + x^2 - (1-x)y$. The system has no equilibrium points, so by NODE, Theorem 3.1 every closed path has index zero. Therefore there are no closed paths since any closed path has the index 1.

(ii) $\dot{x} = -(1-x)^3 + xy^2$, $\dot{y} = y + y^3$. The system has one equilibrium point at $(1,0)$, but $y = 0$ consists of two phase path through this equilibrium point. Hence there can be no closed paths surrounding the equilibrium point.

(iii) $\dot{x} = 2xy + x^3$, $\dot{y} = -x^2 + y - y^2 + y^3$. Note first that $x = 0$, $y > 0$ and $x = 0$, $y < 0$ are two phase paths in opposite directions. Equilibrium occurs where

$$x(2y + x^2) = 0, \tag{i}$$

$$-x^2 + y - y^2 + y^3 = 0. \tag{ii}$$

If $x = 0$ to satisfy (i), then either $y = 0$ or $y^2 - y + 1 = 0$ from (ii). However, the quadratic equation has no real solutions, which leaves the equilibrium point $(0,0)$. Alternatively, if $y = -\frac{1}{2}x^2$ in (i), then (ii) becomes $x^2(x^4 + 2x^2 + 12) = 0$ which has only the solution $x = 0$. Hence the only equilibrium point occurs at the origin, but it lies on a phase path. Hence the system has no closed phase paths.

(iv) $\dot{x} = x$, $\dot{y} = 1 + x + y^2$. Note that $x = 0$ is a phase path. However, $x = 0$ and $1 + x + y^2 = 0$ have no real solutions for x and y. Hence the system cannot have a periodic solution.

(v) $\dot{x} = y$, $\dot{y} = -1 - x^2$. The system has no equilibrium points, and therefore no periodic solutions.

(vi) $\dot{x} = 1 - x^3 + y^2$, $\dot{y} = 2xy$. Note that $y = 0$, $x > 1$ and $y = 0$, $x < 1$ are two phase paths in opposite directions. Equilibrium occurs where $xy = 0$ and $1 - x^3 + y^2 = 0$. Then $x = 0$ leads to no real y, whilst $y = 0$ leads to $x = 1$. Hence there is one equilibrium point, at $(1,0)$, but this lies on $y = 0$. Therefore there can no periodic solutions.

(vii) $\dot{x} = X(x, y) = y$, $\dot{y} = Y(x, y) = (1 + x^2)y + x^3$. Use NODE, Theorem 3.5 (Bendixson):

$$\frac{\partial X}{\partial x} + \frac{\partial Y}{\partial y} = \frac{\partial}{\partial x}(y) + \frac{\partial}{\partial y}[(1 + x^2)y + x^3] = 1 + x^2 > 0,$$

for all x, y. Therefore the system can have no periodic solutions.

- 3.30 Let D be a doubly connected region in the x,y plane. Show that, if $\rho(x,y)$ has contiuous first partial derivatives and div (ρX) is of constant sign in D, then the system has not more than one closed path in D. (An extension of Dulac's test Problem 3.23.)

3.30. Figure 3.18 shows a doubly connected region D. Suppose that L_1 and L_2 are two closed paths in D. Obviously they cannot intersect. Join L_1 and L_2 by two coincident paths AB and BA as shown. This creates a closed path L_1, AB, L_2, BA (call it C) which bounds a simply connected region say S. Apply Green's Theorem in the plane to this curve C and the vector field ρX. Then (as in NODE, Theorem 3.5)

$$\iint_S \operatorname{div}(\rho X)\,dx\,dy = \oint_C X \cdot \mathbf{n}\,ds, \tag{i}$$

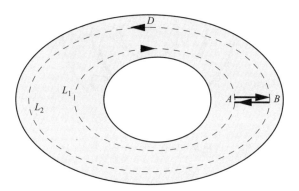

Figure 3.18 Problem 3.30: The dashed curves L_1 and L_2 are two closed phase paths in D.

where $\mathbf{n}$ is the outward normal to C. On L_1 and L_2, X is perpendicular to $\mathbf{n}$ so that $X \cdot \mathbf{n} = 0$, whilst the contributions from AB and BA cancel. Hence the value of the line integral on the right of (i) is zero. Therefore

$$\iint_S \operatorname{div}(\rho X)\,dx\,dy = 0,$$

but this contradicts the requirement that div (ρX) is on one sign in D. Hence D can contain at most one closed path.

- 3.31 A system has exactly two limit cycles with one lying interior to the other and with no equilibrium points between them. Can the limit cycles be described in opposite senses? Obtain the equations of the phase paths of the system
$$\dot{r} = \sin \pi r, \quad \dot{\theta} = \cos \pi r$$
as described in polar coordinates (r,θ). Sketch the phase diagram.

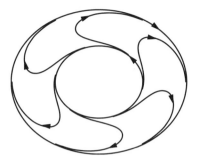

Figure 3.19 Problem 3.31:

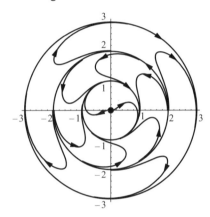

Figure 3.20 Problem 3.31:

3.31. Figure 3.19 shows a phase diagram with an outer counter-clockwise limit cycle and an inner clockwise limit cycle. Between them there are no equilibrium points and phase paths reverse direction, approaching the outer cycle as $t \to \infty$ and the inner cycle as $t \to -\infty$. This configuration shows that limit cycles in opposite senses are possible.

For the polar system $\dot{r} = \sin \pi r$, $\dot{\theta} = \cos \pi r$, the phase paths are given by

$$\frac{dr}{d\theta} = \tan \pi r.$$

This separable equation has the solution

$$\int \cot \pi r \, dr = \int d\theta,$$

that is,

$$\frac{1}{\pi} \ln |\sin \pi r| = \theta + C, \quad \text{or} \quad \sin \pi r = A e^{\pi \theta}.$$

The system has one equilibrium point at $r = 0$. Also $r = 1, 2, 3, \ldots$, with θ arbitrary, are particular solutions. A part of the phase diagram is shown in Figure 3.20. The periodic solutions alternate in direction with the limit cycles $r = 1, 3, 5, \ldots$ stable and $r = 2, 4, 6, \ldots$ unstable.

3: Geometrical aspects of plane autonomous systems

- **3.32** Using Bendixson's theorem (Section 3.4) show that the response amplitudes a, b for the van der Pol equation in the 'van der Pol plane' (this will be discussed later in Chapter 7), described by the equations

$$\dot{a} = \tfrac{1}{2}\varepsilon\left(1 - \tfrac{1}{4}r^2\right)a - \frac{\omega^2-1}{2\omega}b, \quad \dot{b} = \tfrac{1}{2}\varepsilon\left(1 - \tfrac{1}{4}r^2\right)b + \frac{\omega^2-1}{2\omega}a + \frac{\Gamma}{2\omega}$$

have no closed paths in the circle $r < \sqrt{2}$.

3.32. The van der Pol equation is given by

$$\dot{a} = X(a,b) = \tfrac{1}{2}\varepsilon\left(1 - \tfrac{1}{4}r^2\right)a - \frac{\omega^2-1}{2\omega}b,$$

$$\dot{b} = Y(a,b) = \tfrac{1}{2}\varepsilon\left(1 - \tfrac{1}{4}r^2\right)b + \frac{\omega^2-1}{2\omega}a + \frac{\Gamma}{2\omega},$$

where $r = \sqrt{(a^2 + b^2)}$. Use Bendixson's criterion (Theorem 3.5). Then

$$\frac{\partial X}{\partial a} + \frac{\partial Y}{\partial b} = \frac{\partial}{\partial a}\left[\tfrac{1}{2}\varepsilon\left(1 - \tfrac{1}{4}r^2\right)a - \frac{\omega^2-1}{2\omega}b\right]$$

$$+ \frac{\partial}{\partial b}\left[\tfrac{1}{2}\varepsilon\left(1 - \tfrac{1}{4}r^2\right)b + \frac{\omega^2-1}{2\omega}a + \frac{\Gamma}{2\omega}\right]$$

$$= \varepsilon(1 - \tfrac{1}{2}r^2),$$

which takes the sign of ε for $r < \sqrt{2}$. Therefore there can be no closed phase paths within this circle in the (a, b) plane.

- **3.33** Let C be a closed path for the system $\dot{\mathbf{x}} = \mathbf{X}(\mathbf{x})$, having $\mathcal{D}$ as its interior. Show that

$$\iint_{\mathcal{D}} \text{div}\,(\mathbf{X})\,dxdy = 0.$$

3.33. Suppose that C is a closed path for the system $\dot{\mathbf{x}} = \mathbf{X}(\mathbf{x})$, and that $\mathcal{D}$ is the interior of C. By the Divergence Theorem in two dimensions,

$$\int_{\mathcal{D}} \text{div}\,\mathbf{X}\,dxdy = \int_C \mathbf{X}\cdot\mathbf{n}\,ds.$$

This is zero because $\mathbf{X}$ is tangential to C at all points on it, and $\mathbf{n}$ is a unit normal to C.

- **3.34** Assume that van der Pol's equation in the phase plane
$$\dot{x} = y, \quad \dot{y} = -\varepsilon(x^2 - 1)y - x$$
has a single closed path, which, for ε small, is approximately a circle, centre the origin of radius a. Use the result of Problem 3.33 to show that approximately
$$\int_{-a}^{a} \int_{-\sqrt{(a^2-x^2)}}^{\sqrt{(a^2-x^2)}} (x^2 - 1) dy dx = 0,$$
and so deduce a.

3.34. The van der Pol system is
$$\dot{x} = X(x, y) = y, \quad \dot{y} = Y(x, y) = -\varepsilon(x^2 - 1)y - x.$$

We are given that the system has a closed path which is approximately a circle of radius a. In this example, div $X = -\varepsilon(x^2 - 1)$. We can say, approximately, (but not rigorously) that, using Problem 3.33, the double integral of div X over the interior of the circle is zero, that is,
$$J = \int_{-a}^{a} \int_{-\sqrt{(a^2-x^2)}}^{\sqrt{(a^2-x^2)}} (x^2 - 1) dy dx = 0.$$

Integrating as a repeated integral
$$J = 2 \int_{-a}^{a} (x^2 - 1) \sqrt{(a^2 - x^2)} dx$$
$$= 2a^2 \int_{-\frac{1}{2}\pi}^{\frac{1}{2}\pi} (a^2 \sin^2 \theta \cos^2 \theta - \cos^2 \theta) d\theta, \quad (x = a \sin \theta)$$
$$= \frac{a^2}{4} \int_{-\frac{1}{2}\pi}^{\frac{1}{2}\pi} [a^2(1 - \cos 4\theta) - 4(1 + \cos 2\theta)] d\theta$$
$$= \frac{a^2 \pi}{4} (a^2 - 4).$$

Hence we deduce $a \approx 2$.

- **3.35** Following Problems 3.33 and 3.34, deduce a condition on the amplitudes of periodic solutions of
$$\ddot{x} + \varepsilon h(x, \dot{x}) \dot{x} + x = 0, \quad |\varepsilon| \ll 1.$$

3.35 Express the equation $\ddot{x} + \varepsilon h(x, \dot{x})\dot{x} + x = 0$ in the form

$$\dot{x} = X(x, y) = y, \quad \dot{y} = Y(x, y) = -\varepsilon h(x, y)y - x.$$

The system has one equilibrium point, at $(0,0)$, so that any closed phase path must enclose the origin. For small $\varepsilon = 0$, the equation reduces to that for simple harmonic motion with unit frequency so that the centre at the origin is a nest of concentric circles. For small $|\varepsilon|$ any possible closed path will be close to one of these circles. Suppose that C is a closed phase path in the (x, y) plane. In the notation of Problem 3.33, $\mathrm{div}\, \mathbf{X} = -\varepsilon h(x, y)$, so that

$$\iint_{\mathcal{D}} [h(x, y) + h_y(x, y)y] dx dy = 0,$$

which will determine approximately the amplitude of C.

- 3.36 For the system
$$\ddot{x} + \varepsilon h(x, \dot{x})\dot{x} + g(x) = 0,$$
suppose that $g(0) = 0$ and $g'(x) > 0$. Let C be a closed path in the phase plane (as all paths must be) for the equations $\ddot{x} + g(x) = 0$ having interior $\mathcal{D}$. Use the result of Problem 3.33 to deduce that for small ε, C approximately satisfies

$$\int_{\mathcal{D}} \{h(x, y) + h_y(x, y)y\} dx dy = 0.$$

Adapt this result to the equation
$$\ddot{x} + \varepsilon(x^2 - \alpha)\dot{x} + \sin x = 0,$$
with ε small, $0 < \alpha \ll 1$, and $|x| < \frac{1}{2}\pi$. Show that a closed path (a limit cycle) is given by
$$y^2 = 2A + 2\cos x$$
where A satisfies
$$\int_{-\cos^{-1}(-A)}^{\cos^{-1}(-A)} (x^2 - \alpha)\sqrt{(2A + 2\cos x)} dx = 0.$$

3.36. Express the equation $\ddot{x} + \varepsilon h(x, \dot{x})\dot{x} + g(x) = 0$ in the form

$$\dot{x} = X(x, y) = y, \quad \dot{y} = Y(x, y) = -\varepsilon h(x, y)y - g(x) = 0.$$

Since $g(0) = 0$ and $g'(x) > 0$, the system has only one equilibrium point, at $(0, 0)$. In particular for $\varepsilon = 0$ the origin is a centre. The result follows as in the previous problem. The amplitude of any closed path is given approximately by

$$\iint_{\mathcal{D}} [h(x, y) + h_y(x, y)y] dx dy = 0. \qquad (i)$$

Consider the equation $\ddot{x} + \varepsilon(x^2 - \alpha)\dot{x} + \sin x = 0$. In this example

$$X(x, y) = y, \quad Y(x, y) = -\varepsilon(x^2 - \alpha)y - \sin x.$$

If $\varepsilon = 0$, then the equation of the phase paths of the centre at $(0, 0)$ is given by

$$\frac{dy}{dx} = -\frac{\sin x}{y},$$

which has the general solution $y^2 = 2A + 2\cos x$ as required. We assume that any closed path of the original equation is close to one of these. The region $\mathcal{D}$ for eqn (i) is bounded by the curves $y = \pm\sqrt{(2A + 2\cos x)}$ for $-\cos^{-1}(-A) \leq x \leq \cos^{-1}(-A)$. Equation (i) becomes

$$\int_{-\cos^{-1}(A)}^{\cos^{-1}(A)} \int_{-\sqrt{(2A+2\cos x)}}^{\sqrt{(2A+2\cos x)}} (x^2 - \alpha) \, dy \, dx = 0.$$

Integration with respect to y leads to the equation

$$\int_{-\cos^{-1}(A)}^{\cos^{-1}(A)} (x^2 - \alpha)\sqrt{(2A + 2\cos x)} \, dx = 0$$

for the amplitude A.

- **3.37** Consider the system
$\dot{x} = X(x, y) = -(x^2 + y^2)y, \quad \dot{y} = Y(x, y) = bx + (1 - x^2 - y^2)y.$
Let $\mathcal{C}$ be the circle $x^2 + y^2 = a^2$ with interior $\mathcal{R}$. Show that

$$\iint_{\mathcal{R}} \text{div}(X, Y) \, dx \, dy = 0$$

only if $a = 1$. Is $\mathcal{C}$ a phase path (compare Problem 3.33)?

3.37. For the system

$$\dot{x} = X(x, y) = -(x^2 + y^2)y, \quad \dot{y} = Y(x, y) = bx + (1 - x^2 - y^2)y \quad (b > 0)$$

$\text{div}(X, Y) = -2xy + 1 - x^2 - 3y^2$. Let $\mathcal{C}$ be the circle $x^2 + y^2 = a^2$ with interior denoted by $\mathcal{R}$. Then

$$J(a, b) = \iint_{\mathcal{R}} \text{div}(X, Y) \, dx \, dy = \iint_{\mathcal{R}} [-2xy + (1 - x^2 - 3y^2)] \, dx \, dy.$$

In polar coordinates

$$J(a,b) = \int_0^a \int_0^{2\pi} [-2r^2 \sin\theta \cos\theta + (1 - r^2 \cos^2\theta - 3r^2 \sin^2\theta)] r \, dr \, d\theta$$

becomes

$$J(a,b) = \int_0^a \int_0^{2\pi} [-2r^2 \sin\theta \cos\theta + (1 - r^2 \cos^2\theta - 3r^2 \sin^2\theta)] r \, dr \, d\theta$$

$$= a^2(a^2 - 1)\pi.$$

It follows that

$$J(a,b) = \int\int_{\mathcal{R}} \operatorname{div}(X, Y) dx dy = 0,$$

if $a = 1$. However it does not follow (see Problem 3.33) that $\mathcal{C}$ is a phase path. The converse is true, that if $\mathcal{C}$ is a phase path then $J(a, b) = 0$.

Describe the circle parametrically by $x = a \cos t$, $y = a \sin t$. Then, if $a = 1$,

$$\dot{x} + x(x^2 + y^2) = -a \sin t + a^3 \sin t = a(a^2 - 1) \sin t = 0,$$

$$\dot{y} - bx - (1 - x^2 - y^2)y = a \cos t - b \cos t - b^2(1 - a^2) \sin t = (1 - b) \cos t.$$

Therefore, $\mathcal{C}$ is only a phase path if $b = 1$.

This problem confirms that no conclusions can be made about $\mathcal{C}$ as a phase path if

$$\int\int_{\mathcal{R}} \operatorname{div}(X, Y) dx dy = 0.$$

- **3.38** The equation $\ddot{x} + F_0 \tanh k(\dot{x} - 1) + x = 0$, $F_0 > 0$, $k \gg 1$, can be thought of as a plausible continuous representation of the type of Coulomb friction problem of Section 1.6. Show, however, that the only equilibrium point is a stable spiral, and that there are no periodic solutions.

3.38. The friction equation is

$$\ddot{x} + F_0 \tanh k(\dot{x} - 1) + x = 0, \quad (F_0 > 0, \; k \gg 1).$$

In the usual phase plane, let

$$\dot{x} = X(x, y) = y, \quad \dot{y} = Y(x, y) = -F_0 \tanh k(y - 1) - x.$$

Equilibrium only occurs where $y = 0$, $x = -F_0 \tanh(-k) = F_0 \tanh k \approx F_0$ for large k. Let $x = F_0 \tanh k + \xi$. Then the linear approximations become

$$\dot{\xi} = y, \quad \dot{y} \approx -\xi - F_0 k y \operatorname{sech}^2 k.$$

The eigenvalues are given by

$$\begin{vmatrix} -\lambda & 1 \\ -1 & -F_0 k \operatorname{sech}^2 k - \lambda \end{vmatrix} = \lambda^2 + \lambda F_0 k \operatorname{sech}^2 k + 1 = 0.$$

The eigenvalues are

$$\lambda = \tfrac{1}{2}[F_0 k \operatorname{sech}^2 k \pm \sqrt{(F_0^2 k^2 \operatorname{sech}^4 k - 4)}].$$

For k large the roots are complex with negative real part. Hence the equilibrium point is a stable spiral.

For the non-existence of periodic solutions, use Bendixson's Theorem 3.5 (in NODE). Thus

$$\operatorname{div}(X, Y) = \frac{\partial Y}{\partial y} = -F_0 k \operatorname{sech}^2 k (y - 1) < 0, \text{ for all } y.$$

Therefore the system has no periodic solutions.

- **3.39** Show that the third-order system
$$\dot{x}_1 = x_2, \quad \dot{x}_2 = -x_1, \quad \dot{x}_3 = 1 - (x_1^2 + x_2^2)$$
has no equilibrium points but nevertheless has closed paths (periodic solutions).

3.39. The third-order system is

$$\dot{x}_1 = x_2, \quad \dot{x}_2 = -x_1, \quad \dot{x}_3 = 1 - (x_1^2 + x_2^2).$$

Equilibrium points are given by $\dot{x}_1 = \dot{x}_2 = \dot{x}_3 = 0$, that is,

$$x_2 = x_1 = 0, \quad 1 - (x_1^2 + x_2^2) = 0,$$

which are clearly inconsistent.

From the first two equations

$$\frac{dx_2}{dx_1} = -\frac{x_1}{x_2},$$

which can be integrated to give $x_1^2 + x_2^2 = c^2$. This means that all phase paths lie on coaxial circular cylinders with axis the x_3 axis in the (x_1, x_2, x_3) space. The third equation now becomes

$$\dot{x}_3 = 1 - (x_1^2 + x_2^2) = 1 - c^2.$$

Integration gives $x_3 = (1-c^2)t + b$. Generally phase paths are helices on the circular cylinders. They are only periodic if $c = 1$ in which case $x_3 = b$, a constant. To summarize, the phase paths are given by

$$x_1^2 + x_2^2 = 1, \quad x_3 = b,$$

for any value of the constant b.

This example shows that the relation between closed paths and equilibrium points does not immediately generalize to higher dimensions.

- 3.40 Sketch the phase diagram for the quadratic system $\dot{x} = 2xy$, $\dot{y} = y^2 - x^2$.

3.40. The system $\dot{x} = 2xy$, $\dot{y} = y^2 - x^2$ has one equilibrium point, at the origin, but the linear approximation there is not helpful. The phase paths are given by

$$\frac{dy}{dx} = \frac{y^2 - x^2}{2xy}.$$

This is a first-order equation of homogeneous type. Therefore let $y = vx$, so that the equation becomes

$$x \frac{dv}{dx} + v = \frac{v^2 - 1}{2v}, \quad \text{or} \quad \frac{dv}{dx} = -\frac{1 + v^2}{2vx}.$$

This can be integrated to give the general solution $x^2 + y^2 = Ax$. Therefore all phase paths are circles which pass through the origin as shown in Figure 3.21.

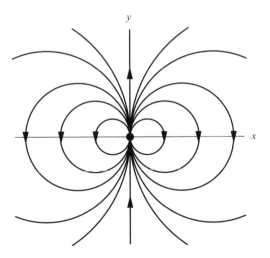

Figure 3.21 Problem 3.40:

- **3.41** Locate the equilibrium points of the system
$$\dot{x} = x(x^2 + y^2 - 1), \quad \dot{y} = y(x^2 + y^2 - 1),$$
and sketch the phase diagram.

3.41. Consider the system
$$\dot{x} = x(x^2 + y^2 - 1), \quad \dot{y} = y(x^2 + y^2 - 1).$$

Equilibrium occurs where
$$x(x^2 + y^2 - 1) = 0, \quad y(x^2 + y^2 - 1) = 0.$$

All points on the circle $x^2 + y^2 = 1$ in equilibrium points, and the origin is also an equilibrium point. Phase paths are given by
$$\frac{dy}{dx} = \frac{y}{x} \quad \Rightarrow \quad y = Cx.$$

The phase diagram is shown in Figure 3.22.

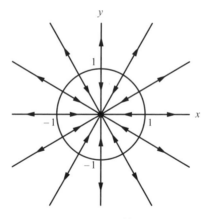

Figure 3.22 Problem 3.41:

- **3.42** Find the equilibrium points of the system
$$\dot{x} = x(1 - y^2), \quad \dot{y} = x - (1 - e^x)y.$$
Show that the system has no closed paths.

3.42. The system
$$\dot{x} = X(x, y) = x(1 - y^2), \quad \dot{y} = Y(x, y) = x - (1 + e^x)y$$

is in equilibrium where
$$x(1 - y^2) = 0, \quad x - (1 + e^x)y = 0.$$

Clearly $(0,0)$ is an equilibrium point (a saddle). If $y=1$, then $x-1-e^x<-1$ for all x, and therefore is never zero. If $y = -1$, then $h(x) = x + 1 + e^x$ is zero for only one value (since $h'(x)$ is always positive and $h(x) \to \infty$ as $x \to \infty$ and $h(x) \to -\infty$ as $x \to -\infty$) at $x = -1.27$ approximately.

To show that there are no closed paths, apply Bendixson's Theorem 3.5. Then
$$\text{div}(X, Y) = \frac{\partial X}{\partial x} + \frac{\partial Y}{\partial y} = 1 - y^2 - 1 - e^x = -y^2 - e^x < 0,$$

which is of one sign for all x, y.

- 3.43 Show, using Bendixson's theorem, that the system
$$\dot{x} = x^2 + y^2, \quad \dot{y} = y^2 + x^2 e^x$$
has no closed paths in $x + y > 0$ or $x + y < 0$. Explain why the system has no closed paths in the x, y plane.

3.43. Apply Bendixson's Theorem 3.5 to the system
$$\dot{x} = X(x, y) = x^2 + y^2, \quad \dot{y} = Y(x, y) = y^2 + x^2 e^x.$$

Then
$$\text{div}(X, Y) = 2x + 2y$$

which is positive in $x + y > 0$, and negative in $x + y < 0$. Therefore, by Bendixson's Theorem, the system can have no closed paths in $x + y > 0$, nor in $x + y < 0$. The system has an equilibrium point at $(0, 0)$ so it is possible that a closed path surrounds the origin. However, $dy/dx > 0$ (except at $(0,0)$). This means that there is no isocline of zero slope, which would be required of any closed path.

- 3.44 Plot the phase diagram, showing the main features of the phase plane, for the equation
$$\ddot{x} + \varepsilon(1 - x^2 - \dot{x}^2)\dot{x} + x = 0$$
using $\dot{x} = y$, for $\varepsilon = 0.1$ and $\varepsilon = 5$.

3.44. The equation is
$$\ddot{x} + \varepsilon(1 - x^2 - \dot{x}^2)\dot{x} + x = 0.$$

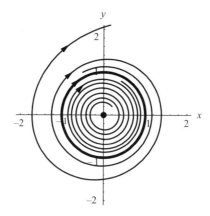

Figure 3.23 Problem 3.44: Phase diagram two phase paths and the limit cycle for $\ddot{x} + \varepsilon(1 - x^2 - \dot{x}^2)\dot{x} + x = 0$ with $\varepsilon = 0.1$.

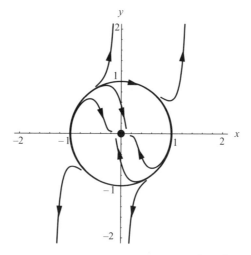

Figure 3.24 Problem 3.44: Phase diagram for $\ddot{x} + \varepsilon(1 - x^2 - \dot{x}^2)\dot{x} + x = 0$ with $\varepsilon = 5$.

Assume that $\dot{x} = y$. Note that the only equilibrium point is at the origin. Near the origin

$$\dot{x} = y, \quad \dot{y} \approx -\varepsilon y - x,$$

which implies that the origin is stable, a spiral if $\varepsilon < 2$ and a node if $\varepsilon > 2$. Note also that the circle $x^2 + y^2 = 1$ is an unstable limit cycles. Phase diagrams for the cases $\varepsilon = 0.1$ and $\varepsilon = 5$ are shown in Figures 3.23 and 3.24.

- **3.45** Plot a phase diagram for the damped pendulum equation $\ddot{x} + 0.15\dot{x} + \sin x = 0$.

3.45. The phase diagram for $\ddot{x} + 0.15\dot{x} + \sin x = 0$ is shown in Figure 3.31 in NODE.

- **3.46** The system

$$\dot{x} = -\frac{1}{2\omega}y\left\{(\omega^2 - 1) - \frac{3}{4}\beta(x^2 + y^2)\right\},$$

$$\dot{y} = \frac{1}{2\omega}x\left\{(\omega^2 - 1) - \frac{3}{4}\beta(x^2 + y^2)\right\} + \frac{\Gamma}{2\omega},$$

occurs in the theory of the forced oscillations of a pendulum. Obtain the phase diagram when $\omega = 0.975$, $\Gamma = 0.005$, $\beta = -1/6$.

3.46. The autonomous system

$$\dot{x} = -\frac{1}{2\omega}y\left\{(\omega^2 - 1) - \frac{3}{4}\beta(x^2 + y^2)\right\},$$

$$\dot{y} = \frac{1}{2\omega}x\left\{(\omega^2 - 1) - \frac{3}{4}\beta(x^2 + y^2)\right\} + \frac{\Gamma}{2\omega},$$

arises from the van der Pol plane for forced oscillations in NODE, Section 7.2. Equilibrium can only occur where $y = 0$, in which case x must be obtained by numerical solution of the cubic

$$\frac{3}{4}\beta x^3 - (\omega^2 - 1)x - \frac{\Gamma}{2\omega} = 0.$$

The system has three equilibrium points: at $(-0.673, 0)$, $(0.104, 0)$ and $(0.569, 0)$. The phase diagram is shown in Figure 7.3 in NODE.

- **3.47** A population of rabbits $R(t)$ and foxes $F(t)$ live together in a certain territory. The combined birth and death rate of the rabbits due to 'natural' causes is $\alpha_1 > 0$, and the additional deaths due to their being eaten by foxes is introduced through an 'encounter factor' β_1, so that

$$\frac{dR}{dt} = \alpha_1 R - \beta_1 RF.$$

The foxes die of old age with death rate $\beta_2 > 0$, and the live birth rate is sustained through an encounter factor α_2, so that (compare Example 2.3)

$$\frac{dF}{dt} = \alpha_2 RF - \beta_2 F.$$

Plot the phase diagram, when $\alpha_1 = 10$, $\beta_1 = 0.2$, $\alpha_2 = 4 \times 10^{-5}$, $\beta_2 = 0.2$. Also plot typical solution curves $R(t)$ and $F(t)$ (these are oscillatory, having the same period but different phase).

3.47. The rabbit $R(t)$ and fox $F(t)$ populations satisfy the differential equations

$$\frac{dR}{dt} = \alpha_1 R - \beta_1 RF, \quad \frac{dF}{dt} = \alpha_2 RF - \beta_2 F.$$

The populations are in equilibrium at $(R, F) = (0, 0)$ and at $(R, F) = (\beta_2/\alpha_2, \alpha_1/\beta_1)$. Note also that $R = 0$ and $F = 0$ are solutions, and also that $R \geq 0$ and $F \geq 0$. With the parameters $\alpha_1 = 10$, $\beta_1 = 0.2$, $\alpha_2 = 4 \times 10^{-5}$, $\beta_2 = 0.2$. Hence the non-zero equilibrium point is at $(5000, 50)$. Some typical closed phase paths are shown in Figure 3.25, and solutions in Figure 3.26.

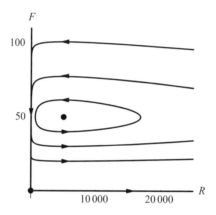

Figure 3.25 Problem 3.47: Phase diagram for $dR/dt = \alpha_1 R - \beta_1 RF$, $dF/dt = \alpha_2 RF - \beta_2 F$ with $\alpha_1 = 10$, $\beta_1 = 0.2$, $\alpha_2 = 4 \times 10^{-5}$, $\beta_2 = 0.2$.

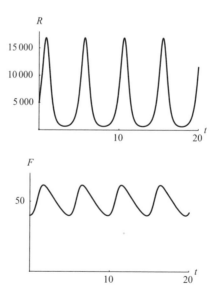

Figure 3.26 Problem 3.47: Solutions for $R(t)$ and $F(t)$ with $R(0) = 5000$ and $F(0) = 40$.

- 3.48 The system

$$\dot{x} = \frac{1}{2}\alpha\left(1 - \frac{1}{4}r^2\right)x - \frac{\omega^2 - 1}{2\omega}y,$$

$$\dot{y} = \frac{\omega^2 - 1}{2\omega}x + \frac{1}{2}\alpha\left(1 - \frac{1}{4}r^2\right)y + \frac{\Gamma}{2\omega},$$

occurs in the theory of forced oscillations of the van der Pol equation (NODE, Section 7.4, and see also Problem 3.32). Plot phase diagrams for the cases:

(i) $\alpha = 1$, $\Gamma = 0.75$, $\omega = 1.2$;
(ii) $\alpha = 1$, $\Gamma = 2.0$, $\omega = 1.6$.

3.48. The equations in the van der Pol plane for the van der Pol equation are (see Section 7.4) are

$$\dot{x} = \frac{1}{2}\alpha\left(1 - \frac{1}{4}r^2\right)x - \frac{\omega^2 - 1}{2\omega}y, \quad \dot{y} = \frac{\omega^2 - 1}{2\omega}x + \frac{1}{2}\alpha\left(1 - \frac{1}{4}r^2\right)y + \frac{\Gamma}{2\omega},$$

where $r^2 = x^2 + y^2$.

(i) Parameters $\alpha = 1$, $\Gamma = 0.75$, $\omega = 1.2$. The system has one equilibrium point at $(-0.245, -0.598)$ found numerically. Some phase paths are shown in Figure 3.27, and the phase diagram indicates an unstable equilibrium point and a stable limit cycle.

(ii) Parameters $\alpha = 1$, $\Gamma = 2$, $\omega = 1.6$. The system has one equilibrium point, at $(-0.814, -0.617)$, and a stable limit cycle. Some phase paths are shown in Figure 3.28.

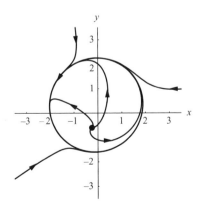

Figure 3.27 Problem 3.48(i): Phase diagram with $\alpha = 1$, $\Gamma = 0.75$, $\omega = 1.2$.

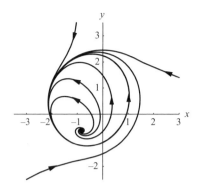

Figure 3.28 Problem 3.48(ii): Phase diagram with $\alpha = 1$, $\Gamma = 2.0$, $\omega = 1.6$.

- 3.49 The equation for a tidal bore on a shallow stream is

$$\varepsilon \frac{d^2\eta}{d\xi^2} - \frac{d\eta}{d\xi} + \eta^2 - \eta = 0.$$

where (in appropriate dimensions), η is the height of the free surface, and $\xi = x - ct$ where c is the wave speed. For $0 < \varepsilon \ll 1$, find the equilibrium points of the equation and classify them according to their linear approximations.
Plot the phase paths in the plane of η, w, where

$$\frac{d\eta}{d\xi} = w, \quad \varepsilon \frac{dw}{d\xi} = \eta + w - \eta^2$$

and show that a separatrix from a saddle point at the origin reaches the other equilibrium point. Interpret this observation in terms of the shape of the wave.

3.49. The tidal bore equation is

$$\varepsilon \frac{d^2\eta}{d\xi^2} - \frac{d\eta}{d\xi} + \eta^2 - \eta = 0.$$

Let the phase plane be (w, η) where

$$\frac{d\eta}{d\xi} = w, \quad \varepsilon \frac{dw}{d\xi} = \eta + w - \eta^2.$$

Equilibrium occurs at the points $(0, 0)$ and $(0, 1)$.

- $(0, 0)$. The linearized equations are $w' \approx (w + \eta)/\varepsilon$, $\eta' = w$, so that the origin is a saddle point.
- $(0, 1)$. Let $\eta = 1 + \eta_1$. Then the linearized equations are

$$w' = (1 + \eta_1 + w - (1 + \eta_1)^2)/\varepsilon \approx (ws - \eta_1)/\varepsilon, \quad \eta_1' = w.$$

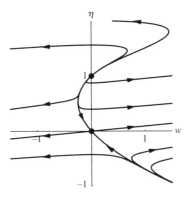

Figure 3.29 Problem 3.49: Phase diagram for $w' = (\eta + w - \eta^2)/\varepsilon$, $\eta' = w$ with $\varepsilon = 0.1$.

The usual parameters are $p = 1/\varepsilon > 0$, $q = 1/\varepsilon > 0$, $\Delta = p^2 - 4q = (1 - 4\varepsilon)/\varepsilon^2 > 0$ for $\varepsilon < \frac{1}{4}$. Therefore $\varepsilon > 0$ small $(0, 1)$ is an unstable node.

A phase diagram with the parameter $\varepsilon = 0.1$ is shown in Figure 3.29. Note that the phase paths asymptotically approach the the parabola $\eta^2 = \eta + w$, obtained by putting $\varepsilon = 0$ in the equations. A separatrix is shown joining the node to the saddle. The solution corresponding to this path represents the bore.

- **3.50** Determine the nature of the equilibrium point, and compute the phase diagrams for the Coulomb friction type problem $\ddot{x} + x = F(\dot{x})$, where

$$F(y) = \begin{cases} -6.0(y-1), & |y-1| \leq 0.4 \\ -[1 + 1.4\exp\{-0.5|y-1| + 0.2\}]\text{sgn}(y-1), & |y-1| \geq 0.4 \end{cases}$$

(See Figure 3.32 in NODE, and compare the simpler case shown in Section 1.6.)

3.50 Equilibrium occurs where $x = x_0 = F(0) = (1 + 1.4e^{1/2} + 0.2) \approx 2.04$. Let $x = x_0 + \xi$. Then, with $\dot{x} = y$,

$$\dot{y} = -x_0 - \xi + F(y) = -x_0 - \xi + [1 + 1.4e^{-0.3+0.5y}] \approx -\xi + 0.5y.$$

Therefore the equilibrium point at $(x_0, 0)$ is an unstable spiral.
The phase diagram is shown in Figure 3.32 in NODE.

- **3.51** Compute the phase diagrams for the system whose polar representations is
$\dot{r} = r(1-r)$, $\dot{\theta} = \sin^2(\frac{1}{2}\theta)$.

3.51. The polar equations are

$$\dot{r} = r(1-r), \quad \dot{\theta} = \sin^2(\tfrac{1}{2}\theta).$$

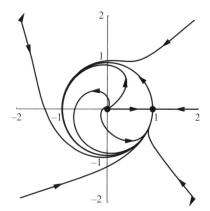

Figure 3.30 Problem 3.51:

Equilibrium occurs at $(r,\theta) = (0,0)$ and $(r,\theta) = (1,0)$. Note that $r = 1$ and $\theta = 0$, $(r > 0)$ are solutions. The phase diagram is shown in Figure 3.30.

- 3.52 Compute the phase diagrams for the following systems: (i) $\dot{x} = 2xy$, $\dot{y} = y^2 + x^2$; (ii) $\dot{x} = 2xy$, $\dot{y} = x^2 - y^2$; (iii) $\dot{x} = x^3 - 2x^2y$, $\dot{y} = 2xy^2 - y^3$.

3.52. (i) $\dot{x} = 2xy$, $\dot{y} = y^2 + x^2$. The system has one equilibrium point at the origin which is a higher-order point. The lines $x = 0$ and $y = \pm x$ are solutions. The phase diagram is shown in Figure 3.31. The origin is a hybrid node/saddle point.
(ii) $\dot{x} = 2xy$, $\dot{y} = x^2 - y^2$. The system has one equilibrium point, at the origin. The phase paths are given by

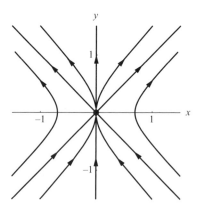

Figure 3.31 Problem 3.52(i): Phase paths for $\dot{x} = 2xy$, $\dot{y} = y^2 + x^2$.

$$\frac{dy}{dx} = \frac{x^2 - y^2}{2xy}.$$

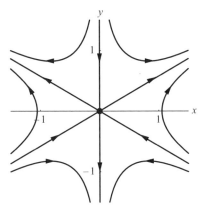

Figure 3.32 Problem 3.52(ii): Phase paths for $\dot{x} = 2xy$, $\dot{y} = x^2 - y^2$.

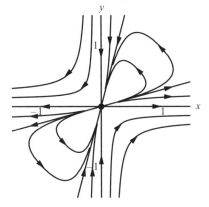

Figure 3.33 Problem 3.52(iii): Phase paths for $\dot{x} = x^3 - 2x^2y$, $\dot{y} = 2xy^2 - y^3$.

Since the equation is homogeneous try the solution $y = kx$, so that

$$k = \frac{1-k^2}{2k}, \quad \text{or } 3k^2 = 1.$$

Hence the lines $y = \pm x/\sqrt{3}$ are phase paths: the axis $x = 0$ is also a path. The computed phase diagram is shown in Figure 3.32: it can be seen that the origin is a higher-order saddle point.

(iii) $\dot{x} = x^3 - 2x^2y$, $\dot{y} = 2xy^2 - y^3$. This system has one equilibrium point, at the origin. The phase paths are given by

$$\frac{dy}{dx} = \frac{2xy^2 - y^3}{x^3 - 2x^2y}.$$

Since the equation is of first-order homogeneous type, we can try solutions of the form $y = kx$, where

$$k = \frac{2k^2 - k^3}{1 - 2k}, \quad \text{or } k^3 - 4k^2 + k = 0.$$

Hence $k = 0$ and $k = 2 \pm \sqrt{3}$. Therefore phase paths lie on the straight lines $y = 0$, $y = (2 \pm \sqrt{3})$: phase paths also lie on the x axis. Some paths in the phase diagram are shown in Figure 3.33.

- **3.53** Obtain the heteroclinic phase paths for the system $\dot{x} = y$, $\dot{y} = -x + x^3$. Show that their time solutions are given by $x = \pm \tanh\left[\frac{1}{2}\sqrt{(2)}(t - t_0)\right]$.

3.53. The system is $\dot{x} = y$, $\dot{y} = -x + x^3$. Equilibrium points occur at $(0,0)$ (a centre) and $(\pm 1, 0)$ (saddle points). A heteroclinic path is a phase path which joins one equilibrium point to another. The phase paths are given by

$$\frac{dy}{dx} = \frac{x^3 - x}{y},$$

which is a separable equation with general solution

$$\tfrac{1}{2}y^2 = -\tfrac{1}{2}x^2 + \tfrac{1}{4}x^4 + C.$$

Since there are only two saddle points and a centre, the only possible heteroclinic paths are ones which link the saddles. A phase path ends at $(1, 0)$ if $y = 0$ where $x = 1$. Therefore $C = \tfrac{1}{4}$ so that the phase paths are given by

$$y = \pm \tfrac{1}{\sqrt{2}}(1 - x^2),$$

which clearly also start at $x = -1$. They are symmetric heteroclinic paths.

The time solutions can be obtained by solving the equations

$$\frac{dx}{dt} = \pm \frac{1}{\sqrt{2}}(1 - x^2).$$

Hence

$$\int \frac{dx}{1 - x^2} = \pm \frac{1}{\sqrt{2}} \int dt = \pm \frac{1}{\sqrt{2}}(t - t_0),$$

so that (for $|x| < 1$)

$$\ln\left(\frac{1+x}{1-x}\right) = \pm \sqrt{2}(t - t_0),$$

or

$$x = \tanh\left[\pm \tfrac{1}{2}\sqrt{2}(t - t_0)\right] = \pm \tanh\left[\tfrac{1}{2}\sqrt{2}(t - t_0)\right].$$

- **3.54** Obtain the heteroclinic phase paths of $\ddot{x} + \sin x = 0$, $\dot{x} = y$. (This is a periodic differential equation in x. If the phase diagram is viewed on a cylinder of circumference 2π, then heteroclinic paths would appear to be homoclinic.)

3.54. The system is $\dot{x} = y$, $\dot{y} = -\sin x$. The equilibrium points are at $(n\pi, 0)$, $(n = 0, \pm 1, \pm 2, \ldots)$ of which centres occur at $(2m\pi, 0)$ and saddles occur at $((2m+1)\pi, 0)$ where $(m = 0, \pm 1, \pm 2, \ldots)$. The phase paths are give by

$$\frac{dy}{dx} = -\frac{\sin x}{y},$$

which can be integrated with the result

$$\tfrac{1}{2} y^2 = \cos x + C.$$

We put $y = 0$ where $x = (2m+1)\pi$ so that $C = 1$ for all such paths. Hence all heteroclinic paths are given by

$$y = \pm\sqrt{2}\sqrt{(1 + \cos x)}, \quad ((2m-1)\pi \le x \le (2m+1)\pi).$$

- **3.55** Find the homoclinic paths of $\ddot{x} - x + 3x^5 = 0$, $\dot{x} = y$. Show that the time solutions are given by $x = \pm\sqrt{[\text{sech}\,(t-t_0)]}$.

3.55. The system is $\dot{x} = y$, $\dot{y} = x - 3x^5$ has an equilibrium point at $(0,0)$ (a saddle point), and at $(\pm 3^{1/4}, 0)$ (centres). The differential equation for the phase paths is

$$\frac{dy}{dx} = \frac{x - 3x^5}{y},$$

which can be integrated to give the phase paths

$$\tfrac{1}{2} y^2 = \tfrac{1}{2} x^2 - \tfrac{1}{2} x^6 + C.$$

Homoclinic paths are given by choosing $C = 0$, that is,

$$y^2 = x^2 - x^6,$$

and they intersect the x axis at $x = \pm 1$.
Time solutions can be found by integrating

$$y = \frac{dx}{dt} = \pm x\sqrt{(1-x^4)},$$

which separates into

$$\int \frac{dx}{x\sqrt{(1-x^4)}} = \pm \int dt = \pm(t - t_0).$$

Using the substitution $u = 1/x^2$, the integral becomes

$$-\int \frac{du}{2\sqrt{(u^2-1)}} = \pm(t-t_0).$$

Hence, for $x > 0$,

$$-\cosh^{-1} u = \pm(t-t_0).$$

Both signs give the same result $u = \cosh[2(t-t_0)]$. For $x > 0$ the homoclinic path is given by

$$x = \sqrt{\{\text{sech}\,[2(t-t_0)]\}}.$$

In a similar manner, the homoclinic path for $x < 0$ is

$$x = -\sqrt{\{\text{sech}\,[2(t-t_0)]\}}.$$

- **3.56** Find all heteroclinic phase paths of $\dot{x} = y(1-x^2)$, $\dot{y} = -x(1-y^2)$ (see NODE, Example 2.1).

3.56. As in Example 2.1, the system is $\dot{x} = y(1-x^2)$, $\dot{y} = -x(1-y^2)$. It was shown that the equations have five equilibrium points, at $(0,0)$ and $(\pm 1, \pm 1)$. From Figure 2.1 (in NODE), it can be seen that $(0,0)$ is a centre, and the four points $(\pm 1, \pm 1)$ are all saddle points. The equation defining the phase paths is $(1-x^2)(1-y^2) = C$. The heteroclinic paths are the straight lines

$$x = \pm 1, \; (-1 \le y \le 1), \quad y = \pm 1, \; (-1 \le x \le 1).$$

- **3.57** The problem of the bead sliding on a rotating wire was discussed in Example 1.12, where it was shown that the equation of motion of the bead is

$$a\ddot{\theta} = g(\lambda \cos\theta - 1)\sin\theta.$$

Find the equations of all homoclinic and heteroclinic paths, carefully distinguishing the cases $0 < \lambda < 1$, $\lambda = 1$ and $\lambda > 1$.

3.57. The angle θ giving the inclination of a bead sliding on a rotating wire is given by

$$a\ddot{\theta} = g(\lambda \cos\theta - 1)\sin\theta,$$

(see NODE, Example 1.12). Equilibrium points occur where

- for $\lambda \le 1$: at $\theta = n\pi$, $(n = 0, \pm 1, \pm 2, \ldots)$, which are centres for $n = 0, \pm 2, \pm 4, \ldots$, and saddles for $n = \pm 1, \pm 3, \ldots$;
- for $\lambda > 1$: at $\theta = n\pi$, $(n = 0, \pm 1, \pm 2, \ldots)$, which are saddles for $n = 0, \pm 2, \pm 4, \ldots$, and centres for $n = \pm 1, \pm 3, \ldots$; also there are centres where $\cos\theta = 1/\lambda$.

From Example 1.12, the phase paths are given by

$$\tfrac{1}{2}a\dot\theta^2 = g(1 - \tfrac{1}{2}\lambda\cos\theta)\cos\theta + C. \tag{i}$$

The phase diagrams shown in Figure 1.30 (in NODE) should be consulted.

- $\lambda < 1$. There are no homoclinic paths, but there are heteroclinic paths which connect two saddle points on either side of a centre. Hence the heteroclinic paths must pass through $\dot\theta = 0$ at $\theta = n\pi$ where $n = \pm 1, \pm 3, \dots$. Hence, from (i),

$$0 = -g(1 + \tfrac{1}{2}\lambda) + C, \text{ so that } C = g(1 + \tfrac{1}{2}\lambda).$$

The heteroclinic paths are given by

$$\tfrac{1}{2}a\dot\theta^2 = g(1 - \tfrac{1}{2}\lambda\cos\theta)\cos\theta + g(1 + \tfrac{1}{2}\lambda).$$

- $\lambda = 1$. There are no homoclinic paths, whilst the heteroclinic paths pass through $\dot\theta = 0$ at $\theta = n\pi$, where $n = \pm 1, \pm 3, \dots$. Hence, from (i),

$$0 = -\tfrac{3}{2}g + C, \text{ so that } C = \tfrac{3}{2}g.$$

The heteroclinic paths are given by

$$\tfrac{1}{2}a\dot\theta^2 = g(1 - \tfrac{1}{2}\cos\theta)\cos\theta + \tfrac{3}{2}g.$$

- $\lambda > 1$. Homoclinic paths are given by putting $\dot\theta = 0$ at $\theta = n\pi$ for $n = 0, \pm 2, \pm 4, \dots$. Hence

$$0 = g(1 - \tfrac{1}{2}\lambda\cos\theta)\cos\theta + C, \text{ so that } C = -g(1 - \tfrac{1}{2}\lambda).$$

The homoclinic paths are given by

$$\tfrac{1}{2}a\dot\theta^2 = g(1 - \tfrac{1}{2}\lambda\cos\theta)\cos\theta - g(1 - \tfrac{1}{2}\lambda).$$

Heteroclinic paths connect the saddles at $\cos\theta = 1/\lambda$. Therefore from (i) $C = -g/(2\lambda)$. The heteroclinic paths are given by

$$\tfrac{1}{2}a\dot\theta^2 = g\left(1 - \tfrac{1}{2}\lambda\cos\theta\right)\cos\theta - \frac{g}{2\lambda}.$$

- **3.58** Consider the equation $\ddot x - x(x-a)(x-b) = 0$, $0 < a < b$. Find the equation of its phase paths. Show that a heteroclinic bifurcation occurs in the neighbourhood of $b = 2a$. Draw sketches showing the homoclinic paths for $b < 2a$ and $b > 2a$.

Show that the time solution for the heteroclinic path ($b = 2a$) is

$$x = \frac{2a}{1 + e^{-a\sqrt{2}(t-t_0)}}.$$

3.58. Consider the system

$$\ddot{x} - x(x-a)(x-b) = 0, \quad 0 < a < b.$$

Assuming that $\dot{x} = y$, the system has equilibrium points at $(0,0)$ (a saddle), $(a,0)$ (a centre) and $(b, 0)$ (a saddle). Phase paths are given by

$$\tfrac{1}{2}y^2 - \int [x^3 - (a+b)x^2 + abx]dx = C,$$

or

$$\tfrac{1}{2}y^2 - \tfrac{1}{4}x^4 + \tfrac{1}{3}(a+b)x^3 - \tfrac{1}{2}abx^2 = C.$$

For the separatrices through the saddle $(0,0)$, $C = 0$, so that they are given by

$$\tfrac{1}{2}y^2 - \tfrac{1}{4}x^4 + \tfrac{1}{3}(a+b)x^3 - \tfrac{1}{2}abx^2 = 0. \tag{i}$$

This is generally a homoclinic path, but is heteroclinic if it connects with the saddle at $(b, 0)$. This occurs if the point $(b, 0)$ also lies on (i), that is, if

$$-\tfrac{1}{4}b^4 + \tfrac{1}{3}(a+b)b^3 - \tfrac{1}{2}ab^3 = 0,$$

which implies that $b = 2a$. The equation of the heteroclinic path is

$$\tfrac{1}{2}y^2 - \tfrac{1}{4}x^4 + ax^3 - a^2x^2 = 0, \quad \text{or } y^2 - \tfrac{1}{2}x^2(x - 2a)^2 = 0, \tag{ii}$$

which is shown in Figure 3.35. For $b > 2a$ the path (i) is homoclinic to the origin as shown in Figure 3.34, but is not homoclinic there if $b \geq 2a$. However, if $b \leq 2a$ there is a path which is homoclinic to the saddle point at $(b, 0)$ as illustrated in Figure 3.36. From (ii), the equation for the heteroclinic solutions is

$$\frac{dx}{dt} = \pm \frac{1}{\sqrt{2}} x(x - 2a).$$

The equation with the minus sign applies to the heteroclinic path in $y > 0$. Thus

$$\int \frac{dx}{x(2a-x)} = \frac{1}{\sqrt{2}} \int dt,$$

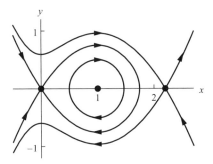

Figure 3.34 Problem 3.58: phase diagram for $\dot{x} = y$, $\dot{y} = x(x-a)(x-b)$ with $a = 1, b = 2.2$.

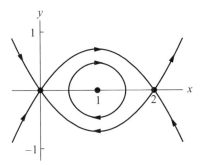

Figure 3.35 Problem 3.58: Phase diagram for $\dot{x} = y$, $\dot{y} = x(x-a)(x-b)$ with $a = 1, b = 2$.

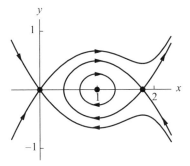

Figure 3.36 Problem 3.58: Phase diagram for $\dot{x} = y$, $\dot{y} = x(x-a)(x-b)$ with $a = 1, b = 1.8$.

which after integration becomes

$$\frac{1}{2a}\ln\left(\frac{x}{2a-x}\right) = \frac{1}{\sqrt{2}}(t-t_0).$$

Solving for x:

$$x = \frac{2a}{1+e^{-a\sqrt{2}(t-t_0)}}.$$

- **3.59** Show that
$$\dot{x} = 4(x^2 + y^2)y - 6xy, \quad \dot{y} = 3y^2 - 3x^2 - 4x(x^2 + y^2)$$
has a higher-order saddle at the origin (neglect the cubic terms for $\dot{x}$ and $\dot{y}$, and show that near the origin the saddle has solutions in the directions of the straight lines $y = \pm x/\sqrt{3}$, $x = 0$. Confirm that the phase paths through the origin are given by
$$(x^2 + y^2)^2 = x(3y^2 - x^2).$$
By plotting this curve, convince yourself that three homoclinic paths are associated with the saddle point at the origin.

3.59 The system is
$$\dot{x} = 4(x^2 + y^2)y - 6xy, \quad \dot{y} = 3y^2 - 3x^2 - 4x(x^2 + y^2). \tag{i}$$

Equilibrium occurs where
$$4(x^2 + y^2)y - 6xy = 0, \text{ and } 3y^2 - 3x^2 - 4x(x^2 + y^2) = 0.$$

Switch to polar coordinates (r, θ), so that the equations become
$$r^2 \sin\theta (2r - 3\cos\theta) = 0, \quad 3r^2(\sin^2\theta - \cos^2\theta) - 4r^3 \cos\theta = 0.$$

From the first equation either $r = 0$ (the origin) or $\sin\theta = 0$ or $r = \frac{3}{2}\cos\theta$. Since $r = 0$ also satisfies the second equation, then $(0, 0)$ is an equilibrium point. If $\sin\theta = 0$, then $\theta = 0$ or $\theta = \pi$, but only $\theta = \pi$ gives a positive value $\frac{3}{4}$ for r. Substitute $r = \frac{3}{2}\cos\theta$ into the second equation so that
$$(\sin^2\theta - \cos^2\theta) - 2\cos^2\theta = 0, \text{ or } \cos\theta = \frac{1}{2},$$
for r to be positive. Hence $\theta = \frac{1}{3}\pi$ or $\theta = \frac{5}{3}\pi$. For these angles $r = \frac{3}{2}\cos\theta = \frac{3}{4}$. To summarize, the system has the equilibrium points $(0,0)$, $(-\frac{3}{4}, 0)$, $(\frac{3}{8}, \frac{3\sqrt{3}}{8})$ and $(\frac{3}{8}, -\frac{3\sqrt{3}}{8})$

Near the origin
$$\dot{x} \approx -6xy, \quad \dot{y} \approx 3y^2 - 3x^2.$$
One solution is $x = 0$. Put $y = kx$; then
$$\frac{dy}{dx} = \frac{\dot{y}}{\dot{x}} = -\frac{3y^2 - 3x^2}{6xy},$$
becomes
$$k = -\frac{3k^2 - 3}{6k},$$
from which it follows that $k = \pm 1/\sqrt{3}$. Hence locally the separatrices of the origin are in the directions of the lines $x = 0$ and $y = \pm x/\sqrt{3}$. The phase diagram displayed in Figure 3.37 shows

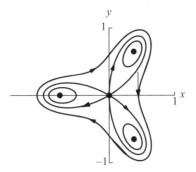

Figure 3.37 Problem 3.59: Phase diagram for $\dot{x} = 4(x^2 + y^2)y - 6xy$, $\dot{y} = 3y^2 - 3x^2 - 4x(x^2 + y^2)$.

three homoclinic paths starting from the origin, each surrounding a centre. The phase diagram was computed by numerical solution of the differential equations although the equation can be solved as follows. From (i)

$$\frac{dy}{dx} = \frac{3y^2 - 3x^2 - 4x(x^2 + y^2)}{4(x^2 + y^2)y - 6xy},$$

which can be expressed in the form

$$4y(x^2 + y^2)\frac{dy}{dx} + 4x(x^2 + y^2) = 6xy\frac{dy}{dx} + 3y^2 - 3x^2.$$

This an exact differential equation equivalent to

$$\frac{d}{dx}[(x^2 + y^2)^2] = \frac{d}{dx}(3xy^2 - x^3).$$

Integrating the phase paths are given by

$$(x^2 + y^2)^2 = 3xy^2 - x^3 + C.$$

- 3.60 Investigate the equilibrium points of
$\dot{x} = y[16(2x^2 + 2y^2 - x) - 1]$, $\dot{y} = x - (2x^2 + 2y^2 - x)(16x - 4)$,
and classify them according to their linear approximations. Show that homoclinic paths through $(0, 0)$ are given by
$(x^2 + y^2 - \frac{1}{2}x)^2 - \frac{1}{16}(x^2 + y^2) = 0$,
and that one homoclinic path lies within the other.

3.60. The system is

$$\dot{x} = P(x, y) = y[16(2x^2 + 2y^2 - x) - 1],$$
$$\dot{y} = Q(x, y) = x - (2x^2 + 2y^2 - x)(16x - 4),$$

say. The equilibrium points are given by

$$y[16(2x^2 + 2y^2 - x) - 1] = 0, \quad x - (2x^2 + 2y^2 - x)(16x - 4) = 0. \tag{i}$$

From the first equation either $y = 0$ or $16(2x^2 + 2y^2 - x) = 1$. Substituted into the second equation, $y = 0$ leads to

$$x(32x^2 - 24x + 3)x = 0.$$

The solutions are $x = 0$, $x = (3 \pm \sqrt{3})/8$. The other equation is inconsistent with the second equation in (i). Therefore, there are three equilibrium points at $(0, 0)$, $((3 + \sqrt{3})/8, 0)$ and $((3 - \sqrt{3})/8, 0)$.

The phase paths are given by the differential equation

$$\frac{dy}{dx} = \frac{Q(x, y)}{P(x, y)} = \frac{x - (2x^2 + 2y^2 - x)(16x - 4)}{y[16(2x^2 + 2y^2 - x) - 1]}.$$

This is an exact equation since it can be verified that $\partial Q/\partial y = -\partial P/\partial x$. Hence there exists a (Hamiltonian) function $H(x, y)$ such that

$$P(x, y) = y[16(2x^2 + 2y^2 - x) - 1] = \frac{\partial H}{\partial y},$$

$$Q(x, y) = x - (2x^2 + 2y^2 - x)(16x - 4) = -\frac{\partial H}{\partial x}.$$

Integration with respect to y and x of these partial derivatives leads to

$$H(x, y) = 16x^2 y^2 + 8y^4 - 8xy^2 - \tfrac{1}{2}y^2 + f(x),$$

and

$$H(x, y) = 8x^4 + 16x^2 y^2 - 8x^3 - 8xy^2 + \tfrac{3}{2}x^2 + g(y).$$

These equations match if $f(x) = -8x^3 + \tfrac{3}{2}x^2$ and $g(y) = -8y^4 - y$, so that

$$H(x, y) = 16x^2 y^2 - 8xy^2 + 8y^4 - \tfrac{1}{2}y^2 + 8x^4 - 8x^3 + \tfrac{3}{2}x^2$$
$$= 8(x^2 + y^2 - \tfrac{1}{2}x)^2 - \tfrac{1}{2}(x^2 + y^2).$$

The phase paths are given by $H(x, y) = C$, a constant.

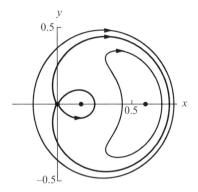

Figure 3.38 Problem 3.60: Phase diagram for $\dot{x} = P(x, y) = y[16(2x^2 + 2y^2 - x) - 1]$, $\dot{y} = Q(x, y) = x - (2x^2 + 2y^2 - x)(16x - 4)$.

For the phase paths through the origin, the constant $C = 0$, and the homoclinic paths are given by
$$16(x^2 + y^2 - \tfrac{1}{2}x)^2 = x^2 + y^2.$$
The paths are symmetric about the x axis and meet it where $y = 0$, that is, where
$$16(x^2 - \tfrac{1}{2}x)^2 = x^2,$$
or, where
$$16x^4 - 16x^3 + 3x^2 = 0.$$
The points where the homoclinic paths intersect the x axis are at $x = 0$, $x = \tfrac{1}{4}$ and $x = \tfrac{3}{4}$, the latter two values being positive, which means that one homoclinic path is within the other as shown in Figure 3.38.

- **3.61** The following model differential equation exhibits two limit cycles bifurcating through homoclinic paths into a single limit cycle of larger amplitude as the parameter ε decreases through zero:
$$\ddot{x} + (\dot{x}^2 - x^2 + \tfrac{1}{2}x^4 + \varepsilon)\dot{x} - x + x^3 = 0.$$
Let $|\varepsilon| < \tfrac{1}{2}$.

(a) Find and classify the equilibrium points of the equation.

(b) Confirm that the equation has phase paths given by
$$y^2 = x^2 - \tfrac{1}{2}x^4 - \varepsilon, \quad y = \dot{x}.$$
Find where the paths cut the x axis.

(c) As ε decreases through zero what happens to the limit cycles which surround the equilibrium point at $x = \pm 1$? (It could be quite helpful to plot phase paths numerically for a sample of ε values.) Are they all stable?

3.61. The equation

$$\ddot{x} + (\dot{x}^2 - x^2 + \tfrac{1}{2}x^4 + \varepsilon)\dot{x} - x + x^3 = 0$$

has a nonlinear friction term and a nonlinear restoring action. Assuming that $\dot{x} = y$, the system has equilibrium points at $(0,0)$ and $(\pm 1, 0)$.

(a) Classification of equilibrium points for $|\varepsilon| < \tfrac{1}{2}$.

- $(0,0)$. The linearized approximation is $\dot{x} = y$, $\dot{y} \approx x - \varepsilon y$. The parameters are (in the notation of Section 2.5)

$$p = \varepsilon, \quad q = -1 < 0, \quad \Delta = \varepsilon^2 + 4 > 0.$$

Therefore $(0,0)$ is a saddle point.

- $(-1, 0)$. Let $x = -1 + \xi$. Then the linearized approximation is

$$\dot{\xi} = y, \quad \dot{y} \approx -2\xi + (\tfrac{1}{2} - \varepsilon)y.$$

The parameters are

$$p = \tfrac{1}{2} - \varepsilon > 0, \quad q = 2 > 0, \quad \Delta = (\tfrac{1}{2} - \varepsilon)^2 - 8 < 0.$$

Hence $(-1, 0)$ is an unstable spiral.

- $(1, 0)$. Let $x = 1 + \xi$. Then the linearized approximation is

$$\dot{x} = y, \quad \dot{y} \approx -2\xi + (\tfrac{1}{2} - \varepsilon)y,$$

as in the previous case. Hence $(1, 0)$ is also an unstable spiral.

(b) If $y^2 = x^2 - \tfrac{1}{2}x^4 - \varepsilon$, then the coefficient of $\dot{x}$ in the differential equation is zero. Also, differentiation with respect to t gives

$$2\dot{x}\ddot{x} = 2x\dot{x} - 2x^3\dot{x},$$

or $\ddot{x} = x - x^3$, $(\dot{x} \neq 0)$. Hence $y^2 = x^2 - \tfrac{1}{2}x^4 - \varepsilon$ is a particular solution. When $y = 0$,

$$x^4 - 2x^2 + 2\varepsilon = 0,$$

which has the solutions

$$x^2 = 1 \pm \sqrt{(1 - 2\varepsilon)}.$$

The equation has four real solutions if $0 < \varepsilon < \tfrac{1}{2}$, and two real solutions if $-\tfrac{1}{2} < \varepsilon \leq 0$.

(c) If $\varepsilon = 0$, then the equation becomes

$$\ddot{x} + \left(\dot{x}^2 - x^2 + \tfrac{1}{2}x^4\right)\dot{x} - x + x^3 = 0,$$

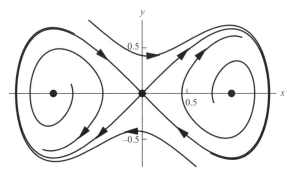

Figure 3.39 Problem 3.61: Phase diagram for $\dot{x} = y$, $\dot{y} = -(y^2 - x^2 + \frac{1}{2}x^4 + \varepsilon)y + x - x^3$ with $\varepsilon = 0$.

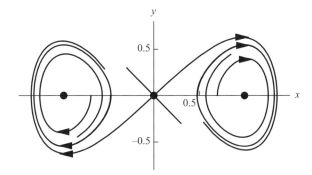

Figure 3.40 Problem 3.61: Phase diagram for $\dot{x} = y$, $\dot{y} = -(y^2 - x^2 + \frac{1}{2}x^4 + \varepsilon)y + x - x^3$ with $\varepsilon = 0.2$.

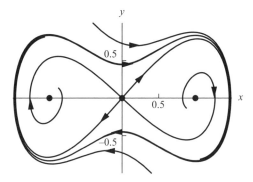

Figure 3.41 Problem 3.61: Phase diagram for $\dot{x} = y$, $\dot{y} = -(y^2 - x^2 + \frac{1}{2}x^4 + \varepsilon)y + x - x^3$ with $\varepsilon = -0.2$.

which has the particular phase paths given by $y^2 = x^2 - \frac{1}{2}x^4$: these are homoclinic paths through the origin a shown in Figure 3.39.

If ε increases from zero, then periodic orbits around the equilibrium points at $(-1, 0)$ and $(1, 0)$ develop in this process of bifurcation. Stable closed paths for $\varepsilon = 0.2$ are shown in Figure 3.40.

If ε decreases from zero, then a stable closed path around both spirals and the saddle point at the origin bifurcates from the homoclinic paths as shown in Figure 3.41 with $\varepsilon = -0.2$.

- **3.62** Classify the equilibrium points of $\ddot{x} = x - 3x^2$, $\dot{x} = y$. Show that the equation has one homoclinic path given by $y^2 = x^2 - 2x^3$. Solve this equation to obtain the (x,t) solution for the homoclinic path.

3.62. The system is $\dot{x} = y$, $\dot{y} = x - 3x^2$. It has two equilibrium points, at $(0,0)$ and $(\frac{1}{3}, 0)$.

- $(0,0)$. The linearized approximation is $\dot{x} = y$, $\dot{y} = x$ which implies that $(0,0)$ is a saddle point.
- $(\frac{1}{3}, 0)$. Let $x = \frac{1}{3} + \xi$. Then the linearized approximation is

$$\dot{\xi} = y, \quad \dot{y} = \frac{1}{3} + \xi - 3(\frac{1}{3} + \xi)^2 \approx -\xi.$$

Hence $(\frac{1}{3}, 0)$ is a centre.

The equation for the phase paths is

$$y \frac{dy}{dx} = x - 3x^2,$$

which can be integrated to give the family of phase paths as

$$y^2 = x^2 - 2x^3 + C.$$

For the paths through the saddle point at the origin, $C = 0$, so that the homoclinic path is given by

$$y^2 = x^2 - 2x^3,$$

which lies in $x > 0$.

The time solution satisfies

$$\frac{dx}{dt} = x\sqrt{(1 - 2x)}.$$

Let $u^2 = 1 - 2x$. Then the differential equation becomes

$$-u \frac{du}{dt} = \frac{u}{\sqrt{2}} (1 - u^2).$$

Separating the variables and integrating,

$$-\int \frac{du}{1 - u^2} = \frac{1}{\sqrt{2}} dt = \frac{t}{\sqrt{2}} + C.$$

3 : Geometrical aspects of plane autonomous systems   193

Hence
$$-\frac{1}{2}\ln\left(\frac{1+u}{1-u}\right) = \frac{1}{\sqrt{2}}t + C$$

or
$$u = \frac{1-e^{\sqrt{2}t}}{1+e^{\sqrt{2}t}} = -\tanh\left(\frac{1}{2}\sqrt{2}t\right).$$

Finally,
$$x = \tfrac{1}{2}(1-u^2) = \tfrac{1}{2}[1-\tanh^2(\tfrac{1}{2}\sqrt{2}t)] = \tfrac{1}{2}\operatorname{sech}^2(\tfrac{1}{2}\sqrt{2}t).$$

- **3.63** Classify all equilibrium points of $\dot{x} = y(2y^2 - 1)$, $\dot{y} = x(x^2 - 1)$ according to their linear approximations. Show that the homoclinic paths are given by $2y^2(y^2 - 1) = x^2(x^2 - 2)$, and that the heteroclinic paths lie on the ellipse $x^2 + \sqrt{2}y^2 = \tfrac{1}{2}(2+\sqrt{2})$ and the hyperbola $x^2 - \sqrt{2}y^2 = \tfrac{1}{2}(2-\sqrt{2})$. Sketch the phase diagram.

3.63. The system is $\dot{x} = X(x,y) = y(2y^2 - 1)$, $\dot{y} = Y(x,y) = x(x^2 - 1)$. There are nine equilibrium points at $(0,0)$, $(0,\pm 1/\sqrt{2})$, $(\pm 1,0)$, $(\pm 1, \pm 1/\sqrt{2})$. Since

$$\frac{\partial X}{\partial x} + \frac{\partial Y}{\partial y} = 0,$$

the system is Hamiltonian, which implies that the equilibrium points will be either centres or saddle points (Section 2.8).

The classification of the equilibrium points is as follows.

- $(0,0)$. The linearized approximation is $\dot{x} = -y$, $\dot{y} = -x$. Hence $(0,0)$ is a saddle point.
- $(0, \pm 1/\sqrt{2})$. Let $y = \pm\frac{1}{\sqrt{2}} + \eta$. Then the linearized approximation is $\dot{x} = 2\eta$, $\dot{\eta} = -x$, so that $(0, \pm 1/\sqrt{2})$ are centres.
- $(\pm 1, 0)$. Let $x = \pm 1 + \xi$. Then the linearized approximation is $\dot{\xi} = -y$, $\dot{y} = 2\xi$. Therefore $(\pm 1, 0)$ are centres.
- $(\pm 1, \pm 1/\sqrt{2})$ (all combinations of signs). Let $x = \pm 1 + \xi$ and $y = \pm\frac{1}{\sqrt{2}} + \eta$. Then the linearized approximation is

$$\dot{\xi} = 2\eta, \quad \dot{\eta} = 2\xi.$$

in each case. Elimination of (say) η leads to $\ddot{\xi} - 4\xi = 0$ in all cases. Hence all these points are saddles.

Since the Hamiltonian $H(x,y)$ satisfies both $\partial H/\partial x = -x(x^2 - 1)$ and $\partial H/\partial y = y(2y^2 - 1)$, it is obvious that
$$H(x,y) = -\tfrac{1}{4}x^4 + \tfrac{1}{2}x^2 + \tfrac{1}{2}y^4 - \tfrac{1}{2}y^2.$$

The general equation of the phase paths is therefore
$$-x^4 + 2x^2 + 2y^4 - 2y^2 = C.$$

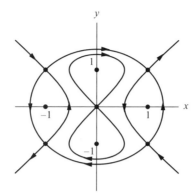

Figure 3.42 Problem 3.63: Heteroclinic and homoclinic paths for $\dot{x} = y(2y^2 - 1)$, $\dot{y} = x(x^2 - 1)$.

The saddle point at the origin has homoclinic paths given by $C = 0$, or

$$2y^2(y^2 - 1) = x^2(x^2 - 2).$$

The other saddle points at $(\pm 1, \pm 1/\sqrt{2})$ have the heteroclinic paths with $C = \frac{1}{2}$, or

$$-x^4 + 2x^2 + 2y^4 - 2y^2 = \frac{1}{2},$$

which factorizes as

$$[x^2 + \sqrt{2}y^2 - \tfrac{1}{2}(\sqrt{2} + 2)][x^2 - \sqrt{2}y^2 - \tfrac{1}{2}(2 - \sqrt{2})] = 0.$$

The heteroclinic paths lie on the ellipse

$$x^2 + \sqrt{2}y^2 = \tfrac{1}{2}(\sqrt{2} + 2),$$

and the hyperbola

$$x^2 - \sqrt{2}y^2 = \tfrac{1}{2}(2 - \sqrt{2}).$$

The paths are shown in Figure 3.42: there are centres at the enclosed equilibrium points.

- **3.64** A dry friction model has the equation of motion $\ddot{x} + x = F(\dot{x})$ where

$$F(y) = \begin{cases} -\mu(y - 1) & |y - 1| \leq \varepsilon \\ -\mu\varepsilon\,\text{sgn}\,(y - 1) & |y - 1| > \varepsilon, \end{cases}$$

where $0 < \varepsilon < 1$ (see Figure 3.33 in NODE). Find the equations of the phase paths in each of the regions $y > 1 + \varepsilon$, $1 - \varepsilon \leq y \leq 1 + \varepsilon$, $y < 1 - \varepsilon$.

3.64. The dry friction has the equation of motion

$$\ddot{x} + x = F(\dot{x}),$$

where

$$F(y) = \begin{cases} -\mu(y-1) & |y-1| \leq \varepsilon \\ -\mu\varepsilon\,\text{sgn}\,(y-1) & |y-1| > \varepsilon, \end{cases}$$

where $0 < \varepsilon < 1$. Equilibrium occurs where $x = F(0)$, that is, where $x = \mu\varepsilon$.
The equations of the phase paths are as follows:

- $y > 1+\varepsilon$. The equation of motion is $\ddot{x}+x = -\mu\varepsilon$, so that the phase paths are solutions of

$$y\frac{dy}{dx} = -x - \mu\varepsilon.$$

Hence the phase paths are arcs of circles given by

$$y^2 + (x + \mu\varepsilon)^2 = A.$$

- $1-\varepsilon \leq y \leq 1+\varepsilon$. The equation of motion is $\ddot{x}+x = -\mu(\dot{x}-1)$, or,

$$\ddot{x} + \mu\dot{x} + x = \mu.$$

The solution of this equation is

$$x = Ae^{m_1 t} + Be^{m_2 t} + \mu,$$

where

$$m_1, m_2 = \tfrac{1}{2}[-\mu \pm \sqrt{(\mu^2 - 4)}].$$

Hence the phase diagram is that of a node, if $\mu > 2$, or a spiral if $0 < \mu < 2$, both stable and centred at $(\mu, 0)$.

- $y < 1-\varepsilon$. The equation of motion is $\ddot{x}+x = \mu\varepsilon$, so that the phase paths are solutions of

$$y\frac{dy}{dx} = -x + \mu\varepsilon.$$

Hence the phase paths are circles and arcs of circles given by

$$y^2 + (x - \mu\varepsilon)^2 = B.$$

The matching of the three phase diagrams is shown in Figure 3.43.

- **3.65** Locate and classify the equilibrium points of
$$\dot{x} = x^2 - 1, \quad \dot{y} = -xy + \varepsilon(x^2 - 1).$$
Find the equations of all phase paths. Show that the separatrices in $|x| < 1$ which approach $x = \pm 1$ are given by $y = \varepsilon\left[\tfrac{1}{2}x\sqrt{(1-x^2)} + \tfrac{1}{2}\sin^{-1}x \mp \tfrac{1}{4}\pi\right]/\sqrt{(1-x^2)}$.

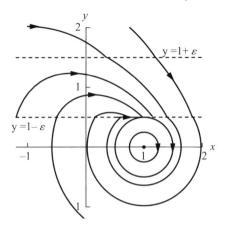

Figure 3.43 Problem 3.64: Dry friction phase diagram with $\varepsilon = 0.5$ and $\mu = 1$.

Sketch typical solutions for $\varepsilon >, =, < 0$, and confirm that a heteroclinic bifurcation occurs at $\varepsilon = 0$.

Show that the displacement $d(x)$ in the y direction between the separatrices for $-1 < x < 1$ is given by

$$d(x) = \frac{\pi\varepsilon}{2\sqrt{(1-x^2)}}.$$

(This displacement is zero when $\varepsilon = 0$ which shows that the separatrices become a heteroclinic path joining $(1, 0)$ and $(-1, 0)$ at this value of ε. This separatrix method is the basis of Melnikov's perturbation method in Chapter 13 for detecting homoclinic and heteroclinic bifurcations.)

3.65. There are two equilibrium points of

$$\dot{x} = x^2 - 1, \quad \dot{y} = -xy + \varepsilon(x^2 - 1),$$

at $(-1, 0)$ and $(1, 0)$.

- $(-1, 0)$. Let $x = -1 + \xi$. Then the linear approximation is

$$\dot{\xi} = (-1+\xi)^2 - 1 \approx -2\xi, \quad \dot{y} = -(-1+\xi)y + \varepsilon[(-1+\xi)^2 - 1] \approx -2\varepsilon x + y.$$

Since the parameter $q = -2$ (Section 2.5), $(-1, 0)$ is a saddle point.
- $(1, 0)$. Let $x = 1 + \xi$. Then the linear approximation is

$$\dot{\xi} = (1+\xi)^2 - 1 \approx 2\xi, \quad \dot{y} = -(1+\xi)y + \varepsilon[(1+\xi)^2 - 1] \approx 2\varepsilon x - y.$$

Since the parameter $q = -2$ (Section 2.5), $(1, 0)$ is also a saddle point.

The equation for the phase paths is

$$\frac{dy}{dx} = \frac{-xy + \varepsilon(x^2 - 1)}{x^2 - 1} = -\frac{x}{x^2 - 1}y + \varepsilon.$$

This is a first-order equation with integrating factor

$$\exp\left[\int \frac{x\,dx}{x^2 - 1}\right] = \exp\left[\frac{1}{2}\ln|x^2 - 1|\right] = \sqrt{(1 - x^2)},$$

for $|x| < 1$. Hence it can be expressed in the form

$$\frac{d}{dx}[y\sqrt{(1 - x^2)}] = \varepsilon\sqrt{(1 - x^2)}.$$

Integrating

$$y\sqrt{(1 - x^2)} = \varepsilon \int \sqrt{(1 - x^2)}\,dx = \frac{1}{2}\varepsilon[x\sqrt{(1 - x^2)} + \arcsin x] + A. \qquad \text{(i)}$$

If $|x| > 1$, the integrating factor is $\sqrt{(x^2 - 1)}$, by similar arguments. The general solution is given by

$$y\sqrt{(x^2 - 1)} = \varepsilon \int \sqrt{(x^2 - 1)}\,dx = \frac{1}{2}\varepsilon[x\sqrt{(x^2 - 1)} - \cosh^{-1} x] + B. \qquad \text{(ii)}$$

Note also that $x = \pm 1$ are particular solutions which are separatrices of the saddle points. For $|x| < 1$, for the other separatrix through $(-1, 0)$, $A = \frac{1}{4}\varepsilon\pi$ in (i) so that it has the equation

$$y_1(x)\sqrt{(1 - x^2)} = \frac{1}{2}\varepsilon[x\sqrt{(1 - x^2)} + \arcsin x] + \frac{1}{4}\varepsilon\pi.$$

By a similar argument the separatrix through $(1, 0)$ is, for $|x| < 1$ is

$$y_2(x)\sqrt{(1 - x^2)} = \frac{1}{2}\varepsilon[x\sqrt{(1 - x^2)} + \arcsin x] - \frac{1}{4}\varepsilon\pi.$$

If $\varepsilon = 0$, then $y = 0$ is a solution for all x, and this solution is a heteroclinic path since it connects the two saddle points.

The displacement $d(x)$ between the separatrices $y_1(x)$ and $y_2(x)$ in the y direction is

$$d(x) = y_1(x) - y_2(x) = \frac{\varepsilon\pi}{2\sqrt{(1-x^2)}}.$$

The heteroclinic bifurcation between the equilibrium points at $(-1,0)$ and $(1,0)$ is shown in the sequence of Figures 3.44, 3.45, 3.46 as ε decreases through zero.

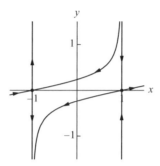

Figure 3.44 Problem 3.65: This shows the separatrices for $\varepsilon = 0.3$.

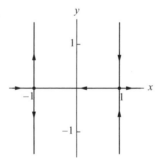

Figure 3.45 Problem 3.65: This shows the heteroclinic path for $\varepsilon = 0$.

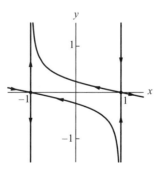

Figure 3.46 Problem 3.65: This shows the separatrices for $\varepsilon = -0.3$.

3: Geometrical aspects of plane autonomous systems

• **3.66** Classify the equilibrium points of the system $\dot{x} = y$, $\dot{y} = x(1-x^2) + ky^2$, according to their linear approximations. Find the equations of the phase paths, and show that, if $k = -\sqrt{(3/2)}$, then there exists a homoclinic path given by $y^2 = x^2(1 - \sqrt{(2/3)}x)$ in $x > 0$. Show that the time solution is given by $x = \sqrt{(\tfrac{3}{2})}\operatorname{sech}^2 \tfrac{1}{2}(t-t_0)$.

3.66. The system $\dot{x} = y$, $\dot{y} = x(1-x^2) + ky^2$ has equilibrium points at $(0,0)$, $(1,0)$ and $(-1,0)$.

- $(0,0)$. The linear approximation is $\dot{x} = y$, $\dot{y} = x$. Hence the origin is a saddle point.
- $(1,0)$. Let $x = 1 + \xi$. Then the linear approximation is

$$\dot{\xi} = y, \quad \dot{y} = (1+\xi)[1 - (1+\xi)^2] + ky^2 \approx -2\xi,$$

for small $|\xi|$. Hence $(1,0)$ is a centre.

- $(-1,0)$. Let $x = -1 + \xi$. Then the linear approximation is

$$\dot{\xi} = y, \quad \dot{y} = (-1+\xi)[1 - (-1+\xi)^2] + ky^2 \approx -2\xi,$$

so that $(-1,0)$ is also a centre.

The differential equation for the phase paths is given by

$$\frac{dy}{dx} = \frac{x(1-x^2) + ky^2}{y},$$

or

$$y\frac{dy}{dx} - ky^2 = x(1-x^2).$$

This first-order equation of integrating-factor type is equivalent to

$$\frac{d}{dx}(e^{-2kx}y^2) = 2x(1-x^2)e^{-2kx},$$

which can be separated, and integrated to give the general solution

$$y^2 e^{-2kx} = 2\int x(1-x^2)e^{-2kx}dx + C$$

$$= 2e^{-2kx}\left[\frac{3 + 6kx + k^2(6x^2 - 2) + 4k^3x(x^2 - 1)}{4k^4}\right] + C.$$

or

$$y^2 = \frac{1}{2k^4}[3 + 6kx + k^2(6x^2 - 2) + 4k^3x(x^2-1)] + Ce^{2kx}.$$

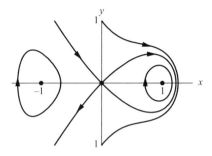

Figure 3.47 Problem 3.66: Phase diagram for $\dot{x} = y$, $\dot{y} = x(1-x^2)+ky^2$ with $k = -\sqrt{(3/2)}$ showing the homoclinic path in $x > 0$.

The origin, being a saddle, is the only equilibrium point with which homoclinic paths can be associated. Paths through the origin are given by the choice

$$C = -\frac{3 - 2k^2}{2k^4}.$$

If $k = \sqrt{(3/2)}$, then $C = 0$, and the corresponding phase path is

$$y^2 = x^2[1 - \sqrt{(\tfrac{2}{3})}x].$$

The homoclinic path is shown in Figure 3.47. For $x > 0$,

$$\frac{dx}{dt} = x\sqrt{[1 - \sqrt{(2/3)}x]}.$$

Separating the variables and integrating

$$\int \frac{dx}{x\sqrt{[1 - \sqrt{(2/3)x}]}} = \int dt + B = t + B,$$

or

$$-2\tanh^{-1}\sqrt{[1 - \sqrt{(2/3)}x]} = t + B = t - t_0,$$

say. Hence

$$x = \sqrt{(\tfrac{3}{2})}\mathrm{sech}^2[\tfrac{1}{2}(t - t_0)].$$

- **3.67** An oscillator has an equation of motion given by $\ddot{x} + f(x) = 0$, where $f(x)$ is a piecewise linear restoring force defined by

$$f(x) = \begin{cases} -x & |x| \le a \\ b(x\,\mathrm{sgn}\,(x) - a) - a & |x| > a \end{cases}.$$

where $a, b > 0$. Find the equations of the homoclinic paths in the phase plane.

3.67. An oscillator has the equation of motion $\ddot{x} + f(x) = 0$, where

$$f(x) = \begin{cases} -x & |x| \le a \\ b(x\,\text{sgn}(x) - a) - a & |x| > a \end{cases}.$$

The system has equilibrium points at $(0,0)$ and $[\pm a(b+1)/b, 0]$. The origin is a saddle point. For $|x| \le a$, the differential equation is $\ddot{x} - x = 0$ which is that for the linear saddle point with separatrices $y = \pm x$ in the phase plane.
For $x > a$, the differential equation is

$$\ddot{x} + b(x - a) - a = 0.$$

The phase paths are given by the equation

$$\frac{dy}{dx} = \frac{a(b+1) - bx}{y},$$

with general solution

$$y^2 + b\left(x - \frac{a(b+1)}{b}\right)^2 = C,$$

which are ellipses centred at $x = a(b+1)/b$, $y = 0$. The particular ellipse which links with the separatrices in $0 < x < a$ at $(a, 0)$ and $(-a, 0)$ has the constant C defined by

$$a^2 + b\left(a - a - \frac{a}{b}\right)^2 = C,$$

that is, $C = a^2(1+b)/b$. The separatrices join the ellipse

$$y^2 + b\left(x - \frac{a(b+1)}{b}\right)^2 = \frac{a^2(1+b)}{b}.$$

Similarly, for $x < -a$, the matching ellipse is

$$y^2 + b\left(x + \frac{a(b+1)}{b}\right)^2 = \frac{a^2(1+b)}{b}.$$

The homoclinic paths for $a = b = 1$ are shown in Figure 3.48.

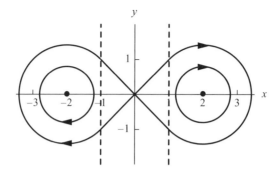

Figure 3.48 Problem 3.67: Homoclinic paths with $a = 1, b = 1$.

- **3.68** Consider the system
$$\dot{x} = y(2y^2 - 3x^2 + \tfrac{19}{9}x^4),$$
$$\dot{y} = y^2(3x - \tfrac{38}{9}x^3) - (4x^3 - \tfrac{28}{3}x^5 + \tfrac{40}{9}x^7).$$
Find the locations of its equilibrium points. Verify that the system has four homoclinic paths given by
$$y^2 = x^2 - x^4 \quad \text{and} \quad y^2 = 2x^2 - \tfrac{10}{9}x^4.$$
Show also that the origin is a higher-order saddle with separatrices in the directions with slopes ± 1 and $\pm\sqrt{2}$.

3.68. The system is
$$\dot{x} = X(x, y) = y\left(2y^2 - 3x^2 + \frac{19}{9}x^4\right) \qquad (i)$$

$$\dot{y} = Y(x, y) = y^2\left(3x - \frac{38}{9}x^3\right) - \left(4x^3 - \frac{28}{3}x^5 + \frac{40}{9}x^7\right). \qquad (ii)$$

First observe that
$$\frac{\partial X}{\partial x} + \frac{\partial Y}{\partial y} = 0,$$
which means that the system is Hamiltonian (see NODE, Section 2.8). A consequence is that equilibrium points are either centres or saddle points. From (i), either $y = 0$ or $y^2 = \tfrac{1}{2}x^2\left(3 - \tfrac{19}{9}x^2\right)$.

(a) $y = 0$. Equation (ii) implies
$$x^3(10x^4 - 21x^2 + 9) = 0.$$

Hence $x = 0$, or $x = \pm\sqrt{\frac{3}{2}}$, or $x = \pm\sqrt{\frac{3}{5}}$. There are five equilibrium points

$$(0,0), \quad \left(\pm\sqrt{\frac{3}{2}}, 0\right), \quad \left(\pm\sqrt{\frac{3}{5}}, 0\right).$$

(b) $y^2 = \frac{1}{2}x^2\left(3 - \frac{19}{9}x^2\right)$. Equation (ii) implies

$$\frac{1}{2}x^2\left(3 - \frac{19}{9}\right)\left(3x - \frac{38}{9}x^3\right) - \left(4x^3 - \frac{28}{3}x^5 + \frac{40}{9}x^7\right) = 0,$$

or

$$x^3(2x^4 - 27x^2 + 81) = 0.$$

The solutions of this equation $x = 0$, or $x = \pm 3$, or $x = \pm\frac{3}{\sqrt{2}}$, but there are corresponding real values for y if $x = \pm 3$, or $x = \pm\frac{3}{\sqrt{2}}$.

The Hamiltonian $H(x, y)$ satisfies

$$\frac{\partial H}{\partial y} = y\left(2y^2 - 3x^2 + \frac{19}{9}x^4\right).$$

Therefore

$$H(x, y) = \left(\frac{1}{2}y^4 - \frac{3}{2}x^2y^2 + \frac{19}{9}x^4y^2\right) + q(x),$$

where

$$q'(x) = -4x^3 + \frac{28}{3}x^5 - \frac{40}{9}x^7.$$

Finally, after integrating this equation, the Hamilitonian is

$$H(x, y) = \frac{1}{2}y^4 - \frac{3}{2}x^2y^2 + \frac{19}{18}x^4y^2 + x^4 - \frac{14}{9}x^6 + \frac{5}{9}x^8,$$

so that the phase paths are given by $H(x, y) = C$. The Hamiltonian can be factorized (use computer algebra such as *Mathematica*) into

$$H(x, y) = \frac{1}{18}(-x^2 + x^4 + y^2)(-18x^2 + 10x^4 + 9y^2).$$

If the constant $C = 0$, then paths through the origin are

$$y^2 = x^2 - x^4, \quad y^2 = 2x^2 - \frac{10}{9}x^4. \tag{iii}$$

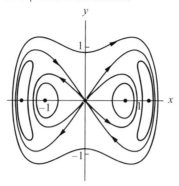

Figure 3.49 Problem 3.68:

These are homoclinic paths associated with the origin since they pass through the origin, are reflected in the x axis, and intersect the x axis so that they are bounded in the x direction. Since there are four homoclinic paths associated with the origin this indicates that the origin is a higher-order saddle point. Near the origin the directions of the homoclinic paths are given approximately by

$$y^2 \approx x^2, \; y^2 \approx 2x^2; \; \text{or} \; y \approx \pm x, \; y \approx \pm\sqrt{2}x,$$

respectively. The phase diagram is shown in Figure 3.49.

- 3.69 Find and classify the equilibrium points of $\dot{x} = a - x^2$, $\dot{y} = -y + (x^2 - a)(1 - 2x)$ for all a. Show that as a decreases through zero, a saddle point and a node coalesce at $a = 0$ after which the equilibrium points disappear. Using the substitution $y = z + x^2 - a$, determine the equations of the phase paths. Show that the phase path connecting the saddle point and the node is $y = x^2 - a$ for $a > 0$, Compute phase diagrams for $a = 0$ and $a = \pm\tfrac{1}{4}$.

3.69. The equilibrium points of

$$\dot{x} = a - x^2, \quad \dot{y} = -y + (x^2 - a)(1 - 2x)$$

occur where

$$a - x^2 = 0, \quad -y + (x^2 - a)(1 - 2x) = 0.$$

Hence equilibrium can only occur where $x^2 = a$ and $y = 0$. Thus, if

- $a < 0$, there are no equilibrium points;
- $a = 0$, there is one equilibrium point at $(0,0)$;
- $a > 0$, there are two equilibrium points at $(\pm\sqrt{a}, 0)$.

Assume that $a > 0$. Let $x = \sqrt{a} + \xi$. Then the linearized approximations are

$$\dot{\xi} = a - (\sqrt{a} + \xi)^2 \approx -2\sqrt{a}\xi,$$

$$\dot{y} = -y + [(\sqrt{a} + \xi)^2 - a][1 - 2(\sqrt{a} + \xi)] \approx 2\sqrt{a}(1 - 2\sqrt{a})\xi - y.$$

3: Geometrical aspects of plane autonomous systems

The parameters associated with approximation are

$$p = -2\sqrt{a} - 1 < 0, \quad q = 2\sqrt{a} > 0, \quad \Delta = (2\sqrt{a} - 1)^2 > 0.$$

Therefore $(\sqrt{a}, 0)$ is an unstable node.
For the other equilibrium point, let $x = -\sqrt{a} + \xi$. Then the linearized approximations are

$$\dot{\xi} = 2\sqrt{a}\xi, \quad \dot{y} = -2\sqrt{a}(1 - 2\sqrt{a})\xi - y.$$

It follows that $q = -2\sqrt{a} < 0$, so that $(-\sqrt{a}, 0)$ is a saddle point.
The differential equation for the phase paths is

$$\frac{dy}{dx} = \frac{-y + (x^2 - a)(1 - 2x)}{a - x^2} = \frac{y}{x^2 - a} + (2x - 1).$$

Let $y = z + x^2 - a$. Then the equation becomes

$$\frac{dz}{dx} = \frac{z}{x^2 - a}. \tag{i}$$

- $a > 0$. The separable first-order equation (i) has the solution

$$\ln|z| = \int \frac{dx}{x^2 - a} = \frac{1}{2\sqrt{a}} \ln\left|\frac{x - \sqrt{a}}{x + \sqrt{a}}\right| + B.$$

or

$$|y - x^2 + a|^{2\sqrt{a}}|x + \sqrt{a}| = C|x - \sqrt{a}|.$$

- $a = 0$. Equation (i) becomes

$$\frac{dz}{dx} = \frac{z}{x^2}.$$

The general solution is $z = y - x^2 + a = De^{-1/x}$.
- $a < 0$. Equation (i) has the general solution

$$\ln|z| = \ln|y - x^2 - a| = \frac{1}{\sqrt{-a}} \tan^{-1}\left[\frac{x}{\sqrt{-a}}\right] + E.$$

Equation (i) also has the singular solution $z = 0$, or $y = x^2 - a$, which joins the equilibrium points $(\pm\sqrt{a}, 0)$ for $a > 0$. Some typical phase paths are shown in Figures 3.50, 3.51, 3.52 for the cases $a > 0$, $a < 0$ and $a = 0$.

206 Nonlinear ordinary differential equations: problems and solutions

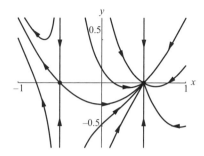

Figure 3.50 Problem 3.69: Phase diagram for $\dot{x} = a - x^2$, $\dot{y} = -y + (x^2 - a)(1 - 2x)$ with $a = \frac{1}{4}$.

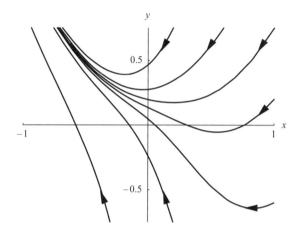

Figure 3.51 Problem 3.69: Phase diagram for $\dot{x} = a - x^2$, $\dot{y} = -y + (x^2 - a)(1 - 2x)$ with $a = -\frac{1}{4}$.

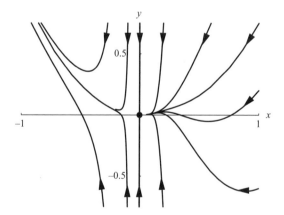

Figure 3.52 Problem 3.69: Phase diagram for $\dot{x} = a - x^2$, $\dot{y} = -y + (x^2 - a)(1 - 2x)$ with $a = 0$.

- **3.70** Locate and classify the equilibrium points of
$$\dot{x} = 1 - x^2, \quad \dot{y} = -(y + x^2 - 1)x^2 - 2x(1 - x^2)$$
according to their linear approximations. *Verify* that the phase diagram has a saddle-node connection given by $y = 1 - x^2$. Find the time solutions $x(t)$, $y(t)$ for this connection. Sketch the phase diagram.

3.70. The system
$$\dot{x} = 1 - x^2, \quad \dot{y} = -(y + x^2 - 1)x^2 - 2x(1 - x^2).$$

has two equilibrium points, at $(1, 0)$ and $(-1, 0)$.

- At $(1, 0)$. Let $x = 1 + \xi$. Then the linear approximation is given by
$$\dot{\xi} = 1 - (1 + \xi)^2 \approx -2\xi,$$
$$\dot{y} = -[y + (1 + \xi)^2 - 1](1 + \xi)^2 - 2(1 + \xi)[1 - (1 + \xi)^2] \approx 2\xi - y.$$
Hence $(1, 0)$ is a stable node.
- At $(-1, 0)$. Let $x = -1 + \xi$. Then the linear approximation is given by
$$\dot{\xi} = 1 - (-1 + \xi)^2 \approx 2\xi,$$
$$\dot{y} = -[y + (-1 + \xi)^2 - 1](-1 + \xi)^2 - 2(-1 + \xi)[1 - (-1 + \xi)^2] \approx -2\xi - y.$$
Hence $(-1, 0)$ is a saddle point.

The phase paths are given by the differential equation
$$\frac{dy}{dx} = \frac{-(y + x^2 - 1)x^2 - 2x(1 - x^2)}{1 - x^2} = -\frac{x^2 y}{1 - x^2} - 2x + x^2.$$

It can be verified that $y = 1 - x^2$ satisfies the differential equation above. It also joins the two equilibrium points, and is, therefore, a saddle–node connection. On this path
$$y = \frac{dx}{dt} = 1 - x^2,$$
which has the required time-solution
$$x = \frac{1 - e^{-2(t - t_0)}}{1 + e^{-2(t - t_0)}}.$$

208 Nonlinear ordinary differential equations: problems and solutions

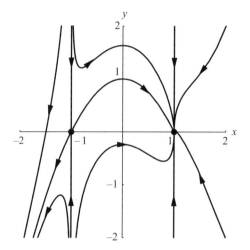

Figure 3.53 Problem 3.70: Phase diagram for $\dot{x} = 1 - x^2$, $\dot{y} = -(y + x^2 - 1)x^2 - 2x(1 - x^2)$.

It follows that
$$y = \dot{x} = \frac{4e^{-2(t-t_0)}}{1 + e^{-2(t-t_0)}}.$$

The phase diagram showing the saddle node connection is shown in Figure 3.53.

- **3.71** Consider the piecewise linear system
$$\dot{x} = x, \qquad \dot{y} = -y, \qquad |x - y| \leq 1,$$
$$\dot{x} = y + 1, \quad \dot{y} = 1 - x, \quad x - y \geq 1,$$
$$\dot{x} = y - 1, \quad \dot{y} = -1 - x, \quad x - y \leq -1.$$
Locate and classify the equilibrium points of the system. By solving the linear equations in each region and matching separatrices, show that the origin has two homoclinic paths.

3.71. The piecewise linear system is
$$\dot{x} = x, \qquad \dot{y} = -y, \qquad |x - y| \leq 1,$$
$$\dot{x} = y + 1, \quad \dot{y} = 1 - x, \quad x - y \geq 1$$
$$\dot{x} = y - 1, \quad \dot{y} = 1 - x, \quad x - y \leq -1.$$

The system has three equilibrium points: at $(0, 0)$ (a saddle point), at $(1, -1)$ (a centre) and at $(-1, 1)$ (a centre).

- In the region $|x - y| \leq 1$, the phase paths are given by the hyperbolas $xy = A$: the separatrices of the saddle point are $x = 0$ and $y = 0$.
- In the region $x - y \geq 1$, the phase paths are given by the circles
$$(x - 1)^2 + (y + 1)^2 = B.$$

3: Geometrical aspects of plane autonomous systems

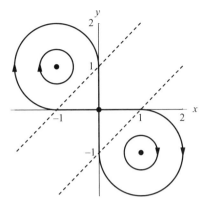

Figure 3.54 Problem 3.71: Phase diagram showing the homoclinic paths associated with the origin.

- In the region $x - y \leq -1$, the phase paths are given by the circles

$$(x + 1)^2 + (y - 1)^2 = C.$$

The homoclinic paths can be constructed by matching circles in $-x + y \geq 1$ and $-x - y \leq -1$ with the separatrices $x = 0$, $y = 0$ on the discontinuity lines. Thus the circle $(x-1)^2 + (y+1)^2 = 1$ joins the separatrices of the origin at the points $(1, 0)$ and $(0, -1)$ as shown in Figure 3.54. Similarly the circles $(x + 1)^2 + (y - 1)^2 = 1$ matches the separatrices at $(-1, 0)$ and $(1, 1)$ to create a second homoclinic path.

- **3.72** Obtain the differential equations for the linear system
$$\dot{x} = ax + by, \quad \dot{y} = cx + dy, \quad (ad \neq bc),$$
in the $\mathcal{U}$-plane (see Figure 3.16 in NODE) using the transformation $x = 1/z$, $y = u/z$. Under what conditions on $\Delta = p^2 - 4q$, $p = a + d$, $q = ad - bc$ does the system on the $\mathcal{U}$-plane have no equilibrium points?

3.72. Apply the transformation $x = 1/z$, $y = u/z$ (see Section 3.3) to the linear system

$$\dot{x} = ax + by, \quad \dot{y} = cx + dy, \quad (ad \neq bc).$$

Then

$$\dot{x} = -\frac{\dot{z}}{z^2}, \quad \dot{y} = \frac{\dot{u}}{z} - \frac{u\dot{z}}{z^2},$$

so that the equations become

$$-\frac{\dot{z}}{z^2} = \frac{a}{z} + \frac{bu}{z}, \quad \frac{\dot{u}}{z} - \frac{u\dot{z}}{z^2} = \frac{c}{z} + \frac{du}{z},$$

or
$$\dot{z} = -z(a+bu), \quad \dot{u} = u(d-a) + c - bu^2.$$

Equilibrium points occur where
$$z(a+bu) = 0, \quad bu^2 + (a-d)u - c = 0.$$

The second equation has the solutions
$$u = \frac{1}{2b}\left\{-(a-d) \pm \sqrt{[(a-d)^2 + 4bc]}\right\}.$$

This will only have real solutions if
$$(a-d)^2 + 4bc \geq 0, \quad \text{or} \quad (a+d)^2 \geq 4(ad-bc).$$

This is equivalent to
$$p^2 \geq 4q, \quad \text{or} \quad \Delta \geq 0.$$

The corresponding real solutions for u are only consistent with $z = 0$.

- **3.73** Classify all the equilibrium points of the system
$$\dot{x} = X(x,y) = (1-x^2)(x+2y), \quad \dot{y} = Y(x,y) = (1-y^2)(-2x+y).$$
Draw the isoclines $X(x,y) = 0$ and $Y(x,y) = 0$, and sketch the phase diagram for the system. A phase path starts near (but not at) the origin. How does its path evolve as t increases? If, on this path, the system experiences small disturbances which cause it to jump to nearby neighbouring paths, what will eventually happen to the system?

3.73. The system
$$\dot{x} = X(x,y) = (1-x^2)(x+2y), \quad \dot{y} = Y(x,y) = (1-y^2)(-2x+y)$$
has nine equilibrium points, at
$$(1,1), (1,-1), (1,2); \ (-1,1), (-1,-1), (-1,-2); \ (-2,1), (2,-1); \ (0,0).$$

The linear classification is as follows.

- $(1,1)$. Let $x = 1+\xi$, $y = 1+\eta$. Then the linear approximation is
$$\dot{\xi} = -4\xi, \quad \dot{\eta} = 2\eta.$$

Hence $(1,1)$ is a saddle point.
- $(1,-1), (-1,1), (-1,-1)$ are also saddle points.

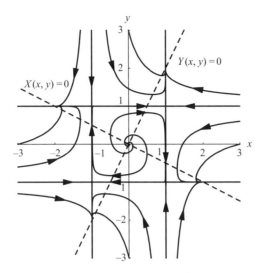

Figure 3.55 Problem 3.73: Phase diagram for $\dot{x} = (1 - x^2)(x + 2y)$, $\dot{y} = (1 - y^2)(-2x + y)$.

- $(1, 2)$. Let $x = 1 + \zeta$, $y = 2 + \eta$. Then the linear approximation is

$$\dot{\zeta} = -10\zeta, \quad \dot{\eta} = 6\zeta - 3\eta.$$

Therefore $(1, 2)$ is a stable node.
- $(-2, 1)$, $(-1, -2)$, $(2, -1)$ are also stable nodes.
- $(0, 0)$. The linear approximations are

$$\dot{x} = x + 2y, \quad \dot{y} = -2x + y.$$

Hence the origin is an unstable spiral.

Note also that the straight lines $x = \pm 1$ and $y = \pm 1$ consist of segments of phase paths. These phase paths are also isoclines with infinite and zero slopes respectively. A further isocline with zero slope is the line $y = 2x$, and a further isocline with infinite slope is the line $x = -2y$.

A phase path starting close to the origin will spiral out and approach asymptotically the square with sides $x = \pm 1$, $y = \pm 1$. This path will be increasingly unstable such that a small disturbance outwards could cause it to jump on to a stable path approaching one of the four nodes outside the square, as shown in Figure 3.55.

4 Periodic solutions; averaging methods

- 4.1 By transforming to polar coordinates, find the limit cycles of the systems
  (i) $\dot{x} = y + x(1 - x^2 - y^2)$, $\dot{y} = -x + y(1 - x^2 - y^2)$;
  (ii) $\dot{x} = (x^2 + y^2 - 1)x - y\sqrt{(x^2 + y^2)}$, $\dot{y} = (x^2 + y^2 - 1)y + x\sqrt{(x^2 + y^2)}$.

  and investigate their stability.

4.1. (i) The system is

$$\dot{x} = y + x(1 - x^2 - y^2), \quad \dot{y} = -x + y(1 - x^2 - y^2). \tag{i}$$

Let $x = r\cos\theta$, $y = r\sin\theta$. Then, differentiating with respect to t, we have

$$\dot{x} = \dot{r}\cos\theta - r\dot{\theta}\sin\theta, \quad \dot{y} = \dot{r}\sin\theta + r\dot{\theta}\cos\theta.$$

Solve these equations for $\dot{r}$ and $\dot{\theta}$ so that

$$\dot{r} = \dot{x}\cos\theta + \dot{y}\sin\theta, \quad \dot{\theta} = -\frac{\dot{x}}{r}\sin\theta + \frac{\dot{y}}{r}\cos\theta.$$

Substitution for $\dot{x}$ and $\dot{y}$ from (i) leads to the polar equations

$$\dot{r} = [r\sin\theta + r(1 - r^2)\cos\theta]\cos\theta + [-r\cos\theta + r(1 - r^2)\sin\theta]\sin\theta = r(1 - r^2), \tag{ii}$$

$$\dot{\theta} = -[\sin\theta + (1 - r^2)\cos\theta]\sin\theta + [-\cos\theta + (1 - r^2)\sin\theta] = -1. \tag{iii}$$

The system (i) has one equilibrium point at the origin. From (ii) it can be seen that $r = 1$ is a phase path representing a periodic solution taken in the clockwise sense since $\dot{\theta}$ is negative. The path is a stable limit cycle since $\dot{r} > 0$ for $r < 1$ and $\dot{r} < 0$ for $r > 1$. The polar equations for the phase paths can be obtained by integrating

$$\frac{dr}{d\theta} = \frac{\dot{r}}{\dot{\theta}} = \frac{r(1 - r^2)}{-1}.$$

214 Nonlinear ordinary differential equations: problems and solutions

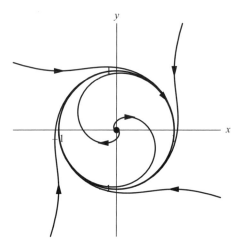

Figure 4.1 Problem 4.1(i): Phase diagram for $\dot{x} = y + x(1 - x^2 - y^2)$, $\dot{y} = -x + y(1 - x^2 - y^2)$.

Hence

$$\ln\left[\frac{r^2}{|1 - r^2|}\right] = -2\theta + C,$$

or

$$r^2 = \frac{Ae^{-2\theta}}{1 + Ae^{-2\theta}}.$$

where A is an arbitrary constant (positive if $r^2 < 1$, negative if $r^2 > 1$). The phase diagram is shown in Figure 4.1.

(ii) The system is

$$\dot{x} = (x^2 + y^2 - 1)x - y\sqrt{(x^2 + y^2)}, \quad \dot{y} = (x^2 + y^2 - 1)y + x\sqrt{(x^2 + y^2)}.$$

As in (i)

$$\dot{r} = \dot{x}\cos\theta + \dot{y}\sin\theta = r^2(r^2 - 1),$$

$$\dot{\theta} = -\frac{\dot{x}}{r}\sin\theta + \frac{\dot{y}}{r}\cos\theta = r.$$

The system has one equilibrium point, at the origin. Also the circle $r = 1$ is a limit cycle. Since $\dot{r} < 0$ for $r < 1$, and $\dot{r} > 0$ for $r > 1$, the limit cycle is unstable. The phase paths are given by the equation

$$\frac{dr}{d\theta} = \frac{\dot{r}}{\dot{\theta}} = \frac{r(1 - r^2)}{-1},$$

which is the same equation as in (i). Therefore the phase diagram is the same as that shown in Figure 4.1 except that the direction of the phase paths is reversed.

- **4.2** Consider the system $\dot{x} = y + xf(r^2)$, $\dot{y} = -x + yf(r^2)$, where $r^2 = x^2 + y^2$ and $f(u)$ is continuous on $u \geq 0$. Show that r satisfies

$$\frac{d(r^2)}{dt} = 2r^2 f(r^2).$$

If $f(r^2)$ has n zeros, at $r = r_k$, $k = 1, 2, \ldots, n$, how many periodic solutions has the system? Discuss their stability in terms of the sign of $f'(r_k^2)$.

4.2. The system is

$$\dot{x} = y + xf(r^2), \quad \dot{y} = -x + yf(r^2),$$

which has one equilibrium point at the origin. From Problem 1(i)

$$\dot{r} = \dot{x}\cos\theta + \dot{y}\sin\theta = rf(r^2), \tag{i}$$

$$\dot{\theta} = -\frac{\dot{x}}{r}\sin\theta + \frac{\dot{y}}{r}\cos\theta = -1. \tag{ii}$$

From (i) it follows that

$$r\frac{dr}{dt} = \frac{1}{2}\frac{d(r^2)}{dt} = r^2 f(r^2) \tag{iii}$$

as required. All solutions of $f(r^2) = 0$ will be concentric circular phase paths of the system. If there are n solutions then there will be n periodic solutions. If $f'(r_k^2) > 0$ then there will be a neighbourhood including the circle in which $f(r_k^2) < 0$ for $r < r_k$, and $f(r_k^2) > 0$ for $r > r_k$. Hence we conclude from (iii) that r is decreasing for $r < r_k$, and increasing for $r > r_k$ implying that $r = r_k$ is unstable. By a similar argument $r = r_k$ is stable if $f'(r_k^2) < 0$.

- **4.3** Apply the energy balance method of NODE, Section 4.1 to each of the following equations where $0 < \varepsilon \ll 1$, and find the amplitude and stability of any limit cycles:
  (i) $\ddot{x} + \varepsilon(x^2 + \dot{x}^2 - 1)\dot{x} + x = 0$;
  (ii) $\ddot{x} + \varepsilon(\frac{1}{3}\dot{x}^3 - \dot{x}) + x = 0$;
  (iii) $\ddot{x} + \varepsilon(x^4 - 1)\dot{x} + x = 0$;
  (iv) $\ddot{x} + \varepsilon\sin(x^2 + \dot{x}^2)\operatorname{sgn}(\dot{x}) + x = 0$;
  (v) $\ddot{x} + \varepsilon(|x| - 1)\dot{x} + x = 0$;
  (vi) $\ddot{x} + \varepsilon(\dot{x} - 3)(\dot{x} + 1)\dot{x} + x = 0$;
  (vii) $\ddot{x} + \varepsilon(x - 3)(x + 1)\dot{x} + x = 0$.

4.3. The method can be applied to equations of the form (see Section 4.1)

$$\ddot{x} + \varepsilon h(x, \dot{x}) + x = 0,$$

where $0 < \varepsilon \ll 1$. An approximate solution $x(t) \approx a\cos t$, $y = \dot{x} \approx -a\sin t$, corresponding to the unperturbed centre ($\varepsilon = 0$) $\ddot{x} + x = 0$, is substituted into the energy-balance equation

$$g(a) = \varepsilon \int_0^{2\pi} h(x(t), y(t))y(t))dt = 0$$

to determine any solutions for the amplitude a. All problems have one equilibrium point, at the origin.

(i) $\ddot{x} + \varepsilon(x^2 + \dot{x}^2 - 1)\dot{x} + x = 0$. In this case $h(x, y) = (x^2 + y^2 - 1)y$. Therefore

$$g(a) = -a^2\varepsilon \int_0^{2\pi}(a^2 - 1)\sin^2 t\, dt = a^2\varepsilon\pi(1 - a^2).$$

The equation has one non-zero positive solution $a = 1$ which will be the amplitude, for ε small, of a limit cycle: note that in this particular example the solution is exact. To investigate its stability we consider the sign of $g'(1)$. Thus

$$g'(a) = \frac{d}{da}[\varepsilon(a^2 - a^4)] = \varepsilon(2a - 4a^3).$$

Therefore, $g'(1) = -2\varepsilon\pi < 0$, which implies that the limit cycle is stable
(ii) $\ddot{x} + \varepsilon(\frac{1}{3}\dot{x}^3 - \dot{x}) + x = 0$. In this example $h(x, y) = (\frac{1}{3}y^3 - y)$. Therefore

$$g(a) = a\varepsilon \int_0^{2\pi}\left(-\frac{1}{3}a^3\sin^3 t + a\sin t\right)\sin t\, dt$$

$$= a\varepsilon\left[-\frac{1}{3}a^3\int_0^{2\pi}\sin^4 t\, dt + \int_0^{2\pi}\sin^2 t\, dt\right]$$

$$= a\varepsilon\left[-\frac{a^3}{6}\int_0^{2\pi}(1 + \cos 2t)^2 dt + \frac{a}{2}\int_0^{2\pi}(1 + \cos 2t)dt\right]$$

$$= a\varepsilon\pi[-\tfrac{1}{2}a^3 + a].$$

Hence the system has a periodic solution of amplitude $a = \sqrt{2}$ approximately. The derivative

$$g'(a) = \varepsilon\pi(-2a^3 + 2a),$$

so that $g'(\sqrt{2}) = -2\varepsilon\pi\sqrt{2} < 0$. The limit cycle is stable.
(iii) $\ddot{x} + \varepsilon(x^4 - 1)\dot{x} + x = 0$. In this case $h(x, y) = (x^4 - 1)y$. Then

$$g(a) = \varepsilon a^2 \int_0^{2\pi}(1 - a^4\cos^4 t)\sin^2 t\, dt$$

$$= \varepsilon a^2 \int_0^{2\pi}\left[\frac{1}{2}(1 - \sin 2t) - \frac{a^4}{8}(1 + \cos 2t)\sin^2 2t\right]dt$$

$$= \varepsilon a^2 \int_0^{2\pi} \left[\frac{1}{2}(1 - \sin 2t) - \frac{a^4}{16}(1 + \cos 2t)(1 - \cos 4t) \right] dt$$

$$= \varepsilon a^2 \left[-\frac{1}{16}(a^4 - 8)t - \frac{1}{64}(16 + a^4) \sin 2t + \frac{1}{64} a^4 \sin 4t + \frac{1}{192} a^4 \sin 6t \right]_0^{2\pi}$$

$$= \varepsilon \frac{a^2 \pi}{8}(8 - a^4)$$

Therefore the system has a periodic solution with amplitude $a = 2^{3/4}$ approximately. The derivative

$$g'(a) = \frac{\varepsilon \pi}{8}(16a - 6a^5),$$

so that $g'(2^{3/4}) = -4\pi \varepsilon 2^{3/4} < 0$. The limit cycle is stable.

(iv) $\ddot{x} + \varepsilon \sin(x^2 + \dot{x}^2)\operatorname{sgn}(\dot{x}) + x = 0$. In this problem $h(x, y) = \sin(x^2 + y^2)\operatorname{sgn}(y)$. Then

$$g(a) = a\varepsilon \int_0^{2\pi} \sin[a^2(\cos^2 t + \sin^2 t)]\operatorname{sgn}(-a \sin t) \sin t \, dt$$

$$= a\varepsilon \sin(a^2) \left[\int_0^{\pi} (-\sin t) dt + \int_{\pi}^{2\pi} \sin t \, dt \right]$$

$$= -2a\varepsilon \sin(a^2)$$

The system has an infinite set of limit cycles, of radius $a = a_n = \sqrt{n\pi}$, $(n = 1, 2, 3, \ldots)$. The derivative

$$g'(a) = -2\varepsilon[\sin(a^2) + 2a^2 \cos(a^2)],$$

so that $g'(a_n) = -4\varepsilon n\pi \cos(n\pi)$, which implies that the limit cycle $a = a_n$ is unstable if n is odd, and stable if n is even.

(v) $\ddot{x} + \varepsilon(|x| - 1)\dot{x} + x = 0$. In this problem $h(x, y) = (|x| - 1)y$. Then

$$g(a) = -a\varepsilon \int_0^{2\pi} (a|\cos t| - 1) \sin^2 t \, dt$$

$$= -a\varepsilon \left[\int_0^{\frac{1}{2}\pi} a \cos t \sin^2 t \, dt - \int_{\frac{1}{2}\pi}^{\frac{3}{2}\pi} a \cos t \sin^2 t \, dt \right.$$

$$\left. + \int_{\frac{3}{2}\pi}^{2\pi} a \cos t \sin^2 t \, dt - \int_0^{2\pi} \sin^2 t \, dt \right]$$

$$= -a\varepsilon \left[\frac{a}{3} - \left\{-\frac{a}{3} - \frac{a}{3}\right\} + \frac{a}{3} - \pi\right]$$

$$= -a\varepsilon \left[\frac{4a}{3} - \pi\right]$$

The system has one limit cycle of approximately radius $a = 3\pi/4$. The derivative

$$g'(a) = \varepsilon \left(-\tfrac{8}{3}a + \pi\right),$$

so that $g'(3\pi/4) = -\pi\varepsilon < 0$ which implies the limit cycle is stable.

(vi) $\ddot{x} + \varepsilon(\dot{x} - 3)(\dot{x} + 1)\dot{x} + x = 0$. In this case $h(x, y) = (y - 3)(y + 1)y$. Then

$$g(a) = a^2\varepsilon \int_0^{2\pi} (a\sin t + 3)(-a\sin t + 1)\sin^2 t \, dt$$

$$= a^2\varepsilon \int_0^{2\pi} (a\sin^4 t - 2a\sin^3 t + 3\sin^2 t) dt$$

$$= a^2\varepsilon \left[\tfrac{3}{4}a^2\pi + 0 + 3\pi\right]$$

$$= \tfrac{3}{4}a^2\varepsilon(a^2 + 4)$$

There are no non-zero real solutions of $g(a) = 0$: hence energy-balance suggests that the system has no limit cycles

(vii) $\ddot{x} + \varepsilon(x - 3)(x + 1)\dot{x} + x = 0$. In this case $h(x, y) = (x - 3)(x + 1)y$. Then

$$g(a) = -a^2\varepsilon \int_0^{2\pi} (a\cos t - 3)(a\cos t + 1) dt$$

$$= -a^2\varepsilon \int_0^{2\pi} (a^2 \cos^2 t \sin^2 t - 2a \cos t \sin^2 t - 3\sin^2 t) dt$$

$$= -a^2\varepsilon \left[\tfrac{1}{4}a^2\pi + 0 - 3\pi\right] = \tfrac{1}{4}a^2\varepsilon\pi(12 - a^2).$$

The system has one limit cycle of radius $a = 2\sqrt{3}$. The derivative

$$g'(a) = \varepsilon\pi a(6 - a^2),$$

so that $g'(2\sqrt{3}) = -12\varepsilon\sqrt{3} < 0$. Hence the limit cycle is stable.

- **4.4** For the equation $\ddot{x}+\varepsilon(x^2+\dot{x}^2-4)\dot{x}+x=0$, the solution $x=2\cos t$ is a limit cycle. Test its stability, using the method of NODE, Section 4.1, and obtain an approximation to the paths close to the limit cycle by the method of Section 4.3.

4.4. It can be verified that the equation

$$\ddot{x}+\varepsilon(x^2+\dot{x}^2-4)\dot{x}+x=0$$

has the exact periodic solution $x=2\cos t$. In this problem

$$h(x,y)=(x^2+y^2-4)y$$

and (see eqn (4.8))

$$g(a)=\varepsilon a\int_0^{2\pi}h(a\cos t,-a\sin t)\sin t\,dt$$

$$=-\varepsilon a^2\int_0^{2\pi}(a\cos^2 t+a\sin^2 t-4)\sin^2 t\,dt$$

$$=-\varepsilon a^2(a^2-4)\int_0^{2\pi}\sin^2 t\,dt$$

$$=\varepsilon\pi a^2(4-a^2).$$

Thus $g(a)=0$ for $a=2$, predicts that the exact solution (above) is the only periodic solution, and therefore is a limit cycle. The derivative

$$g'(a)=\varepsilon\pi(8a-4a^3),$$

so that $g'(2)=-16\pi\varepsilon<0$. Hence the limit cycle is stable.

From NODE, Section 4.3, the amplitude $a(\theta)$ of paths close to the limit cycle are given approximately by the differential equation

$$\frac{da}{d\theta}=\varepsilon p_0(a),$$

where (see eqns (4.27a,b) in NODE)

$$p_0(a)=\frac{1}{2\pi}\int_0^{2\pi}h\{a\cos u,a\sin u\}\sin u\,du$$

$$=\frac{1}{2\pi}(a^2-4)a\int_0^{2\pi}\sin u\,du$$

$$=\tfrac{1}{2}(a^2-4)a.$$

Hence the equation for a becomes the separable equation

$$\frac{da}{d\theta} = \varepsilon p_0(a) = \frac{1}{2}(a^2 - 4)a.$$

Separating the variables and integrating

$$\int \frac{da}{a(a^2 - 4)} = \frac{1}{2}\varepsilon \int d\theta = \frac{1}{2}\varepsilon\theta + C,$$

or

$$\frac{1}{8} \ln \left| \frac{a^2 - 4}{a^2} \right| = \frac{1}{2}\varepsilon\theta + C.$$

Hence

$$1 - \frac{4}{a^2} = A e^{4\varepsilon\theta} = \left(1 - \frac{4}{a_1^2}\right) e^{4\varepsilon\theta},$$

assuming that $a = a_1$ for $\theta = 0$. Finally the polar equation of the phase paths close to the limit cycle is

$$a^2 = \frac{4a_1^2}{a_1^2 - (a_1^2 - 4)e^{4\varepsilon\theta}}.$$

By NODE, (4.21), the period of the limit cycle is given approximately by

$$T \approx 2\pi - \frac{\varepsilon}{a_0} \int_0^{2\pi} h(a_0 \cos\theta, a_0 \sin\theta) \cos\theta \, d\theta,$$

where a_0 is the amplitude of the limit cycle. Hence the approximate theory predicts that

$$T \approx 2\pi + \varepsilon \int_0^{2\pi} (a_0^2 - 4)a_0 \sin u \cos u \, du + O(\varepsilon^2) = 2\pi + O(\varepsilon^2),$$

since $a_0 = 2$. The time solution $x = 2\cos t$ has period 2π exactly, showing that the error in the approximation has magnitude of order ε^2.

- **4.5** For the equation $\ddot{x} + \varepsilon(|x| - 1)\dot{x} + x = 0$, find approximately the amplitude of the limit cycle and its period, and the polar equations for the phase paths near the limit cycle.

4.5. For the equation

$$\ddot{x} + \varepsilon(|x| - 1)\dot{x} + x = 0,$$

4: Periodic solutions; averaging methods

$h(x, y) = (|x| - 1)y$. The unperturbed equation ($\varepsilon = 0$) has the solution $x = a \cos t$. Then the function $g(a)$ (Section 4.1) is given by

$$g(a) = \varepsilon a \int_0^{2\pi} h(a\cos t, -a\sin t) \sin t \, dt$$

$$= -\varepsilon a^2 \int_0^{2\pi} (|a\cos t| - 1) \sin^2 t \, dt$$

$$= -4\varepsilon a^2 \int_0^{\frac{1}{2}\pi} (a\cos t \sin^2 t - \sin^2 t) dt$$

$$= -4\varepsilon a^2 \left(\frac{1}{3}a - \frac{1}{4}\pi\right).$$

Hence the amplitude of the limit cycle is approximately $a = a_0 = \frac{3}{4}\pi$. The derivative of $g(a)$ is given by

$$g'(a) = -4\varepsilon a^2 + 2\varepsilon a \pi,$$

so that $g'(\frac{3}{4}\pi) = -\frac{3}{4}\varepsilon\pi^2 < 0$. Hence the limit cycle is stable.

From NODE, Section 4.3, the amplitude $a(\theta)$ of paths close to the limit cycle are given approximately by the differential equation

$$\frac{da}{d\theta} = \varepsilon p_0(a),$$

where (see eqns (4.27a,b) in NODE)

$$p_0(a) = \frac{1}{2\pi} \int_0^{2\pi} h\{a\cos u, a\sin u\} \sin u \, du$$

$$= \frac{a}{2\pi} \int_0^{2\pi} (a|\cos u| - 1) \sin^2 u \, du$$

$$= \frac{2a}{\pi}\left(\frac{1}{3}a - \frac{1}{4}\pi\right).$$

Hence

$$\frac{da}{d\theta} = \frac{2\varepsilon a}{3\pi}(a - a_0).$$

Separating and integrating

$$\int \frac{da}{a(a-a_0)} = \frac{1}{a_0}\int \left(\frac{1}{a-a_0} - \frac{1}{a}\right) da = \frac{2\varepsilon}{3\pi}\int d\theta = \frac{2\varepsilon\theta}{3\pi} + C.$$

Therefore

$$\ln\left|\frac{a-a_0}{a}\right| = \frac{2a_0\varepsilon\theta}{3\pi} + C,$$

or, if $a = a_1$ where $\theta = 0$,

$$1 - \frac{a_0}{a} = \left(1 - \frac{a_0}{a_1}\right) e^{2a_0 \varepsilon \theta / (3\pi)}.$$

Finally, the polar equation of the phase paths close to the limit cycle are given by

$$a = \frac{a_1 a_0}{a_1 - (a_1 - a_0) e^{2a_0 \varepsilon \theta / (3\pi)}}.$$

The period T is given by

$$T = 2\pi - \frac{\varepsilon}{a_0} \int_0^{2\pi} (a_0 |\cos u| - 1) a_0 \sin u \cos u \, du + O(\varepsilon^2)$$

$$= 2\pi - 0 + O(\varepsilon^2) = 2\pi + O(\varepsilon^2)$$

- **4.6** Repeat Problem 4.5 with Rayleigh's equation, $\ddot{x} + \varepsilon \left(\frac{1}{3}\dot{x}^3 - \dot{x}\right) + x = 0$.

4.6. For Rayleigh's equation

$$\ddot{x} = \varepsilon \left(\frac{1}{3}\dot{x}^3 - \dot{x}\right) + x = 0,$$

$h(x, y) = \frac{1}{3}y^2 - y$. Follow the method given in Problem 4.5. The function $g(a)$ is given by

$$g(a) = -\varepsilon a^2 \int_0^{2\pi} \left(\frac{1}{3} a^2 \sin^4 t - \sin^2 t\right) dt$$

$$= -\varepsilon a^2 \left[\frac{1}{3} a^2 \left(\frac{3}{8} t - \frac{1}{4} \sin 2t + \frac{1}{32} \sin 4t\right) - \frac{1}{2} t + \frac{1}{4} \sin 2t\right]_0^{2\pi}$$

$$= -\varepsilon a^2 \pi (a^2 - 4)$$

Therefore the amplitude of the limit cycle is $a = a_0 = 2$. The derivative

$$g'(a) = -\varepsilon \pi (4a^3 - 8a),$$

so that $g'(2) = -16\varepsilon \pi < 0$. This method implies that the limit cycle is stable.

From NODE, Section 4.3, the amplitude $a(\theta)$ of paths close to the limit cycle are given approximately by the differential equation

$$\frac{da}{d\theta} = \varepsilon p_0(a),$$

where (see eqns (4.27a,b))

$$p_0(a) = \frac{1}{2\pi} \int_0^{2\pi} h\{a\cos u, a\sin u\}\sin u\, du$$

$$= \frac{1}{2\pi} \int_0^{2\pi} \left(\frac{1}{3}a^4 \sin^4 u - a^2 \sin^2 u\right) du$$

$$= \tfrac{1}{8}(a^4 - 4a^2).$$

Hence

$$\frac{da}{d\theta} = \frac{\varepsilon}{8} a^2(a^2 - 4).$$

Separating the variables and integrating

$$\int \frac{da}{a^2(a^2-4)} = \frac{\varepsilon}{8}\int d\theta = \frac{\varepsilon}{8}\theta + C,$$

or

$$\int\left[-\frac{1}{4a^2} + \frac{1}{16}\frac{1}{a-2} - \frac{1}{16}\frac{1}{a+2}\right] da = \frac{\varepsilon}{8}\theta + C.$$

Therefore

$$\frac{1}{4a} + \frac{1}{16}\ln\left[\frac{a-2}{a+2}\right] = \frac{\varepsilon}{8}\theta + C$$

is the polar equation of the spiral phase paths close to the limit cycle
The period T is given by

$$T = 2\pi - \frac{\varepsilon}{2}\int_0^{2\pi}\left(\frac{1}{3}8\sin^3 u - 2\sin u\right)\cos u\, du + O(\varepsilon^2) = 2\pi + O(\varepsilon^2).$$

- 4.7 Find approximately the radius of the limit cycle, and its period, for the equation
$\ddot{x} + \varepsilon(x^2 - 1)\dot{x} + x - \varepsilon x^3 = 0$, $(0 < \varepsilon \ll 1)$.

4.7. The van der Pol equation with nonlinear restoring term is

$$\ddot{x} + \varepsilon(x^2 - 1)\dot{x} + x - \varepsilon x^3 = 0.$$

Assume that $0 < \varepsilon \ll 1$. In the usual notation $h(x, y) = (x^2 - 1)y - x^3$. For the approximate solution $x = a\cos t$, $y = \dot{x} = -a\sin t$, the energy change equation (4.8) becomes

$$g(a) = \varepsilon a \int_0^{2\pi} h(a\cos t, -a\sin t)\sin t\, dt$$

$$= \varepsilon a \int_0^{2\pi} [-a(a^2\cos^2 t - 1)\sin t - a^3\cos^3 t]\sin t\, dt$$

$$= \varepsilon a^2 \int_0^{2\pi} [-a^2\cos^2 t\sin^2 t + \sin^2 t - a^2\cos^3 t\sin t]\, dt$$

$$= \varepsilon a^2 \left[-\frac{1}{4}a^2\pi + \pi - 0\right] = \frac{1}{4}\varepsilon a^2 (4 - a^2)$$

Then $g(a) = 0$ where $a = a_0 = 2$, which is the approximate amplitude of the limit cycle. The derivative is

$$g'(a) = \varepsilon a(2 - a^2),$$

so that $g'(2) = -4\varepsilon < 0$. Therefore the limit cycle is stable.
From NODE, (4.21), the period T is given by

$$T = 2\pi - \frac{\varepsilon}{a_0} \int_0^{2\pi} h(a_0\cos\theta, a_0\sin\theta)\cos\theta\, d\theta$$

$$= 2\pi - \frac{\varepsilon}{2} \int_0^{2\pi} [a^3\cos^3\theta\sin\theta - a\sin\theta\cos\theta - a^3\cos^4\theta]\, d\theta$$

$$= 2\pi - \frac{1}{2}\varepsilon \int_0^{2\pi} \left[0 - 0 - a^3\cos^4\theta\right] d\theta$$

$$= 2\pi + \frac{1}{8}\varepsilon a^3 \int_0^{2\pi} (1 + 2\cos 2\theta + \cos^2 2\theta)d\theta$$

$$= 2\pi + \frac{3}{8}\pi a^3\varepsilon + O(\varepsilon^2).$$

- **4.8** Show that the frequency-amplitude relation for the pendulum equation, $\ddot{x} + \sin x = 0$, is $\omega^2 = 2J_1(a)/a$, using the methods of NODE, Section 4.4 or 4.5. (J_1 is the Bessel function of order 1, with representations

$$J_1(a) = \frac{2}{\pi}\int_0^{\frac{1}{2}\pi} \sin(a\cos u)\cos u\, du = \sum_{n=0}^{\infty} \frac{(-1)^n (1/2a)^{2n+1}}{n!(n+1)!} \right).$$

Show that, for small amplitudes, $\omega = 1 - \frac{1}{16}a^2$.

4.8. For the pendulum equation $\ddot{x} + \sin x = 0$, assume a solution of the form $x = a\cos\omega t$. Expand $\sin(a\cos\omega t)$ as a Fourier series of period $2\pi/\omega$. Thus

$$\sin(a\cos\omega t) = \tfrac{1}{2}a_0 + a_1\cos\omega t + b_1\sin\omega t + \cdots, \qquad (i)$$

where

$$a_0 = \frac{\omega}{\pi}\int_0^{2\pi/\omega}\sin(a\cos\omega t)\,dt = 0,$$

$$b_1 = \frac{\omega}{\pi}\int_0^{2\pi/\omega}\sin(a\cos\omega t)\sin\omega t\,dt = 0,$$

$$a_1 = \frac{\omega}{\pi}\int_0^{2\pi/\omega}\sin(a\cos\omega t)\cos\omega t\,dt$$

$$= \frac{1}{\pi}\int_0^{2\pi}\sin(a\cos u)\cos u\,du$$

$$= \frac{4}{\pi}\int_0^{\frac{1}{2}\pi}\sin(a\cos u)\cos u\,du = 2J_1(a), \qquad (ii)$$

where $J_1(a)$ is the Bessel function of order 1. The Bessel function has the power series expansion

$$J_1(a) = \sum_{n=0}^{\infty}\frac{(-1)^n(1/2a)^{2n+1}}{n!(n+1)!}$$

(see G. N. Watson: *A Treatise on the Theory of Bessel Functions*, Cambridge University Press (1966), Ch. 2). Therefore, from (i) and (ii), $\sin(a\cos\omega t) \approx 2J_1(a)\cos\omega t$. The equivalent linear equation becomes

$$\ddot{x} + \frac{2J_1(a)}{a}x = 0,$$

which has the angular frequency ω where

$$\omega = \sqrt{2J_1(a)} \approx 1 - \tfrac{1}{16}a^2$$

for small amplitude a, using the power series for the Bessel function.

- **4.9** In the equation $\ddot{x} + \varepsilon h(x,\dot{x}) + g(x) = 0$, suppose that $g(0) = 0$, and that in some interval $|x| < \delta$, g is continuous and strictly increasing. Show that the origin for the equation $\ddot{x} + g(x) = 0$ is a centre. Let $\zeta(t,a)$ represent its periodic solutions near the origin, where a is a parameter which distinguishes the solutions, say the amplitude. Also, let $T(a)$ be the corresponding period.

By using an energy balance method show that the periodic solutions of the original equation satisfy
$$\int_0^{T(a)} h(\zeta,\dot\zeta)\dot\zeta\,dt = 0.$$
Apply this equation to obtain the amplitude of the limit cycle of the equation
$$\ddot x + \varepsilon(x^2-1)\dot x + v^2 x = 0.$$

4.9. The conservative system $\ddot x + g(x) = 0$ has one equilibrium point, at the origin. The associated potential energy can be expressed in the form
$$\mathcal{V}(x) = \int_0^x g(u)\,du$$

(see NODE, Section 1.3). Since $g(x)$ is continuous and strictly increasing the origin is a minimum value of the potential energy so that the origin is a centre covering the entire phase plane for this conservative system.

Let the general periodic solution of this equation be $x = \zeta(t,a)$, where a is its amplitude. For the full equation,
$$\ddot x + \varepsilon h(x,\dot x) + g(x) = 0,$$
the energy change over one period is, as in Sections 1.5 and 4.1 given by,
$$\mathcal{E}(T(a)) - \mathcal{E}(0) = -\varepsilon \int_0^{T(a)} h(\zeta(t,a),\dot\zeta(t,a))\dot\zeta(t,a)\,dt.$$

The energy change is zero if
$$\int_0^{T(a)} h(\zeta(t,a),\dot\zeta(t,a))\dot\zeta(t,a)\,dt = 0,$$

which determines the parameter a.

In the application, $g(x) = v^2 x$. Therefore $\zeta(t) = a\cos vt$, $T(a) = 2\pi/v$ and $h(x,y) = (x^2-1)y$. The energy balance equation above becomes
$$\int_0^{2\pi/v} (a^2\cos^2 vt - 1)\sin^2 vt\,dt = \frac{\pi(a^2-1)}{4v} = 0.$$

Therefore, approximately, the amplitude of the limit cycle is given by $a = 2$.

- 4.10 For the following equations, show that, for small ε the amplitude $a(t)$ satisfies approximately the equation given.

 (i) $\ddot{x} + \varepsilon(x^4 - 1)\dot{x} + x = 0$, $\quad 16\dot{a} = -\varepsilon a(a^4 - 16)$;
 (ii) $\ddot{x} + \varepsilon \sin(x^2 + \dot{x}^2)\text{sgn}(\dot{x}) + x = 0$, $\quad \pi\dot{a} = -\varepsilon 2a \sin(a^2)$;
 (iii) $\ddot{x} + \varepsilon(x^2 - 1)\dot{x}^3 + x = 0$, $\quad 16\dot{a} = -\varepsilon a^3(a^2 - 6)$.

4.10. Use eqns (4.28), (4.24) in NODE, namely

$$\frac{da}{dt} = -\varepsilon p_0(a), \qquad (i)$$

$$p_0(a) = \frac{1}{2\pi} \int_0^{2\pi} h(a\cos u, a\sin u) \sin u \, du. \qquad (ii)$$

(i) $\ddot{x} + \varepsilon(x^4 -)\dot{x} + x = 0$. In this problem, $h(x, y) = (x^4 - 2)y$. Therefore (ii) becomes

$$p_0(a) = \frac{a}{2\pi} \int_0^{2\pi} (a^2 \cos^4 u - 2) \sin^2 u \, du = \frac{a}{16}(a^4 - 16).$$

Hence the differential equation for a is

$$\frac{da}{dt} = -\varepsilon a(a^4 - 16),$$

close to the limit cycle, which has amplitude 2.

(ii) $\ddot{x} + \varepsilon \sin(x^2 + \dot{x}^2)\text{sgn}(\dot{x}) + x = 0$. In this example, $h(x, y) = \sin(x^2 + y^2)\text{sgn}(y)$. Therefore (ii) becomes

$$p_0(a) = \frac{a}{2\pi} \int_0^{2\pi} \sin(a^2 \cos^2 u + a^2 \sin^2 u) \sin u \, \text{sgn}(-a \sin u) du$$

$$= -\frac{a \sin(a^2)}{2\pi} \int_0^{2\pi} \sin u \, \text{sgn}(a \sin u) du$$

$$= -\frac{2}{\pi} a^2 \sin(a^2)$$

Therefore the differential equation for a is

$$\pi \frac{da}{dt} = -2a\varepsilon \sin(a^2).$$

The system has an infinite set of limit cycles with amplitudes $a = \sqrt{n\pi}$, $(n = 1, 2, 3, \ldots)$.

(iii) $\ddot{x} + \varepsilon(x^2 - 1)\dot{x}^3 + x = 0$. In this example, $h(x, y) = (x^2 - 1)y^3$. Therefore (ii) becomes

$$p_0(a) = \frac{a^3}{2\pi} \int_0^{2\pi} (a^2 \cos^2 u - 1) \sin^4 u \, du$$

$$= \frac{a^3}{16}(a^2 - 6)$$

Therefore the differential equation for a is

$$\frac{da}{dt} = -\frac{a^3 \varepsilon}{16}(a^2 - 6).$$

The limit cycle has amplitude $\sqrt{3}$.

- **4.11** Verify that the equation $\ddot{x} + \varepsilon h(x^2 + \dot{x}^2 - 1)\dot{x} + x = 0$ where $h(u)$ is differentiable and strictly increasing for all u, and $h(0) = 0$, has the periodic $x = \cos(t + \alpha)$ for any α. Using the method of slowly varying amplitude show that this solution is a stable limit cycle when $\varepsilon > 0$.

4.11. The system

$$\ddot{x} + \varepsilon(x^2 + \dot{x}^2 - 1)\dot{x} + x = 0$$

where $h(u)$ is strictly increasing and $h(0) = 0$, has one equilibrium point, at $(x, y) = (0, 0)$ in the usual phase plane. That the equation has the periodic solution $x = \cos(t + \alpha)$, where α is arbitrary, can be verified by direct substitution. Since the system is autonomous, we can put $\alpha = 0$ without loss. From NODE, eqn (4.8), $g(a)$ is given by

$$g(a) = -\varepsilon a^2 \int_0^{2\pi} h(a^2 \cos^2 t + a^2 \sin^2 t - 1) \sin^2 t \, dt$$

$$= -\varepsilon a^2 h(a^2 - 1) \int_0^{2\pi} \sin^2 t \, dt$$

$$= -\varepsilon \pi a^2 h(a^2 - 1)$$

Its derivative is

$$g'(a) = -2a\varepsilon \pi h(a^2 - 1) - 2a^3 \varepsilon \pi h'(a^2 - 1).$$

Therefore $g'(1) = -2\varepsilon \pi h'(0) < 0$, which implies that the limit cycle is stable.

- 4.12 Find, by the method of NODE, Section 4.5, the equivalent linear equation for
$$\ddot{x} + \varepsilon(x^2 + \dot{x}^2 - 1)\dot{x} + x = 0.$$
Show that it gives the limit cycle exactly. Obtain from the linear equation the equations of the nearby spiral paths.

4.12. Consider the equation
$$\ddot{x} + \varepsilon(x^2 + \dot{x}^2 - 1)\dot{x} + x = 0, \tag{i}$$

Suppose $x \approx a \cos \omega t$ (the equation is autonomous, so the phase is immaterial). For this solution the damping term

$$\varepsilon(x^2 + \dot{x}^2 - 1)\dot{x} = -\varepsilon(a^2 \cos^2 \omega t + a^2\omega^2 \sin^2 \omega t - 1) a\omega \sin \omega t$$
$$= \tfrac{1}{4} a\omega\varepsilon[(4 - a^2 - 3a^2\omega^2)\sin \omega t + (a^2\omega^2 - a^2)\sin 3\omega t],$$

in terms of multiple angles (really a Fourier series expansion). Neglecting the higher harmonic (it turns out later that its coefficient is zero anyhow), and using $\dot{x} = -a\omega \sin \omega t$, the damping term is equivalent to

$$\varepsilon(x^2 + \dot{x}^2 - 1)\dot{x} \approx \tfrac{1}{4} a\omega\varepsilon(4 - a^2 - 3a^2\omega^2)\sin \omega t$$
$$= -\tfrac{1}{4}\varepsilon(4 - a^2 - 3a^2\omega^2)\dot{x}.$$

Hence the equivalent linear equation is

$$\ddot{x} - \tfrac{1}{4}\varepsilon(4 - a^2 - 3a^2\omega^2)\dot{x} + x = 0. \tag{ii}$$

The damping term vanishes if $4 - a^2 - 3a^2\omega^2 = 0$ leaving the simple harmonic equation $\ddot{x} + x = 0$, which has frequency $\omega = 1$. Hence the amplitude of the limit cycle is $a = 1$. Since eqn (i) has the exact solution $x = \cos t$, the equivalent linear equation gives the exact periodic solution in this case.

Equation (ii) becomes
$$\ddot{x} - \varepsilon(1 - a^2)\dot{x} + x = 0. \tag{iii}$$

Consider a solution on a nearby phase path for which $x(0) = a_0$, $\dot{x}(0) = 0$, where $|a_0 - 1|$ is small. Put $a = a_0$ into differential equation (iii). The characteristic equation of this linear damped equation is

$$m^2 - \varepsilon(1 - a_0^2)\dot{x} + x = 0, \tag{iv}$$

which has the solutions

$$\left.\begin{array}{c}m_1\\m_2\end{array}\right\} = \frac{1}{2}\left[\varepsilon(1-a_0^2) \pm i\sqrt{\{4-\varepsilon^2(1-a_0^2)^2\}}\right] = \alpha \pm i\beta,$$

say. The general solution of (iv) is

$$x = e^{\alpha t}[A\cos\beta t + B\sin\beta t],$$

for which the initial conditions imply

$$a_0 = A, \quad 0 = \alpha A + \beta B.$$

Hence $A = a_0$ and $B = -a_0\alpha/\beta$, so that the required solution is

$$x = \frac{a_0}{\beta}e^{\alpha t}[\beta\cos\beta t - \alpha\sin\beta t].$$

- **4.13** Use the method of equivalent linearization to find the amplitude and frequency of the limit cycle of the equation
$$\ddot{x} + \varepsilon(x^2 - 1)\dot{x} + x + \varepsilon x^3 = 0, \quad 0 < \varepsilon \ll 1.$$
Write down the equivalent linear equation.

4.13. (See NODE, Section 4.5.) Consider the equation

$$\ddot{x} + \varepsilon(x^2 - 1)\dot{x} + x + \varepsilon x^3 = 0. \tag{i}$$

Since the system is autonomous, we need only consider the solution $x \approx a\cos\omega t$. Substitute into (i) so that

$$\ddot{x} + \varepsilon(x^2 - 1)\dot{x} + x + \varepsilon x^3 = \tfrac{1}{4}a(4 - 4\omega^2 + 3a^2\varepsilon)\cos\omega t$$
$$+ \tfrac{1}{4}a\omega\varepsilon(4 - a^2)\sin\omega t + \text{higher harmonics}$$

The coefficients of the first harmonics vanish if

$$4 - 4\omega^2 + 3a^2\varepsilon = 0, \tag{ii}$$

$$4 - a^2 = 0. \tag{iii}$$

Hence from (iii), the amplitude of the periodic solution is approximately $a = 2$ and from (ii) its frequency is

$$\omega^2 = 1 + 3a^2\varepsilon = 1 + 3\varepsilon.$$

For small ε, $\omega \approx 1 + \frac{3}{2}\varepsilon$. The equation has a periodic solution given approximately by

$$x = 2\cos\left(1 + \frac{3}{2}\varepsilon\right)t.$$

Putting $x = a\cos\omega t$, nonlinear terms in (i) can be expressed as follows:

$$\varepsilon(x^2 - 1)\dot{x} = -\varepsilon a\omega(a^2\cos^2\omega t - 1)\sin\omega t = \tfrac{1}{4}a\omega\varepsilon(a^2 - 4)\sin\omega t + \text{(higher harmonics)},$$

$$\approx \tfrac{1}{4}\varepsilon(a^2 - 4)\dot{x}$$

$$\varepsilon x^3 = \varepsilon a^3\cos^3\omega t = \tfrac{3}{4}\varepsilon a^3\cos\omega t + \text{(higher harmonics)} \approx \tfrac{3}{4}\varepsilon a^2 x.$$

Finally the equivalent linear equation is

$$\ddot{x} + \tfrac{1}{4}\varepsilon(4 - a^2)\dot{x} + \left(1 + \tfrac{3}{4}\varepsilon a^2\right)x = 0.$$

- **4.14** The equation $\ddot{x} + x^3 = 0$ has a centre at the origin in the phase plane (with $\dot{x} = y$)
(i) Substitute $x = a\cos\omega t$ to find by the harmonic balance method the frequency–amplitude relation $\omega = \sqrt{3}a/2$.
(ii) Construct, by the method of equivalent linearization, the associated linear equation, and show how the processes (i) and (ii) are equivalent.

4.14. The equation $\ddot{x} + x^3 = 0$ has one equilibrium point at the origin, which is a centre.

(i) Let $x \approx a\cos\omega t$, where a and ω are constants. Then

$$\ddot{x} + x^3 = -a\omega^2\cos\omega t + \tfrac{1}{4}(3a^3\cos\omega t + a^3\cos 3\omega t)$$

$$= \left(-a\omega^2 + \tfrac{3}{4}a^3\right)\cos\omega t + \text{higher harmonic}.$$

The coefficient of $\cos\omega t$ is zero if $\omega = \tfrac{1}{2}\sqrt{3}a$, which gives the approximate relation between amplitude and frequency.

(ii) Since, if $x = a\cos\omega t$,

$$x^3 = \tfrac{1}{4}(3a^3\cos\omega t + \text{higher harmonic}),$$

we replace the cube term by $\tfrac{3}{4}a^2 x$. Hence the equivalent linear equation is

$$\ddot{x} + \tfrac{3}{4}a^2 x = 0.$$

From the coefficient of x, it can be confirmed that $\omega = \tfrac{1}{2}\sqrt{3}a$ as in (i).

- **4.15** The displacement x of relativistic oscillator satisfies
$$m_0\ddot{x} + k(1 - (\dot{x}/c)^2)^{3/2}x = 0.$$
Show that the equation becomes $\ddot{x} + (\alpha/a)x = 0$ when linearized with respect to the approximate solution $x = a\cos\omega t$ by the method of equivalent linearization, where
$$\alpha = \frac{1}{\pi}\int_0^{2\pi} \frac{ka}{m_0}\cos^2\theta \left(1 - \frac{a^2\omega^2}{c^2}\sin^2\theta\right)^{3/2} d\theta.$$

Confirm that, when $a^2\omega^2/c^2$ is small, the period of the oscillations is given approximately by
$$2\pi\sqrt{\left(\frac{m_0}{k}\right)}\left(1 + \frac{3a^2k}{16m_0c^2}\right).$$

4.15. The relativistic oscillator has the equation
$$m_0\ddot{x} + k\left[1 - \left(\frac{\dot{x}}{c}\right)^2\right]^{3/2} = 0.$$

If $x = a\cos\omega t$ we require the first cosine term, namely $\alpha\cos\omega t$, of the Fourier series for
$$q(t) = \frac{k}{m_0}\left[1 - \left(\frac{a\omega\sin\omega t}{c}\right)^2\right]^{3/2}\cos\omega t.$$

Thus
$$\alpha = \frac{\omega a k}{\pi m_0}\int_0^{2\pi/\omega}\left[1 - \frac{a^2\omega^2}{c^2}\sin^2\omega t\right]^{3/2}\cos^2\omega t\, dt$$
$$= \frac{ka}{\pi m_0}\int_0^{2\pi}\left[1 - \frac{a^2\omega^2}{c^2}\sin^2\theta\right]^{3/2}\cos^2\theta\, d\theta \quad (\text{putting } \omega t = \theta)$$

The equivalent linear equation becomes
$$\ddot{x} + \left(\frac{\alpha}{a}\right)x = 0,$$
as required, so that $\omega^2 = \alpha/a$.
For $a^2\omega^2/c^2$ small,
$$\left[1 - \frac{a^2\omega^2}{c^2}\sin^2\theta\right]^{3/2} = 1 - \frac{3a^2\omega^2}{2c^2}\sin^2\theta + \cdots.$$

Therefore

$$\alpha \approx \frac{ka}{\pi m_0} \int_0^{2\pi} \left[\cos^2\theta - \frac{3a^2\omega^2}{2c^2} \sin^2\theta \cos^2\theta \right] d\theta$$

$$= \frac{ka}{\pi m_0} \left[\pi - \frac{3a^2\omega^2}{8c^2} \int_0^{2\pi} \sin^2 2\theta\, d\theta \right]$$

$$= \frac{ka}{m_0} \left[1 - \frac{3a^2\omega^2}{8c^2} \right].$$

The period T is given approximately by

$$T = \frac{2\pi}{\omega} = 2\pi\sqrt{\frac{a}{\alpha}} = 2\pi\sqrt{\frac{m_0}{k}} \left(1 - \frac{3a^2\omega^2}{8c^2} \right)^{-1/2}$$

$$\approx 2\pi\sqrt{\frac{m_0}{k}} \left(1 + \frac{3a^2\omega^2}{16c^2} \right)$$

using the binomial expansion.

- **4.16** Show that the phase paths $\ddot{x} + (x^2 + \dot{x}^2)x = 0$, $\dot{x} = y$, are given by $e^{-x^2}(y^2 + x^2 - 1) = $ constant.
Show that the surface $e^{-x^2}(y^2 + x^2 - 1) = z$ has a maximum at the origin, and deduce that the origin is a centre.
Use the method of harmonic balance to obtain the frequency–amplitude relation $\omega^2 = 3a^2/(4-a^2)$ for $a < 2$, assuming solutions of the approximate form $a\cos\omega t$. Verify that $\cos t$ is an exact solution, and that $\omega = 1$, $a = 1$ is predicted by harmonic balance.
Plot some exact phase paths to indicate where the harmonic balance methods likely to be unreliable.

4.16. The phase paths of the equation

$$\ddot{x} + (x^2 + \dot{x}^2)x = 0, \tag{i}$$

are given by the differential equation

$$\frac{dy}{dx} = -\frac{(x^2 + y^2)x}{y}.$$

The equation can be reorganized into

$$\frac{d(y^2)}{dx} + 2xy^2 = -2x^3,$$

which is of integrating-factor type. Hence

$$\frac{d(y^2 e^{x^2})}{dx} = -2x^3 e^{x^2},$$

which can be integrated as follows:

$$y^2 e^{x^2} = -2 \int x^3 e^{x^2} dx = (1 - x^2) e^{x^2} + C.$$

Hence the phase paths are given by

$$e^{-x^2}(y^2 + x^2 - 1) = \text{constant}.$$

The system has one equilibrium point at the origin.
Let $z = e^{-x^2}(y^2 + x^2 - 1)$. Since

$$\frac{\partial z}{\partial x} = 2e^{-x^2}(2 - x^2 - y^2), \quad \frac{\partial z}{\partial y} = 2y e^{-x^2}.$$

Clearly z has a stationary point at $(0, 0, -1)$. Near the origin

$$z = e^{-x^2}(y^2 + x^2 - 1) \approx (1 - x^2)(y^2 + x^2 - 1) \approx -1 + 2x^2 + y^2 > -1$$

for $0 < |x|, |y| \ll 1$. Hence z has a minimum at $(0, 0, 1)$, which means that locally the phase paths are closed about the equilibrium point implying that the origin is a centre.

Suppose that $x(t)$ is approximated by its first harmonic, $x \approx a \cos t$, where a, ω are constant, with initial conditions $x(0) = a > 0$, $\dot{x}(0) = 0$. Then

$$(x^2 + \dot{x}^2)x = a^3(\cos^2 \omega t - \omega^2 \sin^2 \omega t) \cos \omega t$$
$$= \tfrac{1}{4} a^3 (3 + \omega^2) \cos \omega t + \text{higher harmonics} \qquad \text{(ii)}$$

and

$$\ddot{x} = -a\omega^2 \cos \omega t + \text{higher harmonics}. \qquad \text{(iii)}$$

Neglecting the higher harmonics, eqn (i) becomes

$$\left\{ -a\omega^2 + \tfrac{1}{4} a^3 (3 + \omega^2) \right\} \cos \omega t = 0$$

for all t. Therefore the relation between the amplitude a and the circular frequency ω on a particular path is

$$a^2 = \frac{4\omega^2}{3 + \omega^2} \quad \text{or} \quad \omega^2 = \frac{3a^2}{4 - a^2}. \qquad \text{(iv)}$$

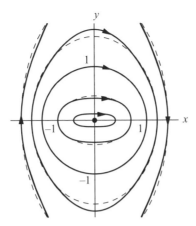

Figure 4.2 Problem 4.16: Phase diagram for $\dot{x} = y$, $\dot{y} = -(x^2 + y^2)x$ is given by the solid curves: the dashed curves represent paths obtained for the equivalent linear equation (v).

In terms of t we than have, from (iv)

$$x(t) \approx \frac{2\omega}{(3+\omega^2)^{1/2}} \cos \omega t,$$

(or in terms of amplitude, $x(t) \approx a \cos [\sqrt{3}at/(\sqrt{(4-a^2)})]$).
Alternatively, we may use (ii) along with $x = a \cos \omega t$, giving

$$(x^2 + \dot{x}^2)x = \tfrac{1}{4}a^2(3+\omega^2)x,$$

to approximate to (i) by the equivalent linear equation

$$\ddot{x} + \tfrac{1}{4}a^2(3+\omega^2)x = 0. \tag{v}$$

The solutions of (v) take the required form $x(t) = a \cos \omega t$ only if ω and a are related by $a^2 = 4\omega^2/(3+\omega^2)$, which is consistent with (iv). The exact phase paths are shown in Figure 4.2, which can be compared with the dashed lines given by the equivalent linear equation (v). Inaccuracies grow for amplitudes greater than about 1.2.

- 4.17 Show, by the method of harmonic balance, that the frequency–amplitude relation for the periodic solutions of the approximate form $a \cos \omega t$, for

$$\ddot{x} - x + \alpha x^3 = 0, \quad \alpha > 0,$$

is $\omega^2 = \tfrac{3}{4}\alpha a^2 - 1$.

By analysing the phase diagram, explain the lower bound $2/\sqrt{(3\alpha)}$ for the amplitude of periodic motion. Find the equation of the phase paths, and compare where the separatrix cuts the x-axis. with the amplitude $2/\sqrt{(3\alpha)}$.

4.17. The system

$$\ddot{x} - x + \alpha x^3 = 0, \quad \alpha > 0$$

has equilibrium points at $x = 0$ (saddle point), $\pm 1/\sqrt{\alpha}$ (centres). Let $x = a \cos \omega t$. Then

$$\alpha x^3 = \alpha a^3 \cos^3 \omega t = \tfrac{3}{4}\alpha a^3 \cos \omega t + \text{higher harmonic}.$$

The equivalent linear equation becomes

$$\ddot{x} + (\tfrac{3}{4}\alpha a^2 - 1)x = 0. \tag{i}$$

Hence the frequency amplitude relation is

$$\omega^2 = \tfrac{3}{4}\alpha a^2 - 1. \tag{ii}$$

The actual phase paths are given by solutions of the equation

$$\frac{dy}{dx} = \frac{x - \alpha x^3}{y}, \tag{iii}$$

which is a separable equation with general solution

$$y^2 = x^2 - \tfrac{1}{2}x^4 + C.$$

Paths through the saddle point at the origin occur for $C = 0$ given by $y^2 = x^2 - \tfrac{1}{2}\alpha x^4$. These are two homoclinic paths each surrounding a centre as shown in Figure 4.3, and they intersect the x axis at $x = \pm\sqrt{(2\alpha)}$. There are periodic solutions which surround these homoclinic paths, and it is these which are approximated to by the harmonic balance above. Since ω^2 must be positive, eqn (ii) implies that these amplitudes must not fall below $2/\sqrt{(3\alpha)}$. There is some

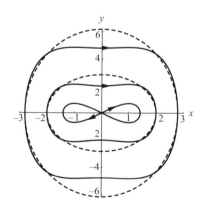

Figure 4.3 Problem 4.17: Phase diagram for $\dot{x} = y$, $\dot{y} = x - \alpha x^3$, with $\alpha = 1$, is given by the solid curves: the dashed curves represent paths obtained for the equivalent linear equation (ii).

discrepancy between these numbers since $\sqrt{2} = 1.414\ldots$ and $2/\sqrt{3} = 1.154\ldots$. From (i), the approximate equations for the phase paths are given by the ellipses

$$\frac{x^2}{a^2} + \frac{y^2}{a^2(\frac{3}{4}\alpha a^2 - 1)} = 1. \qquad (iv)$$

Comparison between the exact phase paths and the approximate ones obtained by harmonic balance are shown in Figure 4.3. The accuracy improves for larger amplitudes, but is not good for phase paths outside but near to the homoclinic paths.

Approximations for the phase paths about the centres at $(\pm 1, 0)$ can be found by finding c and a in the approximation $x = c + a\cos\omega t$ applied in harmonic balance.

- **4.18** Apply the method of harmonic balance to the equation $\ddot{x} + x - \alpha x^2 = 0$, $\alpha > 0$, using the approximate form of solution $x = c + a\cos\omega t$ to show that
$$\omega^2 = 1 - 2\alpha c, \quad c = [1 - \sqrt{(1 - 2\alpha^2 a^2)}]/(2\alpha).$$
Deduce the frequency-amplitude relation
$$\omega = (1 - 2\alpha^2 a^2)^{1/4}, \quad a = 1/(\sqrt{2}\alpha).$$
Explain, in general terms, why an upper bound on the amplitude is to be expected.

4.18. The system
$$\ddot{x} + x - \alpha x^2 = 0, \quad \dot{x} = y$$
has two equilibrium points, at $(0,0)$ (a centre) and $(1/\alpha, 0)$ (a saddle point). For reasons of lack of symmetry we choose $x = c + a\cos\omega t$, and substitute this into the differential equation so that

$$x'' + x - \alpha x^2 = \tfrac{1}{2}(2c - \alpha a^2 - 2\alpha c^2) + (a - a\omega^2 - 2\alpha ac)\cos\omega t - \tfrac{1}{2}\alpha a^2 \cos 2\omega t.$$

Neglecting the second harmonic, the right-hand side satisfies the differential equation if

$$2c - \alpha a^2 - 2\alpha c^2 = 0, \quad a(1 - \omega^2 - 2\alpha c) = 0.$$

Therefore
$$\omega^2 = 1 - 2\alpha c, \quad c = \frac{1}{2\alpha}[1 \pm \sqrt{(1 - 2\alpha^2 a^2)}].$$

The lower sign has to be chosen to ensure ω^2 positive. Elimination of c between these equations leads to the frequency–amplitude equation

$$\omega = (1 - 2\alpha^2 a^2)^{1/4}.$$

> • 4.19 Apply the method of harmonic balance to the equation $\ddot{x} - x + x^3 = 0$ in the neighbourhood of the centre at $x = 1$, using the approximate form of solution $x = 1 + c + a \cos \omega t$. Deduce that the mean displacement, frequency and amplitude are related by
> $$\omega^2 = 3c^2 + 6c + 2 + \tfrac{3}{4}a^2, \quad 2c^3 + 6c^2 + c(4 + 3a^2) + 3a^2 = 0.$$

4.19. The system
$$\ddot{x} - x + x^3 = 0, \quad \dot{x} = y,$$
has three equilibrium points at $(0, 0)$ (a saddle point) and at $(\pm 1, 0)$ (centres). Consider approximate solutions of the form $x \approx 1 + c + a \cos \omega t$. Then

$$\ddot{x} - x + x^3 = \tfrac{1}{2}(3a^2 + 4c + 3a^2 c + 6c^2 + 2c^3)$$

$$+ \tfrac{1}{4}(8a + 3a^3 + 24ac + 12ac^2 - 4a\omega^2) \cos \omega t$$

$$+ \tfrac{3}{2}a^2(1 + c) \cos 2\omega t + \tfrac{1}{4}a^3 \cos 3\omega t.$$

The constant term and the first harmonic are zero if

$$2c^3 + 6c^2 + c(4 + 3a^2) + 3a^2 = 0,$$

$$\omega^2 = 3c^2 + 6c + 2 + \tfrac{3}{4}a^2,$$

as required.

> • 4.20 Consider the van der Pol equation with nonlinear restoring force
> $$\ddot{x} + \varepsilon(x^2 - 1)\dot{x} + x - \alpha x^2 = 0,$$
> where ε and α are small. By assuming solutions approximately of the form $x = c + a \cos \omega t + b \sin \omega t$, show that the mean displacement, frequency, and amplitude are related by
> $$c = 2\alpha, \quad \omega^2 = 1 - 4\alpha^2, \quad a^2 + b^2 = 4(1 - 4\alpha^2).$$

4.20. The system
$$\ddot{x} + \varepsilon(x^2 - 1)\dot{x} + x - \alpha x^2 = 0,$$
has two equilibrium points at $(0, 0)$ (a centre) and at $(1/\alpha, 0)$ (a saddle point). Substitute into the equation $x = c + a \cos \omega t + b \sin \omega t$ and expand in terms of multiple angles, but retain only

the constant term and the first harmonics. Then

$$\ddot{x} + \varepsilon(x^2 - 1)\dot{x} + x - \alpha x^2 = \tfrac{1}{4}[4c - 2\alpha(a^2 + b^2 + 4c^2)] + \tfrac{1}{4}[4a(1 - \omega^2) - 8\alpha ac$$
$$+ \varepsilon b\omega(-4 + a^2 + b^2 + 4c^2)]\cos \omega t$$
$$+ \tfrac{1}{4}[4b(1 - \omega^2) - 8\alpha bc + \varepsilon a\omega(4 - a^2 - b^2 - 4c^2)]\sin \omega t$$
$$+ \text{higher harmonics}$$

The constant term and the first harmonics vanish if

$$2c - \alpha(a^2 + b^2 + 4c^2) = 0, \tag{i}$$

$$4a(1 - \omega^2) - 8\alpha ac + \varepsilon b\omega(-4 + a^2 + b^2 + 4c^2) = 0, \tag{ii}$$

$$4b(1 - \omega^2) - 8\alpha bc + \varepsilon a\omega(4 - a^2 - b^2 - 4c^2) = 0. \tag{iii}$$

From (ii) and (iii),

$$(1 - \omega^2) = 2\alpha c, \quad a^2 + b^2 + 4c^2 = 4.$$

It follows from (i) therefore that

$$c = 2\alpha, \quad \omega^2 = 1 - 4\alpha^2, \quad a^2 + b^2 = 4(1 - 4\alpha^2).$$

Since the system is autonomous any values of a and b which satisfy $a^2 + b^2 = 4(1 - 4\alpha^2)$ will be sufficient. This confirms that we could have chosen $b = 0$ (say) in our original choice of solution since the system is autonomous.

• **4.21** Suppose that the nonlinear system

$$\dot{\mathbf{x}} = \mathbf{p}(\mathbf{x}), \quad \text{where } \mathbf{x} = \begin{bmatrix} x \\ y \end{bmatrix},$$

has an isolated equilibrium point $\mathbf{x} = \mathbf{0}$, and that solutions exist which are approximately of the form

$$\tilde{\mathbf{x}} = \mathbf{B}\begin{bmatrix} \cos \omega t \\ \sin \omega t \end{bmatrix}, \quad \mathbf{B} = \begin{bmatrix} a & b \\ c & d \end{bmatrix}.$$

Adapt the method of equivalent linearization to this problem by approximating $\mathbf{p}(\tilde{\mathbf{x}})$ by its

first harmonic terms:
$$p\{\tilde{x}(t)\} = C \begin{bmatrix} \cos \omega t \\ \sin \omega t \end{bmatrix},$$

where **C** is matrix of the Fourier coefficients. It is assumed that
$$\int_0^{2\pi/\omega} p\{\tilde{x}(t)\} dt = 0.$$

Substitute in the system to show that
$$\mathbf{BU} = \mathbf{C}, \text{ where } \mathbf{U} = \begin{bmatrix} 0 & -\omega \\ \omega & 0 \end{bmatrix}.$$

Deduce that the equivalent linear system is
$$\dot{\tilde{x}} = \mathbf{BUB}^{-1}\tilde{x} = \mathbf{CUC}^{-1}\tilde{x},$$

when **B** and **C** are non-singular.

4.21. Consider the general system
$$\dot{x} = p(x), \quad x = \begin{bmatrix} x \\ y \end{bmatrix}.$$

Consider an approximate solution
$$\tilde{x} \approx \mathbf{B} \begin{bmatrix} \cos \omega t \\ \sin \omega t \end{bmatrix}, \quad \mathbf{B} = \begin{bmatrix} a & b \\ c & d \end{bmatrix}.$$

Assume that
$$p\{\tilde{x}(t)\} = \mathbf{C} \begin{bmatrix} \cos \omega t \\ \sin \omega t \end{bmatrix} + \text{higher harmonics}$$

Substitute $\tilde{x}$ into the differential equation so that
$$\omega \mathbf{B} \begin{bmatrix} -\sin \omega t \\ \cos \omega t \end{bmatrix} = \mathbf{C} \begin{bmatrix} \cos \omega t \\ \sin \omega t \end{bmatrix} + \text{higher harmonics},$$

or
$$\mathbf{BU} \begin{bmatrix} \cos \omega t \\ \sin \omega t \end{bmatrix} = \mathbf{C} \begin{bmatrix} \cos \omega t \\ \sin \omega t \end{bmatrix} + \text{higher harmonics},$$

where
$$\mathbf{U} = \begin{bmatrix} 0 & -\omega \\ \omega & 0 \end{bmatrix}.$$

Therefore the leading harmonics balance if $\mathbf{BU} = \mathbf{C}$. The equivalent linear system is therefore
$$\dot{\tilde{x}} = \mathbf{CB}^{-1}\tilde{x} = \mathbf{BUB}^{-1}\tilde{x} = \mathbf{CUC}^{-1}\tilde{x}.$$

4 : Periodic solutions; averaging methods

• **4.22** Use the method of Problem 4.21 to construct a linear system equivalent to the van der Pol equation
$$\dot{x} = y, \quad \dot{y} = -x - \varepsilon(x^2 - 1)y.$$

4.22. For the system
$$\dot{x} = y, \quad \dot{y} = -x - \varepsilon(x^2 - 1)y,$$

$$\mathbf{x} = \begin{bmatrix} x \\ y \end{bmatrix}, \quad \mathbf{p}(\mathbf{x}) = \begin{bmatrix} y \\ -x - \varepsilon(x^2 - 1)y \end{bmatrix}.$$

It has one equilibrium point, at $(0,0)$. Let

$$\tilde{\mathbf{x}} = \mathbf{B} \begin{bmatrix} \cos \omega t \\ \sin \omega t \end{bmatrix}$$

in the notation of Problem 4.21. Then,

$$\mathbf{p}(\tilde{\mathbf{x}}) = \begin{bmatrix} c \cos \omega t + d \sin \omega t \\ \frac{1}{4}(-4a + 4\varepsilon c - 3\varepsilon a^2 c - \varepsilon b^2 c - 2\varepsilon abd) \cos \omega t \\ + \frac{1}{4}(-4b - 2\varepsilon abc + 4\varepsilon d - \varepsilon a^2 d - 3\varepsilon b^2 d) \sin \omega t + \\ + \text{(higher harmonics)} \end{bmatrix}.$$

Hence

$$\mathbf{C} = \begin{bmatrix} c & d \\ \frac{1}{4}(-4a + 4\varepsilon c - 3\varepsilon a^2 c - \varepsilon b^2 c - 2\varepsilon abd) & \frac{1}{4}(-4b - 2\varepsilon abc + 4\varepsilon d - \varepsilon a^2 d - 3\varepsilon b^2 d) \end{bmatrix}.$$

From Problem 21, we know that

$$\mathbf{BU} = \begin{bmatrix} a & b \\ c & d \end{bmatrix} \begin{bmatrix} 0 & -\omega \\ \omega & 0 \end{bmatrix} = \begin{bmatrix} b\omega & -a\omega \\ d\omega & -c\omega \end{bmatrix} = \mathbf{C},$$

given above. Therefore $c = b\omega$ and $d = -a\omega$. Eliminating c and d in the matrix $\mathbf{C}$, we have

$$\mathbf{C} = \begin{bmatrix} b\omega & -a\omega \\ \frac{1}{4}(-4a + 4\varepsilon b\omega - \varepsilon b\omega(a^2 + b^2)) & \frac{1}{4}(-4b - 4\varepsilon a\omega + \varepsilon a(a^2 + b^2)) \end{bmatrix}.$$

Finally from the second rows in $\mathbf{BU} = \mathbf{C}$,

$$-4a\omega^2 = -4a + 4\varepsilon b\omega - \varepsilon b\omega(a^2 + b^2),$$

$$-4b\omega^2 = -4b - 4\varepsilon a\omega + \varepsilon a(a^2 + b^2).$$

242 Nonlinear ordinary differential equations: problems and solutions

These two equations imply $\omega = 1$ and $a^2 + b^2 = 4$. As expected for the van der Pol equation the frequency of the limit cycle is 1 and its amplitude is 2.

To find the equivalent linear equation go back to the general equation

$$\mathbf{C} = \mathbf{BU} = \begin{bmatrix} b\omega & -a\omega \\ d\omega & -c\omega \end{bmatrix}.$$

Its inverse is given by

$$\mathbf{C}^{-1} = \frac{1}{\omega(bc - ad)} \begin{bmatrix} c & a \\ d & b \end{bmatrix}.$$

The equivalent linear equation from Problem 2.21 is

$$\dot{\tilde{\mathbf{x}}} = \mathbf{CUC}^{-1}\tilde{\mathbf{x}} = \frac{\omega}{ad - bc} \begin{bmatrix} ac + bd & -(a^2 + b^2) \\ c^2 + d^2 & -(ac + bd) \end{bmatrix} \tilde{\mathbf{x}}.$$

• **4.23** Apply the method of Problem 4.21 to construct a linear system equivalent to

$$\begin{bmatrix} \dot{x} \\ \dot{y} \end{bmatrix} = \begin{bmatrix} \varepsilon & 1 \\ -1 & \varepsilon \end{bmatrix} \begin{bmatrix} x \\ y \end{bmatrix} + \begin{bmatrix} 0 \\ -\varepsilon x^2 y \end{bmatrix},$$

and show that the limit cycle has frequency given by $\omega^2 = 1 - 5\varepsilon^2$ for ε small.

4.23. Consider the system

$$\begin{bmatrix} \dot{x} \\ \dot{y} \end{bmatrix} = \begin{bmatrix} \varepsilon & 1 \\ -1 & \varepsilon \end{bmatrix} \begin{bmatrix} x \\ y \end{bmatrix} + \begin{bmatrix} 0 \\ -\varepsilon x^2 y \end{bmatrix}.$$

Equilibrium occurs where

$$\dot{x} = \varepsilon x + y = 0, \quad \dot{y} = -x + \varepsilon y - \varepsilon x^2 y = 0.$$

For ε small (< 1), the system has one equilibrium point, at the origin. As in Problem 4.21, let

$$\tilde{\mathbf{x}} = \mathbf{B} \begin{bmatrix} \cos \omega t \\ \sin \omega t \end{bmatrix}.$$

Then

$$\mathbf{p}(\tilde{\mathbf{x}}) = \mathbf{C} \begin{bmatrix} \cos \omega t \\ \sin \omega t \end{bmatrix},$$

where

$$\mathbf{C} = \begin{bmatrix} \varepsilon a + c & \varepsilon b + d \\ \frac{1}{4}(-4a + 4\varepsilon c - 3\varepsilon a^2 c & \frac{1}{4}(-4b - 2\varepsilon abc + 4\varepsilon d \\ -\varepsilon b^2 c - 2\varepsilon abd) & -\varepsilon a^2 d - 3\varepsilon b^2 d) \end{bmatrix}.$$

The requirement

$$\mathbf{BU} = \begin{bmatrix} a & b \\ c & d \end{bmatrix} \begin{bmatrix} 0 & -\omega \\ \omega & 0 \end{bmatrix} = \begin{bmatrix} b\omega & -a\omega \\ d\omega & -c\omega \end{bmatrix} = \mathbf{C}$$

implies, from the first row,

$$b\omega = \varepsilon a + c, \quad -a\omega = \varepsilon b + d.$$

Hence $c = -\varepsilon a + b\omega$ and $d = -a\omega - \varepsilon b$. Elimination of c and d in $\mathbf{C}$ leads to

$$\mathbf{C} = \begin{bmatrix} b\omega & -a\omega \\ \tfrac{1}{4}[-4a(1+\varepsilon^2) + 4\varepsilon\omega b & \tfrac{1}{4}[-4b(1+\varepsilon^2) - 4\varepsilon\omega a \\ +3\varepsilon^2 ar^2 - \varepsilon\omega br^2] & +3\varepsilon^2 br^2 + \varepsilon\omega ar^2] \end{bmatrix},$$

where $r^2 = a^2 + b^2$. The second rows in $\mathbf{BU} = \mathbf{C}$ imply

$$-4a\omega^2 - 4\varepsilon\omega b = -4a(1+\varepsilon^2) + 4\varepsilon\omega b + 3\varepsilon^2 ar^2 - \varepsilon\omega br^2,$$

$$4\varepsilon\omega a - 4b\omega^2 = -4b(1+\varepsilon^2) - 4\varepsilon\omega a + 3\varepsilon^2 br^2 + \varepsilon\omega ar^2.$$

Elimination between these equations leads to

$$r^2 = a^2 + b^2 = 8, \quad \omega^2 = 1 - 5\varepsilon^2,$$

provided $\varepsilon < 1/\sqrt{5}$.

We can choose a and b to be convenient values, since the system is autonomous. If $b = 0$, then

$$\mathbf{C} = \begin{bmatrix} 0 & -a\omega \\ \tfrac{1}{4}[-4a(1+\varepsilon^2) + 3\varepsilon^2 a^3] & \tfrac{1}{4}[-4\varepsilon\omega a + \varepsilon\omega a^3] \end{bmatrix}.$$

The equivalent linear equation is

$$\dot{\mathbf{x}} = \mathbf{CUC}^{-1}\mathbf{x}.$$

The inverse of $\mathbf{C}$ is given by

$$\mathbf{C}^{-1} = \begin{bmatrix} -\varepsilon & -1 \\ -\omega & 0 \end{bmatrix} \Big/ [2\sqrt{2}(1-5\varepsilon^2)].$$

Hence

$$\mathbf{CUC}^{-1} = \begin{bmatrix} \varepsilon & 1 \\ -1+4\varepsilon^2 & -\varepsilon \end{bmatrix}.$$

The eigenvalues of $\mathbf{CUC}^{-1}$ are given by

$$\begin{vmatrix} \varepsilon - \lambda & 1 \\ -1 + 4\varepsilon^2 & -\varepsilon - \lambda \end{vmatrix} = 1 - 5\varepsilon^2 + \lambda^2.$$

Hence the eigenfrequency is given by $\omega^2 = 1 - 5\varepsilon^2$, which agrees with the earlier result.

- **4.24** Apply the method of Problem 4.21 to the predator–prey equation (see Section 2.2)
$$\dot{x} = x - xy, \quad \dot{y} = -y + xy,$$
in the neighbourhood of the equilibrium point (1, 1), by using the displaced approximations
$$x = m + a\cos\omega t + b\sin\omega t, \quad y = n + c\cos\omega t + d\sin\omega t.$$
Show that $m = n$, $\omega^2 = 2m - 1$ and $a^2 + b^2 = c^2 + d^2$.

4.24. The predator–prey equations

$$\dot{x} = x - xy, \quad \dot{y} = -y + xy, \quad x, y \geq 0,$$

have equilibrium points at $(0, 0)$ (a saddle point) and at $(1, 1)$ (a centre) (see Example 2.3). The equations may be written

$$\dot{\mathbf{x}} = \mathbf{p}(\mathbf{x}).$$

where

$$\mathbf{p}(\mathbf{x}) = \begin{bmatrix} x - xy \\ -y + xy \end{bmatrix}.$$

Since the equilibrium point $(1, 1)$ is not at the origin, let

$$\tilde{\mathbf{x}} = \begin{bmatrix} m \\ n \end{bmatrix} + \mathbf{B} \begin{bmatrix} \cos\omega t \\ \sin\omega t \end{bmatrix} \quad \text{where } \mathbf{B} = \begin{bmatrix} a & b \\ c & d \end{bmatrix}.$$

Then

$$\mathbf{p}(\tilde{\mathbf{x}}) = \begin{bmatrix} \frac{1}{2}(-ac - bd + 2m - 2mn) \\ \frac{1}{2}(ac + bd - 2n + 2mn) \end{bmatrix}$$

$$+ \begin{bmatrix} a - cm - an & b - dm - bn \\ -c + cm + an & -d + dm + bn \end{bmatrix} \begin{bmatrix} \cos\omega t \\ \sin\omega t \end{bmatrix} + \text{higher harmonics}.$$

Since

$$\dot{\tilde{\mathbf{x}}} = \begin{bmatrix} b\omega & -a\omega \\ d\omega & -c\omega \end{bmatrix} \begin{bmatrix} \cos\omega t \\ \sin\omega t \end{bmatrix},$$

comparison of leading harmonics in the two previous terms implies

$$-ac - bd + 2m - 2mn = 0, \quad ac + bd - 2n + 2mn = 0, \tag{i}$$

$$b\omega = a - cm - an, \quad -a\omega = b - dm - bn, \tag{ii}$$

$$d\omega = -c + cm + an, \quad -c\omega = -d + dm + bn. \tag{iii}$$

From (i) it follows that $m = n$. Equations (ii) and (iii) contain four homogeneous linear equations in a, b, c and d. Non-trivial solutions for these amplitudes exist if, and only if,

$$\Delta = \begin{vmatrix} 1-m & -\omega & -m & 0 \\ \omega & 1-m & 0 & -m \\ m & 0 & -(1-m) & -\omega \\ 0 & m & \omega & -(1-m) \end{vmatrix} = 0,$$

Symbolic computation gives the expansion as

$$\Delta = (1 - 2m + \omega^2)^2.$$

Hence

$$\omega^2 = 2m - 1. \tag{iv}$$

Squaring and adding (ii) and (iii) leads to

$$(r^2 - s^2)(\omega^2 - 1 + 2m) = 0, \tag{v}$$

and

$$(r^2 - s^2)(\omega^2 - 1 + 2m) = 0, \tag{vi}$$

where $r^2 = a^2 + b^2$ and $s^2 = c^2 + d^2$. Hence (v) or (vi) compared with (iv) both imply $r = s$.

- **4.25** Show that the approximation solution for the oscillations of the equation $\ddot{x} = x^2 - x^3$ in the neighbourhood of $x = 1$ is $x = c + a\cos\omega t$, where

$$\omega^2 = \frac{c(15c^2 - 15c + 4)}{2(3c - 1)}, \quad a^2 = \frac{2c^2(1 - c^2)}{3c - 1}.$$

4.25. The system $\ddot{x} = x^2 - x^3$, where $\dot{x} = y$, has equilibrium points at $(0, 0)$ (a higher-order equilibrium point) and at $(1, 0)$ (a centre). Let $x \approx c + a\cos\omega t$. Then

$$\ddot{x} - x^2 + x^3 = \tfrac{1}{2}(-a^2 + 3a^2c - 2c^2 + 2c^3)$$
$$+ \tfrac{1}{4}(3a^3 - 8ac + 12ac^2 - 4a\omega^2)\cos\omega t + \text{higher harmonics}.$$

Hence the constant term and first harmonic vanish if

$$-a^2 + 3a^2c - 2c^2 + 2c^3 = 0,$$

$$a(3a^2 - 8c + 12c^2 - 4\omega^2) = 0.$$

From these equations it follows that ($a \neq 0$),

$$\omega^2 = \tfrac{1}{4}(3a^2 - 8c + 12c^2), \quad a^2 = \frac{2c^2(1-c)}{3c-1}.$$

Elimination of a^2 implies

$$\omega^2 = \frac{c(15c^2 - 15c + 4)}{2(3c-1)}.$$

- 4.26 Use the method of Section 4.2 to obtain approximate solutions of the equation
$\ddot{x} + \varepsilon \dot{x}^3 + x = 0, \quad |\varepsilon| \ll 1.$

4.26. The system

$$\ddot{x} + \varepsilon \dot{x}^2 + x = 0$$

has one equilibrium point, at the origin. Assume a solution of the form

$$x = c + a\cos\omega t.$$

Then

$$\ddot{x} + \varepsilon \dot{x}^2 + x = \tfrac{1}{2}(2c + a^2\omega^2\varepsilon)$$

$$+ a(1-\omega^2)\cos\omega t + \text{higher harmonics}$$

The coefficients of the constant term and the first harmonic vanish if

$$2c + \varepsilon\omega^2 a^2 = 0, \quad a(1-\omega^2) = 0.$$

Hence

$$\omega = 1, \quad c = -\tfrac{1}{2}\varepsilon^2 a^2.$$

Therefore, near the origin the solution by harmonic balance is

$$x = -\tfrac{1}{2}\varepsilon a^2 + a\cos t.$$

- 4.27 Suppose that the equation $\ddot{x} + f(x)\dot{x} + g(x) = 0$ has a periodic solution with phase path C. Represent the equation in the (x, y) phase plane given by
$\dot{x} = y - F(x), \quad \dot{y} = -g(x), \quad \text{where } F(x) = \int_0^x f(u)du$

(this particular phase plane is known as the **Liénard plane**.) Let

$$v(x, y) = \frac{1}{2}y^2 + \int_0^x g(u)du,$$

and by considering dv/dt on the closed path C show that

$$\int_C F(x)dy = 0.$$

On the assumption that van der Pol's equation $\ddot{x} + \varepsilon(x^2 - 1)\dot{x} + x = 0$ has a periodic solution approximately of the form $x = A\cos t$, deduce that and $\omega \approx 1$, $A \approx 2$.

4.27. Consider the equation

$$\ddot{x} + f(x)\dot{x} + g(x) = 0.$$

In the (x, y) phase plane, let

$$y = \dot{x} + F(x), \quad \dot{y} = -g(x),$$

where

$$F(x) = \int_0^x f(u)du.$$

Let

$$v(x, y) = \frac{1}{2}y^2 + \int_0^x g(u)du.$$

Then

$$\frac{dv}{dt} = \frac{d}{dt}\left[\frac{1}{2}(\dot{x} + F(x))^2 + \int_0^x g(u)du\right]$$

$$= (\dot{x} + F(x))(\ddot{x} + f(x)\dot{x}) + g(x)\dot{x}$$

$$= (\dot{x} + F(x))(-g(x)) + g(x)\dot{x}$$

$$= -F(x)g(x) = F(x)\frac{dy}{dt}.$$

Therefore

$$\int_C F(x)dy = \int_C dv = 0. \tag{i}$$

For van der Pol's equation,

$$\ddot{x} + \varepsilon(x^2 - 1)\dot{x} + x = 0,$$

the (x, y) phase plane is defined by

$$\dot{x} = y - \varepsilon \int_0^x (u^2 - 1)du = y\varepsilon(\tfrac{1}{3}x^3 - x), \quad \dot{y} = -x.$$

Equation (i) above applied to this equation with $x = A\cos\omega t$ becomes

$$\int_0^{2\pi/\omega} F(x)\frac{dy}{dt}dt = -\varepsilon\int_0^{2\pi/\omega}(\tfrac{1}{3}A^3\cos^3\omega t - A\cos\omega t)A\cos\omega t\, dt$$

$$= \frac{A^2\pi\varepsilon}{4\omega}(A^2 - 4) = 0.$$

We conclude that the amplitude of the limit cycle is $A = 2$.

To obtain the frequency, observe that the period T can be expressed as

$$T = \int_0^T dt = -\int_0^{2\pi/\omega}\frac{1}{x}\frac{dy}{dt}dt \tag{ii}$$

Now, with $x = 2\cos\omega t$,

$$y = \dot{x} + F(x) = -2\omega\sin\omega t + \varepsilon\left(\frac{8}{3}\cos^3\omega t - 2\cos\omega t\right) \approx -2\omega\sin\omega t,$$

for small ε. Hence (ii) becomes

$$T \approx \int_0^{2\pi/\omega}\frac{2\omega^2\cos\omega t}{2\cos\omega t}dt = \int_0^{2\pi/\omega}\omega^2 dt = 2\pi\omega.$$

Since $T = 2\pi/\omega$, it follows that $\omega = 1$. The limit cycle is therefore given approximately by $x = 2\cos t$.

- 4.28 Apply the slowly varying amplitude method of Section 4.3 to
$$\ddot{x} - \varepsilon\sin\dot{x} + x = 0, \quad (0 < \varepsilon \ll 1),$$
and show that the amplitude a satisfies $\dot{a} = \varepsilon J_1(a)$ approximately. [Use the formula
$$J_1(a) = \frac{1}{\pi}\int_0^\pi \sin(a\sin u)\sin u\, du$$
for the Bessel function $J_1(a)$: see Abramowitz and Stegun (1965, p. 360).]
Find also the approximate differential equation for θ. Using a graph of $J_1(a)$ decide how many limit cycles the system has. Which are stable?

4.28. The slowly varying amplitude method is applied to

$$\ddot{x} - \varepsilon\sin\dot{x} + x = 0, \quad (0 < \varepsilon \ll 1).$$

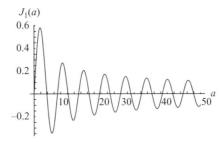

Figure 4.4 Problem 4.28: Bessel function $J_1(a)$ plotted against the amplitude a.

The system has one equilibrium point at the origin which is an unstable spiral. In this problem $h(x, y) = \sin y$. Equation (4.28) (in NODE) becomes

$$\dot a = -\varepsilon p_0(a) = -\frac{\varepsilon}{2\pi}\int_0^{2\pi} \sin(-a \sin u)\sin u\, du$$

$$= \frac{\varepsilon}{2\pi}\int_0^{2\pi} \sin(a \sin u)\sin u\, du$$

$$= \varepsilon J_1(a),$$

where $J_1(a)$ is a Bessel function of order 1. Hence any limit cycles have amplitudes which are the zeros of the Bessel function $J_1(a)$. The graph of $J_1(a)$ versus a displaying its oscillatory character is shown in Figure 4.4. This system has an infinite number of limit cycles. After $a = 0$, the first, third, fifth, etc. zeros correspond to stable limit cycles.

The approximate differential equation for θ is (4.29), namely

$$\dot\theta = -1 - \frac{\varepsilon}{a}r_0(a)$$

$$= -1 - \frac{\varepsilon}{2\pi a}\int_0^{2\pi}\sin(-a\sin u)\cos u\, du$$

$$= -1 + \frac{\varepsilon}{2\pi a}\int_0^{2\pi}\sin(a\sin u)\cos u\, du$$

$$= -1 + \frac{\varepsilon}{2\pi a}\left[-\frac{\cos(a\sin u)}{a}\right]_0^{2\pi}$$

$$= -1$$

Integrating, $\theta = -t + \theta_0$.

5 Perturbation methods

Trigonometric identities

The following identities are useful in the application of perturbation methods.

(A) $\cos^3 t = \frac{3}{4}\cos t + \frac{1}{4}\cos 3t;\ \sin^3 t = \frac{3}{4}\sin t - \frac{1}{4}\sin 3t.$

(B) $(a\cos t + b\sin t)^3 = \frac{3}{4}b(a^2+b^2) + \frac{1}{4}a(a^2+b^2)\sin t - \frac{1}{4}b(3a^2-b^2)\cos 3t$
$+\frac{1}{4}a(3b^2-1)\sin 3t.$

(C) $(c + a\cos t + b\sin t)^3 = \frac{1}{2}c(3a^2+3b^2+2c^2) + \frac{3}{4}b(a^2+b^2+4c^2)\cos t$
$+\frac{3}{4}a(a^2+b^2+4c^2)\sin t + \frac{3}{2}c(b^2-a^2)\cos 2t$
$+3abc\sin 2t + \frac{1}{4}b(b^2-3a^2)\cos 3t$
$+\frac{1}{4}a(3b^2-a^2)\sin 3t.$

- **5.1** Find all the periodic solutions of $\ddot{x} + \Omega^2 x = \Gamma \cos t$ for all values of Ω^2.

5.1. The general solutions of
$$\ddot{x} + \Omega^2 x = \Gamma \cos t$$
are
$$x = A\cos\Omega t + B\sin\Omega t + \frac{\Gamma}{\Omega^2 - 1}\cos t,\quad (\Omega^2 \neq 1),$$
$$x = A\cos\Omega t + B\sin\Omega t + \tfrac{1}{2}\Gamma t\sin t,\quad (\Omega^2 = 1).$$

- $\Omega^2 = 1$. There are no periodic solutions.
- $\Omega = p,\ p(\neq \pm 1)$ an integer. General solution is
$$x = a\cos pt + b\sin pt + \frac{\Gamma}{p^2 - 1}\cos t,$$
which has period 2π.
- $\Omega = 1/q,\ q \neq \pm 1$ an integer. The general solution is
$$x = a\cos(t/q) + b\sin(t/q) + \frac{\Gamma q^2}{1 - q^2}\cos t,$$
which has period $2\pi q$.

- $\Omega = p/q$, p, q integers, $p/q \ne 1$. The general solution is

$$x = a\cos(pt/q) + b\sin(pt/q) + \frac{\Gamma q^2}{p^2 - q^2}\cos t,$$

which has period $2\pi q$.
- Ω = irrational number. The equation has a only one periodic solution

$$x = \frac{\Gamma}{\Omega^2 - 1}\cos t,$$

and this has period 2π.

- **5.2** Find the first harmonics of the solutions of period 2π of the following:
  (i) $\ddot{x} - 0.5x^3 + 0.25x = \cos t$;
  (ii) $\ddot{x} - 0.1x^3 + 0.6x = \cos t$;
  (iii) $\ddot{x} - 0.1\dot{x}^2 + 0.5x = \cos t$.

5.2. These problems are of the form

$$\ddot{x} + \varepsilon h(x, \dot{x}) + \Omega^2 x = \Gamma \cos t.$$

In the direct method (see NODE, Section 5.2) we let

$$x(t) = x_0(t) + \varepsilon x_1(t) + \varepsilon^2 x_2(t) + \cdots .$$

Direct substitution gives the equations

$$\ddot{x}_0 + \Omega^2 x_0 = \Gamma \cos t,$$

$$\ddot{x}_1 + \Omega^2 x_1 = -h(x_0, \dot{x}_0),$$

and so on.

(i) $\ddot{x} - 0.5x^3 + 0.25x = \cos t$. In this problem, $h(x, \dot{x}) = -x^3$, $\Gamma = 1$, $\Omega = \frac{1}{2}$ and $\varepsilon = 0.5$. Therefore x_0 satisfies

$$\ddot{x}_0 + 0.25 x_0 = \cos t.$$

Therefore

$$x_0 = A_0 \cos 0.5t + B_0 \sin 0.5t - \tfrac{4}{3}\cos t.$$

The constants A_0 and B_0 must be put equal to zero since otherwise x_1 will include non-periodic terms. The second term satisfies

$$\ddot{x}_1 + 0.25x_1 = \frac{1}{2}\left(\frac{4}{3}\right)^3 \cos^3 t = \frac{8}{27}(3\cos t + \cos 3t).$$

The general solution is

$$x_1 = A_1 \cos 0.5t + B_1 \sin 0.5t - \frac{32}{27}\cos t - \frac{32}{945}\cos 3t.$$

For the same reason $A_1 = B_1 = 0$ to avoid non-periodic terms. Finally

$$x = -\frac{4}{3}\cos t + 0.5\left[-\frac{32}{27}\cos t - \frac{32}{945}\cos 3t\right] + O(\varepsilon^2)$$

$$= -\frac{52}{27}\cos t - \frac{16}{945}\cos 3t + O(\varepsilon^2).$$

(ii) $\ddot{x} - 0.1x^3 + 0.6x = \cos t$. In this problem, $h(x, \dot{x}) = -x^3$, $\Gamma = 1$, $\Omega^2 = 3/5$ and $\varepsilon = 1/10$. Then x_0 satisfies

$$\ddot{x}_0 + \tfrac{3}{5}x_0 = \cos t.$$

As in (i), complementary functions must be put equal to zero at each stage to eliminate non-periodic terms. The first term in the expansion of the forced solution is

$$x_0 = -\tfrac{5}{2}\cos t.$$

The second term satisfies

$$\ddot{x}_1 + \tfrac{3}{5}x_1 = \frac{1}{10}\left(\frac{5}{2}\right)^3 \cos^3 t = \frac{75}{64}\cos t + \frac{25}{64}\cos 3t.$$

The particular solution is

$$x_1 = -\frac{375}{128}\cos t - \frac{125}{2688}\cos 3t.$$

The expansion is

$$x = -\frac{5}{2}\cos t + \frac{1}{10}\left[-\frac{375}{128}\cos t - \frac{125}{2688}\cos 3t\right] + O(\varepsilon^2) + O(\varepsilon^2)$$

$$= -\frac{765}{256}\cos t - \frac{25}{5376}\cos 3t + O(\varepsilon^2)$$

$$\approx -2.99\cos t - 0.005\cos 3t$$

(iii) $\ddot{x} - 0.1\dot{x}^2 + 0.5x = \cos t$. In this case, $h(x, \dot{x}) = -\dot{x}^3$, $\Gamma = 1$, $\Omega^2 = \frac{1}{2}$ and $\varepsilon = 1/10$. Then x_0 satisfies

$$\ddot{x}_0 + \tfrac{1}{2}x_0 = \cos t.$$

As in (i), complementary functions must be put equal to zero at each stage to eliminate non-periodic terms. The leading term is therefore

$$x_0 = -2\cos t.$$

The second term satisfies

$$\ddot{x}_1 + \tfrac{1}{2}x_1 = \tfrac{1}{10}2^2 \sin^2 t = \tfrac{1}{5}(1 - \cos 2t).$$

Hence

$$x_1 = \tfrac{2}{5} + \tfrac{2}{35}\cos 2t.$$

The expansion is

$$x = -2\cos t + \frac{1}{10}\left[\frac{2}{5} + \frac{2}{35}\cos 2t\right] + O(\varepsilon^2)$$

$$= \frac{1}{25} - 2\cos t + \frac{2}{35}\cos t + \frac{1}{175}\cos 2t + O(\varepsilon^2)$$

- **5.3** Find a first approximation to the limit cycle for Rayleigh's equation
$$\ddot{x} + \varepsilon\left(\tfrac{1}{3}\dot{x}^3 - \dot{x}\right) + x = 0, \quad |\varepsilon| \ll 1,$$
using the method of NODE, Section 5.9 (Lindstedt's method).

5.3. In Rayleigh's equation

$$\ddot{x} + \varepsilon\left(\tfrac{1}{3}\dot{x}^3 - \dot{x}\right) + x = 0, \quad |\varepsilon| \ll 1,$$

apply the change of scale $\tau = \omega t$, so that x satisfies

$$\omega^2 x'' - \varepsilon \left(1 - \tfrac{1}{3}\omega^2 x'^2\right)\omega x' + x = 0,$$

where $x' = dx/d\tau$. Look for periodic solutions which are perturbations of those having period 2π. Let

$$\omega = 1 + \varepsilon\omega_1 + \cdots,$$

$$x(\varepsilon, \tau) = x_0(\tau) + \varepsilon x_1(\tau) + \cdots.$$

Substituting these expansions into the differential equation, we have

$$(1 + 2\omega\varepsilon + \cdots)(x_0'' + \varepsilon x_1'' + \cdots) - \varepsilon\left(1 - \tfrac{1}{3}x_0'^2 + \cdots\right)(x_0' + \cdots) + x_0 + \varepsilon x_1 + \cdots = 0.$$

Equating coefficients of like powers of ε,

$$x_0'' + x_0 = 0, \tag{i}$$

$$x_1'' + x_1 = -2\omega_1 x_0' + \left(1 - \tfrac{1}{3}x_0'^2\right) x_0'. \tag{ii}$$

Without loss of generality, we may assume the boundary conditions $x(0) = a_0$, $\dot{x}(0) = 0$. Applying the expansions to the boundary conditions, we have

$$x_0(0) = a_0; \quad x_1(0) = 0; \ \ldots, \tag{iii}$$

$$\dot{x}_0(0) = 0; \quad \dot{x}_1(0) = 0; \ \ldots, \tag{iv}$$

and so on. The solution of (i) and (iii) is

$$x_0 = a_0 \cos \tau.$$

Equation (iv) for x_1 becomes

$$x_1'' + x_1 = 2\omega_1 a_0 \cos \tau + \left(-a_0 \sin \tau + \tfrac{1}{3}a_0^3 \sin^3 \tau\right)$$

$$= 2\omega_1 a_0 \cos \tau + a_0 \left(\tfrac{1}{4}a_0^2 - 1\right)\sin \tau - \tfrac{1}{12}a_0^3 \sin 3\tau.$$

The coefficients of $\cos \tau$ and $\sin \tau$ must be zero to avoid secular (non-periodic) terms for x_1. Therefore $\omega_1 = 0$ and $a_0 = 2$. The leading term in the expansion is

$$x \approx 2 \cos t.$$

- **5.4** Use the method of Section 5.9 to order ε to obtain solutions of period 2π, and the amplitude–frequency relation, for

  (i) $\ddot{x} - \varepsilon x \dot{x} + x = 0$;

  (ii) $(1 + \varepsilon \dot{x})\ddot{x} + x = 0$.

5.4. (i) In the equation
$$\ddot{x} - \varepsilon x \dot{x} + x = 0,$$
apply the change of scale $\tau = \omega t$, so that
$$\omega^2 x'' - \varepsilon \omega x x' + x = 0.$$

Let
$$\omega = 1 + \varepsilon \omega_1 + \varepsilon^2 \omega_2 + \cdots,$$
$$x(\varepsilon, \tau) = x_0(\tau) + \varepsilon x_1(\tau) + \varepsilon^2 x_2(\tau) + \cdots.$$

Substituting these expansions into the differential equation, it follows that
$$(1 + \varepsilon \omega_1 + \varepsilon^2 \omega_2 + \cdots)^2 (x_0'' + \varepsilon x_1'' + \varepsilon^2 x_2'' + \cdots)$$
$$- \varepsilon (1 + \varepsilon \omega_1 + \varepsilon^2 \omega_2 + \cdots)(x_0 + \varepsilon x_1 + \varepsilon^2 x_2 + \cdots)(x_0' + \varepsilon x_1' + \varepsilon^2 x_2' + \cdots)$$
$$+ (x_0 + \varepsilon x_1 + \varepsilon^2 x_2 + \cdots) = 0.$$

Equating like powers of ε, we obtain the equations

$$x_0'' + x_0 = 0, \qquad \text{(i)}$$
$$x_1'' + x_1 = -2\omega_1 x_0'' + x_0 x_0', \qquad \text{(ii)}$$
$$x_2'' + x_2 = -(\omega_1^2 + 2\omega_2) x_0'' - 2\omega_1 x_1'' + \omega_1 x_0 x_0' + x_0 x_1' + x_0' x_1. \qquad \text{(iii)}$$

Without loss of generality, we may assume the initial conditions $x = a$, $\dot{x} = 0$ when $t = 0$, which become $x_0(0) = a$, $x_i(0) = 0$, $(i = 1, 2, \ldots)$, $x_j'(0) = 0$, $(j = 0, 1, 2, \ldots)$.
Equation (i) has the solution $x_0 = a \cos \tau$, and (ii) becomes
$$x_1'' + x_1 = 2\omega_1 \cos \tau - \tfrac{1}{2} a^2 \sin 2\tau \qquad \text{(iv)}$$

Hence x_1 will only have a periodic solution if the coefficient of $\cos \tau$ is zero. Therefore $\omega_1 = 0$. The remaining eqn (iv) has the general solution
$$x_1 = A \cos \tau + B \sin \tau + \tfrac{1}{6} a^2 \sin 2\tau.$$

From the initial conditions $A = 0$ and $B = -\frac{1}{3}a^2$. The required second term is

$$x_1 = -\frac{1}{3}a^2 \sin\tau + \frac{1}{6}a^2 \sin 2\tau.$$

Now construct the equation for x_2 from (iii), which becomes, on reduction

$$x_2'' + x_2 = \tfrac{1}{12}a(a^2 + 24\omega_2)\cos\tau + \tfrac{1}{3}a^3 \cos 2\tau + \tfrac{1}{4}a^3 \cos 3\tau.$$

Hence x_2 can only be periodic if $\omega_2 = -a^2/24$. The frequency–amplitude relation, to order ε^2 is

$$\omega = 1 + \varepsilon^2 \omega_2 = 1 - \tfrac{1}{24}\varepsilon a^2.$$

The differential equation for x_2 reduces to

$$x_2'' + x_2 = \tfrac{1}{3}a^3 \cos 2\tau + \tfrac{1}{4}a^3 \cos 3\tau,$$

which has the general solution

$$x_2 = C\cos\tau + D\sin\tau - \tfrac{1}{9}\cos 2\tau - \tfrac{1}{32}\cos 3\tau.$$

From the initial conditions $C = 0$ and $D = 91/288$. To order ε^2 the expansion for the periodic solution is

$$x = a\cos\tau + \varepsilon\left(-\tfrac{1}{3}a^2 \sin\tau + \tfrac{1}{6}a^2 \sin 2\tau\right) + \varepsilon^2\left(\tfrac{91}{288}\sin\tau - \tfrac{1}{9}\cos 2\tau - \tfrac{1}{32}\cos 2\tau\right),$$

where $\tau = \omega t = \left(1 - \tfrac{1}{24}\varepsilon^2 a^2\right)t$.

(ii) For the equation

$$(1 + \varepsilon\dot{x})\ddot{x} + x = 0,$$

apply the change of scale $\tau = \omega t$, so that x satisfies

$$(1 + \varepsilon\omega x')\omega^2 x'' + x = 0.$$

Let

$$\omega = 1 + \varepsilon\omega_1 + \varepsilon^2 \omega_2 + \cdots,$$

$$x(\varepsilon, \tau) = x_0(\tau) + \varepsilon x_1(\tau) + \varepsilon^2 x_2(\tau) + \cdots.$$

Substituting these expansions into the differential equation, it follows that

$$[1+\varepsilon(1+\varepsilon w_1+\cdots)(x_0'+\varepsilon x_1'+\cdots)](1+\varepsilon w_1+\varepsilon^2 w_2+\cdots)^2$$
$$\times (x_0''+\varepsilon x_1''+\varepsilon^2 x_2''+\cdots)+(x_0+\varepsilon x_1+\varepsilon x_2+\cdots)=0.$$

Equating like powers of ε we can obtain the differential equations for x_0, x_1 and x_2, namely,

$$x_0''+x_0=0, \tag{i}$$
$$x_1''+x_1=-2w_1 x_0''-x_0' x_0'', \tag{ii}$$
$$x_2''+x_2=-w_1^2 x_0''-2w_2 x_0''-3w_1 x_0' x_0''-x_0'' x_1'-2w_1 x_1''-x_0' x_1''. \tag{iii}$$

Assume the initial conditions $x=a$, $\dot{x}=0$ when $t=0$, which become $x_0(0)=a$, $x_i(0)=0$, $(i=1,2,\ldots)$, $x_j'(0)=0$, $(j=0,1,2,\ldots)$.

From (i) and the initial conditions, $x_0=a\cos\tau$ so that x_1 satisfies

$$x_1''+x_1=-2aw_1\cos\tau-\tfrac{1}{2}a^2\sin 2\tau.$$

To avoid a growth term in x_1 we must put $w_1=0$, leaving the equation

$$x_1''+x_1=-\tfrac{1}{2}a^2\sin 2\tau.$$

The solution satisfying the initial conditions is

$$x_1=a^2\left(-\tfrac{1}{3}\sin\tau+\tfrac{1}{6}\sin 2\tau\right).$$

The equation for x_2 given by (iii) becomes (remember $w_1=0$)

$$x_2''+x_2=\tfrac{1}{6}(12aw_2-a^3)\cos\tau-\tfrac{1}{3}a^3\cos 2\tau+\tfrac{1}{2}a^3\cos 3\tau. \tag{iv}$$

The secular term in x_2 can be eliminated if $w_2=a^2/12$. We can now find x_2 from (iv), which is, subject to the specified initial conditions,

$$x_2=a^3\left(\tfrac{1}{16}\cos\tau-\tfrac{2}{9}\sin\tau+\tfrac{1}{9}\sin 2\tau-\tfrac{1}{16}\cos 3\tau\right).$$

The perturbation solution is therefore

$$x=a\cos\tau+\varepsilon a^2\left(-\tfrac{1}{3}\sin\tau+\tfrac{1}{6}\sin 2\tau\right)$$
$$+\varepsilon^2 a^3\left(\tfrac{1}{16}\cos\tau-\tfrac{2}{9}\sin\tau+\tfrac{1}{9}\sin 2\tau-\tfrac{1}{16}\cos 3\tau\right)$$

where $w=1+\tfrac{1}{12}\varepsilon^2 a^3$.

- 5.5 Apply the perturbation method to the equation $\ddot{x} + \Omega^2 \sin x = \cos t$ by considering
$$\ddot{x} + \Omega^2 x + \varepsilon \Omega^2 (\sin x - x) = \cos t,$$
with $\varepsilon = 1$, and assuming that Ω is not close to an odd integer, to find perod 2π solutions. Use the Fourier expansion
$$\sin(a \cos t) = 2 \sum_{n=0}^{\infty} (-1)^n J_{2n+1}(a) \cos\{(2n+1)t\},$$
where J_{2n+1} is the Bessel function of order $2n+1$. Confirm that the leading terms are given by
$$x = \frac{1}{\Omega^2 - 1} [1 + \Omega^2 - 2\Omega^2 J_1\{1/(\Omega^2 - 1)\}] \cos t + \frac{2}{\Omega^2 - 9} J_3\{1/(\Omega_1^2)\} \cos 3t.$$

5.5. Rewrite the equation
$$\ddot{x} + \Omega^2 \sin x = \cos t \qquad (i)$$
as a member of the family of equations with parameter ε
$$\ddot{x} + \Omega^2 x + \varepsilon \Omega^2 (\sin x - x) = \cos t. \qquad (ii)$$

Substitute the perturbation series
$$x(\varepsilon, t) = x_0(t) + \varepsilon x_1(t) + \cdots$$
into the differential equation (ii) and equate like powers of ε. Then
$$\ddot{x}_0 + \Omega^2 x_0 = \cos t, \qquad (iii)$$
$$\ddot{x}_1 + \Omega^2 x_1 = -\Omega^2 (\sin x_0 - x_0). \qquad (iv)$$
For Ω not an integer, the only periodic solution (of period 2π) of (iii) is the forced solution
$$x_0 = \frac{1}{\Omega^2 - 1} \cos t.$$

Equation (iv) becomes
$$\ddot{x}_1 + \Omega^2 x_1 = -\Omega^2 \left[\sin\left(\frac{\cos t}{\Omega^2 - 1}\right) - \frac{1}{\Omega^2 - 1} \cos t \right].$$

The Fourier series expansion
$$\sin(a \cos t) = 2 \sum_{n=0}^{\infty} (-1)^n J_{2n+1}(a) \cos\{(2n+1)t\}, \qquad (v)$$

where $a = 1/(\Omega^2 - 1)$, is applied to the right-hand side of (v) so that, to leading order,

$$\ddot{x}_1 + \Omega^2 x_1 = -\frac{\Omega^2}{\Omega^2 - 1}\left[2(\Omega^2 - 1)J_1\left(\frac{1}{\Omega^2 - 1}\right) - 1\right]\cos t + 2J_3\left(\frac{1}{\Omega^2 - 1}\right)\cos 3t.$$

The period-2π solution of this equation is

$$x_1 = -\frac{\Omega^2}{(\Omega^2 - 1)^2}\left[2(\Omega^2 - 1)J_1\left(\frac{1}{\Omega^2 - 1}\right) - 1\right]\cos t + \frac{2}{\Omega^2 - 9}J_3\left(\frac{1}{\Omega^2 - 1}\right)\cos 3t.$$

If we now combine the first two terms and put $\varepsilon = 1$, the result is the perturbation

$$x = \frac{1}{\Omega^2 - 1}\left[1 + \Omega^2 - 2\Omega^2 J_1\left(\frac{1}{\Omega^2 - 1}\right)\right]\cos t + \frac{2}{\Omega^2 - 9}J_3\left(\frac{1}{\Omega^2 - 1}\right)\cos 3t.$$

- **5.6** For the equation $\ddot{x} + \Omega^2 x - 0.1x^3 = \cos t$, where Ω is not near $1, 3, 5, \ldots$, find, to order ε, the ratio of the amplitudes of the first two harmonics.

5.6. Consider the family of equations

$$\ddot{x} + \Omega^2 x - \varepsilon x^3 = \cos t,$$

and afterwards put $\varepsilon = 0.1$. We are given that Ω is not close to an odd integer, but it may be close to an even integer. Let

$$x = x_0 + \varepsilon x_1 + \cdots.$$

Substitution into the differential equation leads to

$$\ddot{x}_0 + \Omega^2 x_0 = \cos t,$$
$$\ddot{x}_1 + \Omega^2 x_1 = \varepsilon x_0^3.$$

The 2π-periodic solution for x_0 is

$$x_0 = \frac{1}{\Omega^2 - 1}\cos t.$$

The equation for x_1 is therefore

$$\ddot{x}_1 + \Omega^2 x_1 = \frac{1}{4(\Omega^2 - 1)}(3\cos t + \cos 3t).$$

The 2π-periodic solution of this equation is

$$x_1 = \frac{3}{4(\Omega^2 - 1)^2} \cos t + \frac{1}{4(\Omega^2 - 1)(\Omega^2 - 9)} \cos 3t.$$

to order ε. The approximate solution including the first two harmonics is

$$x_0 + \varepsilon x_1 = \frac{4 + 3\varepsilon}{4(\Omega^2 - 1)} \cos t + \frac{\varepsilon}{4(\Omega^2 - 1)(\Omega^2 - 9)} \cos 3t.$$

If a_3 and a_1 are the coefficients of the first two harmonics, then the ratio of the amplitudes is

$$\frac{|a_3|}{|a_1|} = \frac{\varepsilon}{(4 + 3\varepsilon)(\Omega^2 - 9)}.$$

- 5.7 In the equation $\ddot{x} + \Omega^2 x + \varepsilon f(x) = \Gamma \cos t$, Ω is not close to an odd integer, and $f(x)$ is an odd function of x, with expansion
$$f(a \cos t) = -a_1(a) \cos t - a_3(a) \cos 3t - \cdots.$$
Derive a perturbation solution of period 2π, to order ε.

5.7. In the equation

$$\ddot{x} + \Omega^2 x + \varepsilon f(x) = \Gamma \cos t,$$

let $x = x_0 + \varepsilon x_1 + \cdots$. The two leading terms satisfy

$$\ddot{x}_0 + \Omega^2 x_0 = \Gamma \cos t, \tag{i}$$

$$\ddot{x}_1 + \Omega^2 x_1 = -f(x_0 + \varepsilon x_1 + \cdots) \approx -f(x_0). \tag{ii}$$

The 2π-periodic solution of (i) is

$$x_0 = \frac{\Gamma}{\Omega^2 - 1} \cos t,$$

provided Ω is not close to 1 (or any odd integer for higher-order terms). The next term x_1 in the perturbation satisfies

$$\ddot{x}_1 + \Omega^2 x_1 = -\varepsilon f\left(\frac{\Gamma}{\Omega^2 - 1}\right) = -\varepsilon[a_1(\kappa) \cos t + a_3(\kappa) \cos 3t],$$

where $\kappa = \Gamma/(\Omega^2 - 1)$. The 2π-periodic solution is

$$x_1 = -\frac{a_1(\kappa)}{\Omega^2 - 1} \cos t - \frac{a_3(\kappa)}{\Omega^2 - 9} \cos 3t.$$

Hence to order ε,

$$x = \frac{\Gamma - \varepsilon a_1(\kappa)}{\Omega^2 - 1} \cos t - \frac{\varepsilon a_3(\kappa)}{\Omega^2 - 9} \cos 3t.$$

- **5.8** The Duffing equation near resonance at $\Omega = 3$, with weak excitation, is
$\ddot{x} + 9x = \varepsilon(\gamma \cos t - \beta x + x^3).$
Show that there are solutions of period 2π if the amplitude of the zero-order solution is 0 or $2\sqrt{(\beta/3)}$.

5.8. The undamped Duffing equation near resonance at $\Omega = 3$ with weak excitation is

$$\ddot{x} + 9x = \varepsilon(\gamma \cos t - \beta x + x^3).$$

Let $x = x_0 + \varepsilon x_1 + \cdots$. The equations for x_0 and x_1 are

$$\ddot{x}_0 + 9x_0 = 0, \qquad (i)$$

$$\ddot{x}_1 + 9x_1 = \gamma \cos t - \beta x_0 + x_0^3. \qquad (ii)$$

We are searching for 2π-periodic solutions. Equation (i) has the periodic solution $x_0 = a_0 \cos 3t + b_0 \sin 3t$. Substituting x_0 into (ii) and expanding, x_1 satisfies

$$\ddot{x}_1 + 9x_1 = \gamma \cos t - \beta(a_0 \cos 3t + b_0 \sin 3t) + (a_0 \cos 3t + b_0 \sin 3t)^3$$

$$= \gamma \cos t + \tfrac{1}{4}a_0(3a_0^2 + 3b_0^2 - 4\beta) \cos 3t + \tfrac{1}{4}b_0(3a_0^2 + 3b_0^2 - 4\beta) \sin 3t$$

$$+ \tfrac{1}{4}a_0(a_0^2 - 3b_0^2) \cos 9t + \tfrac{1}{4}b_0(3a_0^2 - b_0^2) \sin 9t.$$

The secular terms can be removed by choosing

$$a_0(3a_0^2 + 3b_0^2 - 4\beta) = 0, \quad b_0(3a_0^2 + 3b_0^2 - 4\beta) = 0.$$

Possible solutions are $a_0 = b_0 = 0$, or

$$r_0 = \sqrt{(a_0^2 + b_0^2)} = 2\sqrt{(\beta/3)}.$$

- 5.9 From eqn (5.40), the amplitude equation for the undamped pendulum is
$$-F = a_0\left(\omega^2 - \omega_0^2 + \tfrac{1}{8}\omega_0^2 a_0^2\right).$$
When ω_0 is given, find for what values of ω there are three possible responses. (Find the stationary values of F with respect to a_0, with ω fixed. These are the points where the response curves of Figure 5.4 (in NODE) turn over.)

5.9. The frequency–amplitude equation is
$$F = -a_0\left(\omega^2 - \omega_0^2 + \tfrac{1}{8}\omega_0^2 a_0^2\right). \tag{i}$$

We find the stationary points of F with respect to a_0 for fixed ω and ω_0. Then
$$\frac{dF}{da_0} = \omega^2 - \omega_0^2 + \tfrac{3}{8}\omega_0^2 a_0^2.$$

Therefore F is stationary where
$$a_0 = \pm\frac{2\sqrt{2}}{\omega_0\sqrt{3}}\sqrt{(\omega_0^2 - \omega^2)},$$

if $\omega^2 \leq \omega_0^2$. If $\omega^2 > \omega_0^2$, there are no stationary values. There are three possible responses if $\omega^2 < \omega_0^2$.

Equation (i) can be expressed in the more convenient form
$$Q = -a_0(a_0^2 - \kappa),$$

where $Q = 8F/\omega_0^2$ and $\kappa = 8(\omega_0^2 - \omega^2)/\omega_0^2$. The surface showing the relation between Q and a_0 and κ is shown in Figure 5.1.

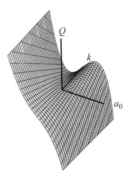

Figure 5.1 Problem 5.9: Surface showing the frequency–amplitude relation between Q, a_0 and κ.

- 5.10 From NODE, eqn (5.42), the amplitude equation for the positively damped pendulum is

$$F^2 = r_0^2 \left\{ k^2\omega^2 + \left(\omega^2 - \omega_0^2 + \tfrac{1}{8}\omega_0^2 r_0^2\right)^2 \right\}.$$

By considering $d(F^2)/d(r_0^2)$, show that if $(\omega^2 - \omega_0^2)^2 \leq 3k^2\omega^2$, then the amplitude equation has only one real root r_0, and three real roots if $(\omega^2 - \omega_0^2)^2 > 3k^2\omega^2$.

5.10. The amplitude equation for the damped pendulum is

$$F^2 = r_0^2 \left\{ k^2\omega^2 + \left(\omega^2 - \omega_0^2 + \tfrac{1}{8}\omega_0^2 r_0^2\right)^2 \right\}.$$

Let

$$G = \frac{F^2 \omega_0^2}{8\omega^3 k^3}, \quad \rho = \frac{\omega_0^2 r_0^2}{8\omega k}, \quad \alpha = \frac{\omega^2 - \omega_0^2}{k\omega},$$

so that

$$G(\rho) = \rho[1 + (\alpha + \rho)^2],$$

We are only interested in the domain $G > 0$, $\rho > 0$.
The derivative of $G(\rho)$ is given by

$$G'(\rho) = 3\rho^2 + 4\alpha\rho + 1 + \alpha^2.$$

The equation $G'(\rho) = 0$ has no real solutions if $\alpha^2 < 3$, one real solution if $\alpha^2 = 3$, and two real solutions $\alpha^2 > 3$. The cases are discussed below.

- $\alpha^2 \leq 3$. This is equivalent to

$$(\omega^2 - \omega_0^2)^2 \leq 3k^2\omega^2 \text{ or } \omega^4 - (2\omega_0^2 + 3k^2)\omega^2 + \omega_0^4 \leq 0.$$

The equation

$$\omega^4 - (2\omega_0^2 + 3k^2)\omega^2 + \omega_0^4 = 0$$

has the solutions

$$\omega^2 = \tfrac{1}{2}[(2\omega_0^2 + 3k^2) \pm \sqrt{(6\omega_0^2 + 9k^2)}],$$

both of which give real solutions for ω (we need only consider positive solutions for ω). Hence the system has just one real solution for the amplitude if

$$\tfrac{1}{2}[(2\omega_0^2 + 3k^2) - \sqrt{(6\omega_0^2 + 9k^2)}] \leq \omega^2 \leq \tfrac{1}{2}[(2\omega_0^2 + 3k^2) + \sqrt{(6\omega_0^2 + 9k^2)}].$$

- $\alpha^2 > 3$. This is equivalent to

$$\omega^4 - (2\omega_0^2 + 3k^2)\omega^2 + \omega_0^4 > 0.$$

By an argument similar to above the amplitude has three solutions if

$$\omega^2 > \tfrac{1}{2}[(2\omega_0^2 + 3k^2) + \sqrt{(6\omega_0^2 + 9k^2)}],$$

or

$$0 < \omega^2 < \tfrac{1}{2}[(2\omega_0^2 - 3k^2) + \sqrt{(6\omega_0^2 + 9k^2)}].$$

- **5.11** Find the equivalent linear form (NODE, Section 4.5) of the expression $\ddot{x} + \Omega^2 x - \varepsilon x^3$, with respect to the periodic form $x = a \cos t$. Use the linear form to obtain the frequency–amplitude relation for the equation
$$\ddot{x} + \Omega^2 x - \varepsilon x^3 = \Gamma \cos t.$$
Solve the equation approximately by assuming that $a = a_0 + \varepsilon a_1$, and show that this agrees with the first harmonic in NODE, eqn (5.23). (Note that there may be three solutions, but that this method of solution shows only the one close to $\{\Gamma/(1 - \Omega^2)\} \cos t$.)

5.11. We require the equivalent linear form (see Section 4.5) of the left-hand side of

$$\ddot{x} + \Omega^2 x - \varepsilon x^3 = \Gamma \cos t.$$

Assume $x \approx a \cos t$. Using the identity

$$\cos^3 t = \tfrac{3}{4} \cos t + \tfrac{1}{4} \cos 3t$$

and neglecting higher harmonics, we replace $-\varepsilon x^3$ by

$$-\varepsilon a^3 \cos^3 t \approx -\tfrac{3}{4}\varepsilon a^3 \cos t = -\tfrac{3}{4}\varepsilon a^2 x.$$

Therefore the equivalent linear equation is

$$\ddot{x} + \left(\Omega^2 - \tfrac{3}{4}\varepsilon a^2\right) x = \Gamma \cos t.$$

Hence the 2π-periodic solution is

$$x = \frac{\Gamma}{\Omega^2 - 1 - \tfrac{3}{4}\varepsilon a^2} \cos t,$$

where the frequency-amplitude equation is

$$a = \frac{\Gamma}{\Omega^2 - 1 - \frac{3}{4}\varepsilon a^2}. \tag{i}$$

Let $a = a_0 + \varepsilon a_1 + \cdots$. Then the expansion of eqn (i) gives

$$a_0 + \varepsilon a_1 + \cdots = \frac{\Gamma}{\Omega^2 - 1 - \frac{3}{4}\varepsilon(a_0 + \varepsilon a_1 + \cdots)^2}$$

$$= \frac{\Gamma}{\Omega^2 - 1}\left[1 + \frac{3}{4}\frac{\varepsilon a_0^2}{\Omega^2 - 1}\right] + O(\varepsilon^2)$$

By comparison of the two sides of this equation we deduce

$$a_0 = \frac{\Gamma}{\Omega^2 - 1}, \quad a_1 = \frac{3}{4}\frac{\Gamma^3}{(\Omega^2 - 1)^4}.$$

Finally to order ε the first harmonic is

$$x = \frac{\Gamma}{\Omega^2 - 1}\cos t + \frac{3}{4}\frac{\varepsilon \Gamma^3}{(\Omega^2 - 1)^4}\cos t,$$

which agrees with (5.23).

- **5.12** Generalize the method of Problem 5.11 for the same equation by putting $x = x^{(0)} + x^{(1)} + \cdots$, where $x^{(0)}$ and $x^{(1)}$ are the first harmonics to order ε, $a\cos t$ and $b\cos 3t$, say, in the expansion of the solution. Show that the linear form equivalent to x^3 is

$$\left(\tfrac{3}{4}a^2 + \tfrac{3}{4}ab + \tfrac{3}{2}b^2\right)x^{(0)} + \left(\tfrac{1}{4}a^3 + \tfrac{3}{2}a^2b + \tfrac{3}{4}b^3\right)x^{(1)}/b.$$

Split the pendulum equation into the two equations

$$\ddot{x}^{(0)} + \left\{\Omega^2 - \varepsilon\left(\tfrac{3}{4}a^2 + \tfrac{3}{4}ab + \tfrac{3}{2}b^2\right)\right\}x^{(0)} = \Gamma\cos t,$$

$$\ddot{x}^{(1)} + \left\{\Omega^2 - \varepsilon\left(\tfrac{1}{4}a^3 + \tfrac{3}{2}a^2b + \tfrac{3}{4}b^3\right)/b\right\}x^{(1)} = 0.$$

Deduce that a and b must satisfy

$$a\left\{\Omega^2 - 1 - \varepsilon\left(\tfrac{3}{4}a^2 + \tfrac{3}{4}ab + \tfrac{3}{2}b^2\right)\right\} = \Gamma,$$

$$b\left\{\Omega^2 - 9 - \varepsilon\left(\tfrac{1}{4}a^3 + \tfrac{3}{2}a^2b + \tfrac{3}{4}b^3\right)\right\} = 0.$$

Assume that $a = a_0 + \varepsilon a_1 + O(\varepsilon^2)$, $b = \varepsilon b_1 + O(\varepsilon^2)$ and obtain a_0, a_1 and b_1.

5.12. In the equation
$$\ddot{x} + \Omega^2 x - \varepsilon x^3 = \Gamma \cos t,$$
let $x \approx x^{(0)} + x^{(1)}$, where $x^{(0)} = a \cos t$ and $x^{(1)} = b \cos 3t$ and a and b are constants. Then

$$-\varepsilon x^3 = -\varepsilon (a \cos t + b \cos 3t)^3$$
$$= -\varepsilon \left[\tfrac{3}{4} a(a^2 + ab + 2b^2) \cos t + \tfrac{1}{4}(a^3 + 6a^2 b + 3b^3) \cos 3t + \text{(higher harmonics)} \right].$$

Therefore, neglecting higher harmonics,

$$-\varepsilon x^3 \approx \tfrac{3}{4}\varepsilon(a^2 + ab + 2b^2) x^{(0)} + \tfrac{1}{4}\varepsilon(a^3 + 6a^2 b + 3b^3) x^{(1)}/b,$$

and the equivalent linear equations are

$$\ddot{x}^{(0)} + \left[\Omega^2 - \tfrac{3}{4}\varepsilon(a^2 + ab + 2b^2) \right] x^{(0)} = \Gamma \cos t,$$

$$\ddot{x}^{(1)} + \left[\Omega^2 - \tfrac{1}{4}\varepsilon(a^3 + 6a^2 b + 3b^3) \right] x^{(1)} = 0.$$

To ensure that $x^{(0)} = a \cos t$ and $x^{(1)} = b \cos 3t$ satisfy these equations only if

$$a \left[\Omega^2 - 1 - \tfrac{3}{4}\varepsilon(a^2 + ab + 2b^2) \right] = \Gamma, \qquad \text{(i)}$$

$$b \left[\Omega^2 - 9 - \tfrac{1}{4}\varepsilon(a^3 + 6a^2 b + 3b^3) \right] = 0. \qquad \text{(ii)}$$

Let $a = a_0 + \varepsilon a_1 + \cdots$ and $b = \varepsilon b_1 + \cdots$. Then (i) and (ii) become

$$(a_0 + \varepsilon a_1 + \cdots) \left[\Omega^2 - 1 - \tfrac{3}{4}\varepsilon \{(a_0 + \cdots)^2 + \cdots \} \right] = \Gamma,$$

$$(\varepsilon b_1 + \cdots) \left[(\Omega^2 - 9) - \tfrac{1}{4}\varepsilon \{(a_0 + \cdots)^3 + \cdots \} \right] = 0.$$

Equating like powers of ε, we have

$$a_0 = \frac{\Gamma}{\Omega^2 - 1}, \qquad a_1 = \tfrac{3}{4} \frac{a_0^3}{\Omega^2 - 1} = \tfrac{3}{4} \frac{\Gamma^3}{(\Omega^2 - 1)^4},$$

$$b_1 = \tfrac{1}{4} \frac{a_0^3}{\Omega^2 - 9} = \tfrac{1}{4} \frac{\Gamma^3}{(\Omega^2 - 1)^3 (\Omega^2 - 1)^3}.$$

- 5.13 Apply the Lindstedt method, Section 5.9, to van der Pol's equation $\ddot{x} + \varepsilon(x^2 - 1)\dot{x} + x = 0$, $|\varepsilon| \ll 1$. Show that the frequency of the limit cycle is given by $\omega = 1 - \frac{1}{16}\varepsilon^2 + O(\varepsilon^3)$.

5.13. Linstedt's method (Section 5.9) is applied to the van der Pol's equation

$$\ddot{x} + \varepsilon(x^2 - 1)\dot{x} + x = 0, \quad |\varepsilon| \ll 1.$$

Let $\tau = \omega t$, so that

$$\omega^2 x'' + \varepsilon\omega(x^2 - 1)x' + x = 0.$$

We now seek solutions of period 2π. Apply the perturbation expansions $x = x_0 + \varepsilon x_1 + \cdots$ and $\omega = \omega_0 + \varepsilon\omega_1 + \cdots$ to this equation:

$$(\omega_0 + \varepsilon\omega_1 + \varepsilon^2\omega_2 + \cdots)^2 (x_0'' + \varepsilon x_1'' + \varepsilon^2 x_2'' + \cdots) + \varepsilon(\omega_0 + \varepsilon\omega_1 + \varepsilon^2\omega_2 + \cdots)$$
$$[(x_0 + \varepsilon x_1 + \varepsilon^2 x_2 + \cdots)^2 - 1](x_0' + \varepsilon x_1' + \varepsilon^2 x_2' \cdots) + (x_0 + \varepsilon x_1 + \varepsilon^2 x_2 + \cdots) = 0$$

Equating powers of ε, we can obtain the differential equations for the leading terms, namely,

$$\omega_0^2 x_0'' + x_0 = 0, \tag{i}$$

$$\omega_0^2 x_1'' + x_1 = -2\omega_0\omega_1 x_0'' - \omega_0(x_0^2 - 1)x_0', \tag{ii}$$

$$\omega_0^2 x_2'' + x_2 = -2\omega_0\omega_1 x_1'' - (2\omega_0\omega_2 + \omega_1^2)x_0'' - (x_0^2 - 1)(\omega_0 x_1' + \omega_1 x_0') - 2\omega_0 x_0 x_0' x_1. \tag{iii}$$

We search for solutions of period 2π; since the system is autonomous we can put $x'(0) = 0$. From (i),

$$x_0 = A_0 \cos[\tau/\omega_0].$$

We must choose, therefore, $\omega_0 = 1$. Substitution of x_0 into (ii) leads to

$$x_1'' + x_1 = 2\omega_1 A_0 \cos\tau + (A_0^2 \cos^2\tau - 1)A_0 \sin\tau$$
$$= 2\omega_1 A_0 \cos\tau - (A_0 - \tfrac{1}{4}A_0^3)\sin\tau + \tfrac{1}{4}A_0^3 \sin 3\tau$$

Any secular term can be removed by making the coefficients of $\cos\tau$ and $\sin\tau$ zero. Therefore we select $A_0 = 2$ and $\omega_1 = 0$. The second term in the expansion is

$$x_1 = A_1 \cos\tau + B_1 \sin\tau - \tfrac{1}{4}\sin 3\tau.$$

Equation (iii) becomes

$$x_2'' + x_2 = \tfrac{1}{4}(1 + 16\omega_2)\cos t + 2A_1 \sin t + \text{higher harmonics}.$$

Hence there are no secular terms if $\omega_2 = -\frac{1}{16}$ and $A_1 = 0$. Finally the frequency of the limit cycle is given by
$$\omega = 1 - \tfrac{1}{16}\varepsilon^2 + O(\varepsilon^3).$$

- **5.14** Investigate the forced periodic solutions of period $\frac{2}{3}\pi$ for the Duffing equation in the form $\ddot{x} + (1+\varepsilon\beta)x - \varepsilon x^3 = \Gamma\cos 3t$.

5.14. The equation
$$\ddot{x} + (1+\varepsilon\beta)x - \varepsilon x^3 = \Gamma\cos 3t.$$
has a forcing term of period $2\pi/3$. Let $x = x_0 + \varepsilon x_1 + \cdots$. Then
$$(\ddot{x}_0 + \varepsilon\ddot{x}_1 + \cdots) + (1+\varepsilon\beta)(x_0 + \varepsilon x_1 + \cdots) - \varepsilon(x_0 + \varepsilon x_1 + \cdots)^3 = \Gamma\cos 3t.$$

Equating powers of ε, the first two terms satisfy
$$\ddot{x}_0 + x_0 = \Gamma\cos 3t, \tag{i}$$
$$\ddot{x}_1 + x_1 = -\beta x_0 + x_0^3. \tag{ii}$$

Only the forced part of the solution (i) is of period $2\pi/3$, namely,
$$x_0 = -\tfrac{1}{8}\Gamma\cos 3t.$$

Equation (ii) becomes
$$\ddot{x}_1 + x_1 = \tfrac{1}{8}\beta\Gamma\cos 3t - \tfrac{1}{8^3}\Gamma^3\cos^3 3t$$
$$= \tfrac{1}{8}\beta\Gamma\cos 3t - \tfrac{1}{8^3}\Gamma^3\left(\tfrac{3}{4}\cos 3t + \cos 9t\right)$$
$$= \tfrac{\Gamma}{8}\left(\beta - \tfrac{3\Gamma^2}{4\times 8^2}\right)\cos 3t + \text{higher harmonics}$$

Therefore the leading harmonic in x_1 is
$$x_1 = -\tfrac{\Gamma}{8^2}\left(\beta - \tfrac{3\Gamma^2}{4\times 8^2}\right)\cos 3t + \cdots.$$

Therefore up to the first harmonic,
$$x = -\tfrac{\Gamma}{8}\left[1 + \tfrac{\varepsilon}{8}\left(\beta - \tfrac{3\Gamma^2}{4\times 8^2}\right)\right]\cos 3t + \cdots.$$

- **5.15** For the equation $\ddot{x} + x + \varepsilon x^3 = 0$, $|\varepsilon| \ll 1$, with $x(0) = a$, $\dot{x}(0) = 0$, assume an expansion of the form $x(t) = x_0(t) + \varepsilon x_1(t) + \cdots$, and carry out the perturbation process without assuming periodicity of the solution. Show that

$$x(t) = a\cos t + \varepsilon a^3 \left\{ -\tfrac{3}{8} t \sin t + \tfrac{1}{32}(\cos 3t - \cos t) \right\} + O(\varepsilon^3).$$

(This expansion is valid, so far as it goes. Why is it not so suitable as those already obtained for describing solutions?)

5.15. Apply the expansion $x = x_0 + \varepsilon x_1 + \cdots$ to the system

$$\ddot{x} + x + \varepsilon x^3 = 0, \quad x(0) = a, \quad \dot{x}(0) = 0.$$

Thus

$$(\ddot{x}_0 + \varepsilon \ddot{x}_1 + \cdots) + (x_0 + \varepsilon x_1 + \cdots) + \varepsilon(x_0 + \cdots)^3 = 0.$$

Equating to zero the coefficients of like powers of ε, we obtain the equations

$$\ddot{x}_0 + x_0 = 0, \qquad (i)$$

$$\ddot{x}_1 + x_1 = -x_0^3, \qquad (ii)$$

subject to the given initial conditions $x_0 = a \cos t$. Equation (ii) then becomes

$$\ddot{x}_1 + x_1 = -x_0^3 = a^3 \cos^3 t$$
$$= \tfrac{3}{4}\cos t + \tfrac{1}{4}\cos 3t.$$

The general solution of this linear second-order equation is

$$x_1 = A_1 \cos t + B_1 \sin t + \tfrac{9}{32}\cos t + \tfrac{3}{8} t \sin t - \tfrac{1}{32}\cos 3t.$$

The expansion of the initial conditions implies $x_1(0) = \dot{x}_1(0) = 0$. Therefore A_1 and B_1 are given by

$$A_1 = -\tfrac{9}{32} + \tfrac{1}{32} = -\tfrac{1}{4}, \quad B_1 = 0.$$

Finally the expansion takes the non-periodic form

$$x \approx x_0 + x_1 = a \cos t + \varepsilon a^3 \left[-\tfrac{3}{8} t \sin t + \tfrac{1}{32}(\cos 3t - \cos t) \right].$$

x_1 has a term with the factor εt (and we would find that x_2 has a term with factor $\varepsilon^2 t^2$, and so on). For any fixed order of approximation the error will increase as t increases, and is unlikely to be small (i.e. the expansion is non-uniform).

- **5.16** Find the first few harmonics in the solution, period 2π, of $\ddot{x}+\Omega^2 x+\varepsilon x^2 = \Gamma \cos t$, by the direct method of Section 5.2. Explain the presence of a constant term in the expansion. For what values of Ω does the expansion fail? Show how, for small values of Γ, an expansion valid near $\Omega = 1$ can be obtained.

5.16. Apply the expansion $x = x_0 + \varepsilon x_1 + \varepsilon^2 x_2 + \cdots$ to

$$\ddot{x} + \Omega^2 x + \varepsilon x^2 = \Gamma \cos t,$$

assuming that $|\varepsilon| \ll 1$. Thus

$$(\ddot{x}_0 + \varepsilon \ddot{x}_1 + \varepsilon^2 \ddot{x}_2 + \cdots) + \Omega^2 (x_0 + \varepsilon x_1 + \varepsilon^2 x_2 + \cdots) + \varepsilon (x_0 + \varepsilon x_1 + \varepsilon^2 x_2 + \cdots)^2 = \Gamma \cos t.$$

The coefficients of the powers ε lead to the perturbation equations

$$\ddot{x}_0 + \Omega^2 x_0 = \Gamma \cos t, \qquad \text{(i)}$$

$$\ddot{x}_1 + \Omega^2 x_1 = -x_0^2, \qquad \text{(ii)}$$

$$\ddot{x}_2 + \Omega^2 x_2 = -2x_0 x_1. \qquad \text{(iii)}$$

The period 2π solution of (i) is

$$x_0 = \frac{\Gamma}{\Omega^2 - 1} \cos t.$$

Equation (ii) then becomes

$$\ddot{x}_1 + \Omega^2 x_1 = -\frac{\Gamma}{2(\Omega^2 - 1)} - \frac{\Gamma}{2(\Omega^2 - 1)} \cos 2t.$$

The period 2π solution of this equation

$$x_1 = -\frac{\Gamma}{2\Omega^2(\Omega^2 - 1)} - \frac{\Gamma}{2(\Omega^2 - 1)(\Omega^2 - 4)} \cos 2t.$$

The equation for x_2 is

$$x_2'' + \Omega^2 x_2 = -2 \left(\frac{\Gamma}{\Omega^2 - 1} \cos t \right) \left(-\frac{\Gamma}{2\Omega^2(\Omega^2 - 1)} - \frac{\Gamma}{2(\Omega^2 - 1)(\Omega^2 - 4)} \cos 2t \right)$$

$$= \frac{\Gamma^2}{2\Omega^2(\Omega^2 - 1)(\Omega^2 - 4)} [(8 - \Omega^2) \cos t + \Omega^2 \cos 3t].$$

The 2π periodic solution for x_2 is

$$x_2 = \frac{\Gamma^2(8-\Omega^2)}{2\Omega^2(\Omega^2-1)^2(\Omega^2-4)}\cos t + \frac{\Gamma^2\Omega^2}{2\Omega^2(\Omega^2-1)(\Omega^2-4)(\Omega^2-9)}\cos 3t.$$

To order ε^2, the 2π periodic solution is

$$x = -\frac{\varepsilon\Gamma}{2(\Omega^2-1)} + \left[\frac{\Gamma}{\Omega^2-1} + \frac{\varepsilon^2\Gamma^2(8-\Omega^2)}{2\Omega^2(\Omega^2-1)^2(\Omega^2-4)}\right]\cos t$$

$$-\frac{\varepsilon\Gamma}{2(\Omega^2-1)(\Omega^2-4)}\cos 2t + \frac{\varepsilon^2\Gamma^2\Omega^2}{2\Omega^2(\Omega^2-1)(\Omega^2-4)(\Omega^2-9)}\cos 3t$$

The appearance of a constant term indicates that the periodic solution is no symmetrically disposed about $x = 0$. The expansion will be unreliable if Ω is close to an integer.

To emphasize that Γ is small, let $\Gamma = \varepsilon\gamma$, so that the differential equation becomes

$$\ddot{x} + \Omega^2 x + \varepsilon x^2 = \varepsilon\gamma\cos t,$$

Let $\Omega = 1 + \varepsilon\Omega_1 + \cdots$ and $x = x_0 + \varepsilon x_1 + \cdots$. Then

$$(\ddot{x}_0 + \varepsilon\ddot{x}_1 + \cdots) + (1 + \varepsilon\Omega_1 + \cdots)^2(x_0 + \varepsilon x_1 + \cdots) + \varepsilon(x_0 + \varepsilon x_1 + \cdots)^2 = \varepsilon\gamma\cos t.$$

Equating powers of ε, we obtain

$$\ddot{x}_0 + x_0 = 0, \qquad\qquad\qquad (i)$$

$$\ddot{x}_1 + x_1 = -2\Omega_1 x_0 - \varepsilon x_0^2 + \gamma\cos t. \qquad (ii)$$

The general solution of (i) is

$$x_0 = A_0 \cos t + B_0 \sin t.$$

Substitute this solution into (ii) which becomes

$$\ddot{x}_1 + x_1 = -2\Omega_1(A_0\cos t + B_0\sin t) - \varepsilon(A_0\cos t + B_0\sin t)^3 + \gamma\cos t$$

$$= -\tfrac{1}{2}\varepsilon(A_0^2 + B_0^2) - (2\Omega_1 A_0 - \gamma)\cos t - 2\Omega_1 B_0 \sin t + \text{higher harmonics}$$

Secular terms can be removed by putting $B_0 = 0$ and $A_0 = \gamma/(2\Omega_1)$, which is the leading frequency–amplitude relation. Further equations relating amplitude and frequency can be obtained by continuing the perturbation.

- **5.17** Use the method of amplitude-phase perturbation (NODE, Section 5.8) to approximate to the solutions, period 2π, of $\ddot{x} + x = \varepsilon(\gamma\cos t - x\dot{x} - \beta x)$.

5.17. In the equation
$$\ddot{x} + x = \varepsilon(\gamma \cos t - x\dot{x} - \beta x),$$
let $s = t + \alpha$ and $X(\varepsilon, s) = x(\varepsilon, t)$ (see Section 5.8). After the change of variable X satisfies
$$X'' + X = \varepsilon[\gamma \cos(s - \alpha) - XX' - \beta X],$$
where $X' = dX/ds$, etc. Let
$$X(\varepsilon, s) = X_0(s) + \varepsilon X_1(s) + \cdots, \qquad \alpha = \alpha_0 + \varepsilon \alpha_1 + \cdots. \tag{i}$$

We are searching for a 2π-period solution for $X(\varepsilon, s)$ and all its coefficients in its perturbation series. We can also assume, without loss of generality, that $X'_i(0) = 0$ for all i. Substitute the series
$$\cos(s - \alpha) = \cos(s - \alpha_0) + \varepsilon \alpha_1 \sin(s - \alpha_0) + \cdots$$
into (i) and equate to zero the coefficients of ε: the result is
$$X''_0 + X_0 = 0, \tag{ii}$$
$$X''_1 + X_1 = \gamma \cos(s - \alpha_0) - X_0 X'_0 - \beta X_0, \tag{iii}$$
$$X''_2 + X_2 = \gamma \alpha_1 \sin(s - \alpha_0) - X_0 X'_1 - X_1 X'_0 - \beta X_1. \tag{iv}$$

From (i), it follows that $X_0(s) = r_0 \cos s$. Equation (ii) now becomes
$$X''_1 + X_1 = \gamma \cos(s - \alpha_0) + r_0^2 \cos s \sin s - \beta r_0 \cos s$$
$$= (\cos \alpha_0 - \beta r_0) \cos s + \gamma \sin \alpha_0 \sin s + \tfrac{1}{2} r_0^2 \sin 2s.$$

We need to eliminate the secular terms on the right, so that
$$\cos \alpha_0 - \beta r_0 = 0, \quad \gamma \sin \alpha_0 = 0.$$

We can choose $\alpha_0 = 0$ so that $r_0 = 1/\beta$. For these values X_1 satisfies
$$X''_1 + X_1 = \frac{1}{2\beta^2} \sin 2s.,$$
which has the general solution
$$X_1 = a_1 \cos s + b_1 \sin s - \frac{1}{6\beta^2} \sin 2s.$$
Using the initial condition, $X'_1(s) = 0$,
$$X_1 = -a_1 \cos s + \frac{1}{3\beta^2} \sin s - \frac{1}{6\beta^2} \sin 2s.$$

Equation (iv) now becomes

$$X_2'' + X_2 = \frac{1}{12\beta^3}(1 + 12a_1\beta^4)\cos s + \frac{1}{3\beta}(3\alpha_1\gamma\beta - 1)\sin s + \text{higher harmonics}.$$

Non-periodic solutions arise unless

$$a_1 = \frac{1}{12\beta^4}, \quad \alpha_1 = \frac{1}{3\gamma\beta}.$$

Finally

$$X_1 = -\frac{1}{12\beta^4}\cos s + \frac{1}{3\beta^2}\sin s - \frac{1}{6\beta^2}\sin 2s.$$

• **5.18** Investigate the solutions, period 2π, of $\ddot{x} + 9x + \varepsilon x^2 = \Gamma\cos t$ obtained by using the direct method of NODE, Section 5.2. If $x = x_0 + \varepsilon x_1 + \cdots$, show that secular terms first appear in x_2.

5.18. We will use the direct method of Section 5.2 to find, approximately, the 2π period solutions of

$$\ddot{x} + 9x + \varepsilon x^2 = \Gamma\cos t.$$

Let $x = x_0 + \varepsilon x_1 + \cdots$. Then

$$(\ddot{x}_0 + \varepsilon\ddot{x}_1 + \cdots) + 9(x_0 + \varepsilon x_1 + \cdots) + \varepsilon(x_0 + \varepsilon x_1 + \cdots)^2 = \Gamma\cos t.$$

Equating like powers of ε, the first few equations are

$$\ddot{x}_0 + 9x_0 = \Gamma\cos t, \tag{i}$$

$$\ddot{x}_1 + 9x_1 = -x_0^2, \tag{ii}$$

$$\ddot{x}_2 + 9x_2 = -2x_0 x_1. \tag{iii}$$

The period 2π-period solution of (i) is

$$x_0 = a_0\cos 3t + b_0\sin 3t + \tfrac{1}{8}\Gamma\cos t.$$

Now substitute this into the equation for x_1:

$$\ddot{x}_1 + 9x_1 = -\frac{1}{2}(a_0^2 + b_0^2) + \frac{\Gamma^2}{128} - \frac{\Gamma}{128}(16a_0 + \Gamma)\cos 2t$$
$$- \tfrac{1}{8}a_0\Gamma\cos 4t + \tfrac{1}{2}(-a_0^2 + b_0^2)\cos 6t - \tfrac{1}{8}b_0\Gamma\sin 2t - \tfrac{1}{8}b_0\Gamma\sin 4t - a_0 b_0\sin 6t$$

Secular terms of the form $t\cos 3t$ and $t\sin 3t$ first appear for x_2 through the forcing term $-2x_0 x_1$.

- **5.19** For the damped pendulum equation with a forcing term
$$\ddot{x} + k\dot{x} + \omega_0^2 x - \tfrac{1}{6}\omega_0^2 x^3 = F\cos\omega t,$$
show that the amplitude–frequency curves have their maxima on $\omega^2 = \omega_0^2\left(1 - \tfrac{1}{8}r_0^2\right) - \tfrac{1}{2}k^2$.

5.19. The damped pendulum equation with a forcing term is
$$\ddot{x} + k\dot{x} + \omega_0^2 x - \tfrac{1}{6}\omega_0^2 x^3 = F\cos\omega t.$$

The amplitude–frequency equation for the equation is

$$r_0^2\left[k^2\omega^2 + \left(\omega^2 - \omega_0^2 + \tfrac{1}{8}\omega_0^2 r_0^2\right)^2\right] = F^2, \tag{i}$$

as given by NODE, (5.42). The family of curves for varying values of F are shown in Figure 5.5 (in NODE). At fixed values of r_0, there will be two solutions for ω below the maximum and one at the maximum. Equation (i) can be rearranged as a quadratic equation in ω^2, namely

$$r_0^2\omega^4 + r_0^2\omega^2\left[k^2 - 2\omega_0^2\left(1 - \tfrac{1}{8}r_0^2\right)\right] + r_0^2\omega_0^4\left(1 - \tfrac{1}{8}r_0^2\right)^2 - F^2 = 0. \tag{ii}$$

This equation has a repeated solution if

$$r_0^4\left[k^2 - 2\omega_0^2\left(1 - \tfrac{1}{8}r_0^2\right)\right]^2 = 4r_0^2\left[r_0^2\omega_0^4\left(1 - \tfrac{1}{8}r_0^2\right)^2 - F^2\right],$$

or, after rearrangement,

$$F^2 = \tfrac{1}{4}k^2 r_0^2\left[2\omega_0^2\left(1 - \tfrac{1}{8}r_0^2\right) - k^2\right]. \tag{iii}$$

The repeated solution of (ii), which locates the position of the maximum, is given by

$$\omega^2 = \omega_0^2\left(1 - \tfrac{1}{8}r_0^2\right) - \tfrac{1}{2}k^2.$$

Equation (iii) identifies the value of F and the particular path in Figure 5.5 (in NODE) on which the maximum occurs.

- **5.20** Show that the first harmonic for the forced van der Pol equation $\ddot{x} + \varepsilon(x^2 - 1)\dot{x} + x = F\cos\omega t$ is the same for both weak and hard excitation, far from resonance.

5.20. In the forced van der Pol equation

$$\ddot{x} + \varepsilon(x^2 - 1)\dot{x} + x = F\cos\omega t,$$

let $\tau = \omega t$. The equation becomes

$$\omega^2 x'' + \omega\varepsilon(x^2 - 1)x' + x = F\cos\tau, \tag{i}$$

where $x' = dx/d\tau$. Hard excitation, far from resonance, is covered in Section 5.10, where it is shown that, in the expansion $x = x_0 + \varepsilon x_1 + \cdots$,

$$x_0 = \frac{F}{1-\omega^2}\cos\tau + O(\varepsilon) = \frac{F}{1-\omega^2}\cos\omega t + O(\varepsilon). \tag{ii}$$

For soft excitation, let $F = \varepsilon F_0$, so that (i) becomes

$$\omega^2 x'' + \omega\varepsilon(x^2 - 1)\dot{x} + x = \varepsilon F_0 \cos\tau,$$

Let $x = x_0 + \varepsilon x_1 + \cdots$ for all ε. Then x_0 and x_1 satisfy the equations

$$\omega^2 x_0'' + x_0 = 0,$$

$$\omega^2 x_1'' + \omega(x_0^2 - 1)x_0' + x_1 = F_0 \cos\tau.$$

Period 2π solutions are

$$x_0 = 0, \quad x_1 = \frac{F_0}{1-\omega^2}\cos\tau.$$

Therefore

$$x = \varepsilon x_1 + O(\varepsilon^2) = \frac{\varepsilon F_0}{1-\omega^2}\cos\tau + O(\varepsilon^2) = \frac{F}{1-\omega^2}\cos\omega t + O(\varepsilon^2),$$

in which the leading term agrees with (ii).

- 5.21 The orbital equation of a planet about the sun is

$$\frac{d^2 u}{d\theta^2} + u = k(1 + \varepsilon u^2),$$

where $u = r^{-1}$ and r, θ, are polar coordinates, $k = \gamma m/h^2$, γ is the gravitational constant, m is the mass of the planet and h is its moment of momentum, a constant, $\varepsilon k u^2$ is the relativistic correction term, where ε is a small constant.

Obtain a perturbation expansion for the solution with initial conditions $u(0) = k(e+1)$, $\dot{u}(0) = 0$. (e is the eccentricity of the unperturbed orbit, and these are initial conditions at the perihelion: the nearest point to the sun on the unperturbed orbit.) Note that the solution of the unperturbed equation is not periodic, and that 'secular' terms cannot be eliminated. Show that the expansion to order ε predicts that in each orbit the perihelion advances by $2k^2\pi\varepsilon$.

5.21. The orbital equation of a planet is

$$u'' + u = k(1 + \varepsilon u^2),$$

where $u = 1/r$ and r, θ are polar coordinates, and ε is a small constant. The initial conditions are

$$u(0) = k(e+1), \quad \dot{u}(0) = 0,$$

where e is the eccentricity of the unperturbed orbit. Let $u = u_0 + \varepsilon u_1 + \cdots$. The leading terms satisfy the differential equations

$$u_0'' + u_0 = k, \qquad \text{(i)}$$
$$u_1'' + u_1 = ku_0^2, \qquad \text{(ii)}$$

subject to the initial conditions $u_0(0) = k(e+1)$, $u_1(0) = 0$, $u_0'(0) = u_1'(0) = 0$. The general solution of (i) is

$$u_0 = A_0 \cos\theta + B_0 \sin\theta + k.$$

From the conditions, $A_0 = ke$, $B_0 = 0$, so that

$$u_0 = k(e\cos\theta + 1).$$

Equation (ii) is now

$$u_1'' + u_1 = k^3(e\cos\theta + 1)^2 = \tfrac{1}{2}k^3[(2 + e^2) + 4e\cos\theta + e^2\cos 2\theta]. \qquad \text{(iii)}$$

Since e is specified, the secular term $2e\cos\theta$ cannot be eliminated, which indicates that the solution will not be periodic. The general solution of this equation is

$$u_1 = A_1 \cos\theta + B_1 \sin\theta + \tfrac{1}{2}k^3(2 + e^2) + ek^3\theta\sin\theta - \tfrac{1}{6}e^2k^3\cos 2\theta.$$

From the initial conditions $A_1 = -\tfrac{1}{3}k^3(3 + e^2)$, $B_1 = 0$. Hence

$$u_1 = \tfrac{1}{2}k^3(2 + e^2) - \tfrac{1}{3}k^3(3 + e^2)\cos\theta + ek^3\theta\sin\theta - \tfrac{1}{6}e^2k^3\cos 2\theta,$$

so that

$$u = k(e\cos\theta + 1) + \varepsilon k^3\left[\tfrac{1}{2}(2 + e^2) - \tfrac{1}{3}(3 + e^2)\cos\theta + e\theta\sin\theta - \tfrac{1}{6}e^2\cos 2\theta\right] + O(\varepsilon^2).$$

Let $\theta = 2\pi + \mu_0\varepsilon$, where $|\mu|$ is small. At the next perihelion,

$$u'(2\pi + \mu_0\varepsilon) = u'_0(2\pi + \mu_0\varepsilon) + \varepsilon u'_1(2\pi + \mu_0\varepsilon) + O(\varepsilon^2)$$
$$= u'_0(2\pi) + \varepsilon\mu_0 u''_0(2\pi) + \varepsilon u'_1(2\pi) + O(\varepsilon^2)$$
$$= \varepsilon[\mu_0 u''_0(2\pi) + u'_1(2\pi)] + O(\varepsilon^2)$$
$$= \varepsilon[\mu_0(-ke) + 2ek^3\pi] + O(\varepsilon^2)$$

The term of order ε vanishes if $\mu_0 = 2k^2\pi$. Hence the perihelion advances by $2k^2\pi\varepsilon$ approximately.

- **5.22** Use the Lindstedt procedure (NODE, Section 5.9) to find the first few terms in the expansion of the periodic solutions of $\ddot{x} + x + \varepsilon x^2 = 0$. Explain the presence of a constant term in the expansion.

5.22. We apply the Lindstedt procedure (Section 5.9) to the equation

$$\ddot{x} + x + \varepsilon x^2 = 0. \tag{i}$$

When $\varepsilon = 0$, the frequency of all solutions is 1. Therefore we let the unknown frequency be

$$\omega = 1 + \varepsilon\omega_1 + \cdots, \quad x(\varepsilon, t) = x_0(t) + \varepsilon x_1(t) + \cdots.$$

Apply the change of variable $\tau = \omega t$, so that (i) becomes

$$\omega^2 x'' + x + \varepsilon x^2 = 0.$$

We are searching for 2π periodic solutions. Substitution of the perturbations leads to

$$(1 + \varepsilon\omega_1 + \cdots)^2(x''_0 + \varepsilon x''_1 + \cdots) + (x_0 + \varepsilon x_1 + \cdots) + \varepsilon(x_0 + \varepsilon x_1 + \cdots)^2 = 0$$

for all ε. Therefore the leading terms satisfy

$$x''_0 + x_0 = 0, \tag{ii}$$
$$x''_1 + x_1 = -2\omega_1 x''_0 - x_0^2 = 0, \tag{iii}$$
$$x''_2 + x_2 = -(\omega_1^2 + 2\omega_2)x''_0 - 2\omega_1 x''_1 - x_0^2 - 2x_0 x_1. \tag{iv}$$

Since it is an autonomous system we can simplify the procedure by assuming $x(\varepsilon, 0) = a$ and $x'(\varepsilon, 0) = 0$, which implies that $x_0 = a$, $x'_0(0) = 0$, and $x_i(0) = x'_i(0) = 0$, ($i = 1, 2, \ldots$). The solution of (i) is

$$x_0 = a \cos \tau,$$

so that x_1 satisfies

$$x''_1 + x_1 = 2\omega_1 a \cos \tau - a^2 \cos^2 \tau = 2\omega_1 a \cos \tau - \tfrac{1}{2}a^2(1 + \cos 2\tau).$$

To eliminate the secular term put $\omega_1 = 0$ ($a_0 = 0$ leads to the solution $x = 0$). By elementary integration follows that

$$x_1 = -\tfrac{1}{2}a^2 + \tfrac{1}{3}a^2 \cos \tau + \tfrac{1}{6}a^2 \cos 2\tau.$$

With $\omega_1 = 0$, eqn (iv) becomes

$$x''_2 + x_2 = -2\omega_2 x''_0 - x_0^2 - 2x_0 x_1$$

$$= 2\omega_2 a \cos \tau - a^2 \cos^2 \tau - 2a \cos \tau \left(-\tfrac{1}{2}a^2 + \tfrac{1}{3}a^2 \cos t + \tfrac{1}{6}a^2 \cos 2t\right)$$

$$= -\tfrac{1}{6}a^2(3 + 2a) + \tfrac{1}{6}a(5a^2 + 24\omega_2) \cos t - \tfrac{1}{6}a^2(3 + 2a) \cos 2t - \tfrac{1}{6}a^3 \cos 3t$$

The secular term vanishes if the coefficient of $\cos t$ is zero, that is, if $\omega_2 = -\tfrac{5}{24}a^2$. Therefore the frequency–amplitude relation is given by

$$\omega = 1 - \frac{5}{24}a^2\varepsilon^2 + O(\varepsilon^3).$$

Finally

$$x = x_0 + \varepsilon x_1 + O(\varepsilon^2)$$

$$= a \cos\left[\left(1 - \frac{5}{24}\varepsilon^2\right)t\right] + \varepsilon\left\{-\tfrac{1}{2}a^2 + \tfrac{1}{3}a^2 \cos\left[\left(1 - \frac{5}{24}\varepsilon^2\right)t\right]\right.$$

$$\left. + \tfrac{1}{6}a^2 \cos\left[2\left(1 - \frac{5}{24}\varepsilon^2\right)t\right]\right\} + O(\varepsilon^2).$$

The constant term indicates that the solution does not have a mean value of zero.

- **5.23** Investigate the forced periodic solutions of period 2π of the equation
$$\ddot{x} + (4+\varepsilon\beta)x - \varepsilon x^3 = \Gamma\cos t$$
where ε is small and β and Γ are not too large. Confirm that there is always a periodic solution of the form $a_0\cos 2t + b_0\sin 2t + \tfrac{1}{3}\Gamma\cos t$, where
$$a_0\left(\tfrac{3}{4}r_0^2 + \tfrac{1}{6}\Gamma^2 - \beta\right) = b_0\left(\tfrac{3}{4}r_0^2 + \tfrac{1}{6}\Gamma^2 - \beta\right) = 0.$$

5.23. In the equation
$$\ddot{x} + 4x + \varepsilon\beta x - \varepsilon x^3 = \Gamma\cos t,$$
let $x = x_0 + \varepsilon x_1 + \cdots$. Then, equating like powers of ε, we have
$$\ddot{x}_0 + 4x_0 = \Gamma\cos t, \qquad\qquad\qquad \text{(i)}$$
$$\ddot{x}_1 + 4x_1 = -\beta x_0 + x_0^3. \qquad\qquad\qquad \text{(ii)}$$

The general solution of (i) is
$$x_0 = a_0\cos 2t + b_0\sin 2t + \tfrac{1}{3}\Gamma\cos t.$$

Substitution of x_0 into (ii) implies
$$\ddot{x}_1 + 4x_1 = -\beta\left(a_0\cos 2t + b_0\sin 2t + \tfrac{1}{3}\Gamma\cos t\right) + \left(a_0\cos 2t + b_0\sin 2t + \tfrac{1}{3}\Gamma\cos t\right)^3$$
$$= \tfrac{1}{36}\Gamma(18r_0^2 - 12\beta + \Gamma^2)\cos t + \tfrac{1}{12}(9r_0^2 - 12\beta + 2\Gamma^2)(a_0\cos 2t + b_0\sin 2t)$$
$$+ \text{ higher harmonics}$$

where $r_0 = \sqrt{(a_0^2 + b_0^2)}$. The secular terms can be eliminated if
$$a_0\left(\tfrac{3}{4}r_0^2 - \beta + \tfrac{1}{6}\Gamma^2\right) = b_0\left(\tfrac{3}{4}r_0^2 - \beta + \tfrac{1}{6}\Gamma^2\right) = 0.$$

The solutions are $a_0 = b_0 = 0$ or any a_0, b_0 which satisfy
$$\tfrac{3}{4}r_0^2 - \beta + \tfrac{1}{6}\Gamma^2 = 0.$$

- 5.24 Investigate the equilibrium points of $\ddot{x} + \varepsilon(\alpha x^4 - \beta)\dot{x} - x + x^3 = 0$, $(\alpha > \beta > 0)$ for $0 < \varepsilon \ll 1$. Use the perturbation method of NODE, Section 5.12 on homoclinic bifurcation to find the approximate value of β/α at which homoclinic paths exist.

5.24. Equilibrium points of

$$\ddot{x} + \varepsilon(\alpha x^4 - \beta)\dot{x} - x + x^3 = 0, \quad \dot{x} = y, \quad (\alpha > \beta > 0), \tag{i}$$

are located at $(0,0)$, $(1,0)$ and $(-1,0)$. Their linear classifications are as follows.

- $(0,0)$. Near the origin

$$\ddot{x} - \beta\varepsilon\dot{x} - x = 0.$$

The solutions of the characteristic equation are

$$m = \tfrac{1}{2}[\varepsilon\beta \pm \sqrt{(\varepsilon^2\beta^2 + 4)}],$$

which are both real but of opposite signs. Hence $(0,0)$ is saddle point.
- $(1,0)$. Let $x = 1 + X$. Then the equation becomes

$$\ddot{X} + \varepsilon[\alpha(1+X)^4 - \beta]\dot{X} - (1+X) + (1+X)^3 = 0.$$

The linear approximation of the equation is

$$\ddot{X} + \varepsilon(\alpha - \beta)\dot{X} + 2X = 0.$$

The solutions of its characteristic equation are

$$m = \tfrac{1}{2}[-\varepsilon(\alpha - \beta) \pm \sqrt{\{\varepsilon^2(\alpha-\beta)^2 - 8\}}].$$

For $\alpha > \beta$ and ε sufficiently small, $(1,0)$ is a stable spiral.
- $(-1,0)$. This has the same linearized equation as that for $(1,0)$. Therefore it is also stable spiral.

Following the method described in Section 5.12, we let $x = x_0 + \varepsilon x_1 + \cdots$ and substitute the series into (i). The differential equations for x_0 and x_1 are

$$\ddot{x}_0 - x_0 + x_0^3 = 0,$$
$$\ddot{x}_1 + (3x_0^2 - 1)x_1 = -(\alpha x_0^4 - \beta)\dot{x}_0.$$

the zero-order equation. As shown in the same section has the homoclinic solution $x_0 = \sqrt{2}\operatorname{sech} t$ for $x > 0$, and $x_0 = -\sqrt{2}\operatorname{sech} t$ for $x < 0$. The condition for a homoclinic path in $x > 0$ is

$$\int_{-\infty}^{\infty} (\beta - \alpha x_0^4(t))\dot{x}_0^2(t) dt = 0$$

as in NODE, eqn (5.105). Since $x_0(t) = \sqrt{2}\operatorname{sech} t$, the condition becomes

$$2\beta \int_{-\infty}^{\infty} \operatorname{sech}^4 t \sinh^2 t \, dt - 8\alpha \int_{-\infty}^{\infty} \operatorname{sech}^8 t \sinh^2 t \, dt = 0. \qquad (ii)$$

Use the substitution $u = \tanh t$. Then $du/dt = \operatorname{sech}^2 t$ and the integrals become

$$\int_{-\infty}^{\infty} \operatorname{sech}^4 t \sinh^2 t \, dt = \int_{-1}^{1} u^2 du = \frac{2}{3},$$

$$\int_{-\infty}^{\infty} \operatorname{sech}^8 t \sinh^2 t \, dt = \int_{-1}^{1} u^2(1-u^2) du = \int_{-1}^{1}(u^2 - 2u^4 + u^6) du = \frac{16}{105}.$$

Hence condition (ii) becomes

$$\frac{4\beta}{3} - \frac{128}{105}\alpha = 0,$$

so that $\beta/\alpha = 32/35$.

- 5.25 Investigate the equilibrium points of $\ddot{x} + \varepsilon(\alpha x^2 - \beta)\dot{x} - x + 3x^5 = 0$, $(\alpha, \beta > 0)$ for $0 < \varepsilon \ll 1$. Confirm that the equation has an unperturbed time solution $x_0 = \sqrt{[\operatorname{sech} 2t]}$ (see Problem 3.55). Use the perturbation method of Section 5.12 to show that a homoclinic bifurcation takes place for $\beta \approx 4\alpha/(3\pi)$.

5.25. Equilibrium points of

$$\ddot{x} + \varepsilon(\alpha x^2 - \beta)\dot{x} - x + 3x^5 = 0 \qquad (i)$$

are located at $(0,0)$, $(3^{-1/4}, 0)$ and $(-3^{-1/4}, 0)$. Their linear classifications are as follows.

- $(0,0)$. Near the origin

$$\ddot{x} - \varepsilon\beta\dot{x} - x = 0.$$

Its characteristic equation has the solutions

$$m = \tfrac{1}{2}[\varepsilon\beta \pm \sqrt{(\varepsilon^2\beta^2 + 4)}].$$

which are real but of opposite signs. Therefore $(0,0)$ is a saddle point.
- $(3^{-1/4}, 0)$. Let $x = 3^{-1/4} + X$. Then the equation becomes

$$\ddot{X} + \varepsilon[\alpha(3^{-1/4} + X)^2 - \beta]\dot{X} - (3^{-1/4} + X) + 3(3^{-1/4} + X)^5 = 0.$$

Its linear approximation

$$\ddot{X} + \varepsilon(3^{-(1/2)}\alpha - \beta)\dot{X} + 4X = 0,$$

which implies that $(3^{-1/4}, 0)$ is a spiral for ε sufficiently small.
- $(-3^{-1/4}, 0)$. By symmetry (replace x by $-x$ in (i)), this equilibrium point is also a spiral.

If $x_0 = \sqrt{(\mathrm{sech}\, 2t)}$, then

$$\dot{x}_0 = -\tfrac{1}{2}\mathrm{sech}^{3/2}t \sinh t,$$

$$\ddot{x}_0 = -2\mathrm{sech}^{1/2}2t + 3\mathrm{sech}^{5/2}2t\sinh^2 2t = \mathrm{sech}^{1/2}2t - 3\mathrm{sech}^{5/2}2t = x_0 - 3x_0^5.$$

The condition for a homoclinic path in $x > 0$ is

$$\int_{-\infty}^{\infty}(\beta - \alpha x_0^4(t))\dot{x}_0^2(t)dt = 0$$

as in eqn (5.15). Since $x_0 = \sqrt{(\mathrm{sech}\, 2t)}$, the condition becomes

$$\beta\int_{-\infty}^{\infty}\mathrm{sech}^3 2t \sinh^2 2t\, dt - \alpha\int_{-\infty}^{\infty}\mathrm{sech}^4 2t \sinh^2 2t\, dt = 0.$$

Use the substitutions $u = \sinh 2t$ in the first integral and $v = \tanh t$ in the second integral, so that the integrals become

$$\frac{\beta}{2}\int_{-\infty}^{\infty}\frac{u^2}{(1+u^2)^2}du - \alpha\int_{-1}^{1}v^2 dv = 0,$$

or

$$\frac{\beta\pi}{4} - \alpha\frac{1}{3} = 0.$$

Therefore for a homoclinic connection we require $\beta \approx 4\alpha/3\pi$.

- **5.26** The equation $\ddot{x} + \varepsilon g(x, \dot{x})\dot{x} + f(x) = 0$, $\dot{x} = y$, $0 < \varepsilon \ll 1$ is known to have a saddle point at $(0, 0)$ with an associated homoclinic trajectory with solution $x = x_0(t)$ for $\varepsilon = 0$. Work through the perturbed method of NODE, Section 5.12, and show that any homoclinic paths of the perturbed system occur where

$$\int_{-\infty}^{\infty} g(x_0, \dot{x}_0)\dot{x}_0^2 dt = 0.$$

If $g(x, \dot{x}) = \beta - \alpha x^2 - \gamma \dot{x}_0^2$ and $f(x) = -x + x^3$, show that homoclinic bifurcation occurs approximately where $\beta = (28\alpha + 12\gamma)/35$ for small ε.

5.26. The equation

$$\ddot{x} + \varepsilon g(x, \dot{x})\dot{x} + f(x) = 0,$$

is known to have a saddle point at the origin, with an associated homoclinic trajectory with solution $x = x_0(t)$ for $\varepsilon = 0$. As in Section 5.12, let $x = x_0 + \varepsilon x_1 + \cdots$. Then the differential equation becomes

$$(\ddot{x}_0 + \varepsilon \ddot{x}_1 + \cdots) + \varepsilon g(x_0 + \varepsilon x_1 + \cdots, \dot{x}_0 + \varepsilon \dot{x}_1 + \cdots)(\dot{x}_0 + \varepsilon \dot{x}_1 + \cdots)$$

$$+ f(x_0 + \varepsilon x_1 + \cdots) = 0.$$

Hence, expanding $g(x_0 + \varepsilon x_1 + \cdots, \dot{x}_0 + \varepsilon \dot{x}_1 + \cdots)$ and $f(x_0 + \varepsilon x_1 + \cdots)$, and equating to zero the coefficients of powers of ε, we have

$$\ddot{x}_0 + f(x_0) = 0, \tag{i}$$

$$\ddot{x}_1 + f'(x_0)x_1 = -g(x_0, \dot{x}_0)\dot{x}_0. \tag{ii}$$

We are given that x_0 satisfies (i) identically. Multiply both sides of (ii) by $\dot{x}_0$ and confirm that

$$\frac{d}{dt}[\dot{x}_1\dot{x}_0 + x_1 f(x_0)] = -g(x_0, \dot{x}_0)\dot{x}_0^2.$$

Integrate over the infinite interval with respect to t, so that

$$[\dot{x}_1\dot{x}_0 + x_1 f(x_0)]_{-\infty}^{\infty} = -\int_{-\infty}^{\infty} g(x_0, \dot{x}_0)\dot{x}_0^2 dt.$$

A necessary condition for x_1 and $\dot{x}_1$ to both approach zero as $t \to \pm\infty$ is that

$$\int_{-\infty}^{\infty} g(x_0, \dot{x}_0)\dot{x}_0^2 dt = 0. \tag{iii}$$

In the application, $g(x, \dot{x}) = \beta - \alpha x^2 - \gamma \dot{x}$ and $f(x) = -x + x^3$. The unperturbed equation is

$$\ddot{x} - x + x^3 = 0,$$

which has the homoclinic solution $x_0 = \sqrt{2}\operatorname{sech} t$ in $x > 0$. The homoclinic path remains for the perturbed system if (see (iii) above)

$$I = \int_{-\infty}^{\infty} (\beta - \alpha x_0^2 - \gamma \dot{x}_0^2) \dot{x}_0^2 dt = 0.$$

Substitution of x_0 leads to

$$I = \int_{-\infty}^{\infty} [2\beta \operatorname{sech}^4 t \sinh^2 t - 4\alpha \operatorname{sech}^6 t \sinh^2 t - 4\gamma \operatorname{sech}^8 \sinh^4 t] dt$$

$$= 2\beta \frac{2}{3} - 4\alpha \frac{4}{15} - 4\gamma \int_{-1}^{1} (1-u^2) u^4 du \quad \text{(using integrals after (5.106))}$$

$$= \frac{4}{3}\beta - \frac{16}{15}\alpha - 4\gamma \left[\frac{2}{5} - \frac{2}{7}\right]$$

$$= \frac{4}{3}\beta - \frac{16}{15}\alpha - \frac{16}{35}\gamma = 0$$

if $\beta = (28\alpha + 12\gamma)/35$.

- **5.27** Apply Lindstedt's method to $\ddot{x} + \varepsilon x \dot{x} + x = 0$, $0 < \varepsilon \ll 1$ where $x(0) = a_0$, $\dot{x}(0) = 0$. Show that the frequency–amplitude relation for periodic solutions is given by $\omega = 1 - \frac{1}{24}a_0^2 \varepsilon^2 + O(\varepsilon^3)$.

5.27. Applying Lindtstedt's (NODE, Section 5.9) method to

$$\ddot{x} + \varepsilon x \dot{x} + x = 0, \quad x(0) = a_0, \quad \dot{x}(0) = 0,$$

let $\tau = \omega t$, $\omega = 1 + \varepsilon \omega_1 + \varepsilon^2 \omega_2 + \cdots$ and $x = x_0 + \varepsilon x_1 + \varepsilon^2 x_2 + \cdots$. The equation becomes

$$\omega^2 x'' + \varepsilon \omega x x' + x = 0,$$

where $x' = dx/d\tau$. Substitution of the series into the equation leads to

$$(1 + \varepsilon \omega_1 + \varepsilon^2 \omega_2 + \cdots)^2 (x_0'' + \varepsilon x_1'' + \varepsilon^2 x_2'' + \cdots)$$
$$+ \varepsilon (1 + \varepsilon \omega_1 + \varepsilon^2 \omega_2 + \cdots)(x_0 + \varepsilon x_1 + \varepsilon^2 x_2 + \cdots)(x_0' + \varepsilon x_1' + \varepsilon^2 x_2' + \cdots)$$
$$+ (x_0 + \varepsilon x_1 + \varepsilon^2 x_2 + \cdots) = 0.$$

Equating to zero the coefficients of the powers of ε, we obtain the differential equations

$$x_0'' + x_0 = 0, \tag{i}$$

$$x_1'' + x_1 = -2\omega_1 x_0'' - x_0 x_0', \tag{ii}$$

$$x_2'' + x_2 = -(\omega_1 x_0 + x_1) x_0' - x_0 x_1' - (\omega_1^2 + 2\omega_2) x_0'' - 2\omega_1 x_1''. \tag{iii}$$

The perturbation initial conditions for $x(0) = a$ and $x'(0) = 0$ become

$$x_0(0) = a_0, \quad x_i(0) = 0, (i = 1, 2, \ldots), \quad x_j'(0) = 0, (j = 0, 1, 2, \ldots).$$

The solution of (i) subject to the initial conditions is $x_0 = a \cos \tau$. Equation (ii) then becomes

$$x_1'' + x_1 = 2\omega_1 a_0 \cos \tau + \tfrac{1}{2} a_0^2 \sin 2\tau.$$

Periodicity for x_1 requires $\omega_1 = 0$, so that x_1 satisfies

$$x_1'' + x_1 = \tfrac{1}{2} a_0^2 \sin 2\tau.$$

The solution subject to $x_1(0) = x_1'(0) = 0$ is

$$x_1 = \tfrac{1}{6} a_0^2 (2 \sin \tau - \sin 2\tau).$$

Substitution of x_0 and x_1 into (iii), the equation for x_2, gives

$$x_2'' + x_2 = \tfrac{1}{12}(a_0^3 + 24 a_0 \omega_2) \cos \tau - 4 a_0^3 \cos 2\tau + 3 a_0^3 \cos 3\tau.$$

Secular terms can be eliminated by choosing $\omega_2 = -a_0^2/24$. Therefore the frequency-amplitude is given by

$$\omega = 1 - \tfrac{1}{24} a_0^2 \varepsilon^2 + O(\varepsilon^3).$$

• **5.28** Find the first three terms in a direct expansion for x in powers of ε for period 2π solutions of the equation $\ddot{x} + \Omega^2 x - \varepsilon \dot{x}^2 = \cos t$, where $0 < \varepsilon \ll 1$ and $\Omega \neq$ an integer.

5.28. In the equation

$$\ddot{x} + \Omega^2 x - \varepsilon \dot{x}^2 = \cos t,$$

let $x = x_0 + \varepsilon x_1 + \varepsilon^2 x_2 + \cdots$. Therefore

$$(\ddot{x}_0 + \varepsilon \ddot{x}_1 + \varepsilon^2 \ddot{x}_2 + \cdots) + \Omega^2(x_0 + \varepsilon x_1 + \varepsilon^2 x_2 + \cdots) - \varepsilon(\dot{x}_0 + \varepsilon \dot{x}_1 + \varepsilon^2 \dot{x}_2 + \cdots)^2 = \cos t.$$

Equating like powers of ε leads to the equations

$$\ddot{x}_0 + \Omega^2 x_0 = \cos t, \qquad \text{(i)}$$

$$\ddot{x}_1 + \Omega^2 x_1 = \dot{x}_0^2, \qquad \text{(ii)}$$

$$\ddot{x}_2 + \Omega^2 x_2 = 2\dot{x}_0 \dot{x}_1. \qquad \text{(iii)}$$

The 2π-periodic solution of (i) is

$$x_0 = \frac{\cos t}{\Omega^2 - 1}. \qquad \text{(iv)}$$

Equation (ii) becomes

$$\ddot{x}_1 + \Omega^2 x_1 = -\frac{\sin^2 t}{(\Omega^2 - 1)^2} = -\frac{1}{2(\Omega^2 - 1)^2}(1 - \cos 2t).$$

The 2π periodic solution is

$$x_1 = \frac{1}{2(\Omega^2 - 1)^2}\left(\frac{1}{\Omega^2} + \frac{\cos 2t}{\Omega^2 - 4}\right). \qquad \text{(v)}$$

Equation (iii) is

$$\ddot{x}_2 + \Omega^2 x_2 = -\frac{\sin t}{2\Omega^2(\Omega^2 - 1)^3} + \frac{\sin t \cos 2t}{2(\Omega^2 - 1)^3(\Omega^2 - 4)}$$

$$= \frac{7 - 2\Omega^2}{4\Omega^2(\Omega^2 - 1)^3(\Omega^2 - 4)}\sin t + \frac{1}{4(\Omega^2 - 1)^3(\Omega^2 - 4)}\sin 3t$$

The 2π periodic solution of this equation is

$$x_2 = \frac{7 - 2\Omega^2}{4\Omega^2(\Omega^2 - 1)^4(\Omega^2 - 4)}\sin t + \frac{1}{4(\Omega^2 - 1)^3(\Omega^2 - 4)(\Omega^2 - 9)}\sin 3t. \qquad \text{(vi)}$$

The periodic solution can be constructed by substituting x, x_1, x_2 from (iv), (v) and (vi) into

$$x = x_0 + \varepsilon x_1 + \varepsilon^2 x_2 + O(\varepsilon^3).$$

6 Singular perturbation methods

• **6.1** Work through the details of NODE, Example 6.1 to obtain an approximate solution of $\ddot{x} + x = \varepsilon x^3$, with $x(\varepsilon, 0) = 1$, $\dot{x}(\varepsilon, 0) = 0$, with error $O(\varepsilon^3)$ uniformly on $t \geq 0$.

6.1. Substitute the expansion

$$x(\varepsilon, t) = x_0(t) + \varepsilon x_1(t) + \varepsilon^2 x_2(t) + \cdots$$

into the equation

$$\ddot{x} + x = \varepsilon x^3,$$

to give

$$(\ddot{x}_0(t) + \varepsilon \ddot{x}_1(t) + \varepsilon^2 \ddot{x}_2(t) + \cdots) + (x_0(t) + \varepsilon x_1(t) + \varepsilon^2 x_2(t) + \cdots) =$$
$$\varepsilon(x_0(t) + \varepsilon x_1(t) + \varepsilon^2 x_2(t) + \cdots)^3.$$

The initial conditions are

$$x_0(0) = 1, \quad x_1(0) = 0, \quad x_2(0) = 0, \ldots;$$
$$\dot{x}_0(0) = 0, \quad \dot{x}_1(0) = 0, \quad \dot{x}_2(0) = 0, \ldots.$$

The terms in the perturbation series satisfy the differential equations

$$\ddot{x}_0 + x_0 = 0, \tag{i}$$
$$\ddot{x}_1 + x_1 = x_0^3, \tag{ii}$$
$$\ddot{x}_2 + x_2 = 3x_0^2 x_1. \tag{iii}$$

The solution of (i) which satisfies the initial conditions is $x_0 = \cos t$. Equation for x_1 given by (ii)

$$\ddot{x}_1 + x_1 = \cos^3 t = \tfrac{3}{4}\cos t + \tfrac{1}{4}\cos 3t.$$

Hence

$$x_1 = A\cos t + B\sin t + \tfrac{3}{8}t\sin t - \tfrac{1}{32}\cos 3t.$$

The initial conditions imply $A = \tfrac{1}{32}$ and $B = 0$, so that

$$x_1 = \tfrac{1}{32}\cos t + \tfrac{3}{8}t\sin t - \tfrac{1}{32}\cos 3t.$$

In (iii)

$$3x_0^2 x_1 = \tfrac{3}{128}(2\cos t - \cos 3t - \cos 5t + 12t\sin t + 12t\sin 3t),$$

so that

$$\ddot{x}_2 + x_2 = 3x_0^2 x_1 = \tfrac{3}{128}(2\cos t - \cos 3t - \cos 5t + 12t\sin t + 12t\sin 3t).$$

The solution satisfying $x_2(0) = \dot{x}_2(0) = 0$ is

$$x_2 = \tfrac{1}{1024}(23\cos t - 72t^2\cos t - 24\cos 3t + \cos 5t + 96t\sin t - 36t\sin 3t).$$

Now let

$$t = \tau + \varepsilon T_1(\tau) + \varepsilon^2 T_2(\tau) + \cdots. \tag{iv}$$

Substitute for t into the expressions for $x_0(t)$, $x_1(t)$ and $x_2(t)$. They

$$x_0 = \cos[\tau + \varepsilon T_1(\tau) + \varepsilon^2 T_2(\tau)] + O(\varepsilon^3)$$

$$= \cos\tau - \varepsilon T_1(\tau)\sin\tau + \varepsilon^2\left(-\tfrac{1}{2}T_1(\tau)^2\cos\tau - T_2(\tau)\sin\tau\right) + O(\varepsilon^3).$$

$$x_1 = \tfrac{1}{32}\cos(\tau + \varepsilon T_1(\tau)) + \tfrac{3}{8}\varepsilon(\tau + \varepsilon T_1(\tau))\sin(\tau + \varepsilon T_1(\tau))$$

$$- \tfrac{1}{32}\cos 3(\tau + \varepsilon T_1(\tau)) = \tfrac{1}{32}[\cos\tau - \cos 3\tau + 12\tau\sin\tau]$$

$$+ \tfrac{1}{32}\varepsilon T_1(\tau)[11\sin\tau + 12\tau\cos\tau + 3\sin 3\tau] + O(\varepsilon^2).$$

$$x_2 = \tfrac{1}{1024}(23\cos\tau - 72\tau^2\cos\tau - 24\cos 3\tau + \cos 5\tau + 96\tau\sin\tau$$

$$- 36\tau\sin 3\tau) + O(\varepsilon).$$

In the expansion $x(\varepsilon,t) = x_0(t) + \varepsilon x_1(t) + \varepsilon^2 x_2(t) + \cdots$, the zero-order coefficient is $\cos \tau$. The coefficient of ε is

$$-T_1(\tau) \sin \tau + \tfrac{1}{32}(\cos \tau - \cos 3\tau + 12\tau \sin \tau).$$

The growth term can be eliminated by choosing

$$T_1(\tau) = \tfrac{3}{8}\tau.$$

The coefficient of ε^2 is

$$-\tfrac{1}{2}T_1(\tau)^2 \cos \tau - T_2(\tau) \sin \tau + \tfrac{1}{32}T_1(\tau)[11 \sin \tau + 12\tau \cos \tau + 3 \sin 3\tau]$$
$$+ \tfrac{1}{1024}(23 \cos \tau - 72\tau^2 \cos \tau - 24 \cos 3\tau + \cos 5\tau + 96\tau \sin \tau - 36\tau \sin 3\tau)$$
$$= -T_2(\tau) \sin \tau + \tfrac{57}{256}\tau \sin \tau + \tfrac{23}{1024} \cos \tau - \tfrac{24}{1024} \cos 3\tau + \tfrac{1}{1024} \cos 5\tau$$

after substituting $T_1(\tau) = \tfrac{3}{8}\tau$. The secular term can be eliminated by choosing $T_2(\tau) = \tfrac{57}{256}\tau$. Hence from (iv) the time perturbation is

$$t = \tau + \tfrac{3}{8}\varepsilon\tau + \tfrac{57}{256}\varepsilon^2\tau + O(\varepsilon^3), \tag{v}$$

and in terms of τ the solution becomes

$$x = \cos \tau + \tfrac{1}{32}\varepsilon(\cos \tau - \cos 3\tau) + \tfrac{1}{1024}\varepsilon^2(23 \cos \tau - 24 \cos 3\tau + \cos 5\tau) + O(\varepsilon^3).$$

The period of this expression in τ is 2π, and the corresponding t-period is obtained from (v).

- **6.2** How does the period obtained by the method of Problem 6.1 compare with that derived in Problem 1.34?

6.2. The equation in Problem 1.34 is, with a change of notation,

$$\ddot{X} + X = \varepsilon' X^3, \tag{i}$$

and, the period T of this pendulum equation (as in Problem 6.1) in terms of the amplitude a was shown to be

$$T = 2\pi \left(1 + \tfrac{3}{8}\varepsilon' a^2 + \tfrac{57}{256}\varepsilon'^2 a^4\right) + O(a^6).$$

In Problem 6.1 the initial condition given is $x(0) = 1$. This can be changed to a by the substitution of $x = X/a$ into the differential equation in Problem 6.1, so that

$$\ddot{X} + X = \frac{\varepsilon}{a^2} X^3$$

and $X(0, \varepsilon) = a$. This is equivalent to (i) if ε' is replaced by ε/a^2. From Problem 6.1, the relation between t and τ is

$$t = \tau + \tfrac{3}{8}\varepsilon\tau + \tfrac{57}{256}\varepsilon^2\tau + O(\varepsilon^3),$$

and the solution is 2π periodic in τ. Hence the period T', say, is given by,

$$T' = 2\pi\left[1 + \tfrac{3}{8}\varepsilon + \tfrac{57}{256}\varepsilon^2 + O(\varepsilon^3)\right]$$
$$= 2\pi\left[1 + \tfrac{3}{8}a\varepsilon' + \tfrac{57}{256}a^2\varepsilon'^2 + O(\varepsilon'^3)\right]$$
$$\approx T.$$

for $a\varepsilon'$ small enough.

- 6.3 Apply the method of Problem 6.1 to the equation $\ddot{x}+x = \varepsilon x^3 + \varepsilon^2 \alpha x^5$ with $x(\varepsilon, 0) = 1$, $\dot{x}(\varepsilon, 0) = 0$. Obtain the period to order ε^3, and confirm that the period is correct for the pendulum when the right-hand side is the first two terms in the expansion of $x - \sin x$. (Compare the result of Problem 1.33. To obtain the required equations simply add the appropriate term to the right-hand side of the equation for x_2 in Example 6.1 in NODE.)

6.3 The equation

$$\ddot{x} + x = \varepsilon x^3 + \varepsilon^2 \alpha x^5,$$

is as in Problem 6.1, but with the additional term $\varepsilon^2 \alpha x^5$ which is of order ε^2. As in Problem 6.1, we substitute the expansion

$$x(\varepsilon, t) = x_0(t) + \varepsilon x_1(t) + \varepsilon^2 x_2(t) + \cdots,$$

into the equation and apply the initial conditions

$$x_0(0) = 1, \quad x_1(0) = 0, \quad x_2(0) = 0, \ldots;$$
$$\dot{x}_0(0) = 0, \quad \dot{x}_1(0) = 0, \quad \dot{x}_2(0) = 0, \ldots.$$

6 : Singular perturbation methods

The order zero and order ε terms will be the same so that, with

$$t = \tau + \varepsilon T_1(\tau) + \varepsilon^2 T_2(\tau) + \cdots,$$

we also find that $x_0 = \cos \tau$, $T_1(\tau) = \frac{3}{8}\tau$ and

$$x_1 = \tfrac{1}{32}(\cos \tau - \cos 3\tau).$$

The term $\varepsilon^2 \alpha x^5$ will first appear in the equation for x_2, which will now become

$$\ddot{x}_2 + x_2 = 3x_0^2 x_1 + \alpha x_0^5$$
$$= \tfrac{1}{128}[(6 + 8\alpha)\cos t + (-3 + 40\alpha)\cos 3t + (-3 + 8\alpha)\cos 5t$$
$$+ 36t \sin t + 36t \sin 3t]$$

using the expansion

$$\cos^5 t = \tfrac{1}{16}(10\cos t + 5 \cos 3t + \cos 5t),$$

The solution of the differential equation satisfying the given initial conditions is

$$x_2 = \tfrac{1}{3072}[(69 + 128\alpha - 216 t^2)\cos t - (72 + 120\alpha)\cos 3t$$
$$+ (3 - 8\alpha)\cos 5t + (288 + 960\alpha)t \sin t - 108 t \sin 3t].$$

Putting $t = \tau + \cdots$ in (i), and using eqn (iv) from Problem 6.1, the coefficient of ε^2 becomes

$$- T_2(\tau) \sin \tau + \tfrac{1}{3072}[(69 + 128\alpha)\cos \tau - (72 + 120\alpha)\cos 3\tau$$
$$+ (3 - 8\alpha)\cos 5\tau + (684 + 960\alpha)\tau \sin \tau]$$

The secular term can be eliminated by choosing

$$T_2(\tau) = \frac{57 + 80\alpha}{256}\tau.$$

We can compare the result with that obtained from the pendulum equation in Problem 1.33. The exact pendulum equation is

$$\ddot{x} + \sin x = 0,$$

which can be approximated by

$$\ddot{x} + x - \tfrac{1}{6}x^3 + \tfrac{1}{120}x^5 = 0.$$

This is the same equation as the one in this problem if $\varepsilon = \tfrac{1}{6}$ and $\varepsilon^2 \alpha = -\tfrac{1}{120}$, so that $\alpha = -\tfrac{3}{10}$. Therefore the timescale becomes

$$t = \tau + \varepsilon T_1(\tau) + \varepsilon^2 T_2(\tau) + \cdots$$

$$= \left[1 + \tfrac{3}{8}\varepsilon + \tfrac{57+80\alpha}{256}\varepsilon^2\right]\tau + O(\varepsilon^3)$$

$$= \left[1 + \tfrac{1}{16} + \tfrac{11}{3072}\right]\tau + O(\varepsilon^3).$$

The period obtained using this timescale agrees with that of Problem 1.33.

- **6.4** Use the substitution to show that the case considered in Problem 6.1 (the equation is $\ddot{x} + x = \varepsilon x^3$) covers all boundary conditions $x(\varepsilon, 0) = a$, $\dot{x}(\varepsilon, 0) = 0$.

6.4. The equation in Problem 6.1 is

$$\ddot{x} + x = \varepsilon x^3.$$

Apply the transformation $x(\varepsilon, t) = X(\varepsilon, t)/a$. Then the equation becomes

$$\ddot{X} + X = \varepsilon X^3/a^2,$$

with the initial conditions $X(\varepsilon, 0) = a$, $\dot{X}(\varepsilon, 0) = 0$. Finally replace ε/a^2 by ε' so that X satisfies

$$\ddot{X} + X = \varepsilon' X^3.$$

This assumes that ε' remains small. Hence solutions for general initial conditions subject to the previous restriction are included in Problem 6.1.

- **6.5** The equation for the relativistic perturbation of a planetary orbit is

$$\frac{d^2 u}{d\theta^2} + u = k(1 + \varepsilon u^2)$$

(see Problem 5.21). Apply the coordinate perturbation technique to eliminate the secular term in $u_1(\theta)$ in the expansion $u(\varepsilon, \theta) = u_0(\theta) + \varepsilon u_1(\theta) + \cdots$, with $\theta = \phi + \varepsilon T_1(\phi) + \cdots$.

Assume the initial conditions $u(0) = k(1 + e)$, $du(0)/d\theta = 0$. Confirm that the perihelion of the orbit advances by approximately $2\pi\varepsilon k^2$ in each planetary year.

6.5. The relativistic equation is

$$\frac{d^2u}{d\theta^2} + u = k(1 + \varepsilon u^2),$$

subject to the initial conditions $u(\varepsilon, 0) = k(1 + e)$ and $x'(\varepsilon, 0) = 0$. Let $u(\varepsilon, \theta) = u_0(\theta) + \varepsilon u_1(\theta) + \cdots$, and substitute this series into the differential equation:

$$\frac{d^2}{d\theta^2}(u_0 + \varepsilon u_1 + \cdots) + (u_0 + \varepsilon u_1 + \cdots) = k[1 + \varepsilon(u_0 + \varepsilon u_1 + \cdots)^2].$$

Equating like powers of ε, we have

$$\frac{d^2 u_0}{d\theta^2} + u_0 = k, \qquad \text{(i)}$$

$$\frac{d^2 u_1}{d\theta^2} + u_1 = ku_0^2. \qquad \text{(ii)}$$

The initial conditions become

$$u_0(0) = k(1 + e), \quad u_i(0) = 0, \ (i = 1, 2, \ldots), \quad u_j'(0) = 0, \ (j = 0, 1, 2, \ldots).$$

Therefore, from (i),

$$u_0 = k(1 + e\cos\theta).$$

Equation (ii) becomes

$$\frac{d^2 u_1}{d\theta^2} + u_1 = k^3(1 + e\cos\theta)^2 = k^3[\tfrac{1}{2}(2 + e^2) + 2e\cos\theta + \tfrac{1}{2}e^2\cos 2\theta].$$

The general solution is

$$u_1(\theta) = A\cos\theta + B\sin\theta + \tfrac{1}{2}k^3(2 + e^2) + k^3 e\theta\sin\theta - \tfrac{1}{6}e^2 k^3\cos 2\theta.$$

The initial conditions $u_1(0) = u_1'(0) = 0$ imply

$$A + \tfrac{1}{2}k^3(e^2 + 2) - \tfrac{1}{6}k^3 e^2 = 0, \quad B = 0.$$

Therefore $A = -\frac{1}{3}k^3(e^2 + 3)$, and the required solution is

$$u_1(\theta) = \frac{1}{2}k^3(2 + e^2) - \frac{1}{3}k^3(e^2 + 3)\cos\theta + k^3 e\theta \sin\theta - \frac{1}{6}e^2 k^3 \cos 2\theta.$$

Finally

$$u(\varepsilon, \theta) = u_0(\theta) + \varepsilon u_1(\theta) \cdots$$
$$= +k(1 + e\cos\theta) + \varepsilon k^3[\tfrac{1}{2}(2 + e^2) - \tfrac{1}{3}(e^2 + 3)\cos\theta$$
$$+ e\theta \sin\theta - \tfrac{1}{6}e^2 \cos 2\theta] + O(\varepsilon^2)$$

Now introduce the expansion for the angle θ:

$$\theta = \phi + \varepsilon T_1(\phi) + \cdots.$$

where ϕ is the strained coordinate. Then

$$u(\varepsilon, \phi + \varepsilon T_1(\phi) + \cdots) = u_0(\phi + \varepsilon T_1(\phi) + \cdots) + \varepsilon u_1(\phi + \varepsilon T_1(\phi) + \cdots) + \cdots$$
$$= k(1 + e\cos\phi - e\varepsilon T_1(\phi)\sin\phi) + \varepsilon k^3[\tfrac{1}{2}(2 + e^2) - \tfrac{1}{3}(e^2 + 3)\cos\phi$$
$$+ e\phi \sin\phi - \tfrac{1}{6}e^2 \cos 2\phi] + O(\varepsilon^2)$$

The non-periodic $\phi \sin\phi$ term can be eliminated by choosing $T_1(\phi) = k^2 \phi$. The period in θ is given by

$$2\pi\theta = 2\pi(1 + \varepsilon k^2 + \cdots).$$

Hence the correction, which is the advance of the perihelion, is approximately $2\pi\varepsilon k^2$.

- **6.6** Apply the multiple-scale method to van der Pol's equation $\ddot{x} + \varepsilon(x^2 - 1)\dot{x} + x = 0$. Show that, if $x(0) = a$ and $\dot{x}(0) = 0$, then for $t = O(\varepsilon^{-1})$,
$$x = 2\{1 + [(4/a^2) - 1]e^{-\varepsilon t}\}^{-(1/2)} \cos t.$$

6.6. The van der Pol equation is

$$\ddot{x} + \varepsilon(x^2 - 1)\dot{x} + x = 0. \tag{i}$$

To apply the multiple-scale method of Section 6.4, let

$$x(t, \varepsilon) = X(t, \eta, \varepsilon) = X_0(t, \eta) + \varepsilon X_1(t, \eta) + O(\varepsilon^2), \quad \eta = \varepsilon t \qquad \text{(ii)}$$

to the equation. The derivatives of $x(\varepsilon, t)$ are given by

$$\frac{dx(t, \varepsilon)}{dt} = \frac{d}{dt} X(t, \varepsilon t, \varepsilon) = \frac{\partial X}{\partial t} + \varepsilon \frac{\partial X}{\partial \eta},$$

$$\frac{d^2 x(t, \varepsilon)}{dt^2} = \frac{\partial^2 X}{\partial t^2} + 2\varepsilon \frac{\partial^2 X}{\partial \eta \partial t} + \varepsilon^2 \frac{\partial^2 X}{\partial \eta^2}.$$

The van der Pol equation is transformed into

$$\frac{\partial^2 X}{\partial t^2} + 2\varepsilon \frac{\partial^2 X}{\partial \eta \partial t} + \varepsilon^2 \frac{\partial^2 X}{\partial \eta^2} + \varepsilon (X^2 - 1) \left[\frac{\partial X}{\partial t} + \varepsilon \frac{\partial X}{\partial \eta} \right] + X = 0. \qquad \text{(iii)}$$

Substitute the series (ii) into (iii), and equate the coefficients of powers of ε to zero. The coefficients X_0 and X_1 satisfy

$$\frac{\partial^2 X_0}{\partial t^2} + X_0 = 0, \qquad \text{(iv)}$$

$$\frac{\partial^2 X_1}{\partial t^2} + X_1 = (1 - X_0^2) \frac{\partial X_0}{\partial t} - 2 \frac{\partial^2 X_0}{\partial t \partial \eta}. \qquad \text{(v)}$$

The initial conditions are (as in eqn (6.53))

$$X(0, 0, \varepsilon) = a, \quad \frac{\partial X}{\partial t}(0, 0, \varepsilon) + \varepsilon \frac{\partial X}{\partial \eta}(0, 0, \varepsilon) = 0.$$

Substitute into these initial conditions the expansion for $x(t, \varepsilon)$ and equate the coefficients of powers of ε. Then

$$X_0(0, 0) = a, \quad X_1(0, 0) = 0, \quad \frac{\partial X_0}{\partial t}(0, 0) = 0, \quad \frac{\partial X_1}{\partial t}(0, 0) + \frac{\partial X_0}{\partial \eta}(0, 0) = 0.$$

The solution of (iv) can be expressed as

$$X_0(t, \eta) = A(\eta) \cos t + B(\eta) \sin t, \quad A(0) = a, \quad B(0) = 0.$$

Equation (v) becomes

$$\frac{\partial^2 X_1}{\partial t^2} + X_1 = [1 - (A(\eta)\cos t + B(\eta)\sin t)^2](-A(\eta)\sin t + B(\eta)\cos t)$$

$$- 2(-A'(\eta)\sin t + B'(\eta)\cos t)$$

$$= \tfrac{1}{4}[4B(\eta) - A(\eta)^2 B(\eta) - B(\eta)^3 - 8B'(\eta)]\cos t$$

$$+ \tfrac{1}{4}[-4A(\eta) + A(\eta)^3 + A(\eta)B(\eta)^2 + 8A'(\eta)]\sin t$$

$$+ \tfrac{1}{4}B(\eta)[B(\eta)^2 - 3A(\eta)^2]\cos 3t + \tfrac{1}{4}A(\eta)[A(\eta)^2 - 3B(\eta)^2]\sin 3t$$

To avoid secular growth the coefficients of $\cos t$ and $\sin t$ are equated to zero so that B and A satisfy the differential equations

$$4B(\eta) - A(\eta)^2 B(\eta) - B(\eta)^3 - 8B'(\eta) = 0,$$

$$-4A(\eta) + A(\eta)^3 + A(\eta)B(\eta)^2 + 8A'(\eta) = 0.$$

The initial condition $B(0) = 0$ in (vi) implies $B(\eta) = 0$ for all η, so that $A(\eta)$ satisfies

$$8A'(\eta) = A(\eta)(4 - A(\eta)^2).$$

This is a separable equation with solution given by

$$8\int \frac{dA}{A(4 - A^2)} = \int d\eta + C = \eta + C,$$

so that

$$\frac{A^2}{|A^2 - 4|} = e^C e^\eta,$$

where C is a constant. By (vi), $A(0) = a$, leading to

$$A^2 = \frac{4K}{K - e^{-\eta}} = \frac{4a^2}{a^2 + (4 - a^2)e^{-\eta}} = \frac{4a^2}{a^2 + (4 - a^2)e^{-\varepsilon t}}.$$

Finally, form (vi), to the first order

$$x = X_0 = A(\eta)\cos t = \frac{2a}{[a^2 + (4 - a^2)e^{-\varepsilon t}]^{1/2}} \cos t.$$

- 6.7 Apply the multiple-scale method to the equation $\ddot{x}+x-\varepsilon x^3 = 0$, with initial conditions $x(0) = a$, $\dot{x}(0) = 0$. Show that, for $t = O(\varepsilon^{-1})$, $x(t) = a\cos\left\{t\left(1-\tfrac{3}{8}\varepsilon a^2\right)\right\}$.

6.7. Apply the multiple-scale method (Section 6.4) to the system

$$\ddot{x} + x - \varepsilon x^3 = 0, \quad x(0) = a, \quad \dot{x}(0) = 0.$$

As in the previous problem let

$$x(t,\varepsilon) = X(t,\eta,\varepsilon) = X_0(t,\eta) + \varepsilon X_1(t,\eta) + O(\varepsilon^2), \quad \eta = \varepsilon t.$$

in which $\eta = \varepsilon t$, so that

$$\frac{dx(t,\varepsilon)}{dt} = \frac{d}{dt}X(t,\varepsilon t,\varepsilon) = \frac{\partial X}{\partial t} + \varepsilon\frac{\partial X}{\partial \eta},$$

$$\frac{d^2 x(t,\varepsilon)}{dt^2} = \frac{\partial^2 X}{\partial t^2} + 2\varepsilon\frac{\partial^2 X}{\partial\eta\partial t} + \varepsilon^2\frac{\partial^2 X}{\partial\eta^2}.$$

The initial conditions are

$$X_0(0,0) = a, \quad X_1(0,0) = 0, \quad \frac{\partial X_0}{\partial t}(0,0) = 0, \quad \frac{\partial X_1}{\partial t}(0,0) + \frac{\partial X_0}{\partial \eta}(0,0) = 0.$$

Hence

$$\frac{\partial^2 X_0}{\partial t^2} + X_0 = 0, \tag{i}$$

$$\frac{\partial^2 X_1}{\partial t^2} + X_1 = -2\frac{\partial^2 X_0}{\partial\eta\partial t} + X_0^3. \tag{ii}$$

In this solution we shall use the alternative approach of complex solutions. Express X_0 in the form

$$X_0 = A_0(\eta)e^{it} + \overline{A}_0(\eta)e^{-it}.$$

Equation (ii) becomes

$$\frac{\partial^2 X_1}{\partial t^2} + X_1 = -2iA_0'(\eta) + 2i\overline{A}_0'(\eta) + [A_0(\eta)e^{it} + \overline{A}_0(\eta)e^{-it}]^3$$

$$= -2iA_0'(\eta)e^{it} + 2i\overline{A}_0'(\eta)e^{-it} + A_0^3(\eta)e^{3it} + 3A_0^2(\eta)\overline{A}_0(\eta)e^{it}$$
$$+ 3A_0(\eta)\overline{A}_0^2(\eta)e^{-it} + \overline{A}_0^3(\eta)e^{-3it}$$

$$= [-2iA_0'(\eta) + 3A_0^2(\eta)\overline{A}_0(\eta)]e^{it} + [2i\overline{A}_0'(\eta) + 3A_0(\eta)\overline{A}_0^2(\eta)]e^{-it}$$
$$+ \text{(higher harmonics)}$$

Secular terms can be removed if

$$-2iA_0'(\eta) + 3A_0^2(\eta)\overline{A}_0(\eta) = 0. \tag{iii}$$

Let $A_0(\eta) = \rho(\eta)e^{i\alpha(\eta)}$. Then (iii) becomes

$$-2i(\rho'e^{i\alpha} + i\rho\alpha'e^{i\alpha}) + 3\rho^3 e^{i\alpha} = 0, \quad \text{or} \quad (2\rho\alpha' + 3\rho^3) - 2i\rho' = 0.$$

The real and imaginary parts of this equation are zero if

$$\rho'(\eta) = 0, \quad 2\alpha'(\eta) + 3\rho(\eta)^2 = 0.$$

Therefore $\rho(\eta) = k$, and $\alpha'(\eta) = -\frac{3}{2}k^2$, which implies $\alpha(\eta) = -\frac{3}{2}k^2\eta + m$. Therefore

$$X_0(t,\eta) = ke^{i(m-(3/2)k^2\eta+t)} + ke^{-i(m-(3/2)k^2\eta+t)} = 2k\cos(m - \tfrac{3}{2}k^2\eta + t).$$

The initial conditions imply $a = 2k\cos m$ and $2k\sin m = 0$, so that $m = 0$ and $k = \frac{1}{2}a$. Finally

$$x \approx a\cos[(1 - \tfrac{3}{8}\varepsilon a^2)t]$$

as required.

- 6.8 Obtain the exact solution of the equation in Example 6.9 namely,

$$\varepsilon\frac{d^2y}{dx^2} + 2\frac{dy}{dx} + y = 0, \quad y(0) = 0, y(1) = 1,$$

and show that it has the first approximation equal to that obtained by the matching method.

6.8. Example 6.9 concerns the approximate solution of the boundary-value problem

$$\varepsilon \frac{d^2 y}{dx^2} + 2\frac{dy}{dx} + y = 0, \quad y(0) = 0, \quad y(1) = 1, \quad 0 < x < 1.$$

This a second-order linear differential equation having the characteristic equation

$$\varepsilon \lambda^2 + 2\lambda + 1 = 0.$$

The solutions of this equation are

$$\lambda_1, \lambda_2 = \frac{1}{2\varepsilon}[-2 \pm \sqrt{(4 - 4\varepsilon)}] = \frac{1}{\varepsilon}[-1 \pm \sqrt{(1 - \varepsilon)}].$$

Subject to the boundary conditions the solution is

$$y(x, \varepsilon) = \frac{e^{\lambda_1 x} - e^{\lambda_2 x}}{e^{\lambda_1} - e^{\lambda_2}}.$$

For small ε,

$$\lambda_1 = \frac{1}{\varepsilon}[-1 + \sqrt{(1 - \varepsilon)}] = \frac{1}{\varepsilon}[-1 + 1 - \tfrac{1}{2}\varepsilon + O(\varepsilon^2)] = -\tfrac{1}{2} + O(\varepsilon).$$

Hence, approximately,

$$y(x, \varepsilon) \approx \frac{e^{-(1/2)x} - e^{-2x/\varepsilon}}{e^{-(1/2)} - e^{-2/\varepsilon}}. \tag{i}$$

Since $x = O(1)$ but not $o(1)$, that is, x is not 'small', the terms $e^{-2x/\varepsilon}$ and $e^{-2/\varepsilon}$ are exponentially small, so that

$$y(x, \varepsilon) \approx e^{(1/2)-(1/2)x} = y_O(\varepsilon, x),$$

the outer solution given by (6.99).

In the boundary layer, let $x = \xi \varepsilon$. Then, from (i),

$$y(x, \varepsilon) \approx \frac{e^{-(1/2)\xi\varepsilon} - e^{-2\xi}}{e^{-(1/2)} - e^{-2/\varepsilon}}$$

$$\approx \frac{e^{-(1/2)\xi\varepsilon} - e^{-2\xi}}{e^{-(1/2)}} \quad \text{(neglecting the exponentially small term)}$$

$$= e^{1/2}[1 - e^{-2\xi}] + O(\varepsilon)$$

Hence

$$y(x,\varepsilon) \approx e^{1/2}[1 - e^{-2\xi}] = e^{1/2}[1 - e^{-2x/\varepsilon}] = y_I(x,\varepsilon),$$

which agrees with the inner approximation given by (6.113) (in NODE).

- **6.9** Consider the problem
  $\varepsilon y'' + y' + y = 0, \quad y(\varepsilon, 0) = 0, \quad y(\varepsilon, 1) = 1,$
  on $0 \le x \le 1$, where ε is small and positive.
  (a) Obtain the outer approximation
  $y(\varepsilon, x) \approx y_O = e^{1-x}, \quad x \text{ fixed}, \quad \varepsilon \to 0+;$
  and the inner approximation
  $y(\varepsilon, x) \approx y_I = C(1 - e^{-x/\varepsilon}), \quad x = O(\varepsilon), \quad \varepsilon \to 0+,$
  where C is a constant.
  (b) Obtain the value of C by matching y_O and y_I in the intermediate region.
  (c) Construct from y_O and y_I a first approximation to the solution which is uniform on $0 \le x \le 1$.

  Compute the exact solution, and show graphically y_O, y_I, the uniform approximation and the exact solution.

6.9. The system is

$$\varepsilon y'' + y' + y = 0, \quad y(\varepsilon, 0) = 0, \quad y(\varepsilon, 1) = 1.$$

(a) Put $\varepsilon = 0$ in the equation so that the outer solution satisfies $y'_O + y_O = 0$ subject to $y_O = 1$ at $x = 1$. Hence

$$y_O = Ae^{-x} = e^{1-x}. \tag{i}$$

Let $x = \xi\varepsilon$. Then eqn (i) is transformed into

$$\frac{d^2 y}{d\xi^2} + \frac{dy}{d\xi} + \varepsilon y = 0$$

For small ε, the inner approximation y_I satisfies

$$\frac{d^2 y}{d\xi^2} + \frac{dy}{d\xi} = 0,$$

for $x = O(\varepsilon)$ and such that $y_I = 0$ at $\xi = 0$. Hence

$$y_I = C + Be^{-\xi} = C(1 - e^{-\xi}) = C(1 - e^{-x/\varepsilon})$$

for $x = O(\varepsilon)$.

(b) The constant C in the previous equation can be found by matching the outer and inner approximations as follows. Let $x = \eta\sqrt{\varepsilon}$. Then

$$y_O = e^{1-x} = e^{1-\eta\sqrt{\varepsilon}} = e + O(\varepsilon), \quad y_I = C(1 - e^{-x/\varepsilon}) = C(1 - e^{-\eta/\sqrt{\varepsilon}})$$
$$= C + o(1),$$

as $\varepsilon \to 0+$. These match to the lowest order if $C = e$. Therefore the inner approximation is

$$y_I = e(1 - e^{-x/\varepsilon}). \tag{ii}$$

(c) Consider

$$q(x, \varepsilon) = y_O + y_I = e^{1-x} + e(1 - e^{-x/\varepsilon}).$$

If $x = O(1)$, then

$$q(x, \varepsilon) = e^{1-x} + e + O(\varepsilon) \tag{iii}$$

as $\varepsilon \to 0$. If $x = \xi\varepsilon$ where $\xi = O(1)$, then

$$q(\xi\varepsilon, \varepsilon) = e^{1-\xi\varepsilon} + e(1 - e^{-\xi}) = e + e(1 - e^{-\xi}) + O(\varepsilon). \tag{iv}$$

Comparison of (i) and (ii) with (iii) and (iv) shows that both contain the unwanted term e. Hence the composite uniform approximation is

$$y_C = y_O + y_I - e = e(e^{-x} - e^{-x/\varepsilon}).$$

- **6.10** Repeat the procedure of Problem 6.9 for the problem $\varepsilon y'' + y' + xy = 0$, $y(0) = 0$, $y(1) = 1$ on $0 \leq x \leq 1$.

6.10. Consider the system

$$\varepsilon y'' + y' + xy = 0, \quad y(0) = 0, \quad y(1) = 1.$$

The outer approximation satisfies $y' + xy = 0$ with $y(1) = 1$, for $y = O(1)$. Therefore

$$y_O = Ae^{-(1/2)x^2} = e^{(1/2)(1-x^2)}.$$

For the inner approximation, let $x = \xi\varepsilon$, so that

$$\frac{d^2y}{d\xi^2} + \frac{dy}{d\xi} + \varepsilon^2 y = 0.$$

For ε small, y satisfies, for $\xi = O(1)$,

$$\frac{d^2y}{d\xi^2} + \frac{dy}{d\xi} = 0, \quad y(0) = 0.$$

Therefore

$$y_I = C + Be^{-\xi} = C(1 - e^{-\xi}) = C(1 - e^{-x/\varepsilon}).$$

To match the outer and inner approximations, let $x = \eta\sqrt{\varepsilon}$. Then

$$y_O = e^{(1/2)(1-\eta^2\varepsilon)} = e^{1/2} + O(\varepsilon),$$

and

$$y_I = C(1 - e^{-\eta/\sqrt{\varepsilon}}) = C + o(1).$$

Hence $C = e^{1/2}$

To obtain uniform approximation, consider

$$q(x,\varepsilon) = y_O + y_I = e^{(1/2)(1-x^2)} + e^{1/2}(1 - e^{-x/\varepsilon}).$$

If $x = O(1)$, then

$$q(x,\varepsilon) = e^{(1/2)(1-x^2)} + e^{1/2} + O(\varepsilon). \tag{i}$$

Let $x = \xi\varepsilon$ where $\xi = O(1)$, then

$$q(\xi\varepsilon) = e^{(1/2)-\xi^2\varepsilon^2} + e^{1/2}(1 - e^{-\xi}) = e^{1/2} + e^{1/2}(1 - e^{-\xi}) + O(\varepsilon). \tag{ii}$$

Both (i) and (ii) contain the unwanted term $e^{1/2}$. Therefore the uniform or composite approximation is

$$y_C = y_O + y_I - e^{1/2} = e^{1/2}(e^{-(1/2)x^2} - e^{-x/\varepsilon}).$$

The inner, outer and composite approximations are shown in Figure 6.1 for $\varepsilon = 0.1$.

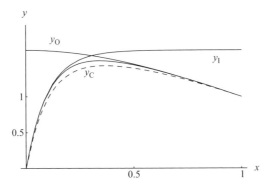

Figure 6.1 Problem 6.10: The diagram shows the outer approximation y_O, the inner approximation y_I and the composite approximation y_C; the dashed line is the exact solution computed numerically.

- **6.11** Find the outer and inner approximations of $\varepsilon y'' + y' + y \sin x = 0$, $y(0) = 0$, $y(\pi) = 1$.

6.11. Consider the system

$$\varepsilon y'' + y' + y \sin x = 0, \quad y(0) = 0, \quad y(\pi) = 1.$$

The outer approximation satisfies $y' + y \sin x = 0$ with $y(\pi) = 1$, for $y = O(1)$. Therefore

$$y_O = A e^{\cos x} = e^{1+\cos x}.$$

For the inner approximation, let $x = \xi \varepsilon$, so that

$$\frac{d^2 y}{d\xi^2} + \frac{dy}{d\xi} + \varepsilon^2 y = 0.$$

For small ε small, y satisfies, for $\xi = O(1)$,

$$\frac{d^2 y}{d\xi^2} + \frac{dy}{d\xi} = 0, \quad y(0) = 0.$$

Therefore

$$y_I = C + B e^{-\xi} = C(1 - e^{-\xi}) = C(1 - e^{-x/\varepsilon}).$$

To match the outer and inner approximations, let $x = \eta \sqrt{\varepsilon}$. Then

$$y_O = e^{1+\cos(\eta\sqrt{\varepsilon})} = e^2 + O(\varepsilon),$$

and

$$y_I = C(1 - e^{-\eta/\sqrt{\varepsilon}}) = C + o(1).$$

Hence $C = e^2$ and the inner solution is given by

$$y_I = e^2(1 - e^{-x/\varepsilon}).$$

• 6.12 By using the method of multiple scales, with variables x and $\xi = x/\varepsilon$, obtain a first approximation uniformly valid on $0 \leq x \leq 1$ to the solution of
$$\varepsilon y'' + y' + xy = 0, \quad y(\varepsilon, 0) = 0, \quad y(\varepsilon, 1) = 1,$$
on $0 \leq x \leq 1$, with $\varepsilon > 0$. Show that the result agrees to order ε with that of Problem 6.10.

6.12. As in Problem 6.10, the system is

$$\varepsilon y'' + y' + xy = 0, \quad y(0) = 0, \quad y(1) = 1. \tag{i}$$

Let $x = \xi\varepsilon$, and

$$y(x, \varepsilon) = Y(x, \xi, \varepsilon) = Y_0(x, \xi) + \varepsilon Y_1(x, \xi) + O(\varepsilon^2). \tag{ii}$$

The derivatives transform into

$$\frac{dy}{dx} = \frac{\partial Y}{\partial x} + \frac{1}{\varepsilon}\frac{\partial Y}{\partial \xi},$$

$$\frac{d^2y}{dx^2} = \frac{\partial^2 Y}{\partial x^2} + \frac{2}{\varepsilon}\frac{\partial^2 Y}{\partial x \partial \xi} + \frac{1}{\varepsilon^2}\frac{\partial^2 Y}{\partial \xi^2}.$$

Equation (i) becomes

$$\varepsilon^2 \frac{\partial^2 Y}{\partial x^2} + 2\varepsilon \frac{\partial^2 Y}{\partial x \partial \xi} + \frac{\partial^2 Y}{\partial \xi^2} + \varepsilon \frac{\partial Y}{\partial x} + \frac{\partial Y}{\partial \xi} + \varepsilon x Y = 0. \tag{iii}$$

Substitute the series (ii) into eqn (iii), and equate to zero the first two coefficients of ε. Hence

$$\frac{\partial^2 Y_0}{\partial \xi^2} + \frac{\partial Y_0}{\partial \xi} = 0, \tag{iv}$$

$$\frac{\partial^2 Y_1}{\partial \xi^2} + \frac{\partial Y_1}{\partial \xi} = -2\frac{\partial^2 Y_0}{\partial x \partial \xi} - \frac{\partial Y_0}{\partial x} - xY_0. \tag{v}$$

The boundary conditions at $x = 0$ translate into

$$Y_0(0,0) = 0, \quad Y_1(0,0) = 0,$$

but the conditions at $x = 1$ are more complicated. They remain as the series

$$Y_0(1, 1/\varepsilon) + \varepsilon Y_1(1, 1/\varepsilon) + \cdots = 0, \tag{vi}$$

since each perturbation contains ε. From (iv) it follows that

$$Y_0 = A_0(x) + B_0(x)e^{-\xi}.$$

From the condition at $x = 0$,

$$A_0(0) + B_0(0) = 0. \tag{vii}$$

From (v), the equation for Y_1 is

$$\frac{\partial^2 Y_1}{\partial \xi^2} + \frac{\partial Y_1}{\partial \xi} = -[A_0'(x) + x A_0(x)] + [B_0'(x) - x B_0(x)]e^{-\xi}.$$

To avoid growth terms in ξ (and consequently x) we put

$$A_0'(x) + x A_0(x) = 0, \text{ and } B_0'(x) - x B_0(x) = 0.$$

Integration of these equations leads to

$$A_0 = a e^{-(1/2)x^2}, \quad B_0 = b e^{(1/2)x^2}.$$

Condition (vii) implies $a + b = 0$, so that

$$Y_0(x, \xi) = a e^{-(1/2)x^2} + (1-a)e^{(1/2)x^2} e^{-\xi}.$$

The boundary condition (vi) becomes

$$a e^{-\frac{1}{2}} + (1-a)e^{1/2} e^{-1/\varepsilon} + O(\varepsilon) = 1.$$

The second term is exponentially small as $\varepsilon \to 0$, so that $a = e^{1/2}$. Finally

$$Y = e^{\frac{1}{2}}(e^{-(1/2)x^2} - e^{(1/2)x^2} e^{-x/\varepsilon}) + O(\varepsilon),$$

which agrees with the approximation obtained in Problem 6.10.

- 6.13 The steady flow of a conducting liquid between insulated parallel plates at $x = \pm 1$ under the influence of a transverse magnetic field satisfies
$$w'' + Mh' = -1, \quad h'' + Mw' = 0, \quad w(\pm 1) = h(\pm 1) = 0,$$
where, in dimensionless form, w is the fluid velocity, h the induced magnetic field, M is the Hartmann number. By putting $p = w + h$ and $q = w - h$, find the exact solution. Plot w and h against x for $M = 10$. The diagram indicates boundary layers adjacent to $x = \pm 1$. From the exact solutions find the outer and inner approximations.

6.13. The steady rectilinear Hartmann flow of a conducting liquid between parallel plates at $x = \pm 1$ satisfies the simultaneous equations

$$w'' + Mh' = -1, \quad h'' + Mw' = 0, \quad w(\pm 1) = h(\pm 1) = 0,$$

where w is the fluid velocity and h the induced magnetic field. Add and subtract the equations to obtain

$$(w + h)'' + M(w + h) = -1, \quad (w - h)'' - M(w - h) = -1.$$

The general solutions of these equations are

$$w + h = A + Be^{-Mx} - \frac{x}{M}, \quad w - h = C + De^{Mx} + \frac{x}{M}.$$

The boundary conditions imply

$$A + Be^{-M} - \frac{1}{M} = A + Be^{M} + \frac{1}{M} = 0,$$

$$C + De^{M} + \frac{1}{M} = C + De^{-M} - \frac{1}{M} = 0.$$

The solutions of these conditions are

$$A = C = \frac{1}{M} \coth M, \quad B = D = -\frac{1}{M \sinh M}.$$

Hence

$$w + h = \frac{1}{M} \coth M - \frac{e^{-Mx}}{M \sinh M} - \frac{x}{M},$$

$$w - h = \frac{1}{M} \coth M - \frac{e^{Mx}}{M \sinh M} + \frac{x}{M}.$$

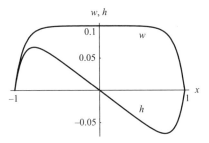

Figure 6.2 Problem 6.13:

Solving these equations, we have

$$w = \frac{1}{M \sinh M}[\cosh M - \cosh Mx], \quad h = \frac{1}{M \sinh M}[\sinh Mx - x \sinh M].$$

Graphs of w and h are shown in Figure 6.2 for $M = 10$, indicating boundary layers for both w and h near the walls at $x = \pm 1$.
Let $M = 1/\varepsilon$, and consider the solutions for w and h not close to $x = \pm 1$. Then

$$w = \varepsilon \left[\coth(1/\varepsilon) - \frac{\cosh(x/\varepsilon)}{\sinh(1/\varepsilon)} \right] \sim \varepsilon,$$

as $\varepsilon \to 0$. Hence $w_O \approx \varepsilon$, which agrees with $w \approx 0.1$ for $M = 10$ in Figure 6.2. For h,

$$h = \varepsilon \left[\frac{\sinh(x/\varepsilon)}{\sinh(1/\varepsilon)} - x \right] \sim -x\varepsilon,$$

as $\varepsilon \to 0$. Hence $h_O \approx -x\varepsilon$ (see Figure 6.2).
For the boundary layer near $x = 1$, let $1 - x = \xi \varepsilon$. Then

$$w = \varepsilon \left[\coth(1/\varepsilon) - \frac{\cosh\{(1-\xi\varepsilon)/\varepsilon\}}{\sinh(1/\varepsilon)} \right] \sim \varepsilon(1 - e^{-\xi})$$

as $\varepsilon \to 0$. Hence the inner solution $w_I \approx \varepsilon(1 - e^{-(1-x)/\varepsilon})$. For h,

$$h = \varepsilon \left[\frac{\sinh\{(1-\xi\varepsilon)/\varepsilon\}}{\sinh(1/\varepsilon)} - (1 - \xi\varepsilon) \right] \sim -\varepsilon(1 - e^{-\xi})$$

as $\varepsilon \to 0$. Hence $h_I \approx -\varepsilon(1 - e^{-(1-x)/\varepsilon})$.
Similar formulas can be found for w and h in the boundary layer close to $x = -1$.

- **6.14** Obtain an approximation, to order ε and for $t = O(\varepsilon^{-1})$, to the solutions of $\ddot{x} + 2\varepsilon \dot{x} + x = 0$, by using the method of multiple scales with the variables t and $\eta = \varepsilon t$.

6.14. For the equation

$$\ddot{x} + 2\varepsilon\dot{x} + x = 0,$$

introduce the variable $\eta = \varepsilon t$ and $x = X(t, \eta, \varepsilon)$, where

$$\dot{x} = \frac{\partial X}{\partial t} + \varepsilon\frac{\partial X}{\partial \eta}, \qquad \ddot{x} = \frac{\partial^2 X}{\partial t^2} + 2\varepsilon\frac{\partial^2 X}{\partial t \partial \eta} + \varepsilon^2\frac{\partial^2 X}{\partial \eta^2}.$$

In terms of X, the differential equation becomes

$$X_{tt} + 2\varepsilon X_{t\eta} + \varepsilon^2 X_{\eta\eta} + 2\varepsilon X_t + 2\varepsilon^2 X_\eta + X = 0. \tag{i}$$

Let $X = X_0 + \varepsilon X_1 + \cdots$, and substitute this series into (i). Putting the coefficients of powers of ε to zero, we have

$$X_{0tt} + X_0 = 0, \tag{ii}$$

$$X_{1tt} + X_1 = -2X_{0t\eta} - 2X_{0t}. \tag{iii}$$

In complex notation, we can express the solution of (ii) as

$$X_0 = A_0(\eta)e^{it} + \bar{A}_0(\eta)e^{-it}.$$

Equation (iii) becomes

$$X_{1tt} + X_1 = [-2iA'_0(\eta) - 2iA_0(\eta)]e^{it} + [2i\bar{A}'_0(\eta) + 2i\bar{A}_0(\eta)]e^{-it}.$$

The terms in square brackets on the right are complex conjugates. Secular terms can be removed if A_0 satisfies

$$A'_0(\eta) + A_0(\eta) = 0,$$

which has the general solution $A_0(\eta) = a_0 e^{-\eta}$. Hence

$$X_0 = a_0 e^{-\eta}e^{it} + \bar{a}_0 e^{-\eta}e^{-it} = 2ae^{-\eta}\cos(t + \alpha),$$

where $a_0 = ae^{i\alpha}$ and a and α are real constants. In terms of x

$$x = 2e^{-\varepsilon t}\cos(t + \alpha).$$

- **6.15** Use the method of multiple scales to obtain a uniform approximation to the solutions of the equation $\ddot{x} + \omega^2 x + \varepsilon x^3 = 0$, in the form

$$x(\varepsilon, t) \approx a_0 \cos\left[\left\{\omega_0 + \frac{3\varepsilon a_0^2}{8\omega_0}\right\} t + \alpha\right],$$

where α is a constant. Explain why the approximation is uniform, and not merely valid for $t = O(\varepsilon^{-1})$.

6.15. In the equation

$$\ddot{x} + \omega^2 x + \varepsilon x^3 = 0,$$

introduce the variable $\eta = \varepsilon t$, and let $x = X(t, \eta, \varepsilon)$. The derivatives are

$$\dot{x} = \frac{\partial X}{\partial t} + \varepsilon \frac{\partial X}{\partial \eta}, \qquad \ddot{x} = \frac{\partial^2 X}{\partial t^2} + 2\varepsilon \frac{\partial^2 X}{\partial t \partial \eta} + \varepsilon^2 \frac{\partial^2 X}{\partial \eta^2}.$$

In terms of X, the differential equation becomes

$$X_{tt} + 2\varepsilon X_{t\eta} + \varepsilon^2 X_{\eta\eta} + \omega^2 X + \varepsilon X^3 = 0. \tag{i}$$

Let $X = X_0 + \varepsilon X_1 + \cdots$, and substitute this series into (i). Putting the coefficients of powers of ε to zero, we have

$$X_{0tt} + \omega^2 X_0 = 0, \tag{ii}$$

$$X_{1tt} + \omega^2 X_1 = -2 X_{0t\eta} - X_0^3. \tag{iii}$$

The general solution of (ii) in complex form is

$$X_0 = A_0(\eta) e^{i\omega t} + \overline{A}_0(\eta) e^{-i\omega t}.$$

Equation (iii) becomes

$$X_{1tt} + \omega^2 X_1 = -2(A_0'(\eta) i\omega e^{i\omega t} - \overline{A}_0'(\eta) i\omega e^{-i\omega t}) - (A_0(\eta) e^{i\omega t} - \overline{A}_0(\eta) e^{-i\omega t})^3$$
$$= [-2A_0'(\eta) i\omega - 3A_0^2(\eta)\overline{A}_0(\eta)] e^{i\omega t} - A_0^3(\eta) e^{3i\omega t} + \text{complex conjugate}$$

Secular terms can be eliminated by putting

$$2A_0'(\eta) i\omega + 3A_0^2(\eta)\overline{A}_0(\eta) = 0.$$

To solve this complex differential equation, let $A_0(\eta) = a(\eta)e^{ib(\eta)}$, where the functions $a(\eta)$ and $b(\eta)$ are real. Then

$$2i\omega(a'(\eta)e^{ib(\eta)} + ia(\eta)b'(\eta)e^{ib(\eta)}) + 3a(\eta)^3 e^{ib(\eta)} = 0,$$

or

$$[2\omega i a'(\eta) + (3a(\eta)^3 - 2a(\eta)\omega b'(\eta))]e^{ib(\eta)} = 0.$$

The real and imaginary parts must be zero, so that

$$a'(\eta) = 0, \quad 3a(\eta)^3 - 2a(\eta)\omega b'(\eta)) = 0.$$

Therefore $a = \tfrac{1}{2}a_0$, a constant, and $b(\eta)$ satisfies

$$b'(\eta) = \frac{3a_0^2}{8\omega}.$$

Hence

$$b(\eta) = \frac{3a_0^2}{8\omega}\eta + \alpha.$$

Finally

$$x(t) \approx X_0(t, \eta) = \tfrac{1}{2}ae^{i[\omega t + b(\eta)]} + \tfrac{1}{2}ae^{-i[\omega t + b(\eta)]}$$

$$= a\cos[\omega t + b(\eta)] = a\cos\left[\left\{\omega + \tfrac{3}{8}(\varepsilon a_0^2/\omega)\right\}t + c\right]$$

Extension of the method to higher-order multiple scales $\eta_1 = \varepsilon t$, $\eta_2 = \varepsilon^2 t$, ... leads to equations in which the derivatives in terms of η_2, ... only appear in higher-order equations. To order ε the result is unaffected.

- **6.16** Use the coordinate perturbation technique to obtain the first approximation $x = \tau^{-1}$, $t = \tau + \tfrac{1}{2}\varepsilon\tau(1 - \tau^{-2})$ to the solution of
$$(t + \varepsilon x)\dot{x} + x = 0, \quad x(\varepsilon, 1) = 1, \quad 0 \le x \le 1.$$
Confirm that the approximation is, in fact, the exact solution, and that an alternative approximation $x = \tau^{-1} + \tfrac{1}{2}\varepsilon\tau^{-1}$, $t = \tau - \tfrac{1}{2}\varepsilon\tau^{-1}$ is correct to order ε, for fixed τ. Construct a graph showing the exact solution and the approximation.

6.16. Use the coordinate perturbation technique to obtain the first approximation of
$$(t+\varepsilon x)\dot{x} + x = 0, \quad x(\varepsilon, 1) = 1, \quad 0 \le x \le 1.$$

Apply the straightforward approximation
$$x(\varepsilon, t) = x_0(t) + \varepsilon x_1(t) + \varepsilon^2 x_2(t) + \cdots, \tag{i}$$

so that
$$t\dot{x}_0 + x_0 = 0, \tag{ii}$$
$$t\dot{x}_1 + x_1 = -x_0\dot{x}_0, \tag{iii}$$
$$t\dot{x}_2 + x_2 = -x_0\dot{x}_1 - x_1\dot{x}_0, \tag{iv}$$

and so on. The boundary condition transform into the sequence of conditions
$$x_0(1) = 1, \quad x_1(1) = 0, \quad x_2(1) = 0, \ldots.$$

Therefore $x_0 = 1/t$. Equation (iii) becomes
$$t\dot{x}_1 + x_1 = \frac{d(tx_1)}{dt} = \frac{1}{t^3},$$

so that
$$x_1 = -\frac{1}{2t^3} + \frac{C_1}{t} = \frac{1}{2t}\left(1 - \frac{1}{t^2}\right).$$

From (iv)
$$t\dot{x}_2 + x_2 = \frac{d(tx_2)}{dt} = -x_0\dot{x}_1 - x_1\dot{x}_0 = -\frac{d(x_0 x_1)}{dt}.$$

Therefore
$$\frac{d}{dt}(tx_2 + x_0 x_1) = 0,$$

so that
$$tx_2 = -x_0 x_1 + C_2 = -\frac{1}{2t^2}\left(1 - \frac{1}{t^2}\right) + C_2 = -\frac{1}{2t^2}\left(1 - \frac{1}{t^2}\right).$$

All these solutions are singular at $t = 0$.

To remove the singularities, let

$$t = \tau + \varepsilon T_1(\tau) + \varepsilon^2 T_2(\tau) + \cdots.$$

We require expansions for inverse powers of t. Using binomial expansions,

$$\frac{1}{t} = \frac{1}{\tau} - \varepsilon \frac{T_1}{\tau^2} + \varepsilon^2 \left(\frac{T_1}{\tau^3} - \frac{T_2}{\tau^2} \right) + O(\varepsilon^3),$$

$$\frac{1}{t^2} = \frac{1}{\tau^2} - 2\varepsilon \frac{T_1}{\tau^3} + \varepsilon^2 \left(\frac{3T_1^2}{\tau^4} - \frac{2T_2}{\tau^3} \right) + O(\varepsilon^3),$$

$$\frac{1}{t^3} = \frac{1}{\tau^3} - 3\varepsilon \frac{T_1}{\tau^4} + 3\varepsilon^2 \left(\frac{2T_1^2}{\tau^5} - \frac{T_2}{\tau^4} \right) + O(\varepsilon^3).$$

In terms of τ,

$$x_0 = \frac{1}{\tau} - \varepsilon \frac{T_1}{\tau^2} + \varepsilon^2 \left(\frac{T_1}{\tau^3} - \frac{T_2}{\tau^2} \right) + O(\varepsilon^3),$$

$$x_1 = \frac{1}{2\tau} - \frac{1}{2\tau^3} = \frac{1}{2}\left[\frac{1}{\tau} - \varepsilon \frac{T_1}{\tau^2} \right] - \frac{1}{2}\left[\frac{1}{\tau^3} - 3\varepsilon \frac{T_1}{\tau^4} \right] + O(\varepsilon^2)$$

$$= \frac{1}{2\tau}\left(1 - \frac{1}{\tau^2} \right) - \frac{\varepsilon T_1}{2\tau^2}\left(1 - \frac{3}{\tau^2} \right) + O(\varepsilon^2).$$

Therefore

$$x_0 + \varepsilon x_1 = \frac{1}{\tau} + \varepsilon \left[\frac{1}{2\tau}\left(1 - \frac{1}{\tau^2} \right) - \frac{T_1}{\tau^2} \right] + O(\varepsilon^2).$$

We can eliminate the $O(\varepsilon)$ term by choosing

$$T_1 = \frac{\tau}{2}\left(1 - \frac{1}{\tau^2} \right),$$

in which case the approximate solution is given by

$$x = \frac{1}{\tau} + O(\varepsilon^2), \quad t = \tau + \frac{\varepsilon \tau}{2}\left(1 - \frac{1}{\tau^2} \right) + O(\varepsilon^2).$$

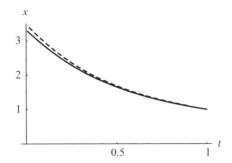

Figure 6.3 Problem 6.16: The solid curve shows the exact solution and the dashed curve shows the approximation given parametrically by $x \approx (1/\tau) + (\varepsilon/(2\tau))$, $t \approx \tau - (\varepsilon/(2\tau))$ with $\varepsilon = 0.2$

We can confirm that the approximation is, in fact, exact by substitution into the differential equation: thus, with $x = x_0 + \varepsilon x_1$,

$$(t + \varepsilon x)\dot{x} = \left(\tau + \frac{\varepsilon\tau}{2} - \frac{\varepsilon}{2\tau} + \frac{\varepsilon}{\tau}\right)\frac{dx\,d\tau}{d\tau\,dt}$$

$$= \left(\tau + \frac{\varepsilon\tau}{2} + \frac{\varepsilon}{2\tau}\right)\left(-\frac{1}{\tau^2}\right) \Big/ \left(1 + \frac{\varepsilon}{2} + \frac{\varepsilon}{2\tau}\right) = -\frac{1}{\tau} = -x$$

as required.

If we put $T_1 = -1/(2\tau)$, then the approximate solution becomes

$$x = x_0 + \varepsilon x_1 + O(\varepsilon^2) = \frac{1}{\tau} + \frac{\varepsilon}{2\tau} + O(\varepsilon^2), \quad t = \tau - \frac{\varepsilon}{2\tau} + O(\varepsilon^2). \tag{v}$$

Figure 6.3 shows the exact solution and the approximation given by (v).

- **6.17** Apply the method of multiple scales, with variables t and $\eta = \varepsilon t$, to van der Pol's equation $\ddot{x} + \varepsilon(x^2 - 1)\dot{x} + x = 0$. Show that, for $t = O(\varepsilon^{-1})$,

$$x(\varepsilon, t) = \frac{2a_0^{1/2}e^{(1/2)\varepsilon t}}{\sqrt{(1 + a_0 e^{\varepsilon t})}}\cos(t + \alpha_0) + O(\varepsilon),$$

where a_0 and α_0 are constants.

6.17. Apply the multiple scale method with $\eta = \varepsilon t$ to the van der Pol equation

$$\ddot{x} + \varepsilon(x^2 - 1)\dot{x} + x = 0.$$

Let $x = X(t, \eta, \varepsilon)$. The derivatives are

$$\dot{x} = \frac{\partial X}{\partial t} + \varepsilon \frac{\partial X}{\partial \eta}, \qquad \ddot{x} = \frac{\partial^2 X}{\partial t^2} + 2\varepsilon \frac{\partial^2 X}{\partial t \partial \eta} + \varepsilon^2 \frac{\partial^2 X}{\partial \eta^2}.$$

In terms of X, van der Pol's equation becomes

$$X_{tt} + 2\varepsilon X_{t\eta} + \varepsilon^2 X_{\eta\eta} + (X^2 - 1)(X_t + \varepsilon X_\eta + X) = 0. \tag{i}$$

Introduce the perturbation series $X = X_0 + \varepsilon X_1 + \cdots$ into (i) so that the first two coefficients lead to

$$X_{0tt} + X_0 = 0, \tag{ii}$$

$$X_{1tt} + X_1 = -2X_{0t\eta} - (X_0^2 - 1)X_{0t}. \tag{iii}$$

In complex notation, the solution of (ii) can be expressed as

$$X_0 = A_0(\eta) e^{it} + \overline{A}_0(\eta) e^{-it}.$$

Equation (iii) is

$$\begin{aligned}X_{1tt} + X_1 &= [-2iA_0'(\eta)e^{it} + 2iA_0(\eta)e^{it}] + [2i\overline{A}_0'(\eta)e^{-it} - i\overline{A}_0(\eta)e^{-it}] \\ &\quad - [(A_0(\eta)e^{it} + \overline{A}_0(\eta)e^{-it})^2 - 1][iA_0(\eta)e^{it} - i\overline{A}_0(\eta)e^{-it}] \\ &= [(-2iA_0'(\eta) + iA_0(\eta) - iA_0^2\overline{A}_0(\eta))e^{it} + \text{(complex conjugate)}] \\ &\quad + \text{(higher harmonics)}\end{aligned}$$

Secular terms can be eliminated if

$$-2A_0'(\eta) + A_0(\eta) - A_0^2 \overline{A}_0(\eta) = 0.$$

To solve this equation, let $A_0(\eta) = a(\eta) e^{i\alpha(\eta)}$, so that

$$-2[a' e^{i\alpha} + a\alpha' i e^{i\alpha}] + a e^{i\alpha} - a^3 e^{i\alpha} = 0,$$

or

$$[-2a' + a - a^3] + i[2a\alpha'] = 0.$$

Therefore

$$\alpha' = 0, \text{ implying } \alpha = \alpha_0, \text{ a constant,}$$

and a satisfies

$$2a' = a - a^3.$$

$$\int \frac{2da}{a(1-a^2)} = \ln\left[\frac{a^2}{|a^2-1|}\right] = \int d\eta + C = \eta + C.$$

Therefore,

$$a = \frac{a_0^{1/2} e^{(1/2)\eta}}{\sqrt{(1+a_0 e^\eta)}},$$

where a_0 is a constant. Finally

$$x(t) = X_0(t, \eta) + O(\varepsilon) = ae^{i(t+\alpha_0)} + ae^{-i(t+\alpha_0)} = 2A_0(\eta)\cos(t+\alpha_0)$$

$$= \frac{2a_0^{1/2} e^{(1/2)\varepsilon t}}{\sqrt{(1+a_0 e^{\varepsilon t})}\cos(t+\alpha_0)}.$$

- **6.18** Use the method of matched approximations to obtain a uniform approximation to the solution of

$$\varepsilon\left(y'' + (2/x)y'\right) - y = 0, \quad y(\varepsilon, 0) = 0, \quad y'(\varepsilon, 1) = 1,$$

($\varepsilon > 0$) on $0 \leq x \leq 1$. Show that there is a boundary layer of thickness $O(\varepsilon^{\frac{1}{2}})$ near $x = 1$ by putting $1 - x = \xi\phi(\varepsilon)$.

6.18. Consider the system

$$\varepsilon(xy'' + 2y') - xy = 0, \quad y(\varepsilon, 0) = 0, \quad y'(\varepsilon, 1) = 1.$$

To obtain the outer expansion for $x = O(1)$, put $\varepsilon = 0$ in the equation, so that $y = 0$, which only agrees with boundary condition at $x = 0$. Hence there must be a boundary layer near $x = 1$ since $y(\varepsilon, 1) = 1$.

To investigate the boundary layer, let $1 - x = \xi\phi(\varepsilon)$. The change of variable leads to

$$\frac{\varepsilon}{\phi^2}\frac{d^2y}{d\xi^2} + \frac{2}{1-\xi\phi}\frac{\varepsilon}{\phi}\frac{dy}{d\xi} - y = 0.$$

The highest derivative is $O(1)$ if we choose $\phi = \sqrt{\varepsilon}$, so that y satisfies

$$\frac{d^2y}{d\xi^2} - y = 0.$$

318 Nonlinear ordinary differential equations: problems and solutions

to the lowest order. Therefore the inner approximation is given by

$$y_I = Ae^{\xi} + Be^{-\xi} = Ae^{(1-x)/\sqrt{\varepsilon}} + Be^{-(1-x)/\sqrt{\varepsilon}}.$$

The inner approximation has to satisfy the boundary condition $y'(\varepsilon, 1) = 1$, so that

$$-\frac{A}{\sqrt{\varepsilon}} + \frac{B}{\sqrt{\varepsilon}} = 1.$$

Hence

$$y_I = Ae^{(1-x)/\sqrt{\varepsilon}} + (A + \sqrt{\varepsilon})e^{-(1-x)/\sqrt{\varepsilon}}.$$

To match the inner and outer approximations, let $1 - x = \eta \psi(\varepsilon)$. then we require

$$\frac{\psi(\varepsilon)}{\phi(\varepsilon)} \to 0,$$

as $\varepsilon \to 0$. The choice $\psi(\varepsilon) = \varepsilon$ will be sufficient for this purpose. Hence

$$y_I = Ae^{\eta\sqrt{\varepsilon}} + (A + \varepsilon)e^{-\eta\sqrt{\varepsilon}} \to 2A + \sqrt{\varepsilon} = 0,$$

as $\varepsilon \to 0$ if $A = -\frac{1}{2}\sqrt{\varepsilon}$. To summarize

$$y_O = 0, \quad y_I = \frac{1}{2}\sqrt{\varepsilon}[e^{-(1-x)/\sqrt{\varepsilon}} - e^{(1-x)/\sqrt{\varepsilon}}] = -\sqrt{\varepsilon}\sinh\left[\frac{1-x}{\sqrt{\varepsilon}}\right].$$

- **6.19** Use the method of matched approximations to obtain a uniform approximation to the solution of the problem
$$\varepsilon(y'' + y) - y = 0, \quad y(\varepsilon, 0) = 1, \quad y(\varepsilon, 1) = 1, \quad (\varepsilon > 0),$$
given that there are boundary layers at $x = 0$ and $x = 1$. Show that both boundary layers have thickness $O(\varepsilon^{1/2})$. Compare with the exact solution.

6.19. We require a uniform approximation to the linear boundary value problem for $y(x, \varepsilon)$

$$\varepsilon(y'' + y') - y = 0, \quad y(\varepsilon, 0) = 1, \quad y(\varepsilon, 1) = 1, \quad 0 \leq x \leq 1$$

Put $\varepsilon = 0$ in the equation, so that the outer approximation in the interval not near the boundaries $x = 0$ and $x = 1$ is given by $y_O = 0$.

For the inner solution y_{I_1} near $x = 0$, put $x = \xi\phi(\varepsilon)$, where $\lim_{\varepsilon\to 0} \phi(\varepsilon) = 0$. Apply the change of variable to the differential equation:

$$\frac{\varepsilon}{\phi^2}\frac{d^2 y}{d\xi^2} + \frac{\varepsilon}{\phi}\frac{dy}{d\xi} - y = 0, \quad y(0, \varepsilon) = 1.$$

We choose $\phi = \sqrt{\varepsilon}$ so that for $\varepsilon \to 0$, y_I satisfies

$$\frac{d^2 y_I}{d\xi^2} - y_I = 0,$$

which has the general solution $y_I = Ae^{\xi} + Be^{-\xi}$. Hence, using the boundary condition,

$$y_{I_1} = Ae^{x/\sqrt{\varepsilon}} + (1-A)e^{-x/\sqrt{\varepsilon}}.$$

To match this solution with the outer approximation, let $x = \eta\varepsilon^{1/4}$ so that

$$y_{I_1} = Ae^{\eta/\varepsilon^{1/4}} + (1-A)e^{-\eta/\varepsilon^{1/4}}.$$

Matching as $\varepsilon \to 0$ gives $A = 0$: therefore

$$y_{I_1} = e^{-x/\sqrt{\varepsilon}}.$$

For the inner approximation near $x = 1$, let $x = 1 - \xi\phi$, so that

$$\frac{\varepsilon}{\phi^2}\frac{d^2 y}{d\xi^2} - \frac{\varepsilon}{\phi}\frac{dy}{d\xi} - y = 0, \quad y(0) = 1.$$

With $\phi = \sqrt{\varepsilon}$, we get the same equation as before, so that, after matching,

$$y_{I_2} = e^{(1-x)/\sqrt{\varepsilon}}.$$

The uniform (or composite) solution is

$$y_C = e^{-x/\sqrt{\varepsilon}} + e^{-(1-x)/\sqrt{\varepsilon}}.$$

The given equation is second-order linear, so we can compare the exact solution y_E with the approximations above. The characteristic equation has the solutions

$$\left.\begin{matrix}\mu_1 \\ \mu_2\end{matrix}\right\} = \begin{cases} -\tfrac{1}{2} + \sqrt{\left(\tfrac{1}{\varepsilon} + \tfrac{1}{4}\right)} \\ -\tfrac{1}{2} - \sqrt{\left(\tfrac{1}{\varepsilon} + \tfrac{1}{4}\right)}. \end{cases}$$

The exact solution is

$$y_E = \frac{1 - e^{\mu_2}}{e^{\mu_1} - e^{\mu_2}} e^{-\mu_1 x} + \frac{e^{\mu_1} - 1}{e^{\mu_1} - e^{\mu_2}} e^{-\mu_2 x}.$$

- **6.20** Obtain a first approximation, uniformly valid on $0 \leq x \leq 1$, to the solution of

$$\varepsilon y'' + \frac{1}{1+x} y' + \varepsilon y = 0, \quad y(\varepsilon, 0) = 0, \quad y(\varepsilon, 1) = 1.$$

6.20. The system is

$$\varepsilon y'' + \frac{1}{1+x} y' + \varepsilon y = 0, \quad y(\varepsilon, 0) = 0, \quad y(\varepsilon, 1) = 1.$$

Put $\varepsilon = 0$ in the equation. It follows that the outer solution $y_O(\varepsilon, x)$ satisfies $y' = 0$ so that $y_O = 1$ for all x from the boundary condition at $x = 1$.

For the inner solution, $y_I(\varepsilon, x)$, let $x = \xi \phi(\varepsilon)$ so that the differential equation becomes

$$\frac{\varepsilon}{\phi^2} \frac{d^2 y}{d\xi^2} + \frac{1}{1 + \xi \phi} \frac{1}{\phi} \frac{dy}{d\xi} + \varepsilon y = 0.$$

Choose $\phi = \varepsilon$ and select the dominant terms. Then, the inner solution satisfies

$$\frac{d^2 y_I}{d\xi^2} + \frac{dy_I}{d\xi} = 0,$$

subject to $y(0) = 0$. Hence

$$y_I = A(1 - e^{-\xi}).$$

The outer and inner solutions match if $A = 1$ so that

$$y_I = 1 - e^{-\xi} = 1 - e^{-\xi/\varepsilon}.$$

The uniform approximation is $y_C = 1 - e^{-x/\varepsilon}$.

- **6.21** Apply the Lighthill technique to obtain a uniform approximation to the solution of
$(t + \varepsilon x)\dot{x} + x = 0, \quad x(\varepsilon, 1) = 1, \quad 0 \leq x \leq 1.$
(Compare Problem 6.16.)

6.21. The Lighthill technique is applied to obtain a uniform approximation to the problem

$$(t + \varepsilon x)\dot{x} + x = 0, \quad x(\varepsilon, 1) = 1, \quad (0 \leq x \leq 1).$$

Write

$$x = X_0(\tau) + \varepsilon X_1(\tau) + \cdots,$$
$$t = \tau + \varepsilon T_1(\tau) + \cdots.$$

The differential equation becomes

$$[(\tau + \varepsilon T_1 + \cdots) + \varepsilon(X_0 + \varepsilon X_1 + \cdots)](X_0' + \varepsilon X_1' + \cdots)(1 - \varepsilon T_1' + \cdots)^{-1}$$
$$+ (X_0 + \varepsilon X_1 + \cdots) = 0.$$

Therefore the perturbations satisfy

$$\tau \frac{dX_0}{d\tau} + X_0 = 0, \tag{i}$$

$$\tau X_1' + X_1 = -X_0'(T_1 + X_0) + \tau X_0' T_1'. \tag{ii}$$

As in Section 2.4, the boundary condition becomes

$$X_0(1) = 1, \quad X_1(1) = T_1(1) X_0'(1). \tag{iii}$$

From (i) it follows that $X_0 = 1/\tau$. Equation (ii) is then

$$\tau X_1' + X_1 = \frac{1}{\tau^2} T_1 + \frac{1}{\tau^3} - \frac{1}{\tau} T_1'.$$

Put the right-hand side equal to zero to remove singularities so that

$$\tau^2 T_1' - \tau T_1 - 1 = 0.$$

The general solution of this equation is

$$T_1 = A\tau - \frac{1}{2\tau}.$$

In this case the solution of (ii) is $X_1 = C/\tau$. The second boundary condition in (iii) implies

$$C - (A - \tfrac{1}{2})(-1) = 0,$$

so that $C = -(A - \frac{1}{2})$. Hence the general formula for the solution is

$$x = \frac{1}{\tau} - \frac{\varepsilon}{\tau}\left(A - \frac{1}{2}\right)\varepsilon + O(\varepsilon^2), \quad t = \tau + \left(A\tau - \frac{1}{2\tau}\right) + O(\varepsilon^2).$$

By choosing $A = \frac{1}{2}$, we can eliminate the $O(\varepsilon)$ in x so that

$$x = \frac{1}{\tau} + O(\varepsilon^2), \quad t = \tau + \frac{1}{2}\left(\tau - \frac{1}{2\tau}\right)\varepsilon + O(\varepsilon^2).$$

- **6.22** Obtain a first approximation, uniform on $0 \le x \le 1$, to the solution of $\varepsilon y' + y = x$, $y(\varepsilon, 0) = 1$, using inner and outer approximations. Compare the exact solution and explain geometrically why the outer approximation is independent of the boundary conditions.

6.22. The system is

$$\varepsilon y' + y = x, \quad y(\varepsilon, 0) = 1.$$

The outer approximation is obtained by putting $\varepsilon = 0$, giving $y_O = x$.
To derive the inner approximation, let $x = \xi \phi(\varepsilon)$, so that

$$\frac{\varepsilon}{\phi}\frac{dy}{d\xi} + y = \phi\xi.$$

With $\phi(\varepsilon) = \varepsilon$, the equation to first order reduces to

$$\frac{dy}{d\xi} + y = 0.$$

Hence $y = Ae^{-\xi} = e^{-\xi}$ using the boundary condition. Therefore $y_I = e^{-x/\varepsilon}$,
The uniform approximation is

$$y_C = x + e^{-x/\varepsilon}.$$

The exact solution of the equation is

$$y = x - \varepsilon + Be^{-x/\varepsilon} = x - \varepsilon + (1 + \varepsilon)e^{-x/\varepsilon},$$

using the boundary condition $y(\varepsilon, 0) = 1$. If $x = O(1)$, then the solution away from the boundary layer adjacent to $x = 1$ is $x + O(\varepsilon)$ provided also that the constant $A = O(1)$). The boundary layer has the width $O(\varepsilon)$. Some solutions indicating the boundary layer are shown in Figure 6.4.

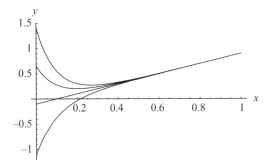

Figure 6.4 Problem 6.22: A selection of exact solutions with $\varepsilon = 0.1$ and various boundary conditions.

• **6.23** Use the method of multiple scales with variables t and $\eta = \varepsilon t$ to show that, to a first approximation, the response of the van der Pol equation to a 'soft' forcing term described by
$$\ddot{x} + \varepsilon(x^2 - 1)\dot{x} + x = \varepsilon\gamma\cos\omega t, \quad \varepsilon > 0,$$
is the same as the unforced response, assuming that $|\omega|$ is not near 1.

6.23. The van der Pol equation with soft forcing is

$$\ddot{x} + x = \varepsilon\{(1 - x^2)\dot{x} + \gamma\cos\omega t\},$$

in the non-resonant case. Let $\eta = \varepsilon t$ and $x = X(t, \eta, \varepsilon)$. Then the derivatives become

$$\dot{x} = \frac{\partial X}{\partial t} + \varepsilon\frac{\partial X}{\partial \eta}, \quad \ddot{x} = \frac{\partial^2 X}{\partial t^2} + 2\varepsilon\frac{\partial^2 X}{\partial t\partial \eta} + \varepsilon^2\frac{\partial^2 X}{\partial \eta^2},$$

and the van der Pol equation is transformed into

$$\frac{\partial^2 X}{\partial t^2} + 2\varepsilon\frac{\partial^2 X}{\partial t\partial \eta} + \varepsilon^2\frac{\partial^2 X}{\partial \eta^2} + X = \varepsilon\left[(1-x^2)\left(\frac{\partial X}{\partial t} + \varepsilon\frac{\partial X}{\partial \eta}\right) + \gamma\cos\omega t\right].$$

Substitute into this equation the series $X = X_0 + \varepsilon X_1 + \cdots$, and equate to zero the coefficients of like powers of ε. The first two equations are

$$X_{0tt} + X_0 = 0, \tag{i}$$

$$X_{1tt} + X_1 = (1 - X_0^2)X_{0t} + \gamma\cos\omega t - 2X_{0t\eta}. \tag{ii}$$

Using complex notation,

$$X_0 = A_0(\eta)e^{it} + \bar{A}_0(\eta)e^{-it}.$$

Equation (ii) becomes

$$X_{1tt} + X_1 = i(A_0 - A_0^2\bar{A}_0 - 2A_0')e^{it} - i(\bar{A}_0 - \bar{A}_0^2 A_0 - 2\bar{A}_0')e^{-it}$$
$$- iA_0^3 e^{3it} + i\bar{A}_0^3 e^{-3it} + \gamma \cos \omega t$$

The period 2π terms will be eliminated only if

$$A_0 - A_0^2\bar{A}_0 - 2A_0' = 0 \tag{iii}$$

together with its conjugate. Let $A_0 = \rho(\eta)e^{i\alpha(\eta)}$, and substitute this form into (iii), resulting in

$$-2\rho'(\eta) - 2i\rho(\eta)\alpha'(\eta) + \rho(\eta) - \rho(\eta)^3 = 0.$$

The real and imaginary parts must vanish so that

$$2\rho'(\eta) - \rho(\eta) - \rho(\eta)^3 = 0, \quad \alpha'(\eta) = 0.$$

It follows that $\alpha = \alpha_0$, a constant, and that

$$\ln\left[\frac{r^2}{|1 - r^2|}\right] = \eta + C.$$

Hence

$$A(\eta) = \frac{e^{it+\alpha_0}}{\sqrt{(1 + Ce^{-\eta})}}.$$

Finally

$$x = X_0(t, \eta) + O(\varepsilon) = \frac{2\cos(t + t\varepsilon) + \alpha_0}{\sqrt{(1 + Ce^{-\varepsilon t})}} + O(\varepsilon)$$

Significantly, X_0 is independent of the angular frequency ω and the amplitude γ of the forcing oscillation, which means the response is independent of the forcing term. This will not be the case if $\omega = 1$.

- **6.24** Repeat Problem 6.23 for $\ddot{x} + \varepsilon(x^2 - 1)\dot{x} + x = \Gamma\cos\omega t$, $(\varepsilon > 0)$, where $\Gamma = O(1)$ and $|\omega|$ is not near 1. Show that

$$x(\varepsilon, t) = \frac{\Gamma}{1-\omega^2}\cos\omega t + O(\varepsilon), \quad \Gamma^2 \geq 2(1-\omega^2)^2;$$

and that for $\Gamma^2 < 2(1-\omega^2)^2$,

$$x(\varepsilon, t) = 2\left(1 - \frac{\Gamma^2}{2(1-\omega^2)^2}\right)^{1/2}\cos t + \frac{\Gamma}{1-\omega^2}\cos\omega t + O(\varepsilon).$$

6.24. In this problem the van der Pol equation is

$$\ddot{x} + x = \varepsilon(1 - x^2)\dot{x} + \Gamma\cos\omega t,$$

where $\Gamma = O(1)$. As in the previous problem, let $\eta = \varepsilon t$ and $x = X(t, \eta, \varepsilon)$. The derivatives $\dot{x}$ and $\ddot{x}$ in terms of X are given in the previous problem. The equations for X_0 and X_1 in the expansion $X = X_0 + \varepsilon X_1 + \cdots$ are

$$X_{0tt} + X_0 = \Gamma\cos\omega t, \tag{i}$$

$$X_{1tt} + X_1 = (1 - X_0^2)X_{0t} - 2X_{0t\eta}. \tag{ii}$$

The solution of (i) can be expressed in the forms

$$X_0 = a_0(\eta)\cos t + b_0(\eta)\sin t + \kappa\cos\omega t,$$

where $\kappa = \Gamma/(1-\omega^2)$. Equation (ii) becomes (symbolic computation for trigonometric identities eases the working)

$$X_{1tt} + X_1 = -2\{[a_0'(\eta) + b_0(\eta)]\cos t + [b_0'(\eta) - a_0(\eta)]\sin t\}$$
$$+ 1 - (a_0(\eta)\cos t + b_0(\eta)\sin t + \kappa\cos\omega t)^2]$$
$$\times [-a_0(\eta)\sin t + b_0(\eta)\cos t - \kappa\omega\sin\omega t]$$
$$= \tfrac{1}{4}[(4 - 2\kappa^2)b_0(\eta) - b_0(\eta)(a_0(\eta)^2 + b_0(\eta)^2) - 8b_0'(\eta)]\cos t$$
$$+ \tfrac{1}{4}[-(4 - 2\kappa^2)a_0(\eta) + a_0(\eta)(a_0(\eta)^2 + b_0(\eta)^2) + 8a_0'(\eta)]\sin t$$
$$+ \text{(non-secular periodic terms)}$$

The solution for X_1 is periodic if

$$(4 - 2\kappa^2)b_0(\eta) - b_0(\eta)(a_0(\eta)^2 + b_0(\eta)^2) - 8b_0'(\eta) = 0, \tag{iii}$$

and

$$-(4 - 2\kappa^2)a_0(\eta) + a_0(\eta)(a_0(\eta)^2 + b_0(\eta)^2) + 8a_0'(\eta) = 0. \tag{iv}$$

Let $u_0 = a_0^2 + b_0^2$. Then $a_0 \times$(iv)$-b_0 \times$(iii) leads to the first-order equation

$$4\frac{du_0}{d\eta} + u_0^2 - (4 - 2\kappa^2)u_0 = 0.$$

Note that this equation always has the solution $u_0 = 0$. This equation is separable with solution

$$4\int \frac{du_0}{u_0(u_0 - \lambda)} = -\frac{4}{\lambda}\ln\left|\frac{u_0}{u_0 - \lambda}\right| = -\int d\eta = -\eta + \text{constant},$$

where $\lambda = 4 - 2\kappa^2$. If $u_0 = \beta$ when $t = 0$, then

$$u_0 = \frac{\lambda}{1 + [(\lambda/\beta) - 1]e^{-(1/4)\lambda\eta}}. \tag{v}$$

If $\lambda < 0$, then u_0 will ultimately become negative, which is not possible: therefore the only possible solution in this case is $u_0 = 0$, from which it follows that $a_0 = b_0 = 0$. Hence the forced periodic response is

$$x = X_0 + O(\varepsilon) = \kappa \cos \omega t + O(\varepsilon) = \frac{\Gamma}{1 - \omega^2}\cos \omega t + O(\varepsilon), \quad \Gamma^2 \geq 2(1 - \omega^2)^2$$

If $\lambda > 0$, then $u_0 \to \lambda$ as $t \to \infty$. From (iii), we can find b_0. Using (v),

$$8\frac{db_0}{d\eta} = \lambda b_0 - b_0 u_0 = \frac{[b_0(\lambda/\beta) - 1]e^{-(1/4)\lambda\eta}}{1 + [(\lambda/\beta) - 1]e^{-(1/4)\lambda\eta}}.$$

Integration of this separable equation leads to the solution

$$8\ln|b_0| = \frac{4}{\lambda}\ln\left[\frac{e^{-(1/4)\lambda\eta}}{(\lambda/\beta - 1) + e^{-\frac{1}{4}\lambda\eta}}\right] + C,$$

or

$$b_0^2 = \frac{e^{-(1/4)\eta}}{[(\lambda/\beta - 1) + e^{-(1/4)\lambda\eta}]^{1/\lambda}}.$$

As $t \to \infty$, then $b_0 \to 0$, if $\lambda > 0$. From the behaviour of u_0, it follows that $a_0 \to \lambda$. Hence the forced output of the oscillator is

$$x = X_0 + O(\varepsilon) = 2\left[1 - \frac{\Gamma^2}{2(1 - \omega^2)^2}\right]^{1/2}\cos t + \frac{\Gamma}{1 - \omega^2}\cos \omega t + O(\varepsilon),$$

if $\Gamma^2 < 2(1 - \omega^2)^2$.

If $\lambda = 0$, then, from (v), it follows that $u_0 = 0$, so that $a_0 = b_0 = 0$. Hence the forced periodic response is

$$x = X_0 = \sqrt{2}\cos\omega t + O(\varepsilon).$$

- 6.25 Apply the matching technique to the damped pendulum equation
$$\varepsilon\ddot{x} + \dot{x} + \sin x = 0, \quad x(\varepsilon,0) = 1, \quad \dot{x}(\varepsilon,0) = 0$$
for ε small and positive. Show that the inner and outer approximations are given by
$$x_{\mathrm{I}} = 1, \quad x_{\mathrm{O}} = 2\tan^{-1}\left(e^{-t}\tan\tfrac{1}{2}\right).$$
(The pendulum has strong damping and strong restoring action, but the damping dominates.)

6.25. The equation is the damped pendulum equation
$$\varepsilon\ddot{x} + \dot{x} + \sin x = 0, \quad x(\varepsilon,0) = 1, \quad \dot{x}(\varepsilon,0) = 1.$$

For the outer approximation x_O, put $\varepsilon = 0$ into the equation so that x_O satisfies
$$\dot{x}_\mathrm{O} + \sin x_\mathrm{O} = 0.$$

Hence
$$\int \frac{dx_\mathrm{O}}{\sin x_\mathrm{O}} = -\int dt = -t + C.$$

Integrating
$$\ln\tan(\tfrac{1}{2}x_\mathrm{O}) = -t + C,$$

so that
$$x_\mathrm{O} = 2\tan^{-1}(Ae^{-t}).$$

For the inner solution, use the transformation $t = \varepsilon\xi$. The transformed equation becomes
$$x'' + x' + \varepsilon\sin x = 0.$$

the derivatives being with respect to ξ. Putting $\varepsilon = 0$, the inner solution x_I satisfies
$$x_\mathrm{I}'' + x_\mathrm{I}' = 0.$$

Therefore the inner solution is given by

$$x_{\mathrm{I}} = M + Ne^{-\xi},$$

where $x_{\mathrm{I}}(0) = 1$, $x_{\mathrm{I}}'(0) = 0$. Hence $M + N = 1$ and $N = 0$. Therefore

$$x_{\mathrm{I}} = 1.$$

To match the outer and inner approximations, let $t = \eta\sqrt{\varepsilon}$. Then, expanding both approximations

$$\lim_{\varepsilon \to 0} x_{\mathrm{O}} = 2\tan^{-1} A,$$

$$\lim_{\varepsilon \to 0} x_{\mathrm{I}} = 1.$$

Matching we find that $A = \tan\frac{1}{2}$ and that

$$x_{\mathrm{O}} = 2\tan^{-1}[\tan(\tfrac{1}{2})e^{-t}].$$

- **6.26** The equation for a tidal bore on a shallow stream is

$$\varepsilon\frac{d^2\eta}{d\xi^2} - \frac{d\eta}{d\xi} - \eta + \eta^2 = 0,$$

where (in appropriate dimensions) η is the height of the free surface, and $\xi = x - ct$, where c is the wave speed. For $0 < \varepsilon \ll 1$, find the equilibrium points for the equation and classify them according to their linear approximations. Apply the coordinate perturbation method to the equation for the phase paths,

$$\varepsilon\frac{dw}{d\eta} = \frac{w + \eta - \eta^2}{w}, \quad \text{where} \quad w = \frac{d\eta}{d\xi},$$

and show that

$$w = -\zeta + \zeta^2 + O(\varepsilon^2), \quad \eta = \zeta - \varepsilon(-\zeta + \zeta^2) + O(\varepsilon^2).$$

Confirm that, to this degree of approximation, a separatrix from the origin reaches the other equilibrium point. Interpret the result in terms of the shape of the bore.

6.26. The tidal bore equation is

$$\varepsilon\frac{d^2\eta}{d\xi^2} - \frac{d\eta}{d\xi} - \eta + \eta^2 = 0.$$

Let $w = d\eta/d\xi$. The system has two equilibrium points in the (η, w) plane, at $(0,0)$ and $(1,0)$. Near the origin η satisfies the linear approximation

$$\varepsilon \eta'' - \eta' - \eta = 0,$$

which implies a saddle point. If $\eta = 1 + \bar{\eta}$, then $\bar{\eta}$ has the linear approximation

$$\varepsilon \bar{\eta}'' - \bar{\eta}' + \bar{\eta} = 0$$

near $(1,0)$. Therefore the equilibrium point is an unstable node.
The equation for the phase paths is

$$\varepsilon \frac{dw}{d\eta} = \frac{w^2 + \eta - \eta^2}{w}. \tag{i}$$

Let $w = w_0 + \varepsilon w_1 + \cdots$ and substitute this expansion into (i) so that

$$\varepsilon(w_0 + \varepsilon w_1 + \cdots)(w_0' + \varepsilon w_1' + \cdots) = w_0 + \varepsilon w_1 + \cdots + \eta - \eta^2.$$

Equating coefficients of like powers of ε to zero, we obtain

$$w_0 = \eta^2 - \eta, \quad w_1 = w_0 w_0' = (\eta^2 - \eta)(2\eta - 1).$$

Now let $\eta = \zeta + \varepsilon \zeta_1 + \cdots$. Then

$$w_0 + \varepsilon w_1 = (\zeta + \varepsilon \zeta_1)^2 - (\zeta + \varepsilon \zeta_1) + \varepsilon[(\zeta + \varepsilon \zeta_1)^2 - (\zeta + \varepsilon \zeta_1)][2(\zeta + \varepsilon \zeta_1) - 1]$$
$$= \zeta^2 - \zeta + \varepsilon[2\zeta\zeta_1 - \zeta_1 + (\zeta^2 - \zeta)(2\zeta - 1)] + O(\varepsilon^2)$$
$$= \zeta^2 - \zeta + \varepsilon(2\zeta - 1)(\zeta_1 + \zeta^2 - \zeta) + O(\varepsilon^2)$$

The order ε term in w can be eliminated by putting $\zeta_1 = \zeta - \zeta^2$. We arrive at the approximate solution given parametrically by

$$w = \zeta^2 - \zeta + O(\varepsilon^2), \quad \eta = \zeta - \varepsilon(\zeta^2 - \zeta) + O(\varepsilon^2). \tag{ii}$$

To order ε^2 this solution passes through $(0,0)$, where $\zeta = 0$, and through $(1,0)$, where $\zeta = 1$. In other words a phase path from the unstable node at $(1,1)$ becomes a separatrix of the saddle point at the origin as shown in Figure 6.5 with $\varepsilon = 0.25$. For this value of ε the approximate phase path given by (ii) is virtually indistinguishable form the computed phase path. The bore consists of wave advancing along a dried bed.

330 Nonlinear ordinary differential equations: problems and solutions

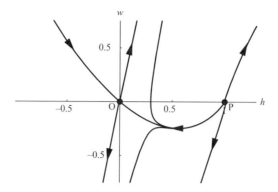

Figure 6.5 Problem 6.26: The computed phase diagram for $\varepsilon\eta'' - \eta' - \eta + \eta^2 = 0$ with $\varepsilon = 0.25$: the separatrix joining the node P to the saddle O is shown.

- **6.27** The function $x(\varepsilon, t)$ satisfies the differential equation $\varepsilon\ddot{x} + x\dot{x} - x = 0$, $(t \geq 0)$ subject to the initial conditions $x(0) = 0$, $\dot{x}(0) = 1/\varepsilon$. To leading order, obtain inner and outer approximations to the solution for small ε. Show that the composite solution is
$$x_C = t + \sqrt{2}\tanh(t/(\varepsilon\sqrt{2})).$$

6.27. The system is
$$\varepsilon\ddot{x} + x\dot{x} - x = 0, \quad x(0) = 0, \quad \dot{x}(0) = 1/\varepsilon.$$

The outer solution x_O satisfies the equation with $\varepsilon = 0$, that is, $x_O\dot{x}_O - x_O = 0$. There are two possible solutions; either
$$x_O = 0, \text{ or } x_O = t + A. \tag{i}$$

For the inner solution, let $t = \varepsilon\xi$, so that
$$x'' + xx' - \varepsilon x = 0.$$

The inner solution x_I therefore satisfies
$$x_I'' + x_I x_I' = 0.$$

Hence
$$\frac{dx_I'}{dx_I} = -x_I, \quad \text{so that} \quad x_I' = -\tfrac{1}{2}x_I^2 + C.$$

Since $x'_I(0) = 1$ when $x_I = 0$, it follows that $C = 1$. Separating the variables, we have

$$\int \frac{2\mathrm{d}x_I}{2 - x_I^2} = \int \mathrm{d}\xi = \xi + B,$$

that is,

$$\frac{1}{\sqrt{2}} \ln\left[\frac{\sqrt{2} + x_I}{\sqrt{2} - x_I}\right] = \xi + B = \xi$$

using the initial condition $x_I(0) = 0$. Therefore, the inner approximation is

$$x_I = \sqrt{2} \tanh\left(\frac{\xi}{\sqrt{2}}\right) = \sqrt{2} \tanh\left(\frac{t}{\sqrt{\varepsilon}\sqrt{2}}\right). \tag{ii}$$

To match the outer and inner approximations, let $t = \eta\sqrt{\varepsilon}$. In (i) try the solution $x_O = t + A$. Then

$$x_O = \eta\sqrt{\varepsilon} + A. \tag{iii}$$

Apply the same transformation to (ii), so that

$$x_I = \sqrt{2} \tanh\left(\frac{\eta\varepsilon}{\sqrt{2}}\right) = \eta\sqrt{\varepsilon} + O(\varepsilon) \tag{iv}$$

for $\eta = O(1)$. Expansions (iii) and (iv) match to leading order if $A = 0$. Hence the composite approximation is

$$x_C = x_O + x_I = t + \sqrt{2} \tanh\left(\frac{t}{\varepsilon\sqrt{2}}\right).$$

- **6.28** Consider the initial-value problem $\varepsilon\ddot{x} + \dot{x} = e^{-t}$, $x(0) = 0$, $\dot{x}(0) = 1/\varepsilon$, $(0 < \varepsilon \ll 1)$. Find inner and outer expansions for x, and confirm that the outer expansion to two terms is $x_O = 2 - e^{-t} - \varepsilon e^{-t}$.
  Compare computed graphs of the composite expansion and the exact solution of the differential equation for $\varepsilon = 0.1$ and for $\varepsilon = 0.25$.

6.28. The initial-value problem is

$$\varepsilon\ddot{x} + \dot{x} = e^{-t}, \quad x(0) = 0, \quad \dot{x}(0) = 1/\varepsilon. \tag{i}$$

For the outer expansion, let $x_O = f_0 + \varepsilon f_1 + \cdots$. Substitute this expansion into the equation, obtaining

$$\varepsilon(\ddot{f}_0 + \varepsilon \ddot{f}_1 + \cdots) + (\dot{f}_0 + \dot{f}_1 \varepsilon + \cdots) = e^{-t}.$$

The first two coefficients satisfy

$$\dot{f}_0 = e^{-t}, \quad \dot{f}_1 = -\ddot{f}_0.$$

The solutions of these equations are

$$f_0 = -e^{-t} + C_0, \quad f_1 = -e^{-t} + C_1.$$

The outer expansion is therefore of the form

$$x_O = (-e^{-t} + C_0) + \varepsilon(-e^{-t} + C_1) + O(\varepsilon^2).$$

For the inner expansion x_I, let $t = \varepsilon \tau$. Equation (i) becomes

$$x_I'' + x_I' = \varepsilon e^{-\varepsilon \tau}.$$

Let $x_I = g_0 + \varepsilon g_1 + \cdots$ and expand $e^{\varepsilon \tau}$ in powers of ε:

$$(g_0'' + \varepsilon g_1'' + \cdots) + (g_0' + \varepsilon g_1' + \cdots) = \varepsilon(1 - \varepsilon \tau + \cdots).$$

The coefficients of the series satisfy

$$g_0'' + g_0' = 0, \quad g_1'' + g_1' = 1.$$

Hence

$$g_0 = A_0 + B_0 e^{-\tau}, \quad g_1 = \tau + A_1 + B_1 e^{-\tau}.$$

The initial conditions in (i) become the sequence of conditions

$$g_i(0) = 0 \ (i = 0, 1, \ldots), \quad g'(0) = 1, \ g_j'(0) = 0, \ (j = 1, 2, \ldots).$$

Therefore

$$A_0 + B_0 = 0, \quad -B_0 = 1, \quad A_1 + B_1 = 0, \quad -B_1 + 1 = 0,$$

which results in $A_0 = 1$, $B_0 = -1$, $A_1 = -1$, $B_1 = 1$. The inner expansion becomes

$$x_I = g_0 + \varepsilon g_1 + O(\varepsilon^2) = (1 - e^{-\tau}) + \varepsilon(\tau - 1 + e^{-\tau}) + O(\varepsilon^2).$$

To match the outer and inner expansions, let $t = \varepsilon^{1/2}\eta$ so that $\tau = t/\varepsilon = \varepsilon^{-(1/2)}\eta$, where $\eta = O(1)$. Then

$$x_O = [C_0 + \exp(-\varepsilon^{1/2}\eta)] + \varepsilon[C_1 + \exp(-\varepsilon^{1/2}\eta)] + \cdots$$
$$= (C_0 - 1) + \varepsilon^{1/2}\eta + \varepsilon(C_1 - 1) + \cdots \qquad \text{(ii)}$$

Also

$$x_I = [1 - \exp(-\varepsilon^{-(1/2)}\eta)] + \varepsilon[\varepsilon^{-(1/2)}\eta - 1 + \exp(-\varepsilon^{-(1/2)}\eta)] + \cdots$$
$$= 1 + \varepsilon^{1/2}\eta - \varepsilon + \cdots, \qquad \text{(iii)}$$

the exponential terms being negligible as $\varepsilon \to 0$. Comparison of (ii) with (iii) shows that they match to the lowest orders if $C_0 = 2$ and $C_1 = 0$. To summarize

$$x_O = (2 - e^{-t}) - \varepsilon e^{-t} + \cdots, \qquad \text{(iv)}$$
$$x_I = (1 + t - e^{-t/\varepsilon}) - \varepsilon(1 - e^{-t/\varepsilon}) + \cdots. \qquad \text{(v)}$$

For the composite solution, form

$$x_O + x_I = (2 - e^{-t}) - \varepsilon e^{-t} + (1 + t - e^{-t/\varepsilon}) - \varepsilon(1 - e^{-t/\varepsilon}) + \cdots$$

If $t = O(1)$, then, expanding in powers of ε,

$$x_O + x_I = (3 + t - e^{-t}) + \varepsilon(-1 - e^{-t}) + O(\varepsilon^2), \qquad \text{(vi)}$$

neglecting the exponential terms $\exp(-t/\varepsilon)$. Comparison of (vi) with (iv) implies that the zero-order term has the unwanted term $1 + t$, and that the first-order term has the unwanted term -1. Now put $t = \varepsilon\tau$ in (vi) so that

$$x_O + x_I = (2 - e^{-\varepsilon\tau}) - \varepsilon e^{-\varepsilon\tau} + (1 + \varepsilon\tau - e^{-\tau}) - \varepsilon(1 - e^{-\tau}).$$

Now assume that $\tau = O(1)$, and expand in powers of ε, so that

$$x_O + x_I = (2 - e^{-\tau}) + \varepsilon(2\tau - 2 + e^{-\tau}) + \cdots$$
$$= (2 + 2t - e^{-t/\varepsilon}) - \varepsilon(2 - e^{-t/\varepsilon}) + \cdots.$$

Comparison of this expansion with (v) shows the same unwanted terms; of $1 + t$ and -1. Therefore the composite solution is

$$x_C = x_O + x_I - (1 + t) + \varepsilon = (2 - e^{-t} - e^{-\tau}) - \varepsilon(e^{-t} - e^{-\tau}) + \cdots.$$

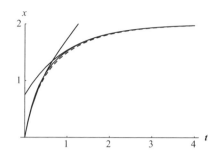

Figure 6.6 Problem 6.28: The graph shows the exact solution (the dashed curve) for $\varepsilon = 0.25$; the outer, inner and composite approximations can be easily identified.

The differential equation is a linear second-order inhomogeneous equation with the exact solution

$$x = 2 - \frac{e^{-t}}{1-\varepsilon} - \frac{1-2\varepsilon}{1-\varepsilon} e^{-t/\varepsilon}.$$

The exact solution and the various approximations are compared in Figure 6.6 only for the case $\varepsilon = 0.25$.

- **6.29** Investigate the solution of the initial/boundary-value problem
$$\varepsilon^3 \dddot{x} + \varepsilon \ddot{x} + \dot{x} + x = 0, \quad 0 < \varepsilon \ll 1,$$
with $x(1) = 1$, $x(0) = 0$, $\dot{x}(0) = 1/\varepsilon^2$ using matched approximations. Start by finding, with a regular expansion, the outer solution x_O and an inner solution x_I using $t = \varepsilon\tau$. Confirm that x_I cannot satisfy the conditions at $t = 0$. The boundary-layer thickness $O(\varepsilon)$ at $t = 0$ is insufficient for this problem. Hence we create an additional boundary layer of thickness $O(\varepsilon^2)$, and a further time scale η where $t = \varepsilon^2\eta$. Show that the leading order equation for the **inner–inner approximation** x_{II} is $x_{II}''' + x_{II}'' = 0$, and confirm that the solution can satisfy the conditions at $t = 0$. Finally match the expansions x_{II} and x_I and the expansions x_I and x_O. Show that the approximations are

$$x_O = e^{1-t}, \quad x_I = e + (1-e)e^{-t/\varepsilon}, \quad x_{II} = 1 - e^{-t/\varepsilon^2}$$

to leading order.
Explain why the composite solution is
$$x_C = e^{1-t} + (1-e)e^{-t/\varepsilon} - e^{-t/\varepsilon^2}.$$
Comparison between the numerical solution of the differential equation and the composite solution is shown in Figure 6.10 in NODE. The composite approximation could be improved by taking all approximations to include $O(\varepsilon)$ terms.

6.29. Consider the third-order system

$$\varepsilon^3 \dddot{x} + \varepsilon \ddot{x} + \dot{x} + x = 0, \quad x(1) = 1, \quad x(0) = 0, \quad \dot{x}(0) = 1/\varepsilon^2.$$

The outer approximation x_O satisfies the equation $\dot{x}_O + x_O = 0$ obtained by putting $\varepsilon = 0$. Therefore using the condition $x(1) = 1$,

$$x_O = A_0 e^{-t} = e^{1-t}. \tag{i}$$

For an inner approximation try $t = \varepsilon\tau$ with $\tau = O(1)$. The differential equation becomes

$$\varepsilon x''' + x'' + x' + \varepsilon x = 0$$

in which the derivatives are with respect to τ. The inner approximation $x_I(\tau)$ therefore satisfies $x_I'' + x_I' = 0$, so that

$$x_I(\tau) = A_1 + B_1 e^{-\tau} = A_1 + B_1 \exp(-t/\varepsilon) \tag{iii}$$

for certain constants A_1, B_1. However, the condition $x(0) = 1/\varepsilon^2$ cannot be satisfied by x_I given by (ii).

This difficulty can be avoided by introducing a further, 'inner–inner', approximation, which will suit the conditions nearer to $t = 0$. To identify this approximation let $t = \varepsilon^2 \eta$, where $\eta = O(1)$. The differential equation becomes

$$x''' + x'' + \varepsilon x' + \varepsilon^3 x = 0$$

in which the derivatives are now with respect to η. The inner–inner approximation $x_{II}(\eta)$ satsifies $x_{II}''' + x_{II}'' = 0$. Therefore

$$x_{II}(\eta) = A_2 + B_2\eta + C_2 e^{-\eta}.$$

The conditions at $t = 0$ ($\eta = 0$) require

$$A_2 + C_2 = 0, \quad B_2 - C_2 = 0,$$

Therefore

$$x_{II} = A_2 + (1 - A_2)\eta - A_2 e^{-\eta}$$
$$= A_2 + (1 - A_2)t/\varepsilon^2 - A_2 \exp(-t/\varepsilon^2).$$

The factor $t/\varepsilon^2 \to \infty$ as $\varepsilon \to 0$ for all positive t, so necessarily $A_2 = 0$, and finally we have

$$x_{II} = 1 - \exp(-t/\varepsilon^2). \tag{iv}$$

We determine the unknown constants in x_I (eqn (iii)) by matching it to x_O and x_{II} over intermediate range of t. To match x_I to x_O put $t = \varepsilon^{1/2} q(t)$ in (i) and (ii), where $q = O(1)$.

Then as $\varepsilon \to 0$

$$x_O = \exp(1 - \varepsilon^{1/2}q) = e + O(\varepsilon^{1/2}),$$
$$x_I = A_1 + B_1 \exp(-q/\varepsilon^{1/2}) = A_1 + o(1).$$

Therefore $A_1 = e$, and in terms of t

$$x_I = e + B_1 \exp(-t/\varepsilon).$$

To match x_I with x_{II} put $t = \varepsilon^{3/2} p(t)$, where $p(t) = O(1)$. The dominant terms match only if $B_1 = 1 - e$. Therefore

$$x_I = e + (1 - e) \exp(-t/\varepsilon). \qquad (v)$$

A composite approximation, valid over the whole interval $0 < t < 1$ can be constructed by considering the sum defined by $x_O + x_I + x_{II}$ on $0 < t < 1$. The following table gives the dominant terms contributed by each of x_O, x_I and x_{II} as $\varepsilon \to 0$, to each of the subintervals considered above.

	$t = O(1)$ but not $O(\varepsilon)$	$t = O(\varepsilon)$ but not $O(\varepsilon^2)$	$t = O(\varepsilon^2)$
$x_O(t)$	e^{1-t}	e	e
$x_I(t)$	e	$e + (1-e)e^{-t/\varepsilon}$	1
$x_{II}(t)$	1	1	$1 - e^{-t/\varepsilon^2}$

Therefore the composite first approximation required is given by

$$x_C = x_O + x_I + x_{II} - (1 + e) = e^{1-t} + e + (1 - e)e^{-t/\varepsilon} + 1 - e^{-t/\varepsilon^2}.$$

- **6.30** Let $y(x, \varepsilon)$ satisfy $\varepsilon y'' + y' = x$, where $y(0, \varepsilon) = 0$, $y(1, \varepsilon) = 1$. Find the inner and outer expansions to order ε using the inner variable $\eta = x/\varepsilon$. Apply the van Dyke matching rule to show that the inner expansion is
$$y_I \approx (\tfrac{1}{2} + \varepsilon)(1 - e^{x/\varepsilon}).$$

6.30. The system is

$$\varepsilon y'' + y' = x, \quad y(0, \varepsilon) = 0, \quad y(1, \varepsilon) = 1.$$

We require the first two terms in the outer and inner expansions. Let $y = f_0 + \varepsilon f_1 + \cdots$. Then

$$\varepsilon(f_0'' + \cdots) + (f_0' + \varepsilon f_1' + \cdots) = x.$$

Equating powers of ε, we have

$$f_0' = x, \quad f_1' = -f_0''.$$

The boundary condition at $x = 1$ becomes $f_0(1) = 1$, $f_i(1) = 0$, $(i = 1, 2, \ldots)$. Therefore

$$f_0 = \tfrac{1}{2}x^2 + A = \tfrac{1}{2}x^2 + \tfrac{1}{2}, \quad f_1 = -x + B = -x + 1.$$

The outer expansion is

$$y_O = (\tfrac{1}{2}x^2 + \tfrac{1}{2}) + \varepsilon(1 - x) + O(\varepsilon^2). \tag{i}$$

For the inner expansion, let $x = \varepsilon\eta$. Then the equation becomes

$$y'' + y' = \varepsilon^2 \eta.$$

Let $y = g_0 + \varepsilon g_1 + \cdots$. Then g_0 and g_1 satisfy

$$g_0'' + g_0' = 0, \quad g_1'' + g_1' = 0,$$

subject to $g_0(0) = 0$, $g_1(0) = 0$, $\ldots$. Hence

$$g_0 = C + De^{-\eta} = C(1 - e^{-\eta}), \quad g_1 = E + Fe^{-\eta} = E(1 - e^{-\eta}),$$

so that the inner expansion is

$$y_I = C(1 - e^{-\eta}) + \varepsilon E(1 - e^{-\eta}) + O(\varepsilon^2). \tag{ii}$$

Put $x = \varepsilon\eta$ in (i) where $\eta(t) = O(1)$:

$$y_O = (\tfrac{1}{2}\varepsilon^2 \eta^2 + \tfrac{1}{2}) + \varepsilon(1 - \varepsilon\eta) + \cdots = \tfrac{1}{2} + \varepsilon + \cdots. \tag{iii}$$

In terms of x, y_I given by (ii) becomes

$$y_I = C(1 - e^{-x/\varepsilon}) + \varepsilon E(1 - e^{-x/\varepsilon}) + \cdots = C + \varepsilon E + \cdots. \tag{iv}$$

Matching of (iii) and (iv) implies $C = \tfrac{1}{2}$ and $E = 1$. Therefore

$$y_I = (\tfrac{1}{2} + \varepsilon)(1 - e^{-x/\varepsilon}) + \cdots.$$

- 6.31 In Example 6.9, a composite solution of
$$\varepsilon\frac{d^2y}{dx^2} + 2\frac{dy}{dx} + y = 0, \quad y(0) = 0, \quad y(1) = 1,$$
valid over the interval $0 \leq x \leq 1$, was found to be (eqn (6.114))
$$y_C = e^{1/2}(e^{-(1/2)x} - e^{-2x/\varepsilon})$$
using matched inner and outer approximations. What linear constant coefficient second-order differential equation and boundary conditions does y_C satisfy exactly?

6.31. The equation
$$\varepsilon y'' + 2y' + y = 0, \quad y(0) = 0, \quad y(1) = 1,$$
has the composite solution
$$y_C = e^{\frac{1}{2}}(e^{-(1/2)x} - e^{-2x/\varepsilon}), \tag{i}$$
obtained by matched inner and outer expansions. The exponents in (i) are $-\frac{1}{2}$ and $-2/\varepsilon$, which could arise from the characteristic equation
$$(m + \tfrac{1}{2})\left(m + \tfrac{2}{\varepsilon}\right) = 0,$$
which defines linearly independent solutions $e^{-(1/2)x}, e^{-2x/\varepsilon}$ of the linear differential equation
$$2\varepsilon y'' + (\varepsilon + 4)y' + 2y = 0.$$

7 Forced oscillations: harmonic and subharmonic response, stability, and entrainment

> • **7.1** Show that eqns (7.16) and (7.17), for the undamped Duffing equation in the van der Pol plane have the exact solution
> $$r^2\{(\omega^2-1)-\tfrac{3}{8}\beta r^2\}+2\Gamma a = \text{constant}, \quad r=\sqrt{(a^2+b^2)}.$$
> Show that these approximate to circles when r is large. Estimate the period on such a path of $a(t), b(t)$.

7.1. Equations (7.16) and (7.17) (in NODE) are

$$\dot a = -\frac{1}{2\omega}b\{(\omega^2-1)-\tfrac{3}{4}\beta(a^2+b^2)\}, \tag{i}$$

$$\dot b = \frac{1}{2\omega}a\{(\omega^2-1)-\tfrac{3}{4}\beta(a^2+b^2)\}+\frac{\Gamma}{2\omega}. \tag{ii}$$

Form $a \times$ (i) $+ b \times$ (ii) which results in

$$\frac{d(r^2)}{dt} = \frac{\Gamma b}{\omega}. \tag{iii}$$

where $r^2 = a^2+b^2$. From (i) and (ii)

$$\frac{d(r^2)}{da} = \frac{d(r^2)}{dt}\bigg/\frac{da}{dt} = -\frac{2\Gamma}{\omega^2-1-\tfrac{3}{4}\beta r^2}.$$

Separating the variables and integrating, we have

$$(\omega^2-1)r^2-\tfrac{3}{8}\beta r^4+2\Gamma a = \text{constant}.$$

From (i) and (ii),

$$\frac{db}{da} = -\frac{a\{(\omega^2-1)-(3/4)\beta(a^2+b^2)\}+\Gamma}{b\{(\omega^2-1)-(3/4)\beta(a^2+b^2)\}}.$$

For r^2 is large in comparison with both $\omega^2 - 1$ and Γ, the differential equation is

$$\frac{db}{da} \approx -\frac{a}{b}.$$

Integration gives the family $a^2 + b^2 = $ constant, which are concentric circles (approximately). Let δs be an increment of length on one these circles of radius r Then $\delta s = \sqrt{[(\delta a)^2 + (\delta b)^2]}$. Hence the period T is given by

$$T \approx \int_0^{2\pi r} t \frac{ds}{\sqrt{(\dot{a}^2 + \dot{b}^2)}}.$$

For large r,

$$\dot{a} \approx \frac{3\beta b}{8\omega} r^2, \quad \dot{b} \approx \frac{3\beta a}{8\omega} r^2,$$

so that

$$T \approx \frac{8\omega}{3\beta r^3} \int_0^{2\pi r} ds = \frac{8\omega}{3\beta r^3} \cdot 2\pi r = \frac{16\pi\omega}{3\beta r^2}.$$

- **7.2** Express eqns (7.16) and (7.17) in polar coordinates. Deduce the approximate period of $a(t)$ and $b(t)$ for large r. Find the approximate equations for these distant paths. Show how frequency modulation occurs, by deriving an expression for $x(t)$.

7.2. Equations (7.16) and (7.17) (in NODE) are

$$\dot{a} = -\frac{1}{2\omega} b\{(\omega^2 - 1) - \tfrac{3}{4}\beta(a^2 + b^2)\}, \tag{i}$$

$$\dot{b} = \frac{1}{2\omega} a\{(\omega^2 - 1) - \tfrac{3}{4}\beta(a^2 + b^2)\} + \frac{\Gamma}{2\omega}. \tag{ii}$$

Let $a = r\cos\theta$ and $b = r\sin\theta$. Then

$$a\dot{a} + b\dot{b} = \frac{\Gamma b}{2\omega},$$

or

$$\frac{dr}{dt} = \frac{\Gamma}{2\omega} \sin\theta. \tag{iii}$$

Also

$$a\dot{b} - b\dot{a} = \frac{1}{2\omega} \left[r^2(\omega^2 - 1) - \frac{3\beta}{4} r^4 + a\Gamma \right].$$

In polar form
$$\frac{d\theta}{dt} = \frac{1}{2\omega r^2}\left[r^2(\omega^2 - 1) - \frac{3\beta}{4}r^4 - a\Gamma\right] \approx -\frac{3\beta r^2}{8\omega} \qquad (iv)$$

for r large. The polar differential equation is, from (iii) and (iv),
$$\frac{dr}{d\theta} = \frac{dr}{dt}\bigg/\frac{d\theta}{dt} = -\frac{4\Gamma \sin\theta}{3\beta r^2}.$$

Therefore
$$3\beta \int r^2 dr = -4\Gamma \int \sin\theta \, d\theta,$$

which can be integrated to give
$$r^3 - \frac{4\Gamma}{\beta}\cos\theta = C, \text{ a constant.}$$

which is the polar equation of phase paths in the van der Pol plane for large r. Since $|\frac{4\Gamma}{\beta}\cos\theta|$ is bounded, the paths will be approximately circles for large r, which agrees with the solution of Problem 7.1.

From (iii) for a fixed and large radius, r will be constant on a path, so that, since θ changes by 2π in one circuit the period is $16\pi\omega/(3\beta r^2)$. Since the frequency is $3\beta r^2/(8\omega)$,

$$x \approx a\cos t + b\sin t = r\left[\cos\left(\frac{3\beta r^2}{8\omega}\right)t\cos t + \sin\left(\frac{3\beta r^2}{8\omega}\right)t\sin t\right]$$

$$= r\cos\left[\left(1 - \frac{3\beta r^2}{8\omega}\right)t\right].$$

The dependence of $[1 - (3\beta r^2/(16\omega))]$ on the radius r indicates frequency modulation of x for large r.

- **7.3** Consider the equation $\ddot{x} + \text{sgn}(x) = \Gamma \cos \omega t$. Assume solutions of the form $x = a\cos\omega t + b\sin\omega t$. Show that solutions of period $2\pi/\omega$ exist when $|\Gamma| \leq 4/\pi$. Show also that
$$a(4 - \pi\omega^2|a|) = \pi\Gamma|a|, \quad b = 0.$$
$$\left(\text{Hint: sgn}\{x(t)\} = \frac{4a}{\pi\sqrt{(a^2+b^2)}}\cos\omega t + \frac{4b}{\pi\sqrt{(a^2+b^2)}}\sin\omega t\right.$$
$$\left. + \text{higher harmonics.}\right)$$

7.3. Consider the equation
$$\ddot{x} + \text{sgn}(x) = \Gamma \cos \omega t.$$
Assume that $x = a \cos \omega t + b \sin \omega t$. We require the leading terms in the Fourier expansion of
$$\text{sgn}\, x = \text{sgn}(a \cos \omega t + b \sin \omega t) = A \cos \omega t + B \sin \omega t + \cdots,$$
say. Then, substituting $\tau = \omega t$,
$$A = \frac{\omega}{\pi} \int_{-\pi/\omega}^{\pi/\omega} \text{sgn}(a \cos \omega t + b \sin \omega t) \cos \omega t \, dt.$$
$$= \frac{1}{\pi} \int_{-\pi}^{\pi} \text{sgn}(a \cos \tau + b \sin \tau) \cos \tau \, d\tau.$$

Let $a = r \cos \phi$, $y = r \sin \phi$ so that
$$A = \frac{1}{\pi} \int_{-\pi}^{\pi} \text{sgn}[r \cos(\tau - \phi)] \cos \tau \, d\tau$$

Now $\cos(\tau - \phi) = 0$ where $\tau = -\tfrac{1}{2}\pi + \phi$ and $\tau = \tfrac{1}{2}\pi + \phi$. Therefore
$$A = \frac{1}{\pi}\left[-\int_{-\pi}^{-\tfrac{1}{2}\pi + \phi} \cos \tau \, d\tau + \int_{-\tfrac{1}{2}\pi+\phi}^{\tfrac{1}{2}\pi+\phi} \cos \tau \, d\tau - \int_{\tfrac{1}{2}\pi+\phi}^{\pi} \cos \tau \, d\tau \right]$$
$$= \frac{1}{\pi}[\cos \phi + 2 \cos \phi + \cos \phi]$$
$$= \frac{4 \cos \phi}{\pi} = \frac{4a}{\pi r}.$$

Similarly
$$B = \frac{1}{\pi} \int_{-\pi}^{\pi} \text{sgn}[r \cos(\tau - \phi)] \sin \tau \, d\tau$$
$$= \frac{1}{\pi}\left[-\int_{-\pi}^{-\tfrac{1}{2}\pi + \phi} \sin \tau \, d\tau + \int_{-\tfrac{1}{2}\pi+\phi}^{\tfrac{1}{2}\pi+\phi} \sin \tau \, d\tau - \int_{\tfrac{1}{2}\pi+\phi}^{\pi} \sin \tau \, d\tau \right]$$
$$= \frac{1}{\pi}[1 + \sin \phi + 2 \sin \phi - 1 + \sin \phi]$$
$$= \frac{4 \sin \phi}{\pi} = \frac{4b}{\pi r}.$$

Hence
$$\text{sgn}(x) \approx \frac{4a}{\pi r} \cos \omega t + \frac{4b}{\pi r} \sin \omega t.$$

Substitute $x = a \cos \omega t + b \sin \omega t$ and $\mathrm{sgn}(x)$ into the differential equation so that

$$-a\omega^2 \cos \omega t - b\omega^2 \sin \omega t + \frac{4a}{\pi r} \cos \omega t + \frac{4b}{\pi r} \sin \omega t \approx \Gamma \cos \omega t.$$

The harmonics balance if

$$-a\omega^2 + \frac{4a}{\pi r} = \Gamma, \quad -b\omega^2 + \frac{4b}{\pi r} = 0.$$

From the second equation $b = 0$ (since $r = 4/(\pi\omega^2)$ is inconsistent with the first equation), so that

$$-a\omega^2 + \frac{4a}{\pi |a|} = \Gamma, \quad \text{or} \quad a(4 - \pi\omega^2|a|) = \pi\Gamma|a|,$$

as required. For $a > 0$,

$$a = \frac{1}{\omega^2}\left(\frac{4}{\pi} - \Gamma\right).$$

Therefore $\Gamma < 4/\pi$. Similarly, if $a < 0$, then $\Gamma > -4/\pi$. The two inequalities can be combined into $|\Gamma| < 4/\pi$.

- 7.4 Show that solutions, period 2π, of the equation $\ddot{x} + x^3 = \Gamma \cos t$ are given approximately by $x = a \cos t$, where a is a solution of $3a^3 - 4a = 4\Gamma$.

7.4. The differential equation is

$$\ddot{x} + x^3 = \Gamma \cos t.$$

Assume that $x \approx a \cos t + b \sin t$. The leading harmonics of x^3 are given by

$$x^3 = \tfrac{3}{4}ar^2 \cos t + \tfrac{3}{4}br^2 \sin t + \cdots,$$

where $r = \sqrt{(a^2 + b^2)}$. The coefficients of $\cos t$ and $\sin t$ are zero if

$$-a + \tfrac{3}{4}ar^2 = \Gamma, \quad -b + \tfrac{3}{4}br^2 = 0.$$

The only solution of these equations is $b = 0$, $3a^3 - 4a = 4\Gamma$.

- 7.5 Show that solutions, period 2π, of $\ddot{x} + k\dot{x} + x + x^3 = \Gamma \cos t$ are given approximately by $x = a \cos t + b \sin t$, where
$$ka - \tfrac{3}{4}br^2 = 0, \quad kb + \tfrac{3}{4}ar^2 = \Gamma, \quad r = \sqrt{(a^2 + b^2)}.$$
Deduce that the response curves are given by $r^2(k^2 + \tfrac{9}{16}r^4) = \Gamma^2$.

7.5. The forced Duffing-type equation is

$$\ddot{x} + k\dot{x} + x + x^3 = \Gamma \cos t. \qquad (i)$$

Let $x \approx a \cos t + b \sin t$, and use the expansion

$$x^3 = (a \cos t + b \sin t)^3 = \tfrac{3}{4}ar^2 \cos t + \tfrac{3}{4}br^2 \sin t + \text{higher harmonics},$$

where $r = \sqrt{(a^2 + b^2)}$. Substituting the expansions into (i), we have

$$(-a \cos t - b \sin t) + k(-a \sin t + b \cos t) + (a \cos t + b \sin t)$$
$$+ (\tfrac{3}{4}ar^2 \cos t + \tfrac{3}{4}br^2 \sin t + \text{higher harmonics}) = \Gamma \cos t.$$

The first harmonics balance if

$$-a + kb + a + \tfrac{3}{4}ar^2 = \Gamma, \quad -b - ka + b + \tfrac{3}{4}br^2 = 0,$$

or

$$bk + \tfrac{3}{4}ar^2 = \Gamma, \quad -ak + \tfrac{3}{4}br^2 = 0.$$

Squaring and adding these equations, we have the response formula

$$k^2 r^2 + \frac{9}{16} r^6 = \Gamma^2.$$

• **7.6** Obtain approximate solutions, period $2\pi/\omega$, of $\ddot{x} + \alpha x + \beta x^2 = \Gamma \cos \omega t$, by assuming the form $x = c + a \cos \omega t$, and deducing equations for c and a.
Show that if β is small, $\Gamma = O(\beta)$, and $\omega^2 - \alpha = O(\beta)$, then there is a solution with $c \approx -\beta a^2/(2\alpha)$ and $a \approx \Gamma/(\alpha - \omega^2)$.

7.6. Let $x \approx c + a \cos \omega t$ in the equation

$$\ddot{x} + \alpha x + \beta x^2 = \Gamma \cos \omega t.$$

Use the identity

$$x^2 = (c + a \cos \omega t)^2 = \tfrac{1}{2}(a^2 + 2c^2 + 4ac \cos \omega t + a^2 \cos 2\omega t).$$

The differential equation becomes

$$-a\omega^2 \cos \omega t + \alpha(c + a \cos \omega t) + \tfrac{1}{2}\beta(a^2 + 2c^2 + 4ac \cos \omega t + \cdots) = \Gamma \cos \omega t,$$

and the translation and first harmonic balance if

$$\alpha c + \tfrac{1}{2}\beta(a^2 + 2c^2) = 0, \quad -a\omega^2 + \alpha a + 2\beta ac = \Gamma. \tag{i}$$

If β is small, $\Gamma = O(\beta)$, and $\omega^2 - \alpha = O(\beta)$, then assume that $a = O(1)$ and that $c = \kappa\beta + O(\beta^2)$. Then

$$\alpha\kappa\beta + \tfrac{1}{2}\beta a^2 \approx 0, \quad -a\omega^2 + \alpha a \approx \Gamma.$$

Therefore $\kappa = -a^2/(2\alpha)$, so that

$$c \approx -\frac{\beta a^2}{2\alpha}, \quad a \approx \frac{\Gamma}{\alpha - \omega^2}.$$

• 7.7 Consider the equation $\ddot{x} + x^3 = \Gamma \cos t$. Substitute $x = a\cos t + b\sin t$, and obtain the solution $x = a\cos t$, where $\tfrac{3}{4}a^3 - a = \Gamma$ (see Problem 7.4).
  Now fit x^3, by a least squares procedure, to a straight line of the form px, where p is a constant on $-A \le x \le A$, so that
$$\int_{-A}^{A}(x^3 - px)^2 dx$$
is a minimum with respect to p. Deduce that this linear approximation to the restoring force is compatible with an oscillation, period 2π, of amplitude A, provided $\tfrac{3}{5}A^3 - A = \Gamma$.

7.7. As in Problem 7.4, the differential equation

$$\ddot{x} + x^3 = \Gamma \cos t,$$

has the approximate solution is $x = a\cos t$, where $\tfrac{3}{4}a^3 - a = \Gamma$.
  In the least squares procedure, the square of the difference between $z = x^3$ and the line $z = px$ over the interval $-A \le x \le A$ is minimized to determine the slope p. The square of the distance is

$$F(p) = \int_{-A}^{A}(x^3 - px)^2 dx = \tfrac{2}{7}A^7 - \tfrac{4}{5}A^5 p + \tfrac{2}{3}A^3 p^2.$$

Since

$$F'(p) = -\tfrac{4}{5}A^5 + \tfrac{4}{3}pA^3,$$

then $F'(p)$ is stationary where $-\tfrac{4}{5}A^2 + \tfrac{4}{3}p = 0$. Therefore $p = \tfrac{3}{5}A^2$ and the best fit is $z = \tfrac{3}{5}A^2 x$. Using this approximation the equivalent linear equation is

$$\ddot{x} + \tfrac{3}{5}A^2 x = \Gamma \cos t.$$

This equation has the solution $x = A\cos t$ if the amplitude A is given by $\tfrac{3}{5}A^3 - A = \Gamma$.

7.8. Consider the equation
$$\ddot{x} + 0.16x^2 = 1 + 0.2\cos t. \tag{i}$$

Without the $0.2\cos t$ term, the unforced system has equilibrium points at $0.16x^2 = 1$, or $x = \pm 2.5$. Let $x = \pm 2.5 + X$, but retain only linear terms in X. Hence X satisfies the approximate equation
$$\ddot{X} \pm 0.8X = 0.2\cos t.$$

This has the periodic solutions $X = K\cos t$, where $K = -1$ or $K = -0.11$ in the two cases. Hence the two modes of oscillation with period 2π are
$$x = 2.5 - \cos t, \quad x = -2.5 - 0.11\cos t.$$

An alternative method assumes that $x = c + a\cos t + b\sin t$, and uses the identity
$$(c + a\cos t + b\sin t)^2 = \tfrac{1}{2}(a^2 + b^2 + 2c^2) + 2ac\cos t + 2bc\sin t + \text{higher harmonics}.$$

Now balance the constant and leading harmonic terms in (i):
$$0.16(a^2 + b^2 + 2c^2) = 2, \quad -a + 0.16 \times 2ac = 0.2, \quad -b + 0.16 \times 2cb = 0.$$

Since $b = 0$ is the only consistent solution of the third equation, it follows that $a = 0.2/(-1 + 0.32c)$ from the second equation. Finally the first equation implies
$$0.16\left(\frac{0.04}{(-1 + 0.32c)^2} + 2c^2\right) = 2,$$

which after expansion becomes
$$0.032768c^4 - 0.2048c^3 + 0.1152c^2 + 1.28c - 1.9936 = 0.$$

Numerical solution gives two real solutions $c = 2.42$ and $c = -2.50$. The corresponding values for a are $a = -0.89$ and $a = -0.11$. Hence the balance method yields the solutions
$$x = 2.42 - 0.89\cos t, \quad x = -2.5 - 0.11\cos t$$

for comparison with the earlier results.

• **7.9** By examining the non-periodic solutions of the linearized equations obtained from the first part of Problem 7.8, show that the two solutions, period 2π, obtained are respectively stable and unstable.

7.9. Refer back to the previous problem and the equation

$$\ddot{x} + 0.16x^2 = 1 + 0.2\cos t.$$

The equivalent linear equations were shown to be, with $x = \pm 2.5 + X$,

$$\ddot{X} \pm 0.8X = 0.2\cos t.$$

Near $x = 2.5$, the transient is

$$X = A\cos(\sqrt{0.8}t + \alpha).$$

which is bounded: hence the solution is stable.

Near $x = -2.5$, the transient is

$$X = Ae^{\sqrt{0.8}t} + Be^{-\sqrt{0.8}t},$$

which is unbounded in general: therefore the solution is unstable.

• **7.10** Show that the equations giving the equilibrium points in the van der Pol plane for solutions period $2\pi/\omega$ for the forced, damped pendulum equation
$$\ddot{x} + k\dot{x} + x - \tfrac{1}{6}x^3 = \Gamma \cos \omega t, \quad k > 0$$
are
$$k\omega a + b\left\{\omega^2 - 1 + \tfrac{1}{8}(a^2+b^2)\right\} = 0, \quad -k\omega b + a\left\{\omega^2 - 1 + \tfrac{1}{8}(a^2+b^2)\right\} = -\Gamma.$$
Deduce that
$$r^2\left(\omega^2 - 1 + \tfrac{1}{8}r^2\right)^2 + \omega^2 k^2 r^2 = \Gamma^2, \quad \omega k r^2 = \Gamma b,$$
where $r = \sqrt{(a^2 + b^2)}$.

7.10. The forced Duffing equation is

$$\ddot{x} + k\dot{x} + x - \tfrac{1}{6}x^3 = \Gamma \cos \omega t, \quad k > 0. \tag{i}$$

Let $x = a\cos\omega t + b\sin\omega t$. Then (see NODE, eqn (7.7))

$$x^3 = \tfrac{3}{4}ar^2\cos\omega t + \tfrac{3}{4}br^2\sin\omega t + \text{higher harmonics}.$$

where $r = \sqrt{(a^2 + b^2)}$. To the order of the first harmonics, eqn (i) becomes

$$(-a\omega^2 + kb\omega + a - \tfrac{1}{8}ar^2)\cos\omega t + (-b\omega^2 - ka\omega + b - \tfrac{1}{8}br^2)\sin\omega t \approx \Gamma\cos\omega t.$$

The harmonics balance if

$$k\omega b + a(1 - \omega^2 - \tfrac{1}{8}r^2) = \Gamma, \qquad \text{(ii)}$$

$$k\omega a - b(1 - \omega^2 - \tfrac{1}{8}r^2) = 0. \qquad \text{(iii)}$$

Square and add (ii) and (iii): the result is

$$\omega^2 k^2 r^2 + r^2(\omega^2 - 1 + \tfrac{1}{8}r^2)^2 = \Gamma^2.$$

Add $b\times$(ii) to $a\times$(iii) to give the second equation

$$\omega k r^2 = \Gamma b.$$

- **7.11** For the equation $\ddot{x} + x - \tfrac{1}{6}x^3 = \Gamma\cos\omega t$, find the frequency–amplitude equations in the van der Pol plane. Show that there are three equilibrium points in the van der Pol plane if $\omega^2 < 1$ and $|\Gamma| > \tfrac{2}{3}\sqrt{(\tfrac{8}{3})}(1 - \omega^2)^{3/2}$, and one otherwise. Investigate their stability.

7.11. The Duffing equation

$$\ddot{x} + x - \tfrac{1}{6}x^3 = \Gamma\cos\omega t, \qquad \text{(i)}$$

is considered in NODE, Section 7.2 with $\beta = -\tfrac{1}{6}$. Using the solution $x = a\cos\omega t + b\sin\omega t$, the frequency–amplitude equations, given by (7.16) and (7.17) are

$$\dot{a} = -\frac{b}{2\omega}\{(\omega^2 - 1) + \tfrac{1}{8}(a^2 + b^2)\}, \qquad \text{(ii)}$$

$$\dot{b} = \frac{a}{2\omega}\{(\omega^2 - 1) + \tfrac{1}{8}(a^2 + b^2)\} + \frac{\Gamma}{2\omega}. \qquad \text{(iii)}$$

Equilibrium points in the van der Pol plane occur where $b = 0$ and a satisfies

$$(\omega^2 - 1)a + \tfrac{1}{8}a^3 + \Gamma = 0.$$

Let

$$z(a) = (\omega^2 - 1)a + \tfrac{1}{8}a^3.$$

Then
$$z'(a) = (\omega^2 - 1) + \tfrac{3}{8}a^2.$$

- $\omega^2 < 1$. The frequency–amplitude curve in the (a, z) plane has two stationary points at $a = \pm\sqrt{\tfrac{8}{3}}\sqrt{(1-\omega^2)}$. Correspondingly, $z = \mp\tfrac{2}{3}\sqrt{\tfrac{8}{3}}(1-\omega^2)^{3/2}$. Therefore there are three equilibrium points if $|\Gamma| < \tfrac{2}{3}\sqrt{\tfrac{8}{3}}(1-\omega^2)^{3/2}$, two equilibrium points if $|\Gamma| = \tfrac{2}{3}\sqrt{\tfrac{8}{3}}(1-\omega^2)^{3/2}$ and one otherwise.
- $\omega^2 > 1$. From (ii), $z(a)$ has no stationary points, but since $z(a) \to \pm\infty$ as $a \to \infty$, there will be just one equilibrium point.

Let $a = a_0 + a_1$, $b = b_1$, where $|a_1|$ and $|b_1|$ are small, and (a_0, b_0) is an equilibrium point. Then the linearized equations derived from (ii) and (iii) are

$$\dot{a}_1 = -\frac{b_1}{2\omega}\left\{(\omega^2 - 1) + \frac{1}{8}a_0^2\right\},$$

$$\dot{b}_1 = \left[\frac{a_0}{2\omega}(\omega^2 - 1) + \frac{1}{16\omega}a_0^3 + \frac{\Gamma}{2\omega}\right] + \frac{a_1}{2\omega}\left[(\omega^2 - 1) + \frac{3}{8}a_0^2\right]$$
$$= \frac{a_1}{2\omega}\left[(\omega^2 - 1) + \frac{3}{8}a_0^2\right].$$

Write these equations as

$$\dot{a}_1 = -\Omega_1 b_1, \quad \Omega_1 = \frac{1}{2\omega}\left\{(\omega^2 - 1) + \frac{1}{8}a_0^2\right\}, \tag{iv}$$

$$\dot{b}_1 = \Omega_2 a_1, \quad \Omega_2 = \frac{1}{2\omega}\left[(\omega^2 - 1) + \frac{3}{8}a_0^2\right] \tag{v}$$

By elimination a_1 satisfies the equation

$$\ddot{a}_1 + \Omega_1 \Omega_2 a_1 = 0. \tag{vi}$$

- $\omega^2 > 1$. From (iv) and (v), $\Omega_1 > 0$ and $\Omega_2 > 0$, so that (vi) implies that the only equilibrium point in the van der Pol plane is a centre. Therefore the corresponding periodic solution is stable.
- $\omega^2 < 1$. The curve in Figure 7.1 typically shows the curve

$$z(a) = (\omega^2 - 1)a + \tfrac{1}{8}a^3$$

for a value of $\omega^2 < 1$. It is helpful to use the figure. The abscissae of the points on the curve are

$$B: a = -\sqrt{8}\sqrt{(1-\omega^2)}, \quad C: a = -\sqrt{\tfrac{8}{3}}\sqrt{(1-\omega^2)},$$
$$D: a = \sqrt{\tfrac{8}{3}}\sqrt{(1-\omega^2)}, \quad E: a = \sqrt{8}\sqrt{(1-\omega^2)}.$$

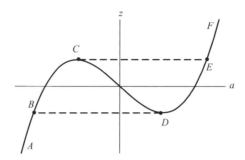

Figure 7.1 Problem 7.11:

From (iv) and (v), we have the following results

$$a_0 < -\sqrt{8}\sqrt{(1-\omega^2)} \qquad \Omega_1\Omega_2 > 0 \quad \text{stable}$$

$$-\sqrt{8}\sqrt{(1-\omega^2)} < a_0 < -\sqrt{\tfrac{8}{3}}\sqrt{(1-\Omega^2)} \qquad \Omega_1\Omega_2 < 0 \quad \text{unstable}$$

$$-\sqrt{\tfrac{8}{3}}\sqrt{(1-\omega^2)} < a_0 < \sqrt{\tfrac{8}{3}}\sqrt{(1-\Omega^2)} \qquad \Omega_1\Omega_2 > 0 \quad \text{stable}$$

$$\sqrt{\tfrac{8}{3}}\sqrt{(1-\omega^2)} < a_0 < \sqrt{8}\sqrt{(1-\Omega^2)} \qquad \Omega_1\Omega_2 < 0 \quad \text{unstable}$$

$$a_0 > \sqrt{8}\sqrt{(1-\omega^2)} \qquad \Omega_1\Omega_2 > 0 \quad \text{stable}$$

To summarize, the periodic solutions are stable in the intervals AE, CD and EF, and unstable in the intervals BC and DE in Figure 7.1.

- **7.12** For the equation $\ddot{x} + \alpha x + \beta x^2 = \Gamma \cos t$, substitute $x = c(t) + a(t)\cos t + b(t)\sin t$, and show that, neglecting $\ddot{a}$ and $\ddot{b}$,
$$\dot{a} = \tfrac{1}{2}b(\alpha - 1 + 2\beta c), \quad \dot{b} = -\tfrac{1}{2}a(\alpha - 1 + 2\beta c) + \Gamma,$$
$$\ddot{c} = -\alpha c - \beta\{c^2 + \tfrac{1}{2}(a^2+b^2)\}.$$
Deduce that if $|\Gamma|$ is large there are no solutions of period 2π, and that if $\alpha < 1$ and Γ is sufficiently small there are two solutions of period 2π.

7.12. Substitute $x = c(t) + a(t)\cos t + b(t)\sin t$ into the differential equation

$$\ddot{x} + \alpha x + \beta x^2 = \Gamma \cos t,$$

and neglect the second derivatives $\ddot{a}$ and $\ddot{b}$. The result is

$$\ddot{c} + (2\dot{b} - a)\cos t - (2\dot{a} + b)\sin t + \alpha c + \alpha a \cos t + \alpha b \sin t$$
$$+ \beta(c^2 + a^2\cos^2 t + b^2\sin^2 t + 2ca\cos t + 2cb\sin t + 2ab\sin t\cos t) = \Gamma \cos t.$$

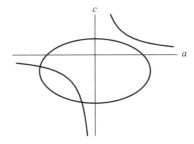

Figure 7.2 Problem 7.12: The diagram illustrates the intersection of an ellipse and a rectangular hyperbola.

Expanding $\cos^2 t$ and $\sin^2 t$, the translation and first harmonics balance if

$$\ddot{c} + \alpha c + \beta(c^2 + \tfrac{1}{2}a^2 + \tfrac{1}{2}b^2) = 0,$$
$$2\dot{b} - a + \alpha a + 2\beta ca = \Gamma,$$
$$-2\dot{a} - b + \alpha b + 2\beta cb = 0.$$

Equilibrium occurs where

$$\alpha c + \beta(c^2 + \tfrac{1}{2}a^2 + \tfrac{1}{2}b^2) = 0, \tag{i}$$
$$a(-1 + \alpha + 2\beta c) = \Gamma, \tag{ii}$$
$$b(-1 + \alpha + 2\beta c) = 0. \tag{iii}$$

Assuming $\Gamma \neq 0$, it follows from (ii) and (iii) that $b = 0$. Therefore (i) becomes

$$\alpha c + \beta\left(c^2 + \frac{1}{2}a^2\right) = 0, \text{ or } \beta a^2 + 2\beta\left(c + \frac{\alpha}{2\beta}\right)^2 = \frac{\alpha^2}{2\beta},$$

which is the equation of an ellipse in the (a, c) plane. Equation (ii) is the equation of a rectangular hyperbola with centre at $[0, (1 - \alpha)/(2\beta)]$. For sufficiently large Γ this hyperbola will not intersect the ellipse, since the ellipse is independent of Γ. Figure 7.2 shows the intersection of one branch of a rectangular hyperbola and an ellipse which occurs for Γ sufficiently small.

- **7.13** Substitute $x = c(t) + a(t)\cos t + b(t)\sin t$ into the equation $\ddot{x} + \alpha x^2 = 1 + \Gamma \cos t$ (compare Problem 7.8), and show that if $\ddot{a}$ and $\ddot{b}$ are neglected, then

$$2\dot{a} = b(2\alpha c - 1), \quad 2\dot{b} = a(1 - 2\alpha c) + \Gamma, \quad \ddot{c} + \alpha(c^2 + \tfrac{1}{2}a^2 + \tfrac{1}{2}b^2) = 1.$$

Use a graphical argument to show that there are two equilibrium points, when $\alpha < \tfrac{1}{4}$ and $\Gamma < \sqrt{(2/\alpha)}$.

7.13. Substitute $x = c(t) + a(t)\cos t + b(t)\sin t$ into the differential equation

$$\ddot{x} + \alpha x^2 = 1 + \Gamma \cos t,$$

and neglect the second derivatives $\ddot{a}$ and $\ddot{b}$. The result is

$$\ddot{c} + (2\dot{b} - a)\cos t - (2\dot{a} + b)\sin t + \alpha(c^2 + a^2\cos^2 t + b^2\sin^2 t + 2ca\cos t + 2cb\sin t$$
$$+ 2ab\sin t\cos t) = 1 + \Gamma\cos t$$

Expanding $\cos^2 t$ and $\sin^2 t$, the translation and first harmonics balance if

$$\ddot{c} + \alpha(c^2 + \tfrac{1}{2}a^2 + \tfrac{1}{2}b^2) = 1,$$
$$2\dot{b} = a - 2\alpha ca + \Gamma,$$
$$2\dot{a} = -b + 2\alpha bc,$$

as required. Equilibrium occurs where

$$\alpha(c^2 + \tfrac{1}{2}a^2 + \tfrac{1}{2}b^2) = 1, \tag{i}$$
$$a - 2\alpha ca + \Gamma = 0 \tag{ii}$$
$$-b + 2\alpha bc = 0. \tag{iii}$$

From (ii) and (iii), $b = 0$ (assuming $\Gamma \neq 0$) so that a and c satisfy

$$\alpha c^2 + \tfrac{1}{2}\alpha a^2 = 1, \tag{iv}$$
$$a(2\alpha c - 1) = \Gamma. \tag{v}$$

Equation (iv) represents an ellipse and (v) a rectangular hyperbola, and any equilibrium points occurs where (if at all) these curves intersect. The ellipse has semi-axes $\sqrt{(2/\alpha)}$ and $1/\sqrt{(\alpha)}$. The horizontal asymptote of the hyperbola is shown in Figure 7.3. If $1/(2\alpha) > 1/\sqrt{\alpha}$, or $\alpha < \tfrac{1}{4}$, then the curves will have at most two intersections. The lower branch of the hyperbola intersects the a axis at $a = -\Gamma$. Therefore there will be two solutions for (a,c), if $\Gamma \leq \sqrt{(2/\alpha)}$.

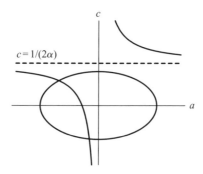

Figure 7.3 Problem 7.13: The diagram illustrates the intersection of an ellipse and a rectangular hyperbola.

7: Forced oscillations

- **7.14** In the forced Duffing equation $\ddot{x} + k\dot{x} + x - \frac{1}{6}x^3 = \Gamma \cos \omega t$, $(k > 0)$, substitute $x = a(t) \cos \omega t + b(t) \sin \omega t$ to investigate the solutions of period $2\pi/\omega$. Assume that a and b are slowly varying and that $k\dot{a}$, $k\dot{b}$ can be neglected. Show that the paths in the van der Pol plane are given by

$$\dot{a} = -\frac{b}{2\omega}\left\{\omega^2 - 1 + \frac{1}{8}(a^2 + b^2)\right\} - \frac{1}{2}ka,$$

$$\dot{b} = \frac{a}{2\omega}\left\{\omega^2 - 1 + \frac{1}{8}(a^2 + b^2)\right\} - \frac{1}{2}kb + \frac{\Gamma}{2\omega}.$$

Show that there is one equilibrium point if $\omega^2 > 1$.

Find the linear approximation in the neighbourhood of the equilibrium point when $\omega^2 > 1$, and show that it is a stable node or spiral when $k > 0$.

7.14. The forced Duffing equation is

$$\ddot{x} + k\dot{x} + x - \frac{1}{6}x^3 = \Gamma \cos \omega t. \tag{i}$$

In this case, let $x = a(t)\cos \omega t + b(t)\sin \omega t$, and assume that $\ddot{a}$, $\ddot{b}$, $k\dot{a}$ and $k\dot{b}$ are small in magnitude. Also, as in Problem 7.10,

$$x^3 = \tfrac{3}{4}ar^2 \cos \omega t + \tfrac{3}{4}br^2 \sin \omega t + \text{higher harmonics}.$$

where $r = \sqrt{(a^2 + b^2)}$. Substitute into (i) and equate to zero the coefficients of the first harmonics: the result is

$$\dot{a} = -\frac{b}{2\omega}\left[\omega^2 - 1 + \frac{1}{8}r^2\right] - \frac{1}{2}ka, \tag{ii}$$

$$\dot{b} = \frac{a}{2\omega}\left[\omega^2 - 1 + \frac{1}{8}r^2\right] - \frac{1}{2}kb + \frac{\Gamma}{2\omega}, \tag{iii}$$

as required.

Equilibrium occurs where

$$b[\omega^2 - 1 + \tfrac{1}{8}r^2] + ka\omega = 0, \tag{iv}$$

$$a[\omega^2 - 1 + \tfrac{1}{8}r^2] - kb\omega = -\Gamma. \tag{v}$$

Square and add these equations:

$$r^2[\omega^2 - 1 + \tfrac{1}{8}r^2]^2 + \omega^2 k^2 r^2 = \Gamma^2.$$

Let

$$f(r) = r^2[\omega^2 - 1 + \tfrac{1}{8}r^2]^2 + \omega^2 k^2 r^2$$

which will only be defined for $r \geq 0$. It follows that

$$f'(r) = 2r[\omega^2 - 1 + \tfrac{1}{8}r^2]^2 = \tfrac{1}{4}r^3[\omega^2 - 1 + \tfrac{1}{8}r^2] + 2\omega^2 k^2 r.$$

For $r > 0$ and $\omega^2 > 1$, $f'(r) > 0$. Therefore, for $\omega^2 > 1$, The equation $f(r) = \Gamma^2$ has only one solution.

Assume $\omega^2 > 1$. To linearize (ii) and (iii), let $a = a_0 + u$ and $b = b_0 + v$, where (a_0, b_0) is the only equilibrium point solution of (iv) and (v). Then the linearized approximations for u and v are

$$\dot{u} = -\frac{u}{2\omega}\left(\frac{1}{4}a_0 b_0 + k\omega\right) - \frac{v}{2\omega}\left(\omega^2 - 1 + \frac{1}{8}r_0^2 + \frac{1}{4}b_0^2\right),$$

$$\dot{v} = \frac{u}{2\omega}\left(\omega^2 - 1 + \frac{1}{8}r_0^2 + \frac{1}{4}a_0^2\right) + \frac{v}{2\omega}\left(\frac{1}{4}a_0 b_0 - k\omega\right).$$

The equilibrium point can be classified by the method of Section 2.5. In the usual notation

$$p = -k < 0, \quad q = \frac{1}{4\omega^2}\left[k^2\omega^2 + (\omega^2 - 1 + \tfrac{1}{8}r_0^2)^2 + \tfrac{1}{4}r_0^2\left(\omega^2 - 1 + \tfrac{1}{8}r_0^2\right)\right] > 0.$$

Hence the equilibrium point is either a stable node or spiral.

- **7.15** For the equation $\ddot{x} + \alpha x + \beta x^3 = \Gamma \cos \omega t$, show that the restoring force $\alpha x + \beta x^3$ is represented in the linear least-squares approximation on $-A \leq x \leq A$ by $(\alpha + \tfrac{3}{5}\beta A^2)x$. Obtain the general solution of the approximating equation corresponding to a solution of amplitude A. Deduce that there may be a subharmonic of order $\tfrac{1}{3}$ if $\alpha + \tfrac{3}{5}\beta A^2 = \tfrac{1}{9}\omega^2$ has a real solution A. Compare NODE, eqn (7.57) for the case when $\Gamma/(8\alpha)$ is small. Deduce that when $\alpha \approx \tfrac{1}{9}\omega^2$ (close to subharmonic resonance), the subharmonic has the approximate form

$$A\cos(\tfrac{1}{3}\omega t + \phi) - \frac{\Gamma}{8\alpha}\cos\omega t,$$

where ϕ is a constant.

(The interpretation is that when $\Gamma/(8\alpha)$ is small enough form the oscillation to lie in $[-A, A]$, A can be adjusted so that the slope of the straight-line fit on $[-A, A]$ is appropriate to the generation of a natural oscillation which is a subharmonic. The phase cannot be determined by this method.)

Show that the amplitude predicted for the equation $\ddot{x} + 0.15x - 0.1x^3 = 0.1\cos t$ is $A = 0.805$.

7.15. The undamped forced Duffing equation is

$$\ddot{x} + \alpha x + \beta x^3 = \Gamma \cos \omega t.$$

Consider the line $z = \mu x$. This becomes the least squares approximation to the restoring term $\alpha x + \beta x^3$ over the interval $(-A, A)$, if μ is given by the stationary value of

$$F(\mu) = \int_{-A}^{A} (\mu x - \alpha x - \beta x^3) dx$$

$$= \int_{-A}^{A} [(\mu - \alpha)^2 x^2 - 2(\mu - \alpha)\beta x^4 + \beta^2 x^6] dx$$

$$= \tfrac{2}{3}(\mu - \alpha)^2 A^3 - \tfrac{4}{5}(\mu - \alpha)\beta A^5 + \tfrac{2}{7}\beta^2 A^7.$$

Differentiating

$$F'(\mu) = \tfrac{4}{3}(\mu - \alpha) A^3 - \tfrac{4}{5} A^5,$$

so that $F'(\mu) = 0$ where $\mu = \tfrac{3}{5} A^2$. The least squares approximation is $z = (\alpha + \tfrac{3}{5}\beta A^2)x$. The equivalent linear equation using this approximation is

$$\ddot{x} + \Omega^2 x = \Gamma \cos \omega t, \quad \Omega = \sqrt{(\alpha + \tfrac{3}{5}\beta A^2)}.$$

The system could have a subharmonic if $\Omega^2 = \tfrac{1}{9}\omega^2$, or

$$\alpha + \tfrac{3}{5}\beta A^2 = \tfrac{1}{9}\omega^2.$$

A will have real solutions if $\alpha < \tfrac{1}{9}\omega^2$. Solving the linear equation, the subharmonic will have the approximate form

$$x = A\cos(\tfrac{1}{3}\omega t + \phi) + \frac{\Gamma}{\Omega^2 - \omega^2} \cos \omega t.$$

If $\alpha \approx \tfrac{1}{9}\omega^2$, then $\tfrac{3}{5}\beta A^2$ is small, and $\Omega^2 \approx \alpha$, so that, approximately,

$$x = A \cos(\tfrac{1}{3}\omega t + \phi) - \frac{\Gamma}{8\alpha} \cos \omega t.$$

The given parameter values are $\alpha = 0.15$, $\beta = -0.1$, $\Gamma = 0.1$ and $\omega = 1$. Then

$$A^2 = \frac{5}{27\beta}(\omega^2 - 9\alpha) = 0.648.$$

Hence $A = 0.805$.

- **7.16** Use the perturbation method to show that $\ddot{x} + k\dot{x} + \alpha x + \beta x^3 = \Gamma \cos \omega t$ has no subharmonic of order $\tfrac{1}{2}$ when β is small and $k = O(\beta)$. (Assume the expansion $(a \cos \tfrac{1}{2}\tau + b \sin \tfrac{1}{2}\tau + c \cos \tau)^3 = \tfrac{3}{4}c(a^2 - b^2) + \tfrac{3}{4}(a^2 + b^2 + 2c^2)$

$(a \cos \tfrac{1}{2}\tau + b \sin \tfrac{1}{2}\tau) +$ higher harmonics.)

7.16. In the Duffing equation

$$\ddot{x} + k\dot{x} + \alpha x + \beta x^3 = \Gamma \cos \omega t,$$

let $k = \kappa \beta$ and $\tau = \omega t$. The equation becomes

$$\omega^2 x'' + \kappa \omega \beta x' + \alpha x + \beta x^3 = \Gamma \cos \tau.$$

Let $x(\tau) = x_0 + \beta x_1 + \cdots$ and $\omega = \omega_0 + \beta \omega_1 + \cdots$, so that the differential equation becomes

$$(\omega_0 + \beta \omega_1 + \cdots)^2 (x_0'' + \beta x_1'' + \cdots) + \kappa \beta (\omega_0 + \cdots)(x_0' + \cdots)$$
$$+ \alpha(x_0 + \beta x_1 + \cdots) + \beta(x_0 + \beta x_1 + \cdots)^3 = \Gamma \cos \tau$$

The perturbation coefficients of β in the equation vanish individually if

$$\omega_0^2 x_0'' + \alpha \beta x_0 = \Gamma \cos \tau, \tag{i}$$

$$\omega_0^2 x_1'' + \alpha \beta x_1 = -2\omega_0 \omega_1 x_0'' - \kappa \omega_0 x_0' - x_0^3, \tag{ii}$$

etc. For a subharmonic of frequency $\tfrac{1}{2}$, it follows from all these equations that $\alpha = \tfrac{1}{4}\omega_0^2$. In this case, the general solution of (i) is

$$x_0 = a_{1/2} \cos \tfrac{1}{2}\tau + b_{1/2} \sin \tfrac{1}{2}\tau - \frac{4\Gamma}{3\omega^2} \cos \tau.$$

Using the identity given in the problem, we have

$$-2\omega_0\omega_1 x_0'' - \kappa \omega_0 x_0' - x_0^3 = -2\omega_0\omega_1 \left(-\tfrac{1}{4}a_{1/2} \cos \tfrac{1}{2}\tau - \tfrac{1}{4}b_{1/2} \sin \tfrac{1}{2}\tau\right)$$

$$- \kappa \omega_0 \left(-\tfrac{1}{2}a_{1/2} \sin \tfrac{1}{2}\tau + \tfrac{1}{2}b_{1/2} \cos \tfrac{1}{2}\tau\right) + \frac{\Gamma}{\omega_0^2}(a_{1/2}^2 - b_{1/2}^2)$$

$$+ \tfrac{3}{4}a_{1/2}\left(a_{1/2}^2 + b_{1/2}^2 + \frac{32\Gamma^2}{9\omega_0^4}\right)\cos \tfrac{1}{2}\tau + \tfrac{3}{4}b_{1/2}\left(a_{1/2}^2 + b_{1/2}^2 + \frac{32\Gamma^2}{9\omega_0^4}\right)$$

$\sin \tfrac{1}{2}\tau +$ (higher harmonics).

The secular term can be removed by putting the coefficients of $\cos \frac{1}{2}\tau$ and $\sin \frac{1}{2}\tau$ in the identity equal to zero, that is,

$$\tfrac{1}{2}\omega_0\omega_1 a_{1/2} - \tfrac{1}{2}\kappa\omega_0 b_{1/2} + \tfrac{3}{4}a_{1/2}\left(a_{1/2}^2 + b_{1/2}^2 + \frac{32\Gamma^2}{9\omega_0^4}\right) = 0, \qquad\text{(iii)}$$

$$\tfrac{1}{2}\omega_0\omega_1 b_{1/2} + \tfrac{1}{2}\kappa\omega_0 a_{1/2} + \tfrac{3}{4}b_{1/2}\left(a_{1/2}^2 + b_{1/2}^2 + \frac{32\Gamma^2}{9\omega_0^4}\right) = 0. \qquad\text{(iv)}$$

The difference $b_{1/2} \times$ (iii) $- a_{1/2} \times$ (iv) leads to

$$\tfrac{1}{2}\kappa\omega_0(b_{1/2}^2 + a_{1/2}^2) = 0,$$

which implies that the coefficients of subharmonic of order $\frac{1}{2}$ are both zero. In which case there can be no subharmonic of this order at least in this approximation.

- 7.17 Use the perturbation method to show that $\ddot{x} + k\dot{x} + \alpha x + \beta x^3 = \Gamma \cos \omega t$ has no subharmonic of order other than $\frac{1}{3}$ when β is small and $k = O(\beta)$.
(Use the identity

$$(a\cos \tfrac{1}{n}\tau + b\sin \tfrac{1}{n}\tau + c\cos\tau)^n = \tfrac{3}{4}(a^2 + b^2 + 2c^2)(a\cos\tau + b\sin\tau)$$

$$+ \text{higher harmonics}$$

for $n \neq 3$.)

7.17. Does the equation

$$\ddot{x} + k\dot{x} + \alpha x + \beta x^3 = \Gamma \cos \omega t$$

have a subharmonic of order other than $\frac{1}{3}$ when β is small and $k = O(\beta)$? Let $k = \kappa\beta$ and $\tau = \omega t$. The differential equation becomes

$$\omega^2 x'' + \kappa\omega\beta x' + \alpha x + \beta x^3 = \Gamma \cos \tau.$$

As in Problem 7.16, let $x(\tau) = x_0 + \beta x_1 + \cdots$ and $\omega = \omega_0 + \beta\omega_1 + \cdots$. Equations (i) and (ii) of Problem 7.16 are

$$\omega_0^2 x_0'' + \alpha x_0 = \Gamma \cos \tau, \qquad\text{(i)}$$

$$\omega_0^2 x_1'' + \alpha x_0 = -2\omega_0\omega_1 x_0'' - \kappa\omega_0 x_0' - x_0^3, \qquad\text{(ii)}$$

Look for subharmonics of order $1/n$, where $n \neq 3$. Let $\alpha = \omega_0^2/n^2$ so that

$$x_0 = a_{1/n} \cos \frac{1}{n}\tau + b_{1/n} \sin \frac{1}{n}\tau - \frac{\Gamma}{(n^2 - 1)\omega_0^2} \sin \tau.$$

Using the identity given, the right-hand side of (ii) becomes

$$-2\omega_0\omega_1 x_0'' - \kappa\omega_0 x_0' - x_0^3$$

$$= -2\omega_0\omega_1\left(-\frac{1}{n^2}a_{1/n}\cos\frac{1}{n}\tau - \frac{1}{n}b_{1/n}\sin\frac{1}{n}\tau\right) - \kappa\omega_0\left(-\frac{1}{n}a_{1/n}\sin\frac{1}{n}\tau\right.$$

$$\left. + \frac{1}{2}b_{1/n}\cos\frac{1}{n}\tau\right) + \frac{3}{4}a_{1/n}\left(a_{1/n}^2 + b_{1/n}^2 + \frac{2n^4\Gamma^2}{(n^2-1)^2\omega_0^4}\right)\cos\frac{1}{n}\tau$$

$$+ \frac{3}{4}b_{1/n}\left(a_{1/n}^2 + b_{1/n}^2 + \frac{2n^4\Gamma^2}{(n^2-1)^2\omega_0^4}\right)\sin\frac{1}{n}\tau + \text{(higher harmonics)}.$$

The secular term can be removed by putting the coefficients of $\cos\frac{1}{n}\tau$ and $\sin\frac{1}{n}\tau$ in the identity equal to zero, that is,

$$\frac{2}{n^2}\omega_0\omega_1 a_{1/n} - \frac{1}{n}\kappa\omega_0 b_{1/n} + \frac{3}{4}a_{1/n}\left(a_{1/n}^2 + b_{1/n}^2 + \frac{2n^4\Gamma^2}{(n^2-1)^2\omega_0^4}\right) = 0, \quad \text{(iii)}$$

$$\frac{2}{n^2}\omega_0\omega_1 b_{1/n} + \frac{1}{n}\kappa\omega_0 a_{1/n} + \frac{3}{4}b_{1/n}\left(a_{1/n}^2 + b_{1/n}^2 + \frac{2n^4\Gamma^2}{(n^2-1)^2\omega_0^4}\right) = 0. \quad \text{(iv)}$$

The difference $b_{1/2} \times$ (iii) $- a_{1/2} \times$ (iv) leads to

$$\tfrac{1}{2}\kappa\omega_0(b_{1/n}^2 + a_{1/n}^2) = 0,$$

which imply that $a_{1/n} = b_{1/n} = 0$. Therefore there are no subharmonics except when $n = 3$.

- **7.18** Look for subharmonics of order $\frac{1}{2}$ for the equation $\ddot{x} + \varepsilon(x^2 - 1)\dot{x} + x = \Gamma\cos\omega t$ using the perturbation method with $\tau = \omega t$.
  If $\omega = \omega_0 + \varepsilon\omega_1 + \cdots$, show that this subharmonic is only possible if $\omega_1 = 0$ and $\Gamma^2 < 18$. (Hint: let $x_0 = a\cos\frac{1}{2}\tau + b\sin\frac{1}{2}\tau - \frac{1}{3}\Gamma\cos\tau$, and use the expansion
  $(x_0^2 - 1)x_0' = \frac{1}{72}[-36 + 9(a^2 + b^2) + 2\Gamma^2](b\cos\frac{1}{2}\tau - a\sin\frac{1}{2}\tau)$
  $+ \text{(higher harmonics)}.)$

7.18. The forced van der Pol equation is

$$\ddot{x} + \varepsilon(x^2 - 1)\dot{x} + x = \Gamma\cos\omega t.$$

Apply the substitution $\omega t = \tau$ so that the differential equation becomes

$$\omega^2 x'' + \omega\varepsilon(x^2 - 1)x' + x = \Gamma \cos \tau.$$

Let $x = a \cos \omega t + b \sin \omega t$. Let $x = x_0 + \varepsilon x_1 + \cdots$ and $\omega = \omega_0 + \varepsilon \omega_1 + \cdots$. The first two terms x_0 and x_1 satisfy

$$\omega_0^2 x_0'' + x_0 = \Gamma \cos \tau, \qquad \text{(i)}$$

$$\omega_0^2 x_1'' + x_1 = -2\omega_0\omega_1 x_0'' - \omega_0(x_0^2 - 1)x_0'. \qquad \text{(ii)}$$

From (i) there could be a subharmonic of order $\tfrac{1}{2}$ if $\omega_0^2 = 4$, in which case the solution of (i) is

$$x_0 = a \cos \tfrac{1}{2}\tau + b \sin \tfrac{1}{2}\tau - \tfrac{1}{3}\Gamma \cos \tau.$$

The right-hand side of (ii) becomes

$$-2\omega_0\omega_1 x_0'' - \omega_0(x_0^2 - 1)x_0' = -2\omega_0\omega_1\left(-\tfrac{1}{4}a \cos \tfrac{1}{2}\tau - \tfrac{1}{4}b \sin \tfrac{1}{2}\tau\right)$$

$$- \omega_0\left\{\frac{b}{72}[-36 + 9(a^2 + b^2) + 2\Gamma^2]\cos \tfrac{1}{2}\tau + \frac{a}{72}[36 - 9(a^2 + b^2)\right.$$

$$\left. -2\Gamma^2]\sin \tfrac{1}{2}\tau\right\} + \text{(higher harmonics)}.$$

To remove secular terms the coefficients of $\cos \tfrac{1}{2}\tau$ and $\sin \tfrac{1}{2}\tau$ must be zero so that

$$a\omega_1 - \frac{b}{36}[-36 + 9(a^2 + b^2) + 2\Gamma^2] = 0,$$

$$b\omega_1 - \frac{a}{36}[36 - 9(a^2 + b^2) - 2\Gamma^2] = 0.$$

Hence the only solution is

$$\omega_1 = 0, \quad 36 - 9(a^2 + b^2) - 2\Gamma^2 = 0),$$

assuming $a^2 + b^2 \neq 0$. The amplitude $a^2 + b^2$ is only real if $\Gamma^2 < 18$.

- **7.19** Extend the analysis of the equation $\ddot{x} + \varepsilon(x^2 - 1)\dot{x} + x = \Gamma \cos \omega t$ in Problem 7.18 by assuming that
$$x = a(t) \cos \tfrac{1}{2}\omega t + b(t) \sin \tfrac{1}{2}\omega t - \tfrac{1}{3}\Gamma \cos \omega t,$$
where a and b are slowly varying. Show that when $\ddot{a}, \ddot{b}, \varepsilon \dot{a}, \varepsilon \dot{b}$, are neglected,
$$\tfrac{1}{2}\omega \dot{a} = (1 - \tfrac{1}{4}\omega^2)b - \tfrac{1}{8}\omega a(a^2 + b^2 + \tfrac{2}{9}\Gamma^2 - 4),$$
$$\tfrac{1}{2}\omega \dot{b} = -(1 - \tfrac{1}{4}\omega^2)a - \tfrac{1}{8}\omega b(a^2 + b^2 + \tfrac{2}{9}\Gamma^2 - 4),$$
in the van der Pol plane for the subharmonic.

By using $\rho = a^2 + b^2$ and ϕ the polar angle on the plane show that
$$\dot{\rho} = \tfrac{1}{4}\varepsilon\rho(\rho + K), \quad \dot{\phi} = -(1 - \tfrac{1}{4}\omega^2)/(2\omega), \quad K = \tfrac{2}{9}\Gamma^2 - 4.$$
Deduce that

(i) When $\omega \neq 2$ and $K \geq 0$, all paths spiral into the origin, which is the only equilibrium point (so no subharmonic exists).

(ii) When $\omega = 2$ and $K \geq 0$, all paths are radial straight lines entering the origin (so there is no subharmonic).

(iii) When $\omega \neq 2$ and $K < 0$, all paths spiral on to a limit cycle, which is a circle, radius $-K$ and centre the origin (so x is not periodic).

(iv) When $\omega = 2$ and $K < 0$, the circle center the origin and radius $-K$ consists entirely of equilibrium points, and all paths are radial straight lines approaching these points (each such point represents a subharmonic).

(Since subharmonics are expected only in case (iv), and for a critical value of ω, entrainment cannot occur. For practical purposes, even if the theory were exact we could never expect to observe the subharmonic, though solutions near to it may occur.)

7.19. The forced van der Pol equation is
$$\ddot{x} + \varepsilon(x^2 - 1)\dot{x} + x = \Gamma \cos \omega t.$$

Let
$$x = a(t) \cos \tfrac{1}{2}\omega t + b(t) \sin \tfrac{1}{2}\omega t - \tfrac{1}{3}\Gamma \cos \omega t.$$

Neglect the terms $\ddot{a}, \ddot{b}, \varepsilon \dot{a}$ and $\varepsilon \dot{b}$. Then
$$\ddot{x} \approx (-\tfrac{1}{4}a\omega^2 + \tfrac{1}{2}\dot{b}\omega) \cos \tfrac{1}{2}\omega t + (-\tfrac{1}{2}\dot{a}\omega - \tfrac{1}{4}b\omega^2) \sin \tfrac{1}{2}\omega t + \tfrac{1}{3}\omega^2 \cos \omega t.$$

Also
$$\dot{x} = (\dot{a} + \tfrac{1}{2}b\omega) \cos \tfrac{1}{2}\omega t + (-\tfrac{1}{2}a\omega + \dot{b}) \sin \tfrac{1}{2}\omega t + \tfrac{1}{3}\omega \Gamma \sin \omega t.$$
$$\approx \tfrac{1}{2}b\omega \cos \tfrac{1}{2}\omega t - \tfrac{1}{2}a\omega \sin \tfrac{1}{2}\omega t + \tfrac{1}{3}\Gamma \omega \sin \omega t$$

in the damping term. Also

$$
\varepsilon(x^2-1)\dot{x} \approx \varepsilon\left(-\frac{1}{2}b\omega + \frac{1}{8}a^2 b\omega + \frac{1}{8}b^3\omega + \frac{1}{36}b\omega\Gamma^2\right)\cos\frac{1}{2}\omega t
$$

$$
+\varepsilon\left(\frac{1}{2}a\omega - \frac{1}{8}a^3\omega - \frac{1}{8}ab^2\omega - \frac{1}{36}a\omega\Gamma^2\right)\sin\frac{1}{2}\omega t
$$

$+$ (higher harmonics)

We can now gather together the coefficients of $\cos\frac{1}{2}\omega t$ and $\sin\frac{1}{2}\omega t$ and put the coefficients equal to zero with the result

$$\tfrac{1}{2}\omega\dot{a} = b(1 - \tfrac{1}{4}\omega^2) - \tfrac{1}{8}\omega a\varepsilon(-4 + a^2 + b^2 + \tfrac{2}{9}\Gamma^2), \tag{i}$$

$$\tfrac{1}{2}\omega\dot{b} = -a(1 - \tfrac{1}{4}\omega^2) - \tfrac{1}{8}\omega b\varepsilon(-4 + a^2 + b^2 + \tfrac{2}{9}\Gamma^2). \tag{ii}$$

Let $\rho = a^2 + b^2$. Then, using (i) and (ii),

$$\dot{\rho} = 2a\dot{a} + 2b\dot{b} = -\tfrac{1}{8}\rho(\rho + K), \tag{iii}$$

where $K = \tfrac{2}{9}\Gamma^2 - 4$. Let $\tan\phi = b/a$. Then

$$\dot{\phi} = \frac{a\dot{b} - b\dot{a}}{\rho} = -\frac{(1 - \tfrac{1}{4}\omega^2)}{2\omega}. \tag{iv}$$

(i) If $\omega \neq 2$ and $K \geq 0$, then the polar form of the equations in the van der Pol plane implies that the origin is the only equilibrium point, which means that there can be no subharmonic in this case. From (iii), $\dot{\rho} < 0$ which implies that the radial distance decreases from any initial radius. For $0 < \omega < 2$, $\dot{\phi} < 0$, whilst for $\omega > 2$, $\dot{\phi} > 0$. Therefore the paths in the van der Pol are spirals into the origin, clockwise if $\omega > 2$ and counterclockwise if $\omega < 2$.

(ii) If $\omega = 2$ and $K \geq 0$, then $\dot{\phi} = 0$ and $\dot{\rho} < 0$. Therefore the polar angles are constant and $\dot{\rho} < 0$. Hence the paths are radial and the phase direction is towards the origin. Again there are no subharmonics.

(iii) If $\omega \neq 2$ and $K < 0$, eqn (iii) has the solution $\rho = -K$ which is a circle and limit cycle in the van der Pol equation. Since $\dot{\rho} < 0$ for $\rho > K$ and $\dot{\rho} > 0$ for $\rho < K$, the limit cycle is stable.

(iv) If $\omega = 2$ and $K < 0$, all points on the circle $\rho = -K$ are stable equilibrium points. Each point corresponds to a subharmonic.

- **7.20** Given eqns (7.34), (7.41), and (7.42) (in NODE) for the response curves and the stability boundaries for the van der Pol's equation (Figure 7.10 in NODE), eliminate r^2 to show that the boundary of the entrainment region in the γ, ν plane is given by

$$\gamma^2 = 8\{1 + 9\nu^2 - (1 - 3\nu^2)^{3/2}\}/27.$$

for $\nu^2 < \tfrac{1}{3}$. Show that, for small ν, $\gamma \approx \pm 2\nu$ or $\gamma \approx \pm\frac{2}{3\sqrt{3}}(1 - \tfrac{9}{8}\nu^2)$.

7.20. The equation and inequalities cited for the forced van der Pol equation are

$$r_0^2\{v^2 + (1 - \tfrac{1}{4}r_0^2)^2\} = \gamma^2, \tag{i}$$

$$\tfrac{3}{16}r_0^4 - r_0^2 + 1 + v^2 > 0, \tag{ii}$$

$$r_0^2 \geq 2. \tag{iii}$$

where (a_0, b_0) is an equilibrium point in the van der Pol plane, and $r_0 = \sqrt{(a_0^2 + b_0^2)}$. The boundary defined by (ii) is given by

$$r_0^2 = \tfrac{4}{3}[2 \pm \sqrt{(1 - 3v^2)}],$$

if $v^2 < \tfrac{1}{3}$. This equation has two positive roots (and therefore real solutions for r_0) say r_1 and r_2 where $0 < r_1^2 < r_2^2$. It follows that inequality (ii) is satisfied if $0 < r_0^2 < r_1^2$ or $r_0^2 > r_2^2$.
For $r_0 = r_2$,

$$\gamma^2 = \tfrac{4}{3}[2 + \sqrt{(1 - 3v^2)}][v^2 + 1 - \tfrac{2}{3}\{2 + \sqrt{(1 - 3v^2)}\} + \tfrac{1}{9}\{2 + \sqrt{(1 - 3v^2)}\}^2]$$

$$= \tfrac{4}{3}[2 + \sqrt{(1 - 3v^2)}][\tfrac{2}{3}v^2 + \tfrac{2}{9} - \tfrac{2}{9}\sqrt{(1 - 3v^2)}]$$

$$= \tfrac{8}{27}[9v^2 + 1 - (1 - 3v^2)^{3/2}].$$

If $r_0 = r_1$, then

$$\gamma^2 = \tfrac{8}{27}[9v^2 + 1 + (1 - 3v^2)^{3/2}].$$

Figure 7.4 shows the entrainment region in the (v, γ). I
If $|v|$ is small, then

$$\gamma^2 \approx \tfrac{8}{27}\left[9v^2 + 1 \pm \left(1 - \tfrac{9}{2}v^2\right)\right],$$

that is

$$\gamma^2 \approx 4v^2, \text{ or } \gamma^2 \approx \tfrac{4}{27}(4 - 9v^2).$$

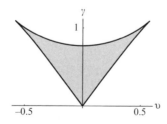

Figure 7.4 Problem 7.20: The shaded area is the entrainment region: the upper boundary is given by $\gamma^2 = \tfrac{8}{27}[9v^2 + 1 + \sqrt{(1 - 3v^2)}]$ and the lower boundary by $\gamma^2 = \tfrac{8}{27}[9v^2 + 1 - \sqrt{(1 - 3v^2)}]$.

Hence
$$\gamma \approx \pm 2v^2, \quad \text{or } \gamma \approx \frac{2}{3\sqrt{3}}\left(1 - \frac{9}{8}v^2\right).$$

• **7.21** Consider the equation $\ddot{x} + \varepsilon(x^2 + \dot{x}^2 - 1)\dot{x} + x = \Gamma \cos \omega t$. To obtain solutions of period $2\pi/\omega$, substitute $x = a(t)\cos \omega t + b(t)\sin \omega t$ and deduce that, if $\ddot{a}, \ddot{b}, \varepsilon\dot{a}, \varepsilon\dot{b}$ can be neglected, then
$\dot{a} = \frac{1}{2}\varepsilon\{a - vb - \frac{1}{4}\mu a(a^2 + b^2)\},$
$\dot{b} = \frac{1}{2}\varepsilon\{va + b - \frac{1}{4}\mu b(a^2 + b^2)\} + \frac{1}{2}\varepsilon\gamma,$
where
$\mu = 1 + 3\omega^2, \quad v = (\omega^2 - 1)/(\varepsilon\omega), \text{ and } \gamma = \Gamma/(\varepsilon\omega).$
Show that the stability boundaries are given by
$1 + v^2 - \mu r^2 + \frac{1}{16}\mu^2 r^4 = 0, \quad 2 - \mu r^2 = 0.$

7.21. The equation is
$$\ddot{x} + \varepsilon(x^2 + \dot{x}^2 - 1)\dot{x} + x = \Gamma \cos \omega t. \tag{i}$$

Let $x = a(t)\cos \omega t + b(t)\sin \omega t$. Then
$$\dot{x} = (\dot{a} + \omega b)\cos \omega t + (\dot{b} - a\omega)\sin \omega t,$$

and, neglecting $\ddot{a}$ and $\ddot{b}$,
$$\ddot{x} \approx (2\omega\dot{b} - a\omega^2)\cos \omega t - (2\omega\dot{a} + b\omega^2)\sin \omega t.$$

Substitution of these derivatives into (i) leads to

$(2\omega\dot{b} - a\omega^2)\cos \omega t - (2\omega\dot{a} + b\omega^2)\sin \omega t + \varepsilon[(a\cos \omega t + b\sin \omega t)^2$
$+ ((\dot{a} + \omega b)\cos \omega t + (\dot{b} - a\omega)\sin \omega t)^2 - 1][(\dot{a} + \omega b)\cos \omega t + (\dot{b} - a\omega)\sin \omega t)]$
$+ a\cos \omega t + b\sin \omega t = \Gamma \cos \omega t$

Neglecting $\varepsilon\dot{x}$ and $\varepsilon\dot{b}$, the equation simplifies to

$(2\omega\dot{b} - a\omega^2)\cos \omega t - (2\omega\dot{a} + b\omega^2)\sin \omega t + \varepsilon[(a\cos \omega t + b\sin \omega t)^2$
$+ (\omega b \cos \omega t - \omega a \sin \omega t)^2 - 1](\omega b\cos \omega t - a\omega \sin \omega t) + a\cos \omega t + b\sin \omega t$
$= \Gamma \cos \omega t.$

The expansion of the ε term is

$$\varepsilon[(a\cos\omega t + b\sin\omega t)^2 + (\omega b\cos\omega t - \omega a\sin\omega t)^2 - 1](\omega b\cos\omega t - a\omega\sin\omega t)$$
$$= \tfrac{1}{4}\varepsilon b\omega[-4\omega + (1+3\omega^2)(a^2+b^2)]\cos\omega t + \tfrac{1}{4}\varepsilon a\omega[4\omega - (1+3\omega^2)(a^2+b^2)]\sin\omega t$$
$$+ \text{(higher harmonics)}.$$

Substitute this into the previous equation and equate the coefficients of $\cos\omega t$ and $\sin\omega t$ to zero, from which the required equations for a and b follow:

$$\dot{a} = \tfrac{1}{2}\varepsilon[a - vb - \tfrac{1}{4}\mu a(a^2+b^2)],$$
$$\dot{b} = \tfrac{1}{2}\varepsilon[va + b - \tfrac{1}{4}\mu b(a^2+b^2)] + \tfrac{1}{2}\varepsilon\gamma,$$

where $\mu = 1 + 3v^2$, $v = (\omega^2 - 1)/(\varepsilon\omega)$ and $\gamma = \Gamma/(\varepsilon\omega)$.
Equilibrium in the van der Pol plane occurs where

$$a - vb - \tfrac{1}{4}\mu a(a^2+b^2) = 0, \qquad \text{(ii)}$$
$$va + b - \tfrac{1}{4}\mu b(a^2+b^2) = -\gamma. \qquad \text{(iii)}$$

The equations can be expressed in the form

$$(1 - \tfrac{1}{4}\mu r^2)a - vb = 0,$$
$$va + (1 - \tfrac{1}{4}\mu r^2)b = -\gamma,$$

where, after squaring and adding, r is given by

$$[(1 - \tfrac{1}{4}\mu r^2)^2 + v^2]r^2 = \gamma^2.$$

Let $a = a_0$ and $b = b_0$ be a solution of (i) and (ii), and consider the perturbation $a = a_0 + a_1$, $b = b_0 + b_1$, where $|a_1|$ and $|b_1|$ are small. Then, to the first order, a_1 and b_1 satisfy

$$\dot{a}_1 = (1 - \tfrac{3}{4}\mu a_0^2 - \tfrac{1}{4}\mu b_0^2)a_1 - v(1 + \tfrac{1}{2}\mu a_0 b_0)b_1,$$
$$\dot{b}_1 = v(1 - \tfrac{1}{2}a_0 b_0)a_1 + (1 - \tfrac{1}{4}\mu a_0^2 - \tfrac{3}{4}b_0^2)b_1.$$

Refer back to Chapter 2, Section 2.5. The solutions of the linearized equations are stable if, (in the notation of Section 2.5),

$$p = 2 - \mu r_0^2 < 0,$$

and

$$q = (1 - \tfrac{3}{4}\mu a_0^2 - \tfrac{1}{4}\mu b_0^2)(1 - \tfrac{1}{4}\mu a_0^2 - \tfrac{3}{4}\mu b_0^2)$$
$$+ (v - \tfrac{1}{2}\mu a_0 b_0)(v + \tfrac{1}{2}\mu a_0 b_0) > 0.$$

The latter inequality simplifies to

$$1 + v^2 - \mu r_0^2 + \frac{3}{16}\mu^2 r_0^4 > 0.$$

• **7.22** Show that the equation $\ddot{x}(1-x\dot{x})+(\dot{x}^2-1)\dot{x}+x = 0$ has an exact periodic solution $x = \cos t$. Show that the corresponding forced equation $\ddot{x}(1-x\dot{x})+(\dot{x}^2-1)\dot{x}+x = \Gamma \cos \omega t$ has an exact solution of the from $a\cos\omega t + b\sin\omega t$, where

$$a(1-\omega^2) - \omega b + \omega^3 b(a^2+b^2) = \Gamma, \quad b(1-\omega^2) + \omega a - \omega^3 a(a^2+b^2) = 0.$$

Deduce that the amplitude $r = \sqrt{(a^2+b^2)}$ satisfies

$$r^2\{(1-\omega^2)^2 + \omega^2(1-r^2\omega^2)^2\} = \Gamma^2.$$

7.22. Let $x = \cos t$. Then

$$L(x) \equiv \ddot{x}(1-x\dot{x}) + (\dot{x}^2-1)\dot{x} + x$$
$$= -\cos t(1+\sin t \cos t) + (\sin^2 t - 1)(-\sin t) + \cos t = 0,$$

which verifies that $x = \cos t$ is an exact solution.
  Consider now the forced equation

$$L(x) = \Gamma \cos \omega t.$$

If $x = a\cos\omega t + b\sin\omega t$, then

$$\begin{aligned}
L(x) - \Gamma \cos \omega t &= \omega^2(a\cos\omega t + b\sin\omega t)\\
&\quad \times [1 - (a\cos\omega t + b\sin\omega t)(-a\sin\omega t + b\cos\omega t)]\\
&\quad + \omega[\omega^2(-a\sin\omega t + b\cos\omega t)^2 - 1](-a\sin\omega t + b\cos\omega t)\\
&\quad + a\cos\omega t + b\sin\omega t - \Gamma \cos \omega t\\
&= (1-\omega^2)(a\cos\omega t + b\sin\omega t) - \omega(-a\sin\omega t + b\cos\omega t)\\
&\quad + \omega^2(\cos\omega t + b\sin\omega t)^2(-a\sin\omega t + b\cos\omega t)\\
&\quad + \omega^3(-\sin\omega t + b\cos\omega t)^3 - \Gamma \cos \omega t\\
&= (1-\omega^2)(a\cos\omega t + b\sin\omega t) - \omega(-a\sin\omega t + b\cos\omega t)\\
&\quad + \omega^3(-a\sin\omega t + b\cos\omega t)[(a\cos\omega t + b\sin\omega t)^2\\
&\quad + (-a\sin\omega t + b\cos\omega t)^2] - \Gamma \cos \omega t\\
&= (1-\omega^2)(a\cos\omega t + b\sin\omega t) - \omega(-a\sin\omega t + b\cos\omega t)\\
&\quad + \omega^3(a^2+b^2)(-a\sin\omega t + b\cos\omega t) - \Gamma \cos \omega t\\
&= 0
\end{aligned}$$

if
$$a(1-\omega^2) - \omega b + \omega^3 b(a^2+b^2) = \Gamma,$$
$$b(1-\omega^2) + \omega a - \omega^3 a(a^2+b^2) = 0.$$

These equations are equivalent to
$$r^2(1-\omega^2) = a\Gamma,$$
and
$$-\omega r^2 + \omega^3(a^2+b^2) = b\Gamma.$$

Square and add these two equations so that
$$r^2[(1-\omega^2)^2 + \omega^2(1-r^2\omega^2)^2] = \Gamma^2.$$

• 7.23 The frequency–amplitude relation for the damped forced pendulum is (eqn (7.23) in NODE), with $\beta = -\frac{1}{6}$) $r^2\{k^2\omega^2 + (\omega^2 - 1 + \frac{1}{8}r^2)^2\} = \Gamma^2$.
Show that the vertex of the cusp bounding the fold in NODE, Figure 7.7 occurs where
$$\omega = \tfrac{1}{2}\{\sqrt{(3k^2+4)} - k\sqrt{3}\}.$$
Find the corresponding value for Γ^2.

7.23. The frequency–amplitude relation for the damped forced pendulum is
$$r^2[k^2\omega^2 + (\omega^2 - 1 + \tfrac{1}{8}r^2)^2] = \Gamma^2.$$

As in NODE, Section 7.3, let $\rho = r^2/6$ and $\gamma = \Gamma/\sqrt{6}$, so that
$$\gamma^2 = G(\rho) = \rho[k^2\omega^2 + (\omega^2 - 1 + \tfrac{3}{4}\rho)^2]$$
$$= \frac{9}{16}\rho^3 + \frac{3}{2}(\omega^2 - 1)\rho^2 + [k^2\omega^2 + (\omega^2-1)^2]\rho$$

Its derivative is
$$G'(\rho) = \frac{27}{16}\rho^2 + 3(\omega^2-1)\rho + k^2\omega^2 + (\omega^2-1)^2. \tag{i}$$

The equation $G(\rho) = \gamma^2$ will have three real roots if $G'(\rho) = 0$ has two roots for $\rho \geq 0$. The solutions of this equation are
$$\rho_1, \rho_2 = \tfrac{8}{9}(1-\omega^2) \pm \tfrac{4}{9}\sqrt{[(1-\omega^2)^2 - 3k^2\omega^2]}.$$

The solutions are real and positive if
$$0 < \omega < 1 \text{ and } (1-\omega^2)^2 > 3k^2\omega^2.$$

The two inequalities are satisfied by

$$0 < \omega < \tfrac{1}{2}[\sqrt{(3k^2+4)} - k\sqrt{3}].$$

The cusp is located at

$$\omega = \tfrac{1}{2}[\sqrt{(3k^2+4)} - k\sqrt{3}].$$

For this value of ω, $\rho = \tfrac{8}{9}(1-\omega^2)$. Therefore

$$\Gamma^2 = 6\gamma^2 = 6G(\rho) = 6\rho[k^2\omega^2 + (\omega^2 - 1 + \tfrac{3}{4}\rho)^2]$$

$$= \frac{16}{3}(1-\omega^2)\left[k^2\omega^2 + \frac{1}{9}(1-\omega^2)^2\right]$$

$$= \frac{64\sqrt{3}}{9}k^3\omega^3,$$

where $\omega = \tfrac{1}{2}[\sqrt{(3k^2+4)} - k\sqrt{3}]$. In terms of k,

$$\Gamma^2 = \frac{8\sqrt{3}}{9}k^3[\sqrt{(3k^2+4)} - \sqrt{3}k]^3.$$

- **7.24** (Combination tones) Consider the equation $\ddot{x} + \alpha x + \beta x^2 = \Gamma_1 \cos\omega_1 t + \Gamma_2 \cos\omega_2 t$, $\alpha > 0$, $|\beta| \ll 1$, where the forcing term contains two distinct frequencies ω_1 and ω_2. To find an approximation to the response, construct the iterative process leading to the sequence of approximations $x^{(0)}(t), x^{(1)}(t), \ldots$, and starting with

$$\ddot{x}^{(0)} + \alpha x^{(0)} = \Gamma_1 \cos\omega_1 t + \Gamma_2 \cos\omega_2 t,$$
$$\ddot{x}^{(1)} + \alpha x^{(1)} = \Gamma_1 \cos\omega_1 t + \Gamma_2 \cos\omega_2 t - \beta(x^{(0)})^2,$$

show that a particular solution is given by approximately by

$$x(t) = -\frac{\beta}{2\alpha}(a^2 + b^2) + a\cos\omega_1 t + b\cos\omega_2 t + \frac{\beta a^2}{2(4\omega_1^2 - \alpha)}\cos 2\omega_1 t$$

$$+ \frac{\beta b^2}{2(4\omega_2^2 - \alpha)}\cos 2\omega_2 t + \frac{\beta ab}{(\omega_1+\omega_2)^2 - \alpha}\cos(\omega_1 + \omega_2)t$$

$$+ \frac{\beta ab}{(\omega_1-\omega_2)^2 - \alpha}\cos(\omega_1 - \omega_2)t,$$

where $a \approx \Gamma_1/(\alpha - \omega_1^2)$, $b \approx \Gamma_2/(\alpha - \omega_2^2)$.
(The presence of 'sum and difference tones' with frequencies $\omega_1 \pm \omega_2$ can be detected in sound resonators having suitable nonlinear characteristics, or as an auditory illusion attributed to the nonlinear detection mechanism in the ear (McLachlan 1956). The iterative method of solution can be adapted to simpler forced oscillation problems involving a single input frequency.)

7.24. Consider the equation

$$\ddot{x} + \alpha x + \beta x^2 = \Gamma_1 \cos \omega_1 t + \Gamma_2 \cos \omega_2 t, \quad \alpha > 0, \quad |\beta| \ll 1.$$

which has two forcing frequencies. Let $x^{(0)}$ be the first approximation, $x^{(1)}$ an improved approximation. Assume that the first approximation satisfies

$$\ddot{x}^{(0)} + \alpha x^{(0)} = \Gamma_1 \cos \omega_1 t + \Gamma_2 \cos \omega_2 t.$$

The forced solution is

$$x^{(0)} = a \cos \omega_1 t + b \cos \omega_2 t,$$

where

$$a = \frac{\Gamma_1}{\alpha - \omega_1^2}, \quad b = \frac{\Gamma_2}{\alpha - \omega_2^2}.$$

Assume that $x^{(0)}$ is an approximation to $x^{(1)}$ and use it in the αx^2 term. Hence $x^{(1)}$ satisfies

$$\ddot{x}^{(1)} + \alpha x^{(1)} = \Gamma_1 \cos \omega_1 t + \Gamma_2 \cos \omega_2 t - \beta x^{(0)^2}$$

$$= \Gamma_1 \cos \omega_1 t + \Gamma_2 \cos \omega_2 t - \beta (a \cos \omega_1 t + b \cos \omega_2 t)^2$$

$$= -\tfrac{1}{2}\beta(a^2 + b^2) + \Gamma_1 \cos \omega_1 t + \Gamma_2 \cos \omega_2 t - \tfrac{1}{2}\beta a^2 \cos 2\omega_1 t$$

$$- \tfrac{1}{2}\beta b^2 \cos 2\omega_2 t - \beta ab \cos(\omega_1 - \omega_2)t - \beta ab \cos(\omega_1 + \omega_2)t$$

This is a standard second-order linear differential equation with a constant and cosine forcing terms. Therefore

$$x^{(1)} = -\frac{\beta}{2\alpha}(a^2 + b^2) + a \cos \omega_1 t + b \cos \omega_2 t + \frac{\beta a^2}{2(4\omega_1^2 - \alpha)} \cos 2\omega_1 t$$

$$+ \frac{\beta b^2}{2(4\omega_2^2 - \alpha)} \cos 2\omega_2 t + \frac{\beta ab}{(\omega_1 + \omega_2)^2 - \alpha} \cos(\omega_1 + \omega_2)t$$

$$+ \frac{\beta ab}{(\omega_1 - \omega_2)^2 - \alpha} \cos(\omega_1 - \omega_2)t,$$

provided that α does not take any of the values $4\omega_1^2$, $4\omega_2^2$, $(\omega_1 + \omega_2)^2$, or $(\omega_1 - \omega_2)^2$.

- 7.25 Apply the method of Problem 7.24 to the Duffing equation
$$\ddot{x} + \alpha x + \beta x^3 = \Gamma_1 \cos \omega_1 t + \Gamma_2 \cos \omega_2 t.$$

7.25. Use the method of the previous problem for the Duffing equation

$$\ddot{x} + \alpha x + \beta x^3 = \Gamma_1 \cos \omega_1 t + \Gamma_2 \cos \omega_2 t.$$

The first approximation $x^{(0)}$ satisfies

$$\ddot{x}^{(0)} + \alpha x^{(0)} = \Gamma_1 \cos \omega_1 t + \Gamma_2 t,$$

which has the forced solution

$$x^0 = a \cos \omega_1 t + b \cos \omega_2 t,$$

where

$$a = \frac{\Gamma_1}{\alpha - \omega_1^2}, \quad b = \frac{\Gamma_2}{\alpha - \omega_2^2}.$$

The equation for the next approximation becomes

$$\ddot{x}^{(1)} + \alpha x^{(1)} = \Gamma_1 \cos \omega_1 t + \Gamma_2 \cos \omega_2 t - \beta(x^0)^3$$

$$= \Gamma_1 \cos \omega_1 t + \Gamma_2 \cos \omega_2 t - \beta(a \cos \omega_1 t + b \cos \omega_2 t)^3$$

$$= (\Gamma_1 - \tfrac{3}{4}\beta^3 - \tfrac{3}{2}\beta ab^2) \cos \omega_1 t + (\Gamma_2 - \tfrac{3}{2}\beta a^2 b - \tfrac{3}{4}\beta b^3) \cos \omega_2 t$$

$$- \tfrac{1}{4}\beta a^3 \cos 3\omega_1 t - \tfrac{1}{4}\beta b^3 \cos 3\omega_2 t - \tfrac{3}{4}\beta ab^2 \cos(\omega_1 - 2\omega_2)t$$

$$- \tfrac{3}{4}\beta a^2 b \cos(2\omega_1 - \omega_2)t - \tfrac{3}{4}\beta a^2 b \cos(2\omega_1 + \omega_2)t$$

$$- \tfrac{3}{4}\beta ab^2 \cos(\omega_1 + 2\omega_2)t.$$

The forced solution is

$$x^{(1)} = A \cos \omega_1 t + B \cos \omega_2 t + C \cos 3\omega_1 t + D \cos 3\omega_2 t + E \cos(\omega_1 - 2\omega_2)t$$
$$+ F \cos(2\omega_1 - \omega_2)t + G \cos(2\omega_1 + \omega_2)t + H \cos(\omega_1 + 2\omega_2)t,$$

where

$$A = \frac{\Gamma_1 - (3/4)\beta^3 - (3/2)\beta ab^2}{\alpha - \omega_1^2}, \quad B = \frac{\Gamma_2 - (3/2)\beta a^2 b - (3/4)\beta b^3}{\alpha - \omega_2},$$

$$C = \frac{\beta a^3}{4(9\omega_1^2 - \alpha)}, \quad D = \frac{\beta b^3}{4(9\omega_2^2 - \alpha)}, \quad E = \frac{3\beta ab^2}{(\omega_1 - 2\omega_2)^2 - \alpha},$$

$$F = \frac{3\beta a^2 b}{(2\omega_1 - \omega_2)^2 - \alpha}, \quad G = \frac{3\beta a^2 b}{(2\omega_1 + \omega_2)^2 - \alpha}, \quad H = \frac{3\beta ab^2}{(\omega_1 + 2\omega_2)^2 - \alpha}.$$

370 Nonlinear ordinary differential equations: problems and solutions

There are obvious conditions on α, ω_1 and ω_2 to avoid zeros of the denominators: in these cases the solutions will be covered by various special solutions.

- **7.26** Investigate the resonant solutions of Duffing's equation in the form
$$\ddot{x} + x + \varepsilon^3 x^3 = \cos t, \quad |\varepsilon| \ll 1,$$
by the method of multiple scales (Section 6.4 in NODE) using a slow time $\eta = \varepsilon t$ and a solution of the form
$$x(\varepsilon, t) = \frac{1}{\varepsilon} X(\varepsilon, t, \eta) = \frac{1}{\varepsilon} \sum_{n=0}^{\infty} \varepsilon^n X_n(t, \eta).$$
show that $X_0 = a_0(\eta) \cos t + b_0(\eta) \sin t$, where
$$8a_0' - 3b_0(a_0^2 + b_0^2) = 0, \quad 8b_0' + 3a_0(a_0^2 + b_0^2) = 4.$$
(This example illustrates that even a small nonlinear term may inhibit the growth of resonant solutions.)

7.26. The Duffing equation is
$$\ddot{x} + x + \varepsilon^3 x^3 = \cos t, \quad |\varepsilon| \ll 1.$$

Use the method of multiple scales and a solution of the form
$$x(\varepsilon, t) = \frac{1}{\varepsilon} X(\varepsilon, t, \eta) = \frac{1}{\varepsilon} \sum_{n=0}^{\infty} \varepsilon^n X_n(t, \eta).$$

In terms of X, the differential equation becomes
$$\frac{\partial^2 X}{\partial t^2} + 2\varepsilon \frac{\partial^2 X}{\partial \eta \partial t} + \varepsilon^2 \frac{\partial^2 X}{\partial \eta^2} + X + \varepsilon X^3 = \varepsilon \cos t.$$

Substitute the series into this equation and equate to zero the coefficients of ε so that X_0 and X_1 satisfy
$$X_{0tt} + X_0 = 0,$$
$$X_{1tt} + X_1 = -2X_{0\eta t} - X_0^3 + \cos t.$$
Therefore
$$X_0 = a_0(\eta) \cos t + b_0(\eta) \sin t.$$
The equation for X_1 becomes
$$X_{1tt} + X_1 = \cos t - 2[-a_0'(\eta) \sin t + b_0'(\eta) \cos t] - (a_0(\eta) \cos t + b_0(\eta) \sin t)^3. \tag{i}$$

Using the identity

$$(a_0 \cos t + b_0 \sin t)^3 = \tfrac{3}{4}a(a^2 + b^2)\cos t + \tfrac{3}{4}b(a^2 + b^2)\sin t + \text{higher harmonics},$$

eqn (i) is

$$X_{1tt} + X_1 = [1 - 2b_0' - \tfrac{3}{4}a_0(a_0^2 + b_0^2)]\cos t + \left[2a_0' - \tfrac{3}{4}b_0(a_0^2 + b_0^2)\right]\sin t$$
$$+ \text{higher harmonics}.$$

Secular terms disappear if the coefficients of $\cos t$ and $\sin t$ are zero, namely

$$1 - 2b_0' - \tfrac{3}{4}a_0(a_0^2 + b_0^2) = 0, \quad 2a_0' - \tfrac{3}{4}b_0(a_0^2 + b_0^2) = 0.$$

- 7.27 Repeat the multiple scale procedure of the previous exercise for the equation
$$\ddot{x} + x + \varepsilon^3 x^2 = \cos t, \quad |\varepsilon| \ll 1,$$
which has an unsymmetrical, quadratic departure from linearity. Use a slow time $\eta = \varepsilon^2 t$ and an expansion
$$x(\varepsilon, t) = \frac{1}{\varepsilon^2}\sum_{n=0}^{\infty} \varepsilon^n X_n(t, \eta).$$

7.27. Repeat the method of the previous problem for the equation

$$\ddot{x} + x\varepsilon^3 x^2 = \cos t.$$

However, in this case the slow time $\eta = \varepsilon^2 t$. Let

$$x(\varepsilon, t) = \frac{1}{\varepsilon^2}\sum_{n=0}^{\infty} \varepsilon^n X_n(t, \eta).$$

In terms of X, the differential equation becomes

$$\frac{\partial^2 X}{\partial t^2} + 2\varepsilon^2 \frac{\partial^2 X}{\partial \eta \partial t} + \varepsilon^4 \frac{\partial^2 X}{\partial \eta^2} + X + \varepsilon X^2 = \varepsilon^2 \cos t.$$

As will become clear, we require the first three terms in the expansion. The equations for X_0, X_1 and X_2 are

$$X_{0tt} + X_0 = 0, \qquad (i)$$

$$X_{1tt} + X_1 = -X_0^2, \qquad (ii)$$

$$X_{2tt} + X_2 = \cos t - 2X_{0tt} - 2X_0 X_1. \qquad (iii)$$

From (i)
$$X_0 = a_0(\eta) \cos t + b_0(\eta) \sin t.$$

Equation (ii) is therefore

$$X_{1tt} + X_1 = -(a_0 \cos t + b_0 \sin t)^2$$
$$= -\tfrac{1}{2}(a_0^2 + b_0^2) - \tfrac{1}{2}(a_0^2 - b_0^2) \cos 2t - a_0 b_0 \sin 2t.$$

The general solution of this equation is

$$X_1 = a_1 \cos t + b_1 \sin t - \tfrac{1}{2}(a_0^2 + b_0^2) + \tfrac{1}{3} a_0 b_0 \sin 2t + \tfrac{1}{6}(a_0^2 - b_0^2) \cos 2t,$$

where a_1 and b_1 are also functions of η.
Equation (iii) now becomes

$$X_{2tt} + X_2 = \cos t - 2(-a_0' \sin t + b_0' \cos t) - 2(a_0 \cos t + b_0 \sin t)$$
$$\times [a_1 \cos t + b_1 \sin t - \tfrac{1}{2}(a_0^2 + b_0^2) + \tfrac{1}{3} a_0 b_0 \sin 2t + \tfrac{1}{6}(a_0^2 - b_0^2) \cos 2t]$$
$$= -(a_0 a_1 + b_0 b_1) + \tfrac{1}{6}(-12 b_0' + 6 + 5 a_0^3 + 5 a_0 b_0^2) \cos t$$
$$+ \tfrac{1}{6}(12 a_0' + 5 a_0^2 b_0 + 5 b_0^3) \sin t + \text{higher harmonics}.$$

Finally secular terms do appear in X_2 if

$$a_0' = -\frac{5}{12} b_0 (a_0^2 + b_0^2), \quad b_0' = \frac{1}{2} + \frac{5}{12} a_0 (a_0^2 + b_0^2).$$

- **7.28** Let $\ddot{x} - x + bx^3 = c \cos t$. Show that this system has an exact subharmonic $k \cos \tfrac{1}{3} t$ if b, c, k satisfy
$$k = \frac{27}{10} c, \quad b = \frac{4c}{k^3}.$$

7.28. We have to show that, for what conditions, does

$$\ddot{x} - x + bx^3 = c \cos t$$

have the exact subharmonic $k \cos \frac{1}{3} t$. Substituting

$$\ddot{x} - x + bx^3 - c \cos t$$

$$= -\tfrac{1}{9} k \cos \tfrac{1}{3} t - k \cos \tfrac{1}{3} t + bk^3 \cos^3 \tfrac{1}{3} t - c \cos t$$

$$= -\tfrac{1}{9} k \cos \tfrac{1}{3} t - k \cos \tfrac{1}{3} t + \tfrac{3}{4} bk^3 \cos \tfrac{1}{3} t + \tfrac{1}{4} k^3 b \cos t - c \cos t$$

$$= 0$$

if

$$-\tfrac{1}{9} k - k + \tfrac{3}{4} bk^3 = 0, \text{ and } \tfrac{1}{4} bk^3 - c = 0.$$

Therefore $k = 27c/10$ and $b = 4c/k^3$.

- **7.29** Noting that $y = 0$ is a solution of the second equation in the forced system
$$\dot{x} = -x(1+y) + \gamma \cos t, \quad \dot{y} = -y(x+1),$$
obtain the forced periodic solution of the system.

7.29. The second equation in

$$\dot{x} = -x(1+y) + \gamma \cos t, \quad \dot{y} = -y(x+1),$$

obviously has the solution $y = 0$. For $y = 0$, the first equation becomes

$$\dot{x} = -x + \gamma \cos t.$$

Let $x = \alpha \cos t + \beta \sin t$. Then the first equation is satisfied if

$$-\alpha \sin t + \beta \cos t = -\alpha \cos t - \beta \sin t + \gamma \cos t,$$

that is, if

$$-\alpha = -\beta, \quad \beta = -\alpha + \gamma.$$

Therefore $\alpha = \beta = \tfrac{1}{2} \gamma$, so that

$$x = \tfrac{1}{2} \gamma (\cos t + \sin t).$$

- **7.30** Show that, if
$$\dot{x} = \alpha y \sin t - (x^2 + y^2 - 1)x, \quad \dot{y} = -\alpha x \sin t - (x^2 + y^2 - 1)y,$$
where $0 < \alpha < \pi$, then $2\dot{r} = (r^2 - 1)r$. Find r as a function of t, and show that $r \to 1$ as $t \to \infty$. Discuss the periodic oscillations which occur on the circle $r = 1$.

7.30. Consider the equations

$$\dot{x} = \alpha y \sin t - (x^2 + y^2 - 1)x,$$

$$\dot{y} = -\alpha x \sin t - (x^2 + y^2 - 1)y,$$

where $0 < \alpha < \pi$. Even though this is a forced system, it has an equilibrium point at the origin in the phase plane. It follows that

$$x\dot{x} + y\dot{y} = 2r\dot{r} = -(x^2 + y^2 - 1)x^2 - (x^2 + y^2 - 1)y^2$$

$$= -(r^2 - 1)r^2.$$

Separation of variables leads to

$$\int \frac{2\mathrm{d}r}{r(r^2-1)} = -\int \mathrm{d}t = -t + C.$$

Routine integration leads to the general solution

$$r^2 = \frac{1}{1 - e^{-(t+C)}} \quad (r^2 > 1), \quad r^2 = \frac{1}{1 + e^{-(t+C)}} \quad (r^2 < 1).$$

As $t \to \infty$, $r \to 1$ in both cases: the circle $r = 1$ is a closed path in the phase plane. If $r = 1$, then

$$\dot{x} = \alpha y \sin t, \quad \dot{y} = -\alpha x \sin t.$$

Let $x = \cos\theta$, $y = \sin\theta$. Then both equations become

$$\dot{\theta} = \alpha \sin,$$

which has the general solution $\theta = \alpha \cos t + B$. Therefore

$$x = \cos(\alpha \cos t + B), \quad y = \sin(\alpha \cos t + B).$$

If $x = 1$ at $t = 0$, then $1 = \cos(\alpha + B)$. Hence $B = -\alpha$. The solutions ar then

$$x = \cos(\alpha \cos t - \alpha), \quad y = \sin(\alpha \sin t - \alpha).$$

Solutions for x and y are shown in Figure 7.5 for the case $a = 1$ and the initial condition $x(0) = 1$.

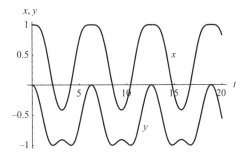

Figure 7.5 Problem 7.30: Solutions for x and y on the circle $r = 1$ with $a = 1$ and $x(0) = 1$ at $t - 0$.

- **7.31** Show that $\ddot{x}+kx+x^2 = \Gamma \cos t$ has an exact subharmonic of the form $x = A+B \cos \frac{1}{2}t$ provided $16k^2 > 1$. Find A and B.

7.31. Substitute $x = A + B \cos \frac{1}{2}t$ into the equation

$$\ddot{x} + kx + x^2 = \Gamma \cos t.$$

Then

$$\ddot{x} + kx + x^3 - \Gamma \cos t = -\tfrac{1}{4}B \cos \tfrac{1}{2}t + k(A + B \cos \tfrac{1}{2}t)$$
$$+(A + B \cos \tfrac{1}{2}t)^2 - \Gamma \cos t$$
$$= (-\tfrac{1}{4}B + Bk) \cos t + kA + A^2 + 2AB \cos \tfrac{1}{2}t$$
$$+\tfrac{1}{2}B^2(1 + \cos t) - \Gamma \cos t$$
$$= (-\tfrac{1}{4}B + Bk + 2AB) \cos \tfrac{1}{2}t + kA + A^2 + \tfrac{1}{2}B^2$$
$$+(\tfrac{1}{2}B^2 - \Gamma) \cos t = 0$$

if

$$B(-\tfrac{1}{4} + k + 2A) = 0, \quad kA + A^2 + \tfrac{1}{2}B^2 = 0, \quad \tfrac{1}{2}B^2 = \Gamma.$$

Therefore $A = \tfrac{1}{8} - \tfrac{1}{2}k$. Eliminating B^2 and A:

$$k(\tfrac{1}{8} - \tfrac{1}{2}k) + (\tfrac{1}{8} - \tfrac{1}{2}k)^2 + \Gamma = 0.$$

Hence $\Gamma = \tfrac{1}{64}(16k^2 - 1)$, and $B^2 = 2\Gamma = \tfrac{1}{32}(16k^2 - 1)$ provided $16k^2 > 1$. Subharmonics are of the form

$$x = (\tfrac{1}{8} - \tfrac{1}{2}k) \pm \tfrac{1}{4}\sqrt{(8k^2 - \tfrac{1}{2})} \cos \tfrac{1}{2}t.$$

- **7.32** Computed solutions of the particular two-parameter Duffing equation
$$\ddot{x} + k\dot{x} + x^3 = \Gamma \cos t$$
have been investigated in considerable detail by Ueda (1980). Using $x = a(t)\cos t + b(t)\sin t$, and assuming that $a(t)$ and $b(t)$ are slowly varying amplitudes, obtain the equations for $\dot{a}(t)$ and $\dot{b}(t)$ as in Section 7.2 (in NODE). Show that the response amplitude, r, and the forcing amplitude, Γ, satisfy $r^2\{k^2 + (1 - \tfrac{3}{4}r^2)^2\} = \Gamma^2$ for 2π-periodic solutions. By investigating the zeros of $d(\Gamma^2)/d(r^2)$, show that there are three response amplitudes if $0 < k < 1/\sqrt{3}$. Sketch this region in the (Γ, k) plane.

7.32. Apply the approximation $x = a(t)\cos t + b(t)\sin t$ to the Duffing equation

$$\ddot{x} + k\dot{x} + x^3 = \Gamma \cos t,$$

assuming that $\ddot{a}$ and $\ddot{b}$ and higher harmonics can be neglected. Using the result (see (7.14))

$$x^3 = \tfrac{3}{4}a(a^2 + b^2)\cos t + \tfrac{3}{4}b(a^2 + b^2)\sin t + \text{higher harmonics},$$

the amplitudes a and b satisfy, approximately,

$$(k\dot{a} + 2\dot{b} - a + bk + \tfrac{3}{4}ar^2)\cos t + (-2\dot{a} + k\dot{b} - ak - b + \tfrac{3}{4}br^2)\sin t = \Gamma \cos t.$$

where $r^2 = a^2 + b^2$. The coefficients of $\cos t$ and $\sin t$ vanish if

$$k\dot{a} + 2\dot{b} - a + bk + \tfrac{3}{4}ar^2 = \Gamma, \quad -2\dot{a} + k\dot{b} - ak - b + \tfrac{3}{4}br^2 = 0.$$

Equilibrium in the van der Pol plane occurs where

$$-a + bk + \tfrac{3}{4}ar^2 = \Gamma, \quad -ak - b + \tfrac{3}{4}br^2 = 0.$$

Square and add these equations to obtain the amplitude equation

$$r^2[k^2 + (1 - \tfrac{3}{4}r^2)^2] = \Gamma^2. \tag{i}$$

This equation expresses the relation between response and forcing frequencies and the damping coefficient. For some values of k and Γ, the response amplitude can take three values. To find where these occur, find where $d(\Gamma)^2/d(r^2) = 0$. Differentiating with respect to r^2,

$$\frac{d(\Gamma^2)}{d(r^2)} = k^2 + \left(1 - \tfrac{3}{4}r^2\right)^2 - \tfrac{3}{2}r^2\left(1 - \tfrac{3}{4}r^2\right),$$

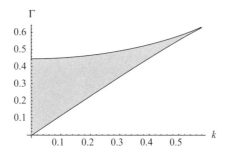

Figure 7.6 Problem 7.32.

which is zero where

$$r^4 - \frac{16}{9}r^2 + \frac{16}{27}(k^2 + 1) = 0.$$

The solutions are given by

$$r^2 = \tfrac{8}{9} \pm \tfrac{4}{9}\sqrt{(1 - 3k^2)}, \qquad \text{(ii)}$$

provided $k^2 \leq \tfrac{1}{3}$. Assuming that r, Γ and k are all positive, the system has three response amplitudes if $0 < k < 1/\sqrt{3}$ and one if $k > 1/\sqrt{3}$. The relations between Γ and k can be found by eliminating r between eqns (i) and (ii). Computed curves are shown in Figure 7.6. The cusp is located at $k = 1/\sqrt{3}$, where $r = \tfrac{2}{3}\sqrt{2}$ and $\Gamma = \tfrac{2\sqrt{2}}{9}\sqrt{(1 + 9k^2)}$. The shaded region indicates the three amplitude responses.

- 7.33 Show that there exists a Hamiltonian
$$H(x, y, t) = \tfrac{1}{2}(x^2 + y^2) - \tfrac{1}{4}\beta x^4 - \Gamma x \cos \omega t$$
for the undamped Duffing equation
$$\ddot{x} + x + \beta x^3 = \Gamma \cos \omega t, \quad \dot{x} = y \quad \text{(see eqn (7.4))}$$
Show also that the autonomous system for the slowly varying amplitudes a and b in the van der Pol plane (eqns (7.16) and (7.17)) is also Hamiltonian (see Section 2.8 in NODE). What are the implications for the types of equilibrium points in the van der Pol plane?

7.33. The Duffing equation can be expressed in the form

$$\dot{x} = X(x, y, t) = y, \quad \dot{y} = Y(x, y, t) = -x - \beta x^3 + \Gamma \cos \omega t,$$

As in Section 2.8 (with the extension to time-dependent functions) we can observe that

$$\frac{\partial X}{\partial x} + \frac{\partial Y}{\partial y} = 0,$$

which implies that the system is Hamiltonian. Since

$$X(x, y, t) = \frac{\partial H}{\partial y} = y,$$

it follows that

$$H(x, y, t) = \tfrac{1}{2}y^2 + F(x, t).$$

Hence

$$\frac{\partial F}{\partial x} = x + \beta x^3 - \Gamma \cos \omega t,$$

so that

$$F(x, t) = \tfrac{1}{2}x^2 + \tfrac{1}{4}\beta x^3 - \Gamma x \cos \omega t.$$

Finally the Hamiltonian is

$$H(x, y, t) = \tfrac{1}{2}(x^2 + y^2) + \tfrac{1}{4}\beta x^4 - \Gamma \cos \omega t.$$

From (7.16) and (7.17) the equations for a and b in the van der Pol plane are

$$\dot{a} = -\frac{b}{2\omega}\{(\omega^2 - 1) - \frac{3}{4}\beta(a^2 + b^2)\} \equiv A(a, b),$$

$$\dot{b} = \frac{a}{2\omega}\{(\omega^2 - 1) - \frac{3}{4}(a^2 + b^2)\} + \frac{\Gamma}{2\omega} \equiv B(a, b).$$

Then

$$\frac{\partial A}{\partial a} + \frac{\partial B}{\partial b} = \frac{3\beta ba}{4\omega} - \frac{3\beta ab}{4\omega} = 0.$$

Therefore the system in the van der Pol plane is also Hamiltonian (see Section 2.8). The implication of this result is that the equilibrium points in the van der Pol plane must be either centres or saddle points, or higher-order versions of centres or saddle points

- **7.34** Show that the exact solution of the equation $\ddot{x} + x = \Gamma \cos \omega t$, $(\omega \neq 1)$ is
$$x(t) = A \cos t + B \sin t + \frac{\Gamma}{1 - \omega^2} \cos \omega t,$$
where A and B are arbitrary constants.
Introduce the van der Pol variables $a(t)$ and $b(t)$ through
$$x(t) = a(t) \cos \omega t + b(t) \sin \omega t,$$
and show that $x(t)$ satisfies the differential equation if $a(t)$ and $b(t)$ satisfy
$$\ddot{a} + 2\omega \dot{b} + (1 - \omega^2)a = \Gamma, \quad \ddot{b} - 2\omega \dot{a} + (1 - \omega^2)b = 0.$$
Solve these equations for a and b by combining them into an equation in $z = a + ib$. Solve this equation, and confirm that, although the equations for a and b contain four constants, these constants combine in such a way that the solution for x still contains just two arbitrary constants.

7.34. The equation
$$\ddot{x} + x = \Gamma \cos \omega t, \quad (\omega \neq 1), \tag{i}$$
has the characteristic equation $m^2 + 1 = 0$, and the complementary function
$$x_f = A \cos t + B \sin t.$$

A particular solution is
$$x_p = \frac{\Gamma}{1 - \omega^2} \cos \omega t.$$

Hence the general solution is
$$x = x_f + x_p = A \cos t + B \sin t + \frac{\Gamma}{1 - \omega^2} \cos \omega t.$$

Let $x(t) = a(t) \cos \omega t + b(t) \sin \omega t$. Then
$$\dot{x} = (\dot{a} + \omega b) \cos \omega t + (\dot{b} - \omega a) \sin \omega t, \tag{ii}$$
$$\ddot{x} = (\ddot{a} + 2\omega \dot{b} - \omega^2 a) \cos \omega t + (\ddot{b} - 2\omega \dot{a} - \omega^2 b) \sin \omega t. \tag{iii}$$

Substitute (ii) and (iii) into (i) and equate to zero the coefficients of $\cos \omega t$ and $\sin \omega t$ with the results
$$\ddot{a} + 2\omega \dot{b} - \omega^2 a + a = \Gamma,$$
$$\ddot{b} - 2\omega \dot{a} - \omega^2 b + b = 0.$$

Let $z = a + ib$ so that z satisfies

$$\ddot{z} - 2\omega i \dot{z} + (1 - \omega^2)z = \Gamma.$$

The characteristic equation

$$\lambda^2 - 2\omega i \lambda + (1 - \omega^2) = 0,$$

has the solutions $\lambda = (\omega \pm 1)i$. A particular solution is

$$z_p = \frac{\Gamma}{1 - \omega^2}.$$

Therefore the general solution is

$$z = a + ib = A_1 e^{(\omega+1)it} + B_1 e^{(\omega-1)it}. \tag{iv}$$

Let $A_1 = \alpha_1 + i\alpha_2$ and $B_1 = \beta_1 + i\beta_2$. From (iv)

$$a = \alpha_1 \cos(\omega + 1)t - \alpha_2 \sin(\omega + 1)t + \beta_1 \cos(\omega - 1)t - \beta_2 \sin(\omega - 1)t + \frac{\Gamma}{1 - \omega^2},$$

$$b = \alpha_2 \cos(\omega + 1)t + \alpha_1 \sin(\omega + 1)t + \beta_2 \cos(\omega - 1)t + \beta_1 \sin(\omega - 1)t.$$

Finally

$$\begin{aligned}
x &= a \cos \omega t + b \sin \omega t \\
&= \alpha_1[\cos(\omega + 1)t \cos \omega t + \sin(\omega + 1)t \sin \omega t] \\
&\quad + \alpha_2[-\sin(\omega + 1)t \cos \omega t + \cos(\omega + 1)t \sin \omega t] \\
&\quad + \beta_1[\cos(\omega - 1)t \cos \omega t + \sin(\omega - 1)t \sin \omega t] \\
&\quad + \beta_2[-\sin(\omega - 1)t \cos \omega t + \cos(\omega - 1)t \sin \omega t] + \frac{\Gamma}{1 - \omega^2} \cos \omega t \\
&= \alpha_1 \cos t - \alpha_2 \sin t + \beta_1 \cos t - \beta_2 \sin t + \frac{\Gamma}{1 - \omega^2} \cos \omega t \\
&= (\alpha_1 + \beta_1) \cos t - (\alpha_2 + \beta_2) \sin t + \frac{\Gamma}{1 - \omega^2} \cos \omega t
\end{aligned}$$

In the final line $\alpha_1 + \beta_1$ are $\alpha_2 + \beta_2$ are the arbitrary constants A and B in the original solution.

• **7.35** Show that the system
$$\ddot{x} + (k - x^2 - \dot{x}^2)\dot{x} + \beta x = \Gamma \cos t, \quad (k, \Gamma > 0, \beta \neq 1),$$
has exact harmonic solutions of the form $x(t) = a \cos t + b \sin t$, if the amplitude $r = \sqrt{(a^2 + b^2)}$ satisfies
$$r^2[(\beta - 1)^2 + (k - r^2)^2] = \Gamma^2.$$
By investigating the solutions of $d(\Gamma^2)/d(r^2) = 0$, show that there are three harmonic solutions for an interval of values of Γ if $k^2 > 3(\beta - 1)^2$. Find this interval if $k = \beta = 2$. Draw the amplitude diagram r against Γ in this case.

7.35. Let $x = a \cos t + b \sin t$. Then

$$\ddot{x} + (k - x^2 - \dot{x}^2)\dot{x} + \beta x - \Gamma \cos t$$
$$= -a \cos t - b \sin t + [k - (a \cos t + b \sin t)^2 - (-a \sin t + b \cos t)^2]$$
$$\times (-a \sin t + b \cos t) + \beta(a \cos t + b \sin t) - \Gamma \cos t$$
$$= [-a + b(k - r^2) + \beta a - \Gamma] \cos t + [-b - a(k - r^2) + \beta b] \sin t = 0$$

if

$$b(k - r^2) + (\beta - 1)a = \Gamma,$$
$$-a(k - r^2) + (\beta - 1)b = 0.$$

Squaring and adding it follows that

$$r^2[(\beta - 1)^2 + (k - r^2)^2] = \Gamma^2,$$

as required.
The derivative

$$\frac{d(\Gamma^2)}{d(r^2)} = (\beta - 1)^2 + (k - r^2)^2 - 2r^2(k - r^2)$$

is zero where

$$3r^4 - 4kr^2 + k^2 + (\beta - 1)^2 = 0.$$

This quadratic equation in r^2 has the solutions

$$r_1^2, r_2^2 = \tfrac{1}{3}[2k \pm \sqrt{\{k^2 - 3(\beta - 1)^2\}}].$$

which are real and positive if $k^2 > 3(\beta - 1)^2$.

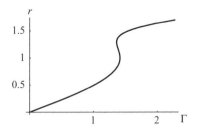

Figure 7.7 Problem 7.35: Showing the graph of $\Gamma^2 = r^2[1 + (2 - r^2)^2]$.

If $k = \beta = 2$, then
$$\Gamma^2 = r^2[1 + (2 - r^2)^2].$$
Figure 7.7 shows the relation between Γ and r.

- 7.36 Show that the equation $\ddot{x} + k\dot{x} - x + \omega^2 x^2 + \dot{x}^2 = \Gamma \cos \omega t$ has exact solutions of the form $x = c + a \cos \omega t + b \sin \omega t$, where the translation c and the amplitude $r = \sqrt{(a^2 + b^2)}$ satisfy
$$\Gamma^2 = r^2[\{1 + \omega^2(1 - 2c)\}^2 + k^2\omega^2] \text{ and } \omega^2 r^2 = c(1 - c\omega^2).$$
Sketch a graph showing response amplitude r against the forcing amplitude Γ.

7.36. Let $x = c + a \cos \omega t + b \sin \omega t$. Then

$$\ddot{x} + k\dot{x} - x + \omega^2 x^2 + \dot{x}^2 - \Gamma \cos \omega t$$
$$= -a\omega^2 \cos \omega t - b\omega^2 \sin \omega t - ka\omega \sin \omega t + kb\omega \cos \omega t$$
$$\quad - c - a \cos \omega t - b \sin \omega t + \omega^2(c + a \cos \omega t + b \sin \omega t)^2$$
$$\quad + (-\omega a \sin \omega t + b \cos \omega t)^2 - \Gamma \cos \omega t$$
$$= (-a\omega^2 + bk\omega - a + 2ac\omega^2 - \Gamma) \cos \omega t + (-b\omega^2 - ak\omega - b + 2bc\omega^2)$$
$$\quad \times \sin \omega t + \omega^2(a^2 + b^2 + c^2) - c = 0$$

if

$$-a[1 + (1 - 2c)\omega^2] + \omega k b = \Gamma, \tag{i}$$
$$-\omega k a - b[1 + (1 - 2c)\omega^2] = 0, \tag{ii}$$
$$\omega^2(r^2 + c^2) - c = 0, \tag{iii}$$

where $r = \sqrt{(a^2 + b^2)}$. Squaring and adding (i) and (ii), we have

$$\Gamma^2 = r^2[\{1 + \omega^2(1 - 2c)\}^2 + k^2\omega^2]. \tag{iv}$$

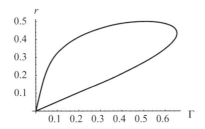

Figure 7.8 Problem 7.36: Amplitude(r)–amplitude(Γ) relation defined by (iv) and (v) with $\omega = 1$ and $k = 0.2$.

From (iii),
$$r = \frac{1}{\omega}\sqrt{[c(1-c\omega^2)]}. \tag{v}$$

which is real for $0 \leq c \leq 1/\omega^2$. Using (iv) and (v) both Γ and r can be expressed in terms of c, which is used to plot r against Γ with c as a parameter in Figure 7.8 with $\omega = 1$ and $k = 0.2$. The figure shows that for small forcing amplitudes, the system has two (exact) forced periodic solutions.

8 Stability

Poincaré or orbital stability is defined by Definition 8.1. A general descriptive or graphical approach is adopted in Problems 1–4, which are concerned with Poincaré stability.

- **8.1** Use the phase diagram for the pendulum equation $\ddot{x} + \sin x = 0$, to say which paths are not Poincaré stable. (See Figure 1.2 in NODE.)

8.1. Figure 8.1 shows the phase diagram for the pendulum equation

$$\ddot{x} + \sin x = 0, \quad \dot{x} = y.$$

Consider the stability of a typical closed path $\mathcal{P}_1$, within any strip bounded by two nearby closed paths (shown shaded). All half-paths starting in the strip remain in it for all time, so $\mathcal{P}_1$ is Pincaré stable. The same applies to any typical path $\mathcal{P}_2$ in the region describing (periodic) whirling motion beyond the separatrices. The separatrices are not Poincaré stable, since there are neighbouring half-paths that deviate unboundedly from any separatrix.

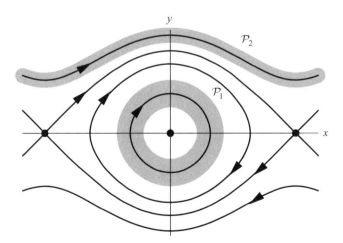

Figure 8.1 Problem 1.1(iv): Phase diagram for the pendulum equation $\ddot{x} + \sin x = 0$.

- 8.2 Show that all the paths of $\dot{x} = x$, $\dot{y} = y$ are Poincaré unstable.

8.2. The phase paths of $\dot{x} = x$, $\dot{y} = y$ are given by $y = Cx$, a family of straight lines through the origin as shown in Figure 8.2. All paths diverge from all neighbouring paths starting from any initial point. Therefore no paths are Poincaré or orbitally stable.

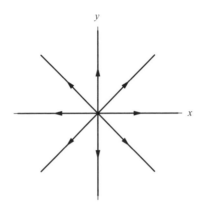

Figure 8.2 Problem 8.2: Phase diagram of $\dot{x} = x$, $\dot{y} = y$.

- 8.3 Find the limit cycles of the system
$$\dot{x} = -y + x \sin r, \quad \dot{y} = x + y \sin r, \quad r = \sqrt{(x^2 + y^2)}.$$
Which cycles are Poincaré stable?

8.3. Express the equations
$$\dot{x} = -y + x \sin r, \quad \dot{y} = x + y \sin r, \quad r = \sqrt{(x^2 + y^2)},$$
in polar coordinates, so that
$$\dot{r} = r \sin r, \quad \dot{\theta} = 1.$$

Limit cycles of the system are given by $r = n\pi$, $(n = 1, 2, \ldots)$ as shown in Figure 8.3. For $(2n - 1)\pi < r < 2n\pi$, $\dot{r} < 0$, which means that r is decreasing: for $2n\pi < r < (2n + 1)\pi$, $\dot{r} > 0$ and r is increasing. Since $\dot{\theta} = 1$ solutions on all paths progress progress at a constant rate in a counterclockwise sense about the origin. Hence all the limit cycles with radius $(2n + 1)\pi$ are Poincaré stable since, for example, with $n = 1$, any path which starts in the circle C will subsequently remain in the shaded strip.

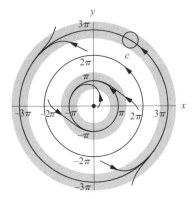

Figure 8.3 Problem 8.3: Phase diagram for $\dot{x} = -y + x\sin r$, $\dot{y} = x + y\sin r$, $r = \sqrt{(x^2 + y^2)}$.

- 8.4 Find the phase paths for $\dot{x} = x$, $\dot{y} = y \ln y$, in the half-plane $y > 0$. Which paths are Poincaré stable?

8.4. The system
$$\dot{x} = x, \quad \dot{y} = y \ln y, \quad (y > 0)$$
has equilibrium points at $(0, 0)$ and $(1, 1)$. The separable differential equation for the phase paths is
$$\frac{dy}{dx} = \frac{y \ln y}{x},$$
which has the general solution
$$\ln|\ln y| = \ln|x| + C, \quad \text{or} \quad y = e^{Ax}.$$

The phase diagram is shown in Figure 8.4. For $A < 0$, the phase paths all approach the x axis as $x \to \infty$, that is they all converge to one another. Hence they are all Poincaré stable. For $A \geq 0$, all the paths diverge: hence these paths are all unstable.

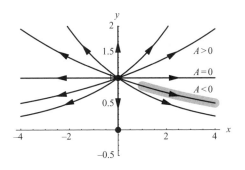

Figure 8.4 Problem 8.4: Phase diagram for $\dot{x} = x$, $\dot{y} = y \ln y$.

- **8.5** Show that every non-zero solution of $\dot{x} = x$ is unbounded and Liapunov unstable, but that every solution of $\dot{x} = 1$ is unbounded and stable.

8.5. Consider the one-dimensional system $\dot{x} = x$. Its general solution is $x = Ae^t$. If $A \neq 0$, the solution is clearly unbounded. Let $x(t) = A_1 e^t$ and $x^*(t) = A_2 e^t$ be two solutions with A_1 and A_2 both non-zero and $A_1 \neq A_2$. Then (adapting NODE, (8.3) to the one-dimensional case)

$$\| x(t) - x^*(t) \| = |x(t) - x^*(t)| = |A_1 - A_2|e^t \to \infty,$$

as $t \to \infty$. Therefore all non-zero solutions are unstable in the Liapunov sense.

The system $\dot{x} = 1$ has the general solution $x(t) = t + x(0)$: all solutions are unbounded. Consider the stability of $x^*(t) = t + x^*(0)$. Then

$$\| x^*(t) - x(t) \| = |x^*(t) - x(t)| = |x^*(0) - x(0)|.$$

Given any $\varepsilon > 0$,

$$|x^*(0) - x(0)| < \varepsilon \Rightarrow |x^*(t) - x(t)| < \varepsilon \text{ for } t > 0.$$

By Definition 8.2 with $\delta = \varepsilon$, all solutions are Liapunov stable.

- **8.6** Show that the solutions of the system $\dot{x} = 1$, $\dot{y} = 0$, are Poincaré and Liapunov stable, but that the system $\dot{x} = y$, $\dot{y} = 0$ is Poincaré but not Liapunov stable.

8.6. The system $\dot{x} = 1$, $\dot{y} = 0$ has the general solution

$$x(t) = t + x(0), \quad y(t) = y(0).$$

The phase diagram is shown in Figure 8.5. Consider the shaded strip which contains one of the solutions. Any neighbouring solution which starts within the shaded region will subsequently

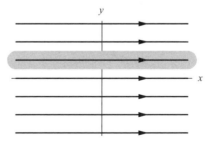

Figure 8.5 Problem 8.6: Phase diagram of $\dot{x} = 1$, $\dot{y} = 0$.

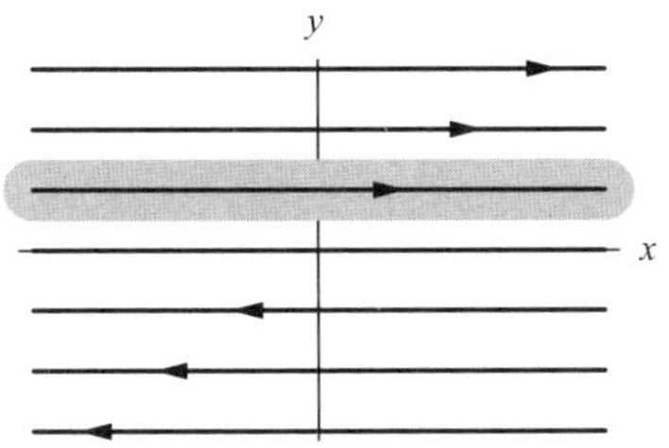

Figure 8.6 Problem 8.6: Phase diagram of $\dot{x} = y$, $\dot{y} = 0$.

stay within the shaded region. Therefore all solutions are Poincaré stable. Consider the stability of $\mathbf{x}^*(t) = [x^*(t), y^*(t)]$. In NODE, Definition 8.2, let $t_0 = 0$ (note that the system is autonomous). Then

$$\| \mathbf{x}(0) - \mathbf{x}^*(0) \| = \| \mathbf{x}(t) - \mathbf{x}^*(t) \| = \sqrt{[(x(0) - x^*(0))^2 + (y(0) - y(0))^2]}.$$

Therefore, given any $\varepsilon > 0$,

$$\| \mathbf{x}^*(0) - \mathbf{x}(0) \| < \varepsilon \Rightarrow \| \mathbf{x}^*(t) - \mathbf{x}(t) \| < \varepsilon \text{ for } t > 0,$$

which implies, with $\delta = \varepsilon$, that all solutions are Liapunov stable.

The system $\dot{x} = y$, $\dot{y} = 0$ has the general solution

$$x(t) = y(0)t + x(0), \quad y(t) = y(0).$$

Also all points on the x axis are equilibrium points. The phase diagram is shown in Figure 8.6. The phase paths are the same as those of the first part of the problem but the sense and phase speed are different. For the same reasons as for the previous system all solutions except the equilibrium states along $y = 0$ are Poincaré stable. For Liapunov stability consider

$$\| \mathbf{x}^*(t) - \mathbf{x}(t) \| = \sqrt{[(y^*(0)t - y(0)t + x^*(0) - x(0))^2 + (y^*(0) - y(0))^2]},$$

which is unbounded in t. Hence all solutions are Liapunov unstable.

- 8.7 Solve the equations $\dot{x} = -y(x^2 + y^2)$, $\dot{y} = x(x^2 + y^2)$, and show that the zero solution is Liapunov stable and that all other solutions are unstable.
  Replace the coordinates x, y by r, ϕ where $x = r\cos(r^2 t + \phi)$, $y = r\sin(r^2 t + \phi)$ and deduce that $\dot{r} = 0$, $\dot{\phi} = 0$. Show that in this coordinate system the solutions are stable. (Change of coordinates can affect the stability of a system. (See Cesari (1971, p. 12).)

8.7. Consider the system

$$\dot{x} = -y(x^2 + y^2), \quad \dot{y} = x(x^2 + y^2).$$

The phase paths are given by
$$\frac{dy}{dx} = -\frac{x}{y},$$
which has the general solution $x^2 + y^2 = c^2$: the phase paths are circles centred at the origin. Note that the origin is an equilibrium point. Substituting back into the system, we have
$$\dot{x} = -c^2 y, \quad \dot{y} = c^2 x.$$
Elimination of y leads to $\ddot{x} + c^4 x = 0$, which has the general solution
$$x = \alpha \cos(c^2 t + \beta), \text{ and } y = \alpha \sin(c^2 t + \beta),$$
where α, c and β are constants. However the three constants are not independent since
$$x^2 + y^2 = \alpha^2 = c^2.$$
Therefore the general solution is given by
$$x(t) = c\cos(c^2 t + \beta), \quad y(t) = c\sin(c^2 t + \beta). \tag{i}$$

Observe that
$$||\mathbf{x}(t)|| = \sqrt{[x^2(t) + y^2(t)]} = c. \tag{ii}$$
The system is autonomous, so we need only consider the solutions for $t \geq t_0$ when $t_0 = 0$ (see Definition 8.2(ii)).

Consider first the stability of the constant solution
$$\mathbf{x}^* = \mathbf{0} = (0, 0).$$
Choose any $\varepsilon > 0$. Then, for all t,
$$||\mathbf{x}^*(t) - \mathbf{x}(t)|| = ||\mathbf{0} - \mathbf{x}(t)|| = c, \tag{iii}$$
and in particular
$$||\mathbf{x}^*(0) - \mathbf{x}(0)|| = c. \tag{iv}$$
Now choose $0 < c < \varepsilon$, and in NODE, Definition 8.2, eqn (8.5), put
$$\delta = \varepsilon \tag{v}$$
for this case, given any $\varepsilon > 0$, we have:
$$\text{if } ||\mathbf{x}^* - \mathbf{x}(0)|| < \delta, \text{ then } ||\mathbf{x}^*(t) - \mathbf{x}(t)|| < \varepsilon \text{ for all } t.$$
This proves the Liapunov stability of the constant solution $(0,0)$.

For other solutions, consider a solution which starts at (x_0, y_0). Then $x_0 = c\cos\varepsilon$ and $y_0 = c\sin\varepsilon$. Hence $c^2 = x_0^2 + y_0^2$. The time taken for the solution to make one circuit of the origin is $2\pi/(x_0^2 + y_0^2)$, which depends on the initial value. Hence there will always be solutions which start close together but do not remain so. Hence all non-zero solutions are not Liapunov stable.

Consider the change of variable $(x, y) \to (r, \phi)$ defined by

$$x = r\cos(r^2 t + \phi), \quad y = r\sin(r^2 t + \phi).$$

Obviously $r^2 = x^2 + y^2$, so that

$$r\dot{r} = x\dot{x} + y\dot{y} = 0,$$

and, since r must be a constant in the equation for $\dot{x}$,

$$-r\sin(r^2 t + \phi)(r^2 + \dot{\phi}) = -r\sin(r^2 t + \phi)r^2, \quad \text{or } \dot{\phi} = 0.$$

The general solution is $r = r_0$, $\phi = \phi_0$. All solutions are Liapunov stable since

$$\| \mathbf{x}^*(t) - \mathbf{x}(t) \| = \text{a constant}.$$

- **8.8** Prove that Liapunov stability of a solution implies Poincaré stability for plane autonomous systems, but not conversely: see Problem 8.6.

8.8. Briefly, NODE, Definition 8.2 for Liapunov stability for plane autonomous systems states that given any $\varepsilon > 0$, there exists a $\delta(\varepsilon) > 0$ such that

$$\| \mathbf{x}(0) - \mathbf{x}^*(0) \| < \delta \Rightarrow \| \mathbf{x}(t) - \mathbf{x}^*(t) \| < \varepsilon, \qquad \text{(i)}$$

for $t \geq 0$, where $\mathbf{x}(t)$ represents any neighbouring solution. In the notation of NODE, Definition 8.1 (for orbital stability), let $\mathbf{a} = \mathbf{x}(0)$, $\mathbf{a}^* = \mathbf{x}^*$ and $\mathcal{H}^*$ be the half-path in the phase plane defined by $\mathbf{x}^*$ for $t \geq 0$. Let

$$q = \max_{\mathbf{x} \in \mathcal{H}} \text{dist}(\mathbf{x}, \mathcal{H}^*).$$

Then in (i),

$$\| \mathbf{x}(t) - \mathbf{x}^*(t) \| \leq q < \varepsilon,$$

which establishes that Liapunov stability implies orbital stability.

The counter-example in Problem 8.6 shows that the converse cannot be true: we can have Poincaré stability without Liapunov stability.

- **8.9** Determine the stability of the solutions of
  (i) $\dot{x}_1 = x_2 \sin t$, $\dot{x}_2 = 0$;
  (ii) $\dot{x}_1 = 0$, $\dot{x}_2 = x_1 + x_2$.

8.9. (i) The equations $\dot{x}_1 = x_2 \sin t$, $\dot{x}_2 = 0$ can be expressed as

$$\begin{bmatrix} \dot{x}_1 \\ \dot{x}_2 \end{bmatrix} = \begin{bmatrix} 0 & \sin t \\ 0 & 0 \end{bmatrix} \begin{bmatrix} x_1 \\ x_2 \end{bmatrix}.$$

The general solution is given by

$$x_1 = -A \cos t + B, \quad x_2 = A,$$

and its norm is clearly bounded. By NODE, Theorems 8.9 and 8.1 the solution is Liapunov stable.

(ii) The equations of the autonomous system $\dot{x}_1 = 0$, $\dot{x}_2 = x_1 + x_2$ has the general solution

$$x_1 = A, \quad x_2 = -A + Be^t.$$

By Theorem 8.1, we need only consider the zero solution $0 = (0,0)$. Consider the solution

$$\mathbf{x}^*(t) = [x_1^*(0), -x_1^*(0) + (x_1^*(0) + x_2^*(0))e^t],$$

which starts close to the origin. Then

$$\| 0 - \mathbf{x}^*(t) \| = \sqrt{[x_1^*(0)]^2 + \{-x_1^*(0) + (x_1^*(0) + x_2^*(0))e^t\}^2]},$$

which is clearly unbounded as $t \to \infty$. Hence the system is not Liapunov stable.

The phase paths are straight lines parallel to the x_2 axis taken in the same sense. Hence the paths are Poincaré stable.

- **8.10** Determine the stability of the solutions of

(i) $\begin{bmatrix} \dot{x}_1 \\ \dot{x}_2 \end{bmatrix} = \begin{bmatrix} -2 & 1 \\ 1 & -2 \end{bmatrix} \begin{bmatrix} x_1 \\ x_2 \end{bmatrix} + \begin{bmatrix} 1 \\ -2 \end{bmatrix} e^t$

(ii) $\ddot{x} + e^{-t}\dot{x} + x = e^t$

8.10. (i) By Theorem 8.1, instead of

$$\begin{bmatrix} \dot{x}_1 \\ \dot{x}_2 \end{bmatrix} = \begin{bmatrix} -2 & 1 \\ 1 & -2 \end{bmatrix} \begin{bmatrix} x_1 \\ x_2 \end{bmatrix} + \begin{bmatrix} 1 \\ -2 \end{bmatrix} e^t,$$

we need only consider

$$\begin{bmatrix} \dot\xi_1 \\ \dot\xi_2 \end{bmatrix} = \begin{bmatrix} -2 & 1 \\ 1 & -2 \end{bmatrix} \begin{bmatrix} \xi_1 \\ \xi_2 \end{bmatrix}.$$

The eigenvalues of the matrix are $\lambda_1 = -1$ and $\lambda_2 = -3$. Therefore all solutions are asymptotically stable, which implies that the original system is also asymptotically stable.

(ii) Express the equation

$$\ddot x + e^{-t}\dot x + x = e^t,$$

in the matrix form

$$\begin{bmatrix} \dot x \\ \dot y \end{bmatrix} = \begin{bmatrix} 0 & 1 \\ -1 & -e^{-t} \end{bmatrix} \begin{bmatrix} x \\ y \end{bmatrix} + \begin{bmatrix} 0 \\ e^{-t} \end{bmatrix}.$$

By NODE, Theorem 8.1, we need only consider the zero solution of

$$\begin{bmatrix} \dot\xi \\ \dot\eta \end{bmatrix} = \begin{bmatrix} 0 & 1 \\ -1 & -e^{-t} \end{bmatrix} \begin{bmatrix} \xi \\ \eta \end{bmatrix}.$$

In the notation of NODE, Theorem 8.15, let

$$\mathbf{A} = \begin{bmatrix} 0 & 1 \\ -1 & 0 \end{bmatrix}, \quad \mathbf{C}(t) = \begin{bmatrix} 0 & 0 \\ 0 & -e^{-t} \end{bmatrix}.$$

The norm of the matrix $\mathbf{C}(t)$ is (see (8.21))

$$\| \mathbf{C}(t) \| = \sqrt{\{e^{-2t}\}} = e^{-t}.$$

Then,

$$\int_{t_0}^{t} \| \mathbf{C}(s) \| \, ds = \int_{t_0}^{t} e^{-s} ds = -e^{-t} + e^{-t_0},$$

which is bounded. Therefore by the Corollary to Theorem 8.15 all solutions are stable.

- **8.11** Show that every solution of the system $\dot x = -t^2 x$, $\dot y = -ty$ is asymptotically stable.

8.11. By Theorem 8.1, the Liapunov stability of any solution of the system

$$\dot x = -t^2 x, \quad \dot y = -ty$$

is the same as that of the zero solution. The general solution of the system is

$$x = Ae^{(1/3)t^3}, \quad y = Be - \tfrac{1}{2}t^2.$$

Then $x \to 0$ and $y \to 0$ as $t \to \infty$. Therefore the origin is asymptotically stable as are all solutions.

- **8.12** The motion of a heavy particle on a smooth surface of revolution with vertical axis z and shape $z = f(r)$ in cylindrical polar coordinates is

$$\frac{1}{r^4}\{1+f'^2(r)\}\frac{d^2r}{d\theta^2} + \frac{1}{r^5}[rf'(r)f''(r) - 2\{1+f'^2(r)\}]\left(\frac{dr}{d\theta}\right)^2 - \frac{1}{r^3} = -\frac{g}{h^2}f'(r),$$

where h is the angular momentum ($h = r^2\dot\theta$). Show that plane, horizontal motion $r = a$, $z = f(a)$, is stable for perturbations leaving h unaltered provided $3 + [af''(a)/f'(a)] > 0$.

8.12. The equation can be rewritten as

$$r\{1+f'^2(r)\}\frac{d^2r}{d\theta^2} + [rf'(r)f''(r) - 2\{1+f'^2(r)\}]\left(\frac{dr}{d\theta}\right)^2 - r^2 = -\frac{g}{h^2}r^5 f'(r).$$

In equilibrium $r = a$, which means that

$$-a^2 = -\frac{g}{h^2}f'(a)a^5 \text{ or } h^2 = ga^3 f'(a),$$

which remains constant. Substitute for h^2, and consider the perturbation $r = a + \rho$ in the differential equation. The linearization leads to

$$(a+\rho)[1+f'^2(a)]\rho'' - a^2 - 2a\rho \approx -\frac{a^2}{f'(a)}[f'(a) + f''(a)\rho],$$

or

$$a[1+f'^2(a)]\rho'' + \rho\left[3a + \frac{a^2 f''(a)}{f'(a)}\right] = 0.$$

The solution for ρ is bounded and therefore stable if

$$3 + \frac{af''(a)}{f'(a)} > 0.$$

- **8.13** Determine the linear dependence or independence of the following:
  (i) $(1, 1, -1), (2, 1, 1), (0, 1, -3)$;
  (ii) $(t, 2t), (3t, 4t), (5t, 6t)$;
  (iii) $(e^t, e^{-t}), (e^{-t}, e^t)$. Could these both be solutions of a 2×2 homogeneous linear systems?

8.13. (i) The vectors $(1, 1, -1), (2, 1, 1)$ and $(0, 1, -3)$ are linearly independent since

$$\begin{vmatrix} 1 & 1 & -1 \\ 2 & 1 & 1 \\ 0 & 1 & -3 \end{vmatrix} = 0.$$

(ii) The equations

$$\alpha_1 \begin{bmatrix} t \\ 2t \end{bmatrix} + \alpha_2 \begin{bmatrix} 3t \\ 4t \end{bmatrix} + \alpha_3 \begin{bmatrix} 5t \\ 6t \end{bmatrix} = 0,$$

have the non-zero solution $\alpha_1 = 2$, $\alpha_2 = 1$ and $\alpha_3 = -1$ for all t. Then by Definition 8.5, the vectors are linearly dependent.

(iii) The only solution of the equations

$$\alpha_1 \begin{bmatrix} e^t \\ e^{-t} \end{bmatrix} + \alpha_2 \begin{bmatrix} e^{-t} \\ e^t \end{bmatrix} = 0$$

is $\alpha_1 = \alpha_2 = 0$. Hence the vectors are linearly independent.

- **8.14** Construct a fundamental matrix Φ for the system $\dot{x} = y$, $\dot{y} = -x - 2y$. Deduce a fundamental matrix Ψ satisfying $\Psi(0) = I$.

8.14. The system $\dot{x} = y$, $\dot{y} = -x - 2y$ is equivalent to

$$\ddot{x} + 2\dot{x} + x = 0.$$

the solution of the characteristic equation is the repeated root $x = -1$. Therefore, a general solution is

$$\begin{bmatrix} x \\ y \end{bmatrix} = \begin{bmatrix} A \\ -A + B \end{bmatrix} e^{-t} + B \begin{bmatrix} 1 \\ -1 \end{bmatrix} te^{-t}.$$

Hence a fundamental matrix is (e.g. put $B = 0$ and put $A = 0$)

$$\Phi(t) = \begin{bmatrix} e^{-t} & te^{-t} \\ -e^{-t} & (1-t)e^{-t} \end{bmatrix}.$$

Use NODE, Theorem 8.6, the required fundamental matrix is

$$\Psi(t) = \Phi(t)\Phi^{-1}(0)$$

$$= \begin{bmatrix} e^{-t} & te^{-t} \\ -e^{-t} & (1-t)e^{-t} \end{bmatrix} \begin{bmatrix} 1 & 0 \\ -1 & 1 \end{bmatrix}^{-1}$$

$$= \begin{bmatrix} e^{-t} & te^{-t} \\ -e^{-t} & (1-t)e^{-t} \end{bmatrix} \begin{bmatrix} 1 & 0 \\ 1 & 1 \end{bmatrix}$$

$$= \begin{bmatrix} (1+t)e^{-t} & te^{-t} \\ -te^{-t} & (1-t)e^{-t} \end{bmatrix}$$

- **8.15** Construct a fundamental matrix for the system $\dot{x}_1 = -x_1$, $\dot{x}_2 = x_1 + x_2 + x_3$, $\dot{x}_3 = -x_2$.

8.15. In matrix form the system can be expressed as $\dot{x}_1 = -x_1$, $\dot{x}_2 = x_1 + x_2 + x_3$, $\dot{x}_3 = -x_2$ as

$$\dot{x} = Ax, \quad A = \begin{bmatrix} -1 & 0 & 0 \\ 1 & 1 & 1 \\ 0 & -1 & 0 \end{bmatrix}.$$

The eigenvalues of A are given by

$$\begin{vmatrix} -1-\lambda & 0 & 0 \\ 1 & 1-\lambda & 1 \\ 0 & -1 & -\lambda \end{vmatrix} = -(\lambda+1)(\lambda^2 - \lambda + 1) = 0.$$

Therefore the eigenvalues are given by $\lambda = -1$ and $\lambda = \frac{1}{2} \pm \frac{\sqrt{3}}{2}i$. Corresponding eigenvectors are given by:

(i) $\lambda = -1$. Let $\mathbf{u} = (u_1\ u_2\ u_3)^T$. Then

$$[A - \lambda I] = \begin{bmatrix} 0 & 0 & 0 \\ 1 & 2 & 1 \\ 0 & -1 & 1 \end{bmatrix} \begin{bmatrix} u_1 \\ u_2 \\ u_3 \end{bmatrix} = 0.$$

Choose the solution $u_1 = -3$, $u_2 = 1$, $u_3 = 1$.

(ii) $\lambda = \frac{1}{2} + \frac{\sqrt{3}}{2}i$. Let $\mathbf{v} = (v_1\ v_2\ v_3)^T$. Then

$$[\mathbf{A} - \lambda \mathbf{I}] = \begin{bmatrix} \frac{3}{2} - \frac{\sqrt{3}}{2}i & 0 & 0 \\ 1 & \frac{1}{2} - \frac{\sqrt{3}}{2}i & 1 \\ 0 & -1 & -\frac{1}{2} - \frac{\sqrt{3}}{2}i \end{bmatrix} \begin{bmatrix} v_1 \\ v_2 \\ v_3 \end{bmatrix} = 0.$$

Choose the solution $v_1 = 0$, $v_2 = \frac{1}{2} + \frac{\sqrt{3}}{2}i$, $v_3 = -1$.

(iii) $\lambda = \frac{1}{2} - \frac{\sqrt{3}}{2}i$. Let $\mathbf{w} = (w_1\ w_2\ w_3)$. We can choose an eigenvector which is the conjugate of $\mathbf{v}$ in (ii), namely $w_1 = 0$, $w_2 = \frac{1}{2} - \frac{\sqrt{3}}{2}i$, $w_3 = -1$.

Finally a fundamental matrix is (see Definition 8.6)

$$\Phi(t) = \begin{bmatrix} e^{-t} & 0 & 0 \\ e^{-t} & \left(\frac{1}{2} + \frac{\sqrt{3}}{2}i\right) e^{\left(\frac{1}{2} + \frac{\sqrt{3}}{2}i\right)t} & \left(\frac{1}{2} - \frac{\sqrt{3}}{2}i\right) e^{\left(\frac{1}{2} - \frac{\sqrt{3}}{2}i\right)t} \\ -3e^{-t} & -e^{\left(\frac{1}{2} + \frac{\sqrt{3}}{2}i\right)t} & -e^{\left(\frac{1}{2} - \frac{\sqrt{3}}{2}i\right)t} \end{bmatrix}.$$

- **8.16** Construct a fundamental matrix for the system $\dot{x}_1 = x_2$, $\dot{x}_2 = x_1$, and deduce the solution satisfying $x_1 = 1$, $\dot{x}_2 = 0$, at $t = 0$.

8.16. The system $\dot{x}_1 = x_2$, $\dot{x}_2 = x_1$ can be expressed in the form

$$\dot{\mathbf{x}} = \mathbf{A}\mathbf{x}, \text{ where } \mathbf{A} = \begin{bmatrix} 0 & 1 \\ 1 & 0 \end{bmatrix}.$$

The eigenvalues of $\mathbf{A}$ are given by $\lambda = \pm 1$. Corresponding eigenvectors are $\mathbf{r} = (1, 1)^T$ and $\mathbf{s} = (1, -1)^T$. Hence a fundamental matrix is

$$\Phi(t) = \begin{bmatrix} e^t & e^{-t} \\ e^t & -e^{-t} \end{bmatrix}.$$

By NODE, Theorem 8.6, the required solution is

$$\mathbf{x}(t) = \Phi(t)\Phi^{-1}(0)\mathbf{x}_0,$$

where $\mathbf{x}_0 = (1, 0)^T$. Therefore

$$\mathbf{x}(t) = \begin{bmatrix} e^t & e^{-t} \\ e^t & -e^{-t} \end{bmatrix} \begin{bmatrix} 1 & 1 \\ 1 & -1 \end{bmatrix}^{-1} \begin{bmatrix} 1 \\ 0 \end{bmatrix}$$

Therefore

$$\mathbf{x}(t) = \begin{bmatrix} e^t & e^{-t} \\ e^t & -e^{-t} \end{bmatrix} \begin{bmatrix} \frac{1}{2} & \frac{1}{2} \\ \frac{1}{2} & -\frac{1}{2} \end{bmatrix}^{-1} \begin{bmatrix} 1 \\ 0 \end{bmatrix} = \begin{bmatrix} \cosh t \\ \sinh t \end{bmatrix}.$$

- **8.17** Construct a fundamental matrix for the system $\dot{x}_1 = x_2$, $\dot{x}_2 = x_3$, $\dot{x}_3 = -2x_1 + x_2 + 2x_3$, and deduce the solution of $\dot{x}_1 = x_2 + e^t$, $\dot{x}_2 = x_3$, $\dot{x}_3 = -2x_1 + x_2 + 2x_3$, with $\mathbf{x}(0) = (1, 0, 0)^T$.

8.17. The matrix of coefficients for the system $\dot{x}_1 = x_2$, $\dot{x}_2 = x_3$, $\dot{x}_3 = -2x_1 + x_2 + 2x_3$ is given by

$$\mathbf{A} = \begin{bmatrix} 0 & 1 & 0 \\ 0 & 0 & 1 \\ -2 & 1 & 2 \end{bmatrix}.$$

The eigenvalues of $\mathbf{A}$ are given by

$$|\mathbf{A} - \lambda \mathbf{I}| = \begin{vmatrix} -\lambda & 1 & 0 \\ 0 & -\lambda & 1 \\ -2 & 1 & 2 - \lambda \end{vmatrix} = -(\lambda - 1)(\lambda + 1)(\lambda - 2) = 0.$$

Corresponding eigenvectors are:

- for $\lambda_1 = 1$, $\mathbf{u} = (1, 1, 1)^T$;
- for $\lambda_2 = -1$, $\mathbf{v} = (1, -1, 1)^T$;
- for $\lambda_3 = 2$, $\mathbf{w} = (1, 2, 4)^T$.

A fundamental matrix for the system is therefore

$$\Phi(t) = \begin{bmatrix} e^t & e^{-t} & e^{2t} \\ e^t & -e^{-t} & 2e^{2t} \\ e^t & e^{-t} & 4e^{2t} \end{bmatrix}.$$

Use NODE, Theorem 8.13 to obtain the solution of the inhomogeneous equation

$$\dot{\mathbf{x}}(t) = \mathbf{A}\mathbf{x}(t) + \mathbf{f}(t),$$

where $\mathbf{f}(t) = (e^t, 0, 0)^T$. We require

$$\Phi(0) = \begin{bmatrix} 1 & 1 & 1 \\ 1 & -1 & 2 \\ 1 & 1 & 4 \end{bmatrix}, \quad \Phi^{-1}(0) = \frac{1}{6}\begin{bmatrix} 6 & 3 & -3 \\ 2 & -3 & 1 \\ -2 & 0 & 2 \end{bmatrix}.$$

Then

$$\mathbf{x}(t) = \Phi(t)\Phi^{-1}(0)(1,0,0)^T + \int_0^t \Phi(t-s)\Phi^{-1}(0)\mathbf{f}(s)ds$$

$$= \begin{bmatrix} e^t & e^{-t} & e^{2t} \\ e^t & -e^{-t} & 2e^{2t} \\ e^t & e^{-t} & 4e^{2t} \end{bmatrix} \begin{bmatrix} 1 \\ \frac{1}{3} \\ -\frac{1}{3} \end{bmatrix}$$

$$+ \int_0^t \begin{bmatrix} e^{t-s} & e^{-t+s} & e^{2t-2s} \\ e^{t-s} & -e^{-t+s} & 2e^{2t-2s} \\ e^{t-s} & e^{-t+s} & 4e^{2t-2s} \end{bmatrix} \begin{bmatrix} e^s \\ \frac{1}{3}e^s \\ -\frac{1}{3}e^s \end{bmatrix} ds$$

$$= \begin{bmatrix} e^t & e^{-t} & e^{2t} \\ e^t & -e^{-t} & 2e^{2t} \\ e^t & e^{-t} & 4e^{2t} \end{bmatrix} \begin{bmatrix} 1 \\ \frac{1}{3} \\ -\frac{1}{3} \end{bmatrix} + \int_0^t \begin{bmatrix} e^t + \frac{1}{3}e^{-t+2s} - \frac{1}{3}e^{2t-s} \\ e^t - \frac{1}{3}e^{-t+2s} - \frac{2}{3}e^{2t-s} \\ e^t + \frac{1}{3}e^{-t+2s} - \frac{4}{3}e^{2t+s} \end{bmatrix} ds$$

$$= \begin{bmatrix} e^t + \frac{1}{3}e^{-t} - \frac{1}{3}e^{2t} + te^t + \frac{1}{6}(e^t - e^{-t}) + \frac{1}{3}(e^t - e^{2t}) \\ e^t - \frac{1}{3}e^{-t} - \frac{2}{3}e^{2t} + te^t - \frac{1}{6}(e^t - e^{-t}) + \frac{2}{3}(e^t - e^{2t}) \\ e^t + \frac{1}{3}e^{-t} - \frac{4}{3}e^{2t} + te^t + \frac{1}{6}(e^t - e^{-t}) - \frac{4}{3}(e^t - e^{2t}) \end{bmatrix}$$

$$= \left[\frac{3}{2}e^t + \frac{1}{6}e^{-t} - \frac{2}{3}e^{2t} + te^t, \frac{3}{2}e^t - \frac{1}{6}e^{-t} - \frac{4}{3}e^{2t} + te^t, -\frac{1}{6}e^t + \frac{1}{6}e^{-t} + te^t \right]^T$$

- **8.18** Show that the differential equation $x^{(n)} + a_1 x^{(n-1)} + \cdots + a_n x = 0$ is equivalent to the system
$$\dot{x}_1 = x_2, \quad \dot{x}_2 = x_3, \cdots, \quad \dot{x}_{n-1} = x_n, \quad \dot{x}_n = -a_n x_1 - \cdots - a_1 x_n,$$
with $x = x_1$. Show that the equation for the eigenvalues is
$$\lambda^n + a_1 \lambda^{n-1} + \cdots + a_n = 0.$$

8.18. The matrix of coefficients for the system

$$\dot{x}_1, \quad \dot{x}_2 = x_3, \cdots, \dot{x}_{n-1} = x_n, \quad \dot{x}_n = -a_n - \cdots - a_1 x_n,$$

is

$$A = \begin{bmatrix} 0 & 1 & 0 & 0 & \cdots & 0 \\ 0 & 0 & 1 & 0 & \cdots & 0 \\ 0 & 0 & 0 & 1 & \cdots & 0 \\ \cdots & \cdots & \cdots & \cdots & \cdots & \cdots \\ 0 & 0 & 0 & 0 & \cdots & 1 \\ -a_n & -a_{n-1} & -a_{n-2} & -a_{n-3} & \cdots & -a_1 \end{bmatrix}.$$

The eigenvalues are given by

$$\begin{vmatrix} -\lambda & 1 & 0 & 0 & \cdots & 0 \\ 0 & -\lambda & 1 & 0 & \cdots & 0 \\ 0 & 0 & -\lambda & 1 & \cdots & 0 \\ \cdots & \cdots & \cdots & \cdots & \cdots & \cdots \\ 0 & 0 & 0 & 0 & \cdots & 1 \\ -a_n & -a_{n-1} & -a_{n-2} & -a_{n-3} & \cdots & -a_1 - \lambda \end{vmatrix} = 0.$$

Let $D_n(\lambda)$ denote the determinant in the previous equation. Then expansion by row 1 leads to

$$D_n(\lambda) = -\lambda D_{n-1}(\lambda) + (-1)^n a_n. \qquad \text{(i)}$$

For decreasing n, we have

$$D_{n-1}(\lambda) = -\lambda D_{n-2}(\lambda) + (-1)^{n-1} a_{n-1}, \qquad \text{(ii)}$$

$$\cdots$$

$$D_2(\lambda) = -\lambda D_1(\lambda) + a_2, \qquad \text{(iii)}$$

where $D_1(\lambda) = -a_1 - \lambda$. Now eliminate $D_{n-1}(\lambda), D_{n-2}(\lambda), \ldots$ from eqns (i) through (iii) by multiplying successive equations by $-\lambda, +\lambda$ and so on, and adding them. The result is

$$D_n(\lambda) = (-1)^n (a_n + a_{n-1}\lambda + \cdots + a_1 \lambda^{n-1} + \lambda^n).$$

The required result follows by equating $D_n(\lambda)$ to zero.

- **8.19** A bird population, $p(t)$, is governed by the differential equation $\dot p = \mu(t)p - kp$, where k is the death rate and $\mu(t)$ represents a variable periodic birth rate with period 1 year. Derive a condition which ensures that the mean annual population remains constant. Assuming that this condition is fulfilled, does it seem likely that, in practice, the average population will remain constant? (This is asking a question about a particular kind of stability.)

8.19. The general solution of $\dot{p} = \mu(t)p - kp$ is

$$p = p(0) \exp\left[\int_0^t \mu(s)dt - ks\right].$$

If T represents the duration of one year, then the population is on average constant if

$$\int_0^T \mu(s)ds = kT,$$

where k is the constant population average.

Consider a particular solution

$$p^*(t) = p^*(0) \exp\left[\int_0^t \mu(s)dt - ks\right].$$

Let $K = \max_{t \in T}[\int_0^T \mu(s)ds - kT]$. Choose any $\varepsilon > 0$, and let $\delta = \varepsilon e^{-K}$. Choose initial conditions such that $|p(0) - p^*(0)| < \delta$. Then

$$\| p(t) - p^*(t) \| = |p(t) - p^*(t)| \le |p(0) - p^*(0)|e^K = \delta e^K = \varepsilon.$$

Hence every solution is Liapunov stable.

• 8.20 Are the periodic solutions of $\ddot{x} + \mathrm{sgn}(x) = 0$, (i) Poincaré stable, (ii) Liapunov stable?

8.20. Consider periodic solutions of $\ddot{x} = \mathrm{sgn}(x)$. The phase diagram of the system is shown in Figure 8.7. There is one equilibrium point, at the origin, and the paths are given by

$$y^2 = C - 2x, \ (x > 0); \quad y^2 = C + 2x, \ (x < 0).$$

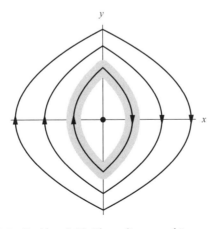

Figure 8.7 Problem 8.20: Phase diagram of $\ddot{x} + \mathrm{sgn}(x) = 0$.

(i) Paths starting in the shaded region remain within it which implies that the phase paths are Poincaré stable.

(ii) The origin is a centre which covers the entire (x, y) plane. Hence by NODE, Section 8.4, all solutions are Liapunov stable.

- **8.21** Give a descriptive argument to show that if the index of an equilibrium point in a plane autonomous system is not unity, then the equilibrium point is not stable.

8.21. For the equilibrium point to be locally Liapunov stable the phase paths must be either closed (as in a centre) or approach the equilibrium point (as for a node or spiral), all of which have index 1. Therefore if the index is not 1, then the equilibrium point must be unstable. Note that the unstable spiral has index 1 but is obviously unstable in the Liapunov sense.

- **8.22** Show that the system
$$\dot{x} = x + y - x(x^2 + y^2), \quad \dot{y} = -x + y - y(x^2 + y^2), \quad \dot{z} = -z,$$
has a limit cycle $x^2 + y^2 = 1$, $z = 0$. Find the linear approximation at the origin and so confirm that the origin is unstable. Use cylindrical polar coordinates $r = \sqrt{(x^2 + y^2)}$, z to show that the limit cycle is stable. Sketch the phase diagram in the x, y, z space.

8.22. The system is given by
$$\dot{x} = x + y - z(x^2 + y^2), \quad \dot{y} = -x + y - y(x^2 + y^2), \quad \dot{z} = -z,$$
which has an equilibrium point at the origin. Apply polar coordinates in the (x, y) plane, so that
$$2r\dot{r} = 2x\dot{x} + 2y\dot{y} = 2[x^2 + y^2 - (x^2 + y^2)^2] = 2r^2(1 - r^2),$$
or
$$\dot{r} = r(1 - r^2),$$
and
$$\dot{\theta} \sec^2\theta = \frac{\dot{y}x - \dot{x}y}{x^2} = -\frac{r^2}{x^2}.$$

Hence $\dot{\theta} = -1$. It is obvious that $r = 1$ is a limit cycle which remains in the (x, y) plane, since it is consistent with the solution $z = 0$ of the third equation.

Near the origin, the linear approximation is

$$\dot{x} \approx x+y, \quad \dot{y} \approx -x+y, \quad \dot{z} = -z,$$

or

$$\begin{bmatrix} \dot{x} \\ \dot{y} \\ \dot{z} \end{bmatrix} = \begin{bmatrix} 1 & 1 & 0 \\ -1 & 1 & 0 \\ 0 & 0 & -1 \end{bmatrix} \begin{bmatrix} x \\ y \\ z \end{bmatrix}.$$

The eigenvalues of the matrix are given by

$$\begin{vmatrix} 1-\lambda & 1 & 0 \\ -1 & 1-\lambda & 0 \\ 0 & 0 & -1-\lambda \end{vmatrix} = 0,$$

which has the solutions

$$\lambda = -1, \quad \lambda = 1 \pm i.$$

Since the complex eigenvalues have positive real parts, the equilibrium point at the origin is unstable.

The equations for the phase paths in cylindrical polar coordinates (r, θ, z) are

$$\dot{r} = r(1-r^2), \quad \dot{\theta} = -1, \quad \dot{z} = -z.$$

The general solutions are

$$r = 1/\sqrt{(1-Ae^{-2t})}, \quad \theta = -t+B, \quad z = Ce^{-t}.$$

Some three-dimensional phase paths are shown in Figure 8.8. The system has a stable limit cycle because all paths approach $r = 1$, $z = 0$ as $t \to \infty$.

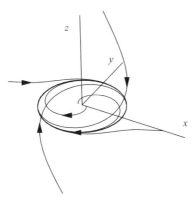

Figure 8.8 Problem 8.22: Phase diagram for $\dot{x} = x+y-z(x^2+y^2)$, $\dot{y} = -x+y-y(x^2+y^2)$, $\dot{z} = -z$.

- **8.23** Show that the nth-order non-autonomous system $\dot{\mathbf{x}} = \mathbf{X}(\mathbf{x},t)$ can be reduced to an $(n+1)$th order autonomous system by introducing a new variable, $x_{n+1} = t$. (The $(n+1)$th order dimensional phase diagram for the modified system is of the type suggested by Figure 8.9 (in NODE). The system has no equilibrium points.)

8.23. For the nth order system $\dot{\mathbf{x}} = \mathbf{X}(\mathbf{x},t)$, let

$$\mathbf{x} = [x_1, x_2, \cdots, x_n]^T, \quad \mathbf{X} = [X_1, X_2, \cdots, X_n]^T.$$

If we introduce a further variable x_{n+1} and put it equal to t. Then if we define

$$\mathbf{y} = [x_1, x_2, \cdots, x_n, x_{n+1}]^T, \quad \mathbf{Y} = [X_1, X_2, \cdots, X_n, 1]^T,$$

then the system is equivalent to the autonomous system

$$\dot{\mathbf{y}} = \mathbf{Y}(\mathbf{y}).$$

Since $\dot{x}_{n+1} = t$, the system can have no equilibrium points.

- **8.24** Show that all phase paths of $\ddot{x} = x - x^3$ are Poincaré stable except the homoclinic paths (see Section 3.6 in NODE).

8.24. The system $\ddot{x} = x - x^3$ has three equilibrium points, at $(0,0)$, $(1,0)$ and $(-1,0)$. The phase paths are given by

$$y^2 = x^2 - \tfrac{1}{2}x^4 + C.$$

The phase diagram is shown in Figure 8.9. All paths except the homoclinic paths are closed, and therefore are Poincaré stable.

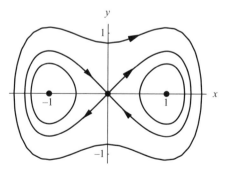

Figure 8.9 Problem 8.24: Phase diagram for $\ddot{x} = x - x^3$.

Consider any initial point on one of the homoclinic paths. Any neighbouring point not on the homoclinic path will lie on a periodic path which is either external or internal to the homoclinic path. In terms of Definition 8.1 a radius δ cannot be found, because as $t \to \infty$ the path periodically departs from the separatrix by a fixed amount.

- **8.25** Investigate the equilibrium points of
$$\dot{x} = y, \quad \dot{y} = z - y - x^3, \quad \dot{z} = y + x - x^3.$$
Confirm that the origin has homoclinic paths given by
$$x = \pm\sqrt{2}\operatorname{sech} t, \quad y = \mp\sqrt{2}\operatorname{sech}^2 t \sinh t, \quad z = \pm\sqrt{2}\operatorname{sech} t \mp \sqrt{2}\operatorname{sech}^2 t \sinh t.$$
In which directions do the solutions approach the origin as $t \to \pm\infty$?

8.25. The third-order system is
$$\dot{x} = y, \quad \dot{y} = z - y - x^3, \quad \dot{z} = y + x - x^3. \tag{i}$$

Equilibrium occurs where
$$y = 0, \quad z - y - x^3 = 0, \quad y + x - x^3 = 0,$$

at the points
$$(0, 0, 0), \quad (1, 0, 1), \quad (-1, 0, -1).$$

- The point $(0,0,0)$. The linear approximation is
$$\begin{bmatrix} \dot{x} \\ \dot{y} \\ \dot{z} \end{bmatrix} = \begin{bmatrix} 0 & 1 & 0 \\ 0 & -1 & 1 \\ 1 & 1 & 0 \end{bmatrix} \begin{bmatrix} x \\ y \\ z \end{bmatrix}.$$

Its eigenvalues are given by
$$\begin{vmatrix} -\lambda & 1 & 0 \\ 0 & -1-\lambda & 1 \\ 1 & 1 & -\lambda \end{vmatrix} = -(\lambda - 1)(\lambda + 1)^2 = 0.$$

Since one eigenvalue is $\lambda = 1$, the origin is an unstable equilibrium point.
- The point $(1, 0, 1)$. Let $x = 1 + u$, $y = v$, $z = 1 + w$. Hence the linear approximation near this point is given by
$$\begin{bmatrix} \dot{u} \\ \dot{v} \\ \dot{w} \end{bmatrix} = \begin{bmatrix} 0 & 1 & 0 \\ -3 & -1 & 1 \\ -2 & 1 & 0 \end{bmatrix} \begin{bmatrix} u \\ v \\ w \end{bmatrix}.$$

Its eigenvalues are given by

$$\begin{vmatrix} -\lambda & 1 & 0 \\ -3 & -1-\lambda & 1 \\ -2 & 1 & -\lambda \end{vmatrix} = -(\lambda-1)(\lambda^2+2) = 0.,$$

that is, -1 and $\pm i\sqrt{2}$. The equilibrium point is stable in the form of a centre/node (but is not asymptotically stable).

- The point $(-1, 0, -1)$. Let $x = -1+u$, $y = v$, $z = -1+w$. The linear approximation is the same as the previous case so that $(-1, 0, -1)$ is also stable.

Note that on any homoclinic path of the origin, $x, y, z \to 0$ as $t \to \pm\infty$. From (i)

$$\dot{z} - \dot{x} - \dot{y} = (y+x-x^3) - y - (z-y-x^3) = -(z-x-y),$$

which has the general solution $z - x - y = Ke^{-t}$. Homoclinicity is only possible if $K = 0$, in which case $z = x + y$. In (i) eliminate z so that x and y satisfy

$$\dot{x} = y, \quad \dot{y} = x - x^3.$$

Therefore, after elimination of y, x satisfies $\ddot{x} = x - x^3$. The homoclinic solutions are

$$x = \pm\sqrt{2}\operatorname{sech} t, \quad y = \dot{x} = \mp\sqrt{2}\operatorname{sech}^2 t \sinh t.$$

Finally

$$z = x + y = \pm\sqrt{2}\operatorname{sech} t[1 - \tanh t].$$

- **8.26** By using linear approximations investigate the equilibrium points of the Lorenz equations

$$\dot{x} = a(y-x), \quad \dot{y} = bx - y - xz, \quad \dot{z} = xy - cz,$$

where $a, b, c > 0$ are constants. Show that if $b \leq 1$, then the origin is the only equilibrium point, and that there are three equilibrium points if $b > 1$. Discuss the stability of the zero solution.

8.26. The Lorenz equations are

$$\dot{x} = a(y-x), \quad \dot{y} = bx - y - xz, \quad \dot{z} = xy - cz.$$

Equilibrium occurs where

$$y - x = 0, \quad bx - y - xz = 0, \quad xy - cz = 0.$$

Elimination of y and z leads to

$$c(b-1)x - x^3 = 0,$$

which has the solutions $x = 0$, $x = \pm\sqrt{[c(b-1)]}$. Therefore the Lorenz equations have up to three equilibrium points, at

$$(0,0,0), \quad (\sqrt{[c(b-1)]}, \sqrt{[c(b-1)]}, b-1), \quad (-\sqrt{[c(b-1)]}, -\sqrt{[c(b-1)]}, b-1).$$

It is clear that, if $b \leq 1$, the origin is the only equilibrium point: if $b > 1$ there are three equilibrium points. Near $(0,0,0)$, the linearized equations are

$$\begin{bmatrix} \dot{x} \\ \dot{y} \\ \dot{z} \end{bmatrix} = \begin{bmatrix} -a & a & 0 \\ b & -1 & 0 \\ 0 & 0 & -c \end{bmatrix} \begin{bmatrix} x \\ y \\ z \end{bmatrix}.$$

Its eigenvalues are given by

$$\begin{vmatrix} -a-\lambda & a & 0 \\ b & -1-\lambda & 0 \\ 0 & 0 & -c-\lambda \end{vmatrix} = -(\lambda + c)(\lambda^2 + a\lambda + \lambda + a - ab) = 0.$$

Therefore the eigenvalues are

$$-c, \quad \tfrac{1}{2}[-(a+1) \pm \sqrt{\{(a-1)^2 + 4ab\}}].$$

The origin is stable if $b \leq 1$, and unstable if $b > 1$.

- 8.27 Test the stability of the linear system
$\dot{x}_1 = t^{-2}x_1 - 4x_2 - 2x_3 + t^2$,
$\dot{x}_2 = -x_1 + t^{-2}x_2 + x_3 + t$,
$\dot{x}_3 = t^{-2}x_1 - 9x_2 - 4x_3 + 1$.

8.27. The system can be expressed in the form $\dot{\mathbf{x}} = \mathbf{A}(t)\mathbf{x} + \mathbf{f}(t)$, where

$$\mathbf{A}(t) = \begin{bmatrix} t^{-2} & -4 & -2 \\ -1 & t^{-2} & 1 \\ t^{-2} & -9 & -4 \end{bmatrix}, \quad \mathbf{f}(t) = \begin{bmatrix} t^2 \\ t \\ 1 \end{bmatrix}.$$

By NODE, Theorem 8.1, the stability of this system will be the same as the zero solution of $\dot{\xi} = A(t)\xi$. With NODE, Theorem 8.15 in view, let $A(t) = B + C(t)$, where

$$B = \begin{bmatrix} 0 & -4 & -2 \\ -1 & 0 & 1 \\ 0 & -9 & -4 \end{bmatrix}, \quad C(t) = \begin{bmatrix} t^{-2} & 0 & 0 \\ 0 & t^{-2} & 0 \\ t^{-2} & 0 & 0 \end{bmatrix}.$$

Hence

$$\|C(t)\| = \sqrt{[t^{-4} + t^{-4} + t^{-4}]} = \frac{\sqrt{3}}{t^2}.$$

For any $t > t_0 > 0$,

$$\int_{t_0}^{t} \|C(s)\| ds = \sqrt{3} \int_{t_0}^{t} \frac{ds}{s^2} = \sqrt{3} \left[\frac{1}{t_0} - \frac{1}{t} \right],$$

which is bounded in t.

Also the eigenvalues of B are given by

$$\begin{vmatrix} -\lambda & -4 & -2 \\ -1 & -\lambda & 1 \\ 0 & -9 & -4-\lambda \end{vmatrix} = -(1+\lambda)^2(2+\lambda) = 0.$$

The eigenvalues are real and negative.

Hence by Theorem 8.15 all solutions ξ and therefore x are asymptotically stable.

- **8.28** Test the stability of the solutions of the linear system
$\dot{x}_1 = 2x_1 + e^{-t}x_2 - 3x_3 + e^t$,
$\dot{x}_2 = -2x_1 + e^{-t}x_2 + x_3 + 1$,
$\dot{x}_3 = (4 + e^{-t})x_1 - x_2 - 4x_3 + e^t$.

8.28. The system can be expressed in the form $\dot{x} = A(t)x + f(t)$, where

$$A(t) = \begin{bmatrix} 2 & e^{-t} & -3 \\ -2 & e^{-t} & 1 \\ (4+e^{-t}) & -1 & -4 \end{bmatrix}, \quad f(t) = \begin{bmatrix} e^t \\ 1 \\ e^t \end{bmatrix}.$$

By NODE, Theorem 8.1, the stability of this system will be the same as the zero solution of $\dot{\xi} = A(t)\xi$. With NODE, Theorem 8.15 in view, let $A(t) = B + C(t)$, where

$$B = \begin{bmatrix} 2 & 0 & -3 \\ -2 & 0 & 1 \\ 4 & -1 & -4 \end{bmatrix}, \quad C(t) = \begin{bmatrix} 0 & e^{-t} & 0 \\ 0 & e^{-t} & 0 \\ e^{-t} & 0 & 0 \end{bmatrix}.$$

Hence
$$\| C(t) \| = \sqrt{[e^{-2t} + e^{-2t} + e^{-2t}]} = \sqrt{3} e^{-t}.$$

Then for any t_0,
$$\int_{t_0}^{t} \| C(s) \| \, ds = \sqrt{3} \int_{t_0}^{t} e^{-s} ds = \sqrt{3} [e^{-t_0} - e^{-t}],$$

which is bounded in t. Also the eigenvalues of $\mathbf{B}$ are given by

$$\begin{vmatrix} 2-\lambda & 0 & -3 \\ -2 & -\lambda & 1 \\ 4 & -1 & -4-\lambda \end{vmatrix} = -(1+\lambda)(4+\lambda+\lambda^2).$$

Therefore the eigenvalues are
$$-1, \quad \tfrac{1}{2}(-1 \pm \sqrt{15}),$$

which all have negative real part.

Hence by Theorem 8.15 all solutions $\boldsymbol{\xi}$ and therefore $\mathbf{x}$ are asymptotically stable.

- 8.29 Test the stability of the zero solution of the system

$$\dot{x} = y + \frac{xy}{1+t^2}, \quad \dot{y} = -x - y + \frac{y^2}{1+t^2}.$$

8.29. Express the system in the form $\dot{\mathbf{x}} = \mathbf{A}\mathbf{x} + \mathbf{h}(\mathbf{x}, t)$, where

$$\mathbf{A} = \begin{bmatrix} 0 & 1 \\ -1 & -1 \end{bmatrix}, \quad \mathbf{h}(\mathbf{x}, t) = \begin{bmatrix} xy/(1+t^2) \\ y^2/(1+t^2) \end{bmatrix},$$

and then apply NODE, Theorem 8.16. The eigenvalues of $\mathbf{A}$ are given by

$$\begin{vmatrix} -\lambda & 1 \\ -1 & -1-\lambda \end{vmatrix} = \lambda^2 + \lambda + 1 = 0.$$

Hence $\lambda = \tfrac{1}{2}(-1 \pm \sqrt{3}i)$. Hence, as required, the solutions of $\dot{\mathbf{x}} = \mathbf{A}\mathbf{x}$ are asymptotically stable. Also

$$\| \mathbf{h}(\mathbf{x}, t) \| = \sqrt{\left[\frac{x^2 y^2 + y^4}{(1+t^2)^2}\right]} \le |y| \| \mathbf{x} \| \le \| \mathbf{x} \|^2.$$

Hence
$$\frac{\| \mathbf{h}(\mathbf{x}, t) \|}{\| \mathbf{x} \|} \le \| \mathbf{x} \| \to 0$$

- **8.30** Test the stability of the zero solution of the system

$$\dot{x}_1 = e^{-x_1-x_2} - 1, \quad \dot{x}_2 = e^{-x_2-x_3} - 1, \quad \dot{x}_3 = -x_3.$$

8.30. The system

$$\dot{x}_1 = e^{-x_1-x_2} - 1, \quad \dot{x}_2 = e^{-x_2-x_3} - 1, \quad \dot{x}_3 = -x_3$$

can be expressed in the form $\dot{x} = Ax + h(x, t)$, where

$$A = \begin{bmatrix} -1 & -1 & 0 \\ 0 & -1 & -1 \\ 0 & 0 & -1 \end{bmatrix}, \quad h(x) = \begin{bmatrix} x_1 + x_2 + e^{-x_1-x_2} - 1 \\ x_2 + x_3 + e^{-x_2-x_3} - 1 \\ 0 \end{bmatrix}.$$

We intend to apply Theorem 8.15. The eigenvalue of A is obviously $\lambda = -1$ (repeated) which satisfies (i) of the theorem.

To evaluate the behaviour of $\| h(x) \|$ note that

$$u - 1 + e^{-u} = \frac{u^2}{2!} - \frac{u^3}{3!} + \frac{u^4}{4!} - \cdots \leq \frac{u^2}{2}, \tag{i}$$

for $0 \leq u < 4$. Let $p = (1, 1, 0)$ and $q = (0, 1, 1)$. Then

$$h(x) = \begin{bmatrix} p \cdot x + e^{-p \cdot x} - 1 \\ q \cdot x + e^{-q \cdot x} - 1 \\ 0 \end{bmatrix}.$$

Using (i),

$$\| h(x) \| = \sqrt{[(p \cdot x + e^{-p \cdot x} - 1)^2 + (q \cdot x + e^{-q \cdot x} - 1)^2]}$$

$$\leq \sqrt{\left[\tfrac{1}{4}(p \cdot x)^4 + \tfrac{1}{4}(q \cdot x)^4\right]}$$

$$\leq \sqrt{[\| x \|^4 + \| x \|^4]}$$

$$= \sqrt{2} \| x \|^2$$

since $\| p \cdot x \| \leq \| p \| \| x \| = \sqrt{2} \| x \|$, etc. Hence condition (ii) of Theorem 15 is also satisfied. It follows that the zero solution of the system is asymptotically stable.

- 8.31 Test the stability of the zero solution of the equation

$$\ddot{x} + \left[\frac{1+(t-1)|\dot{x}|}{1+t|\dot{x}|}\right]\dot{x} + \tfrac{1}{4}x = 0.$$

8.31. The non-autonomous equation

$$\ddot{x} + [\{1 + (t-1)|\dot{x}|\}/\{1 + t|\dot{x}|\}]\dot{x} + \tfrac{1}{4}x = 0$$

can be represented by the system $\dot{\mathbf{x}} = \mathbf{A}\mathbf{x} + \mathbf{h}(\mathbf{x}, t)$, where $\dot{x} = y$, and

$$\mathbf{x} = \begin{bmatrix} x \\ y \end{bmatrix}, \quad \mathbf{A} = \begin{bmatrix} 0 & 1 \\ -\tfrac{1}{4} & -1 \end{bmatrix}$$

$$\mathbf{h}(\mathbf{x}, t) = \begin{bmatrix} 0 \\ -[\{1 + (t-1)|y|\}/\{1 + t|y|\}]y + y \end{bmatrix}.$$

The matrix $\mathbf{A}$ has the repeated eigenvalue $-\tfrac{1}{2}$ so that condition (i) of Theorem 8.16 is satisfied. For the other condition

$$\| \mathbf{h}(\mathbf{x}, t) \| = \left| -\frac{1 + (t-1)|y|}{1 + t|y|}|y| + y \right| = \frac{|y|^2}{|1 + t|y||} \leq |y|^2 \leq \| \mathbf{x} \|^2.$$

Therefore condition (ii) is satisfied so that the zero solution is asymptotically stable.

- 8.32 Consider the restricted three-body problem in planetary dynamics in which one body (possibly a satellite) has negligible mass in comparison with the other two. Suppose that the two massive bodies (gravitational masses μ_1 and μ_2) remain at a fixed distance a apart, so that the line joining them must rotate with spin $\omega = \sqrt{[(\mu_1 + \mu_2)/a^3]}$. It can be shown (see Hill (1964)) that the equations of motion of the third body are given by

$$\ddot{\xi} - 2\omega\dot{\eta} = \frac{\partial U}{\partial \xi}, \quad \ddot{\eta} + 2\omega\dot{\xi} = \frac{\partial U}{\partial \eta},$$

where the gravitational field

$$U(\xi, \eta) = \tfrac{1}{2}\omega^2(\xi^2 + \eta^2) + \frac{\mu_1}{d_1} + \frac{\mu_2}{d_2},$$

and

$$d_1 = \sqrt{\left[\left(\xi + \frac{\mu_1 a}{\mu_1 + \mu_2}\right)^2 + \eta^2\right]}, \quad d_2 = \sqrt{\left[\left(\xi - \frac{\mu_2 a}{\mu_1 + \mu_2}\right)^2 + \eta^2\right]}.$$

The origin of the rotating (ξ, η) plane is at the mass centre with the ξ axis along the common radius of the two massive bodies in the distance of μ_2.

Consider the special case in which $\mu_1 = \mu_2 = \mu$. Show that there are three equilibrium points along the ξ axis (use a computed graph to establish), and two equilibrium points at the triangulation points of μ_1 and μ_2.

8.32. The equations of motion in the restricted three-body problem are

$$\ddot\xi - 2\omega\dot\eta = \frac{\partial U}{\partial \xi}, \quad \ddot\eta + 2\omega\dot\xi = \frac{\partial U}{\partial \eta},$$

where

$$U(\xi, \eta) = \tfrac{1}{2}\omega^2(\xi^2 + \eta^2) + \frac{\mu_1}{d_1} + \frac{\mu_2}{d_2},$$

and

$$d_1 = \sqrt{\left(\xi + \frac{\mu_1 a}{\mu_1 + \mu_2}\right)^2 + \eta^2}, \quad d_2 = \sqrt{\left(\xi + \frac{\mu_2 a}{\mu_1 + \mu_2}\right)^2 + \eta^2}.$$

The coordinate scheme is shown in Figure 8.10. Let $\dot\xi = \sigma\omega$ and $\dot\eta = \rho\omega$. Then

$$\omega\dot\sigma = 2\omega^2\sigma + \frac{\partial U}{\partial \xi}, \quad \omega\dot\rho = -2\omega^2\rho + \frac{\partial U}{\partial \eta}.$$

Therefore the system is fourth-order in $(\xi, \eta, \omega, \eta)$ phase space. Consider the case $\mu_1 = \mu_2 = \mu$. Equilibrium occurs where $\sigma = 0$, $\rho = 0$ and

$$\frac{\partial U}{\partial \xi} = \omega^2\xi - \frac{\mu}{d_1^3}\left(\xi + \tfrac{1}{2}a\right) - \frac{\mu}{d_2^3}\left(\xi - \tfrac{1}{2}a\right) = 0, \quad (i)$$

$$\frac{\partial U}{\partial \eta} = \omega^2\eta - \frac{\mu\eta}{d_1^3} - \frac{\mu\eta}{d_2^3} = 0. \quad (ii)$$

From (ii) $\eta = 0$ is a solution, in which case in (i) ξ satisfies (with $\omega^2 = 2\mu/a^3$),

$$h(q) = 2q|q + \tfrac{1}{2}|^3|q - \tfrac{1}{2}|^3 - \left(q + \tfrac{1}{2}\right)|q - \tfrac{1}{2}|^3 - \left(q - \tfrac{1}{2}\right)|q + \tfrac{1}{2}|^3 = 0. \quad (iii)$$

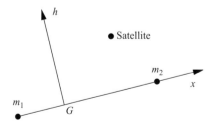

Figure 8.10 Problem 8.32: Coordinate scheme.

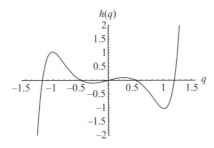

Figure 8.11 Problem 8.32:

where $q = \xi/a$. The graph of $h(q)$ against q is shown in Figure 8.11. The solutions at $q = \pm\frac{1}{2}$ are discounted since they correspond to the locations of the massive bodies. Numerical solution equation of (iii) gives to the two solutions $\xi = \xi_0 = \pm 1.1984$ in addition of course to the origin.

If $\xi = 0$, eqn (i) is satisfied identically whilst (ii) implies $\eta = \pm\frac{1}{2}\sqrt{3}$. Each of these points together with the locations of μ_1 and μ_2 form equilateral triangles. $(0, \pm\frac{1}{2}\sqrt{3})$ are known as triangulation points.

To summarize the five equilibrium points are at $(0, 0)$, $(\pm\xi_0, 0)$ and $(0, \pm\frac{1}{2}\sqrt{3})$.

- **8.33** Express the equations

$$\dot{x} = x[1 - \sqrt{(x^2+y^2)}] - \tfrac{1}{2}y[\sqrt{(x^2+y^2)} - x],$$
$$\dot{y} = y[1 - \sqrt{(x^2+y^2)}] + \tfrac{1}{2}x[\sqrt{(x^2+y^2)} - x]$$

in polar form in terms of r and θ. Show that the system has two equilibrium points at $(0, 0)$ and $(1, 0)$. Solve the equations for the phase paths in terms of r and θ, and confirm that all paths which start at any point other than the origin approach $(1, 0)$ as $t \to \infty$. Sketch the phase diagram for the system.

Consider the half-path which starts at $(0, 1)$. Is this path stable in the Poincaré sense? Is the equilibrium point at $(1, 0)$ stable?

8.33. The equations

$$\dot{x} = x[1 - \sqrt{(x^2+y^2)}] - \tfrac{1}{2}y[\sqrt{(x^2+y^2)} - x],$$

$$\dot{y} = y[1 - \sqrt{(x^2+y^2)}] + \tfrac{1}{2}x[\sqrt{(x^2+y^2)} - x]$$

in polar coordinates (r, θ), where $x = r\cos\theta$ and $y = r\sin\theta$, become

$$\dot{r} = r(1-r), \quad \dot{\theta} = r\sin^2\tfrac{1}{2}\theta.$$

The polar equations have equilibrium points given by

$$r(1-r) = 0, \quad r\sin^2\tfrac{1}{2}\theta = 0.$$

The solutions are $r = 0$, and $r = 1$, $\sin \frac{1}{2}\theta = 0$. In terms of x and y equilibrium occurs at $(0, 0)$ and $(1, 0)$.

The phase paths are given by

$$\frac{dr}{d\theta} = \frac{1-r}{\sin^2 \frac{1}{2}\theta}$$

Hence

$$\int \frac{dr}{1-r} = \int \frac{d\theta}{\sin^2 \frac{1}{2}\theta} = \int \operatorname{cosec}^2 \tfrac{1}{2}\theta \, d\theta,$$

so that

$$\ln|1 - r| = 2\cot\left(\tfrac{1}{2}\theta\right) + C.$$

All solutions are given by

$$r = 1 + Ae^{2\cot[(1/2)\theta]}.$$

As $\theta \to 2\pi$, $\cot \frac{1}{2}\theta \to -\infty$ and $r \to 1$: as $\theta \to 0$, $\cot \frac{1}{2}\theta \to \infty$ and $r \to \infty$ $(A > 0)$ or r stops at $r = 0$ $(A < 0)$ for some value of θ (r cannot be negative). Note that $r = 1$ is a path but not a limit cycle since the path passes through the equilibrium point as shown in the phase diagram in Figure 8.12: it is a separatrix. Also $y = 0$, $x > 0$ contains two phase paths which approach $(1, 0)$. Both equilibrium points are unstable.

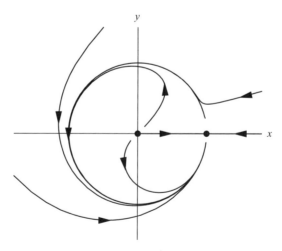

Figure 8.12 Problem 8.33:

● 8.34 Consider the system
$$\dot{x} = -y, \quad \dot{y} = x + \lambda(1 - y^2 - z^2)y, \quad \dot{z} = -y + \mu(1 - x^2 - y^2)z.$$
Classify the linear approximation of the equilibrium point at the origin in terms of the parameters $\lambda \neq 0$ and $\mu \neq 0$. Verify that the system has a periodic solution
$$x = \cos(t - t_0), \quad y = \sin(t - t_0), \quad z = \cos(t - t_0),$$
for any t_0.

8.34. The system
$$\dot{x} = y, \quad \dot{y} = x + \lambda(1 - y^2 - z^2)y, \quad \dot{z} = -y + \mu(1 - x^2 - y^2)z,$$
has one equilibrium point at the origin. Its linear approximation near the origin is
$$\dot{x} = y, \quad \dot{y} \approx x + \lambda y, \quad \dot{z} \approx -y + \mu z,$$

or

$$\begin{bmatrix} \dot{x} \\ \dot{y} \\ \dot{z} \end{bmatrix} = \begin{bmatrix} 0 & -1 & 0 \\ 1 & \lambda & 0 \\ 0 & -1 & \mu \end{bmatrix} \begin{bmatrix} x \\ y \\ z \end{bmatrix}.$$

The eigenvalues of the matrix of coefficients are given by

$$\begin{vmatrix} -m & -1 & 0 \\ 1 & \lambda - m & 0 \\ 0 & -1 & \mu - m \end{vmatrix} = (\mu - m)(m^2 - \lambda m + 1) = 0.$$

Hence the eigenvalues are μ and $\frac{1}{2}\lambda \pm \sqrt{(\lambda^2 - 4)}$. The classification in three dimensions is as follows:

- $\mu < 0, \lambda > 2$. All the eigenvalues are real with 2 positive and 1 negative. The origin is unstable and is a saddle/node.
- $\mu > 0, \lambda > 2$. All the eigenvalues are real and positive. Hence the origin is an unstable node.
- $\mu < 0, \lambda < -2$. All eigenvalues are real and negative so that the origin resembles a stable node.
- $\mu > 0, \lambda < -2$. All eigenvalues are real with 2 negative and 1 positive. The origin is unstable.
- $\mu < 0, 0 < \lambda < 2$. One eigenvalue is negative, and the others are complex conjugates with positive real part so that the origin is an unstable spiral. There are two stable paths which enter the origin.
- $\mu > 0, 0 < \lambda < 2$. One eigenvalue is positive, and the others are complex conjugates with positive real part so that the origin is an unstable spiral.

- $\mu < 0$, $-2 < \lambda < 0$. One eigenvalue is negative, and the others are complex conjugates with negative real part. The origin is stable with spiral paths.
- $\mu > 0$, $-2 < \lambda < 0$. One eigenvalue is positive, and the others are complex conjugates with negative real part. The origin is unstable with spiral paths.

It can be verified that

$$x = \cos(t - t_0), \quad y = \sin(t - t_0), \quad z = \cos(t - t_0),$$

is an exact periodic solution irrespective of the type of equilibrium point at the origin.

9 Stability by solution perturbation: Mathieu's equation

- 9.1 The system
$$\dot{x}_1 = (-\sin 2t)x_1 + (\cos 2t - 1)x_2, \quad \dot{x}_2 = (\cos 2t + 1)x_1 + (\sin 2t)x_2,$$
has a fundamental matrix of normal solutions
$$\Phi(t) = \begin{bmatrix} e^t(\cos t - \sin t) & e^{-t}(\cos t + \sin t) \\ e^t(\cos t + \sin t) & e^{-t}(-\cos t + \sin t) \end{bmatrix}.$$
Obtain the corresponding E matrix (Theorem 9.1 in NODE), the characteristic numbers, and the characteristic exponents.

9.1. The system has the fundamental matrix of normal solutions
$$\Phi(t) = \begin{bmatrix} e^t(\cos t - \sin t) & e^{-t}(\cos t + \sin t) \\ e^t(\cos t + \sin t) & e^{-t}(-\cos t + \sin t) \end{bmatrix}$$

(which can be verified). The matrix E is given by

$$E = \Phi^{-1}(0)\Phi(\pi) = \frac{1}{2}\begin{bmatrix} 1 & 1 \\ 1 & -1 \end{bmatrix}\begin{bmatrix} -e^\pi & -e^{-\pi} \\ -e^\pi & e^{-\pi} \end{bmatrix} = \begin{bmatrix} -e^\pi & 0 \\ 0 & -e^{-\pi} \end{bmatrix}.$$

The characteristic numbers are the eigenvalues of E, which are obviously $\mu_1 = -e^\pi$ and $\mu_2 = -e^{-\pi}$. The corresponding characteristic exponents are $\rho_1 = 1 + i$ and $\rho_2 = -1 + i$.

- 9.2 Let the system $\dot{x} = P(t)x$ have a matrix of coefficients P with minimal period T (and therefore also with periods $2T, 3T, \ldots$). Follow the argument of Theorem 9.1, using period mT, $m > 1$, to show that $\Phi(t + mT) = \Phi(t)E^m$. Assuming that, if the eigenvalues of E are μ_i, then those of E^m are μ_i^m, discuss possible periodic solutions.

9.2. Let $\Phi(t)$ be a fundamental matrix of $\dot{x} = P(t)x$, where P(t) has a minimal period of T. From eqn (9.15)
$$\Phi(t + T) = \Phi(t)E,$$

for all t, where $\mathbf{E}$ is a non-singular constant matrix. By repetition of the result

$$\Phi(t+mT) = \Phi(t+(m-1)T)\mathbf{E} = \Phi(t+(m-2)T)\mathbf{E}^2 = \cdots = \Phi(t)\mathbf{E}^m,$$

where m is any positive integer.

A well-known result in matrix algebra states that if $\mathbf{E}$ has eigenvalues μ_i, then $\mathbf{E}^m$ has eigenvalues μ_i^m. The system will only have a solution of period T if there exists a unit eigenvalue. It will have a solution of period mT if $\mu^m = 1$, which means that μ must be an mth root of unity.

- **9.3** Obtain Wronskians for the following linear systems
  (i) $\dot{x}_1 = x_1 \sin t + x_2 \cos t$, $\dot{x}_2 = -x_1 \cos t + x_2 \sin t$;
  (ii) $\dot{x}_1 = f(t)x_2$, $\dot{x}_2 = g(t)x_1$.

9.3. (i) For the linear system

$$\dot{x}_1 = x_1 \sin t + x_2 \cos t, \quad \dot{x}_2 = -x_1 \cos t + x_2 \sin t,$$

the matrix of coefficients is

$$\mathbf{A}(t) = \begin{bmatrix} \sin t & \cos t \\ -\cos t & \sin t \end{bmatrix}.$$

From NODE, (9.25), the Wronskian $W(t)$ is given by

$$W(t) = W(t_0) \exp\left(\int_{t_0}^{t} \operatorname{tr}\{\mathbf{A}(s)\} ds\right) = W(t_0) \exp\left(\int_{t_0}^{t} 2\sin s \, ds\right)$$
$$= W(t_0) \exp[2(\cos t_0 - \cos t)].$$

(ii) For the system $\dot{x}_1 = f(t)x_2$, $\dot{x}_2 = g(t)x_1$, the matrix of coefficients is

$$\mathbf{A}(t) = \begin{bmatrix} 0 & f(t) \\ g(t) & 0 \end{bmatrix}.$$

The Wronskian is given by

$$W(t) = W(t_0) \exp\left(\int_{t_0}^{t} \operatorname{tr}\{\mathbf{A}(s)\} ds\right) = W(t_0),$$

a constant, since $\operatorname{tr}\{\mathbf{A}(s)\} = 0$.

9 : Stability by solution perturbation: Mathieu's equation   419

• 9.4 By substituting $x = c + a\cos t + b\sin t$ into Mathieu's equation $\ddot{x} + (\alpha + \beta \cos t)x = 0$, obtain by harmonic balance an approximation to the transition curve near $\alpha = 0$, $\beta = 0$ (compare with Section 9.4 in NODE).
By substituting $x = c + a\cos\tfrac{1}{2}t + b\sin\tfrac{1}{2}t$, find the transition curves near $\alpha = \tfrac{1}{4}$, $\beta = 0$.

9.4. Substitute $x = c + a\cos t + b\sin t$ into Mathieu's equation

$$\ddot{x} + (\alpha + \beta \cos t)x = 0.$$

Then

$$\begin{aligned}
\ddot{x} + (\alpha + \beta \cos t)x &= -a\cos t - b\sin t + (\alpha + \beta \cos t)(c + a\cos t + b\sin t)\\
&= -a\cos t - b\sin t + \alpha c + a\alpha \cos t + b\alpha \sin t\\
&\quad + c\beta \cos t + \tfrac{1}{2}a\beta + \text{(higher harmonics)}\\
&= \alpha c + \tfrac{1}{2}a\beta + (-a + a\alpha + c\beta)\cos t + (-b + b\alpha)\sin t\\
&\quad + \text{(higher harmonics)}
\end{aligned}$$

The constant and first harmonic terms vanish if

$$\alpha c + \tfrac{1}{2}a\beta = 0, \quad a(\alpha - 1) + c\beta = 0, \quad b(\alpha - 1) = 0.$$

Hence $b = 0$ and

$$c = -\frac{a\beta}{2\alpha} = -\frac{a(\alpha - 1)}{\beta}.$$

Therefore (with $a \neq 0$), α and β satisfy

$$\alpha^2 - \alpha - \tfrac{1}{2}\beta^2 = 0,$$

so that

$$\alpha = \tfrac{1}{2}[1 \pm \sqrt{(1 + 2\beta^2)}] \approx \begin{cases} 1 + \tfrac{1}{2}\beta^2 \\ -\tfrac{1}{2}\beta^2 \end{cases},$$

for small β. Near the origin in the (α, β) plane the transition curve is given by $\alpha = -\tfrac{1}{2}\beta^2$ which agrees with eqn (9.44).

Substitute $x = c + a\cos\tfrac{1}{2}t + b\sin\tfrac{1}{2}t$ into Mathieu's equation so that

$$\ddot{x} + (\alpha + \beta \cos t)x = -\tfrac{1}{4}a\cos\tfrac{1}{2}t - \tfrac{1}{4}b\sin\tfrac{1}{2}t + (\alpha + \beta \cos t)(c + a\cos\tfrac{1}{2}t + b\sin\tfrac{1}{2}t)$$

$$= \alpha c + a(-\tfrac{1}{4} + \alpha + \tfrac{1}{2}\beta)\cos\tfrac{1}{2}t$$

$$+ b(-\tfrac{1}{4} + \alpha - \tfrac{1}{2}\beta)\sin\tfrac{1}{2}t + \text{(higher harmonics)}$$

The constant term and the harmonics of lowest order vanish if

$$c = 0, \quad a\left(-\tfrac{1}{4} + \alpha + \tfrac{1}{2}\beta\right) = 0, \quad b\left(-\tfrac{1}{4} + \alpha - \tfrac{1}{2}\beta\right) = 0.$$

Hence $b = 0$ leads to $\alpha \approx \tfrac{1}{4} - \tfrac{1}{2}\beta$, and $a = 0$ implies $\alpha \approx \tfrac{1}{4} + \tfrac{1}{2}\beta$. These transition curves near $\alpha = \tfrac{1}{4}$ agree with eqn (9.45) (in NODE).

• **9.5** Figure 9.4 (in NODE) or Figure 9.1 represents a mass m attached to two identical linear strings of stiffness λ and natural length l. The ends of the strings pass through frictionless guides A and B at a distance $2L$, $l < L$, apart. The particle is set into lateral motion at the mid-point, and symmetrical tensions $a + b\cos\omega t$, $a > b$ are imposed on the ends of the string. Show that, for $x \ll L$,

$$\ddot{x} + \left(\frac{2\lambda(L - l + a)}{mL} + \frac{2\lambda b}{mL}\cos\omega t\right)x = 0.$$

Analyse the motion in terms of suitable parameters, using the information of NODE, Sections 9.3 and 9.4 on the growth or decay, periodicity and near periodicity of the solutions of Mathieu's equation in the regions of its parameter plane.

9.5. Let T be the tension in the string, and let x be the displacement of the particle. The transverse equation of motion is

$$-2T\sin\theta = m\ddot{x}.$$

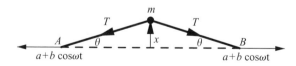

Figure 9.1 Problem 9.5: Transverse oscillations.

Hence

$$m\ddot{x} = -2\lambda[\sqrt{(L^2+x^2)} + a + b\cos\omega t - l]\frac{x}{\sqrt{(L^2+x^2)}}$$

$$= -2\lambda[1 + (a - l + b\cos\omega t)(L^2+x^2)^{-1/2}]x$$

$$= \left[1 + \frac{(a-l+b\cos\omega t)}{L}\left(1 + \frac{x^2}{L^2}\right)^{-1/2}\right]x$$

$$\approx -2\lambda\left[1 + \frac{a-l}{L} + \frac{b\cos\omega t}{L}\right]x \quad (x \ll L)$$

The linearized equation of motion becomes

$$\ddot{x} + \left(\frac{2\lambda(L-l+a)}{mL} + \frac{2\lambda b}{mL}\cos\omega t\right)x = 0.$$

Introduce the timescale τ where $\tau = \omega t$. Hence x satisfies

$$x'' + \left(\frac{2\lambda(L-l+a)}{m\omega^2 L} + \frac{2\lambda b}{m\omega^2 L}\cos\tau\right)x = 0.$$

We can express this equation in the standard form

$$x'' + (\alpha + \beta\cos\tau)x = 0,$$

where

$$\alpha = \frac{2\lambda(L-l+a)}{m\omega^2 L}, \quad \beta = \frac{2\lambda b}{m\omega^2 L}.$$

The stability regions can be seen by consulting Figure 9.3 (in NODE, showing the stability diagram for Mathieu's equation). We must assume that $\alpha = L + a - l > 0$, since otherwise the string would become slack. The critical curves on which period 2π solutions exist pass through the points with $\beta = 0$, $\alpha = n^2$, ($n = 0, 1, 2, \ldots$). In terms of the parameters, $\beta = 0$ corresponds to $b = 0$.

• 9.6 A pendulum with a light, rigid suspension is placed upside-down on end, and the point of suspension is caused to oscillate vertically with displacement y upwards given by $y = \varepsilon\cos\omega t$, $\varepsilon \ll 1$. Show that the equation of motion is

$$\ddot{\theta} + \left(-\frac{g}{a} - \frac{1}{a}\ddot{y}\right)\sin\theta = 0,$$

where a is the length of the pendulum, g is gravitational acceleration and θ the inclination to the vertical. Linearize the equation for small amplitudes and show that the vertical position is stable (i.e. the motion of the pendulum restricts itself to the neighbourhood of the vertical: it does not topple over) provided $\varepsilon^2\omega^2/(2ag) > 1$. For further discussion of the inverted pendulum and its stability see Acheson (1997).

9.6. The pendulum and notation are shown in Figure 9.2. Let R be the stress in the pendulum, and let (x, z) be the coordinates of the bob. The horizontal and vertical equations of motion are given by

$$-R\sin\theta = m\ddot{x} = m\frac{d^2}{dt^2}(a\sin\theta) = m(-a\sin\theta\dot{\theta}^2 + a\cos\theta\ddot{\theta}),$$

$$-mg - R\cos\theta = m\ddot{z} = m\left[\frac{d^2}{dt^2}(a\cos\theta) + \ddot{y}\right] = m(-a\cos\theta\dot{\theta}^2 - a\sin\theta\ddot{\theta} + \ddot{y}).$$

Elimination of R between these equations leads to

$$-g\sin\theta = -a\ddot{\theta} + \ddot{y}\sin\theta,$$

or

$$\ddot{\theta} - \left(\frac{g+\ddot{y}}{a}\right)\sin\theta = \ddot{\theta} - \frac{1}{a}(g - \varepsilon\omega^2\cos\omega t)\sin\theta = 0.$$

For small $|\theta|$, $\sin\theta \approx \theta$ so that the linearized equation is

$$\ddot{\theta} - \frac{1}{a}(g - \varepsilon\omega^2\cos\omega t)\theta = 0.$$

Let $\tau = \omega t$ so that, in standard Mathieu form, the equation is

$$\theta'' + (\alpha + \beta\cos\tau)\theta = 0,$$

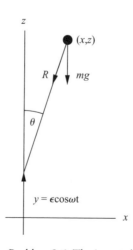

Figure 9.2 Problem 9.6: The inverted pendulum.

where
$$\alpha = -\frac{g}{a\omega^2}, \quad \beta = \frac{\varepsilon}{a}.$$

Consult Figure 9.3 (in NODE). Since $\alpha < 0$, the figure indicates that there is a stability region for small β. The period 2π solution occurs approximately on $\alpha = -\frac{1}{2}\beta^2$ (see eqn (9.44) in NODE). Hence for sufficiently small ε, stability occurs where

$$-\tfrac{1}{2}\beta^2 < \alpha < 0, \text{ or } \varepsilon^2\omega^2 > 2ag.$$

- 9.7 Let $\Phi(t) = [\phi_{ij}(t)]$, $i, j = 1, 2$, be the fundamental matrix for the system $\dot{x}_1 = x_2$, $\dot{x}_2 = -(\alpha + \beta \cos t)x_1$, satisfying $\Phi(0) = I$ (Mathieu's equation). Show that the characteristic numbers μ satisfy the equation
$\mu^2 - \mu\{\phi_{11}(2\pi) + \phi_{22}(2\pi)\} + 1 = 0.$

9.7. Mathieu's equation in the form

$$\dot{x}_1 = x_2, \quad \dot{x}_2 = -(\alpha + \beta \cos t)x_1,$$

is assumed to have the fundamental matrix $\Phi(t)$ satisfying $\Phi(0) = I$. The matrix E is given by

$$E = \Phi^{-1}(0)\Phi(2\pi) = \Phi(2\pi),$$

since $\Phi(0) = I$. Let μ_1 and μ_2 be the characteristic numbers. These satisfy

$$\det(E - \mu I) = \det(\Phi(2\pi) - \mu I)$$

$$= \begin{vmatrix} \phi_{11}(2\pi) - \mu & \phi_{12}(2\pi) \\ \phi_{21}(2\pi) & \phi_{22}(2\pi) - \mu \end{vmatrix}$$

$$= \mu^2 - (\phi_{11}(2\pi) + \phi_{22}(2\pi))\mu + 1 = 0,$$

since $\det(E) = 1$, by NODE, Theorem 9.5.

- **9.8** In Section 9.3, for the transition curves of Mathieu's equation for solutions period 2π, let

$$D_{m,n} = \begin{vmatrix} 1 & \gamma_m & 0 & & & & & \\ \gamma_{m-1} & 1 & \gamma_{m-1} & & & & & \\ 0 & \gamma_{m-2} & 1 & & & & & \\ & & & \ddots & & & & \\ & & & & \gamma_0 & 1 & \gamma_0 & \\ & & & & & & \ddots & \\ & & & & & \gamma_{n-1} & 1 & \gamma_{n-1} \\ & & & & & 0 & \gamma_n & 1 \end{vmatrix}$$

for $m \geq 0$, $n \geq 0$. Show that $D_{m,n} = D_{m-1,n} - \gamma_m \gamma_{m-1} D_{m-2,n}$. Let $E_n = D_{n,n}$, and verify that

$E_0 = 1$, $E_1 = 1 - 2\gamma_0\gamma_1$, $E_2 = (1 - \gamma_1\gamma_2)^2 - 2\gamma_0\gamma_1(1 - \gamma_1\gamma_2)$.

Prove that, for $n \geq 1$,

$$E_{n+2} = (1 - \gamma_{n+1}\gamma_{n+2})E_{n+1} - \gamma_{n+1}\gamma_{n+2}(1 - \gamma_{n+1}\gamma_{n+2})E_n + \gamma_n^2 \gamma_{n+1}^3 \gamma_{n+2} E_{n-1}.$$

9.8. Expansion of the determinant in the problem by its first row gives

$$D_{m,n} = D_{m-1,n} - \gamma_m \gamma_{m-1} D_{m-2,n}. \tag{i}$$

Let $m = n$ and $E_n = D_{n,n}$. Finite approximations for the transition curves are given by $E_n = 0$ for $n = 1, 2, 3, \ldots$. The first three expressions are

$$E_0 = 1, \quad E_1 = \begin{vmatrix} 1 & \gamma_1 & 0 \\ \gamma_0 & 1 & \gamma_0 \\ 0 & \gamma_1 & 1 \end{vmatrix} = 1 - 2\gamma_0\gamma_1,$$

$$E_2 = \begin{vmatrix} 1 & \gamma_2 & 0 & 0 & 0 \\ \gamma_1 & 1 & \gamma_1 & 0 & 0 \\ 0 & \gamma_0 & 1 & \gamma_0 & 0 \\ 0 & 0 & \gamma_1 & 1 & \gamma_1 \\ 0 & 0 & 0 & \gamma_2 & 1 \end{vmatrix} = (1 - \gamma_1\gamma_2)^2 - 2\gamma_0\gamma_1(1 - \gamma_1\gamma_2).$$

Observe that $D_{m,n} = D_{n,m}$. Let $E_n = D_{n,n}$, $P_n = D_{n-1,n}$ and $Q_n = D_{n-2,n}$. Put $m = n, n+1, n+2$ in (i) resulting in the three equations

$$E_n = P_n - \gamma_n \gamma_{n-1} Q_n, \tag{ii}$$

$$P_{n+1} = E_n - \gamma_{n+1} \gamma_n P_n, \tag{iii}$$

$$Q_{n+2} = P_{n+1} - \gamma_{n+2} \gamma_{n+1} E_n. \tag{iv}$$

Eliminate Q_n between (ii) and (iv), so that

$$E_{n+2} = P_{n+2} - \gamma_{n+2}\gamma_{n+1}P_{n+1} + \gamma_{n+2}^2\gamma_{n+1}^2 E_n. \qquad \text{(v)}$$

From (iii) and (v) so that

$$2\gamma_{n+1}\gamma_{n+2}P_{n+1} = E_{n+1} - E_{n+2} + \gamma_{n+1}^2\gamma_{n+2}^2 E_n. \qquad \text{(vi)}$$

Finally substitute P_n from (vi) back into (iii) which leads to the third-order difference equation

$$E_{n+2} = (1 - \gamma_{n+1}\gamma_{n+2})E_{n+1} - \gamma_{n+1}\gamma_{n+2}(1 - \gamma_{n+1}\gamma_{n+2})E_n + \gamma_n^2\gamma_{n+1}^3\gamma_{n+2}E_{n-1}.$$

- 9.9 In eqn (9.38) (in NODE), for the transition curves of Mathieu's equation for solutions of period 4π, let

$$F_{m,n} = \begin{vmatrix} 1 & \delta_m & 0 & & & & & \\ \delta_{m-1} & 1 & \delta_{m-1} & & & & & \\ 0 & \delta_{m-2} & 1 & & & & & \\ & & & \ddots & & & & \\ & & & & \delta_1 & 1 & \delta_1 & 0 \\ & & & & 0 & \delta_1 & 1 & \delta_1 \\ & & & & & & \ddots & \\ & & & & & \delta_{n-1} & 1 & \delta_{n-1} \\ & & & & & 0 & \delta_n & 1 \end{vmatrix}$$

Show as in the last exercise that $G_n = F_{n,n}$ satisfies the same recurrence relation as E_n for $n \geq 2$ (see Problem 9.8). Verify that

$$G_1 = 1 - \delta_1^2, \quad G_2 = (1 - \delta_1\delta_2)^2 - \delta_1^2,$$
$$G_3 = (1 - \delta_1\delta_2 - \delta_2\delta_3)^2 - \delta_1^2(1 - \delta_2\delta_3)^2.$$

9.9. Expansion by the first row gives

$$F_{m,n} = F_{m-1,n} - \delta_m\delta_{m-1}F_{m-2,n},$$

which is essentially the same difference equation as for $D_{m,n}$ in the previous problem. Hence if $G_n = F_{n,n}$, then must satisfy the same difference equation as E_n, that is

$$G_{n+2} = (1 - \delta_{n+1}\delta_{n+2})G_{n+1} - \delta_{n+1}\delta_{n+2}(1 - \delta_{n+1}\delta_{n+2})G_n + \delta_n^2\delta_{n+1}^3\delta_{n+2}G_{n-1}.$$

However, the initial terms will differ. Thus

$$G_1 = \begin{vmatrix} 1 & \delta_1 \\ \delta_1 & 1 \end{vmatrix} = 1 - \delta_1^2, \quad G_2 = \begin{vmatrix} 1 & \delta_2 & 0 & 0 \\ \delta_1 & 1 & \delta_1 & 0 \\ 0 & \delta_1 & 1 & \delta_1 \\ 0 & 0 & \delta_2 & 1 \end{vmatrix} = (1 - \delta_1\delta_2)^2 - \delta_1^2,$$

$$G_3 = \begin{vmatrix} 1 & \delta_3 & 0 & 0 & 0 & 0 \\ \delta_2 & 1 & \delta_2 & 0 & 0 & 0 \\ 0 & \delta_1 & 1 & \delta_1 & 0 & 0 \\ 0 & 0 & \delta_1 & 1 & \delta_1 & 0 \\ 0 & 0 & 0 & \delta_2 & 1 & \delta_2 \\ 0 & 0 & 0 & 0 & \delta_3 & 1 \end{vmatrix} = (1 - \delta_1\delta_2 - \delta_2\delta_3)^2 - \delta_1^2(1 - \delta_2\delta_3)^2.$$

Note that the determinants E_n in Problem 9.7 are determinants of odd order, but that G_n are determinants of even order.

• **9.10** Show, by the perturbation method, that the transition curves for Mathieu's equation
$$\ddot{x} + (\alpha + \beta \cos t)x = 0,$$
near $\alpha = 1$, $\beta = 0$, are given approximately by $\alpha = 1 + \frac{1}{12}\beta^2$, $\alpha = 1 - \frac{5}{12}\beta^2$.

9.10. In Mathieu's equation
$$\ddot{x} + (\alpha + \beta \cos t)x = 0,$$
assume that $|\beta|$ is small, substitute the expansions $\alpha = \alpha_0 + \beta\alpha_1 + \beta^2\alpha_2 + \cdots$ and $x = x_0 + \beta x_1 + \beta^2 x_2 + \cdots$. Therefore

$$(\ddot{x}_0 + \beta\ddot{x}_1 + \beta^2\ddot{x}_2 + \cdots) + [(\alpha_0 + \beta\alpha_1 + \beta^2\alpha_2 + \cdots) + \beta \cos t]$$

$$(x_0 + \beta x_1 + \beta^2 x_2 + \cdots) = 0.$$

Equating the coefficients of powers of β to zero we obtain

$$\ddot{x}_0 + \alpha_0 x_0 = 0, \tag{i}$$

$$\ddot{x}_1 + \alpha_0 x_1 = -\alpha_1 x_0 - x_0 \cos t, \tag{ii}$$

$$\ddot{x}_2 + \alpha_0 x_2 = -\alpha_2 x_0 - \alpha_1 x_1 - x_1 \cos t. \tag{iii}$$

Since $\alpha \approx 1$, we are searching for period 2π solutions. Therefore put $\alpha_0 = 1$, so that (i) implies $x_0 = a_0 \cos t + b_0 \sin t$. Equation (ii) becomes

$$\ddot{x}_1 + x_1 = -\alpha_1 a_0 \cos t - \alpha_1 b_0 \sin t - \tfrac{1}{2}a_0 - \tfrac{1}{2}a_0 \cos 2t - \tfrac{1}{2}b_0 \sin 2t. \tag{iv}$$

Secular terms can only be removed by putting $\alpha_1 = 0$. For this value of α_1, eqn (iv) has the general solution

$$x_1 = a_1 \cos t + b_1 \sin t - \tfrac{1}{2}a_0 + \tfrac{1}{6}a_0 \cos 2t + \tfrac{1}{6}b_0 \sin 2t.$$

Equation (iii) becomes

$$\ddot{x}_2 + x_2 = -\alpha_2 a_0 \cos t - \alpha_2 b_0 \sin t - a_1 \cos t + b_1 \sin t - \tfrac{1}{2}a_0$$
$$+ \tfrac{1}{6}a_0 \cos 2t + \tfrac{1}{6}b_0 \sin 2t \cos t$$
$$= -\alpha_2 a_0 \cos t - \alpha_2 b_0 \sin t + \tfrac{1}{2}a_1 - \tfrac{5}{12}a_0 \cos t + \tfrac{1}{12}b_0 \sin t$$
$$+ \tfrac{1}{2}a_1 \cos 2t + \tfrac{1}{2}b_1 \sin 2t + \tfrac{1}{12}a_0 \cos 3t + \tfrac{1}{12}b_0 \sin 3t$$

Secularity is removed if the coefficients of $\cos t$ and $\sin t$ are zero. Hence α_2 can take two possible values, namely, $\alpha_2 = -\tfrac{5}{12}$ and $\alpha_2 = \tfrac{1}{12}$. Therefore the curves along which period-2π solutions occur are $\alpha \approx 1 - \tfrac{5}{12}\beta^2$ and $\alpha \approx 1 + \tfrac{1}{12}\beta^2$.

- 9.11 Consider Hill's equation $\ddot{x} + f(t)x = 0$, where f has period 2π, and

$$f(t) = \alpha + \sum_{r=1}^{\infty} \beta_r \cos rt$$

is its Fourier expansion, with $\alpha \approx \tfrac{1}{4}$ and $|\beta_r| \ll 1$, $r = 1, 2, \ldots$. Assume an approximate solution $e^{\sigma t}q(t)$, where σ is real and q has period 4π as in (9.34) (in NODE). Show that

$$\ddot{q} + 2\sigma\dot{q} + \left(\sigma^2 + \alpha\sum_{r=1}^{\infty}\beta_r \cos rt\right)q = 0.$$

Take $q \approx \sin(\tfrac{1}{2}t + \gamma)$ as the approximate form for q and match terms in $\sin\tfrac{1}{2}t$, $\cos\tfrac{1}{2}t$, on the assumption that these terms dominate. Deduce that

$$\sigma^2 = -\left(\alpha + \tfrac{1}{4}\right) + \tfrac{1}{2}\sqrt{(4\alpha + \beta_1^2)}$$

and that the transition curves near $\alpha = \tfrac{1}{4}$ are given by $\alpha = \tfrac{1}{4} \pm \tfrac{1}{2}\beta_1$.

9.11. Consider Hill's equation $\ddot{x} + f(t)x = 0$, where

$$f(t) = \alpha + \sum_{r=1}^{\infty} \beta_r \cos rt.$$

Assume $\alpha \approx \tfrac{1}{4}$ and $|\beta_r| \ll 1$, $r = 1, 2, \ldots$. Let $x = e^{\sigma t}q(t)$. Then

$$\dot{x} = e^{\sigma t}(\dot{q} + \sigma q), \quad \ddot{x} = e^{\sigma t}(\ddot{q} + \sigma\dot{q} + \sigma^2 q).$$

Therefore q satisfies

$$\ddot{q} + \sigma\dot{q} + \left(\sigma^2 + \alpha + \sum_{r=1}^{\infty}\beta_r \cos rt\right)q = 0.$$

Let $q \approx \sin(\tfrac{1}{2}t + \gamma)$, and substitute q into Hill's equation so that

$$-\tfrac{1}{4}\sin(\tfrac{1}{2}t + \gamma) + \sigma\cos(\tfrac{1}{2}t + \gamma) + (\sigma^2 + \alpha)\sin(\tfrac{1}{2}t + \gamma)$$
$$+ \tfrac{1}{2}\sum_{r=1}^{\infty}\beta_r[\sin\{(r + \tfrac{1}{2})t + \gamma\} + \sin\{(\tfrac{1}{2} - r)t + \gamma\}] = 0.$$

The leading harmonics vanish if

$$(\sigma^2 + \alpha - \tfrac{1}{4})(\sin\tfrac{1}{2}t\cos\gamma + \cos\tfrac{1}{2}t\sin\gamma) + \sigma(\cos\tfrac{1}{2}t\cos\gamma - \sin\tfrac{1}{2}t\sin\gamma)$$
$$+ \tfrac{1}{2}\beta_1(-\sin\tfrac{1}{2}t\cos\gamma + \cos\tfrac{1}{2}t\sin\gamma) = 0.$$

Coefficients of $\cos\tfrac{1}{2}t$ and $\sin\tfrac{1}{2}t$ imply

$$(\sigma^2 + \alpha - \tfrac{1}{4} - \tfrac{1}{2}\beta_1)\cos\gamma - \sigma\sin\gamma = 0,$$
$$\sigma\cos\gamma + (\sigma^2 + \alpha - \tfrac{1}{4} + \tfrac{1}{2}\beta_1)\sin\gamma = 0.$$

These equations are consistent if

$$(\sigma^2 + \alpha - \tfrac{1}{4} - \tfrac{1}{2}\beta_1)(\sigma^2 + \alpha - \tfrac{1}{4} + \tfrac{1}{2}\beta_1) + \sigma^2 = 0,$$

or

$$(\sigma^2 + \alpha - \tfrac{1}{4})^2 + (\sigma^2 + \alpha - \tfrac{1}{4}) - (\tfrac{1}{4}\beta_1^2 + \alpha - \tfrac{1}{4}) = 0..$$

The solutions of this quadratic equation are

$$\sigma^2 = -(\alpha + \tfrac{1}{4}) \pm \tfrac{1}{2}\sqrt{(\beta_1^2 + 4\alpha)}.$$

For $\alpha = \tfrac{1}{4}$ and $\beta_1 = 0$, $\sigma^2 = -\tfrac{1}{2} \pm \tfrac{1}{2}$ which will only be zero (giving a 4π-periodic solution) if we choose the $+$ sign. Therefore

$$\sigma^2 = -(\alpha + \tfrac{1}{4}) + \tfrac{1}{2}\sqrt{(\beta_1^2 + 4\alpha)}.$$

σ real implies unstable solutions, so unstable solutions require $\sigma^2 > 0$. Hence instability occurs if

$$\tfrac{1}{2}\sqrt{(\beta_1^2 + 4\alpha)} > \alpha + \tfrac{1}{4}, \text{ or } \alpha^2 - \tfrac{1}{2}\alpha + \tfrac{1}{16} - \tfrac{1}{4}\beta_1^2 < 0,$$

or

$$(\alpha - \tfrac{1}{4} - \tfrac{1}{2}\beta_1)(\alpha - \tfrac{1}{4} + \tfrac{1}{2}\beta_1) < 0, \text{ or } \tfrac{1}{2}|\beta_1| > |\alpha - \tfrac{1}{4}|.$$

The stability boundaries ($\sigma = 0$) are $\alpha = \tfrac{1}{4} \pm \tfrac{1}{2}\beta_1$.

- 9.12 Obtain, as in NODE, Section 9.4, the boundary of the stable region in the neighbourhood of $v = 1$, $\beta = 0$ for Mathieu's equation with damping,
$$\ddot{x} + \kappa \dot{x} + (v + \beta \cos t)x = 0,$$
where $\kappa = O(\beta^2)$.

9.12. The Mathieu equation with damping is

$$\ddot{x} + \kappa \dot{x} + (v + \beta \cos t)x = 0. \tag{i}$$

It is assumed that $\kappa = O(\beta^2)$. To remove the damping term, let $x = e^{-\frac{1}{2}\kappa t}\eta(t)$. Then

$$\dot{x} = e^{-(1/2)\kappa t}(\eta' - \tfrac{1}{2}\kappa \eta), \ddot{x} = e^{-(1/2)\kappa t}(\eta'' - \kappa \eta' + \tfrac{1}{4}\kappa^2 \eta).$$

Therefore eqn (i) is transformed into the Mathieu equation

$$\ddot{\eta} + (v + \tfrac{1}{2}\kappa^2 + \beta \cos t)x = 0.$$

Solutions of period 2π exist near critical values of v. In the usual notation $\alpha = v + \tfrac{1}{2}\kappa^2$. If $\phi(\alpha, \beta) > 0$, then

$$\eta = c_1 e^{\sigma t} p_1(t) + c_2 e^{\sigma t} p_2(t),$$

but

$$x = c_1 e^{(\sigma - (1/2)\kappa)t} p_1(t) + c_2 e^{(-\sigma - (1/2)\kappa)t} p_2(t).$$

Therefore the boundary of the 2π periodic solution will be $\sigma - \tfrac{1}{2}\kappa = 0$. Consider the perturbation procedure in which

$$x = x_0 + \beta x_1 + \beta^2 x_2 + \cdots, v = v_0 + v_1 \beta + v_2 \beta^2 + \cdots, \kappa = \kappa_2 \beta^2 + \cdots.$$

Then (i) becomes

$$(\ddot{x}_0 + \beta \ddot{x}_1 + \beta^2 \ddot{x}_2 + \cdots) + (\kappa_2 \beta^2 \dot{x}_0 + \cdots)$$
$$+ (v_0 + \beta v_1 + \beta^2 v_2 + \cdots + \beta \cos t)(x_0 + \beta x_1 + \beta^2 x_2 + \cdots) = 0$$

Hence the perturbation equations are

$$\ddot{x}_0 + v_0 x_0 = 0, \tag{ii}$$
$$\ddot{x}_1 + v_0 x_1 = -(v_1 + \cos t)x_0, \tag{iii}$$
$$\ddot{x}_2 + v_0 x_2 = -\kappa_2 \dot{x}_0 - v_2 x_0 - v_1 x_1 - x_1 \cos t. \tag{iv}$$

For 2π periodicity, $v_0 = 1$: hence (ii) implies

$$x_0 = a_0 \cos t + b_0 \sin t.$$

From (iii), x_1 satisfies

$$\ddot{x}_1 + x_1 = -v_1 a_0 \cos t - v_1 b_0 \sin t - \tfrac{1}{2} a_0 - \tfrac{1}{2} a_0 \cos 2t - \tfrac{1}{2} b_0 \sin 2t.$$

Must choose $v_1 = 0$, since otherwise we can only have $a_0 = b_0 = 0$ which leads to the trivial solution $x = 0$. Hence

$$x_1 = -\tfrac{1}{2} a_0 + \tfrac{1}{6} a_0 \cos 2t + \tfrac{1}{6} \sin 2t.$$

Equation (iv) becomes

$$\ddot{x}_2 + x_2 = \kappa_2 a_0 \sin t - \kappa_2 b_0 \cos t - v_2 a_0 \cos t - v_2 b_0 \sin t +$$

$$\tfrac{1}{2} a_0 \cos t - \tfrac{1}{12} a_0 \cos t - \tfrac{1}{12} b_0 \sin t + \text{(higher harmonics)}$$

$$= \left(-\kappa_2 b_0 - v_2 a_0 + \tfrac{1}{2} a_0 - \tfrac{1}{12} a_0\right) \cos t +$$

$$\left(\kappa_2 a_0 - v_2 b_0 - \tfrac{1}{12} b_0\right) \sin t + \text{(higher harmonics)}$$

To remove secularity, we must put

$$\left(\tfrac{5}{12} - v_2\right) a_0 - \kappa_2 b_0 = 0,$$

$$\kappa_2 a_0 - \left(\tfrac{1}{12} + v_2\right) b_0 = 0.$$

These linear equations have non-trivial solutions if

$$\left(\tfrac{5}{12} - v_2\right)\left(\tfrac{1}{12} + v_2\right) - \kappa_2^2 = 0.$$

Therefore

$$v_2^2 - \tfrac{1}{3} v_2 + \left(\kappa_2^2 - \tfrac{5}{144}\right) = 0,$$

so that

$$v_2 = \tfrac{1}{6} \pm \tfrac{1}{4}\sqrt{(1 - 16\kappa_2^2)},$$

where it is required that $\kappa_2 < \tfrac{1}{4}$. Hence 2π periodic solutions occur on the curves

$$v \approx 1 \pm \tfrac{1}{4}\sqrt{(1 - 16\kappa_2^2)}.$$

9: Stability by solution perturbation: Mathieu's equation

• **9.13** Solve Meissner's equation $\ddot{x} + (\alpha + \beta f(t))x = 0$ where $f(t) = 1, 0 \leq t < \pi$; $f(t) = -1, \pi \leq t < 2\pi$ and $f(t + 2\pi) = f(t)$ for all t. Find the conditions on α, β, for periodic solutions by putting $x(0) = x(2\pi), \dot{x}(0) = \dot{x}(2\pi)$ and by making x and $\dot{x}$ continuous at $t = \pi$. Find a determinant equation for α and β.

9.13. Meissner's equation is
$$\ddot{x} + [\alpha + \beta f(t)]x = 0,$$
where
$$f(t) = f(t + 2\pi), \quad f(t) = \begin{cases} 1 & 0 \leq t < \pi \\ -1 & \pi \leq t < 2\pi \end{cases}.$$

Assume that $\alpha + \beta > 0$ and $\alpha - \beta > 0$. In the interval $(0, \pi)$, Meissner's equation is
$$\ddot{x} + (\alpha + \beta)x = 0,$$
which has the general solution
$$x_1 = A \cos \lambda t + B \sin \lambda t, \lambda = \sqrt{(\alpha + \beta)}.$$

In the interval $(\pi, 2\pi)$, the equation
$$\ddot{x} + (\alpha - \beta)x = 0,$$
has the general solution
$$x_2 = C \cos \mu t + D \sin \mu t, \mu = \sqrt{(\alpha - \beta)}.$$

Periodicity occurs if
$$x_1(\pi) = x_2(\pi), \dot{x}_1(\pi) = \dot{x}_2(\pi), x_1(0) = x_2(2\pi), \dot{x}_1(0) = \dot{x}_2(2\pi).$$

These conditions become
$$A \cos \lambda \pi + B \sin \lambda \pi = C \cos \mu \pi + D \sin \mu \pi,$$
$$-A\lambda \sin \lambda \pi + B\lambda \cos \lambda \pi = -C\mu \sin \mu \pi + D\mu \cos \mu \pi,$$
$$A = C \cos 2\mu \pi + D \sin 2\mu \pi,$$
$$B\lambda = -C\mu \sin 2\mu \pi + D\mu \cos 2\mu \pi.$$

432 Nonlinear Ordinary Differential Equations: Problems and Solutions

These equations have non-trivial solutions for A, B, C, D if

$$\begin{vmatrix} \cos \lambda \pi & \sin \lambda \pi & -\cos \mu \pi & -\sin \mu \pi \\ -\lambda \sin \lambda \pi & \lambda \cos \lambda \pi & \mu \sin \mu \pi & -\mu \cos \mu \pi \\ 1 & 0 & -\cos 2\mu \pi & -\sin 2\mu \pi \\ 0 & \lambda & \mu \sin 2\mu \pi & -\mu \cos 2\mu \pi \end{vmatrix} = 0.$$

Expansion of the determinant leads to

$$2\lambda \mu - 2\lambda \mu \cos \lambda \pi \cos \mu \pi + (\lambda^2 + \mu^2) \sin \lambda \pi \sin \mu \pi = 0,$$

or

$$\sqrt{(\alpha^2 - \beta^2)}[1 - \cos \sqrt{(\alpha + \beta)} \pi \cos \sqrt{(\alpha - \beta)} \pi]$$

$$+ \alpha \sin \sqrt{(\alpha + \beta)} \sin \sqrt{(\alpha - \beta)} \pi = 0. \qquad \text{(i)}$$

If $\beta = 0$, then $\cos 2\sqrt{\alpha} \pi = 1$. Therefore the critical values on the α axis occur at αn^2, ($n = 0, 1, 2, \ldots$). The general solutions of (i) are straight lines $\beta = \pm \alpha$ along which 2π periodic solutions occur (subject to the restriction $\alpha > \beta$).

> **9.14** By using the harmonic balance method of Chapter 4 in NODE, show that the van der Pol equation with parametric excitation,
> $$\ddot{x} + \varepsilon(x^2 - 1)\dot{x} + (1 + \beta \cos t)x = 0$$
> has a 2π-periodic solution with approximately the same amplitude as the unforced van der Pol equation.

9.14. The van der Pol equation with parametric excitation is

$$\ddot{x} + \varepsilon(x^2 - 1)\dot{x} + (1 + \beta \cos t)x = 0.$$

Let $x \approx c + a \cos t + b \sin t$. Then

$$\ddot{x} + \varepsilon(x^2 - 1)\dot{x} + (1 + \beta \cos t)x$$

$$= (-a \cos t - b \sin t) + \varepsilon[(c + a \cos t + b \sin t)^2 - 1](-a \sin t + b \cos t)$$

$$+ (1 + \beta \cos t)(c + a \cos t + b \sin t)$$

$$= (c + \tfrac{1}{2}a\beta) + [c\beta + b\varepsilon(-1 + \tfrac{1}{4}(a^2 + b^2) + c^2)]\cos t$$

$$+ a\varepsilon[1 - \tfrac{1}{4}(a^2 + b^2) - c^2]\sin t + \text{(higher harmonics)}$$

The approximation is a solution if the constant term and the coefficients of $\cos t$ and $\sin t$ are zero, that is, if

$$c + \tfrac{1}{2}a\beta = 0,$$

$$c\beta + b\varepsilon(-1 + \tfrac{1}{4}(a^2 + b^2) + c^2) = 0,$$

$$a\varepsilon[1 - \tfrac{1}{4}(a^2 + b^2) - c^2] = 0.$$

The only non-trivial solution of these equations is $a = c = 0$ and $b^2 = 4$. The solution becomes $x \approx 2\sin t$, which has amplitude 2, the same as that for the unforced van der Pol equation.

• 9.15 The male population M and female population F for a bird community have a constant death rate k and a variable birth rate $\mu(t)$ which has period T, so that
$\dot{M} = -kM + \mu(t)F$, $\dot{F} = -kF + \mu(t)F$.
The births are seasonal, with rate

$$\mu(t) = \begin{cases} \delta, & 0 < t \leq \varepsilon; \\ 0, & \varepsilon < t \leq T \end{cases}.$$

Show that periodic solutions of period T exist for M and F if $kT = \delta\varepsilon$.

9.15. The male (M) and female (F) population sizes satisfy

$$\dot{M} = -kM + \mu(t)F, \quad \dot{F} = -kF + \mu(t)F,$$

where $\mu(t)$, defined by

$$\mu(t) = \begin{cases} \delta & 0 < t \leq \varepsilon \\ 0 & \varepsilon < t \leq T \end{cases},$$

is periodic with period T. The equation for F has the general solution

$$F = \begin{cases} Ae^{(\delta-k)t} & 0 < t \leq \varepsilon \\ Be^{-kt} & \varepsilon < t \leq T \end{cases}.$$

The function is periodic and continuous if it is continuous at $t = \varepsilon$, and if $F(0) = F(T)$. Therefore

$$Ae^{(\delta-k)\varepsilon} = Be^{-k\varepsilon}, \text{ or, } Ae^{\delta\varepsilon} = B, \tag{i}$$

$$A = Be^{-kT}. \tag{ii}$$

From (i) and (ii) if $e^{-kT+\delta\varepsilon} = 1$, or if $kT = \delta\varepsilon$.

The equation for M is

$$\dot{M} + kM = \begin{cases} \delta A e^{(\delta-k)t} & 0 < t \leq \varepsilon \\ 0 & \varepsilon < t \leq T \end{cases}$$

The general solution is

$$M = \begin{cases} Ce^{-kt} + Ae^{(\delta-k)t} & 0 < t \leq \varepsilon \\ De^{-kt} & \varepsilon < t \leq T \end{cases}.$$

Continuity and periodicity imply

$$Ce^{-k\varepsilon} + Ae^{(\delta-k)\varepsilon} = De^{-k\varepsilon}, \tag{iii}$$

$$C + A = De^{-kT}. \tag{iv}$$

If $kT = \delta\varepsilon$, then $C(1 - e^{\delta\varepsilon}) = 0$. Hence $C = 0$, and eqns (iii) and (iv) are satisfied. If $kT = \delta\varepsilon$ then both F and M are periodic.

• **9.16** A pendulum bob is suspended by a light rod of length a, and the support is constrained to move vertically with displacement $\zeta(t)$. Show that the equation of motion is
$a\ddot{\theta} + (g + \ddot{\zeta}(t))\sin\theta = 0$,
where θ is the angle of inclination to the downward vertical. Examine the stability of the motion for the case when $\zeta(t) = c\sin\omega t$, on the assumption that it is permissible to put $\sin\theta \approx \theta$.

9.16. The suspended pendulum is shown in Figure 9.3: the position of the bob is given by the coordinates (x, y), and R is the reaction in the rod. The upward displacement of the support is given by $\zeta(t)$. In terms of θ, $x = a\sin\theta$ and $y = \zeta(t) - a\cos\theta$. The horizontal and vertical equations of motion are

$$-R\sin\theta = m\ddot{x} = m[a\cos\theta\ddot{\theta} - a\sin\theta\dot{\theta}^2],$$

$$R\cos\theta - mg = m\ddot{y} = m[\ddot{\zeta} + a\sin\theta\ddot{\theta} + a\cos\theta\dot{\theta}^2].$$

Elimination of R leads to

$$-g\sin\theta = \ddot{\zeta}\sin\theta + a\ddot{\theta} \quad \text{or} \quad a\ddot{\theta} + (g + \ddot{\zeta})\sin\theta = 0.$$

If $\zeta = c\sin\omega t$ and $\sin\theta \approx \theta$, then

$$a\ddot{\theta} + (g - c\omega^2 \sin\omega t)\theta \approx 0.$$

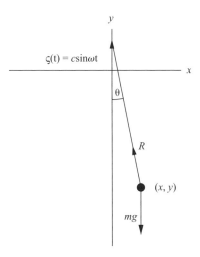

Figure 9.3 Problem 9.16:

To express this equation in standard Mathieu form, let $\omega t = \tau - \tfrac{1}{2}\pi$, where τ is a new variable, so that

$$\theta'' + (\alpha + \beta \cos \tau)\theta = 0, \quad \alpha = \frac{g}{a\omega^2}, \beta = \frac{c}{a}.$$

The stability regions are indicated in the parameter diagram shown in Figure 9.3 (in NODE). The 2π periodic boundaries pass through the points

$$\alpha = \frac{g}{a\omega^2} = n^2, \ \beta = 0, \ (n = 0, 1, 2, \ldots),$$

and the 4π periodic boundaries pass through

$$\alpha = \frac{g}{a\omega^2} = (n + \tfrac{1}{2})^2, \ \beta = 0, \ (n = 0, 1, 2, \ldots).$$

- **9.17** A pendulum, with bob of mass m and rigid suspension of length a, hangs from a support which is constrained to move with vertical and horizontal displacements $\zeta(t)$ and $\eta(t)$ respectively. Show that the inclination θ of the pendulum satisfies the equation $a\ddot\theta + (g + \ddot\zeta) \sin\theta + \ddot\eta \cos\theta = 0$.

Let $\zeta = A \sin \omega t$ and $\eta = B \sin \omega t$, where $\omega = \sqrt{(g/a)}$. Show that after linearizing this equation for small amplitudes, the resulting equation has a solution $\theta = -(8B/A)\cos \omega t$

9.17. The pendulum is shown in Figure 9.4; the position of the bob is given by the coordinates (x, y), and R is the reaction in the rod. The upward displacement of the support is $\zeta(t)$ and its horizontal displacement is $\eta(t)$. In terms of θ, $x = \eta(t) + a \sin \theta$ and $y = \zeta(t) - a \cos \theta$. The

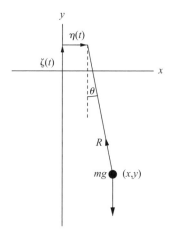

Figure 9.4 Problem 9.17: Pendulum with horizontal forcing.

horizontal and vertical equations of motion are

$$-R\sin\theta = m(\ddot{\eta} + \ddot{x}) = m(\ddot{\eta} - a\sin\theta\dot{\theta}^2 + a\cos\theta\ddot{\theta}),$$

$$R\cos\theta - mg = m(\ddot{\zeta} + \ddot{y}) = m(\ddot{\zeta} + a\cos\theta\dot{\theta}^2 + a\cos\theta\ddot{\theta}).$$

Elimination of R leads to

$$a\ddot{\theta} + (g + \ddot{\zeta})\sin\theta + \ddot{\eta}\cos\theta = 0.$$

Assume that $|\theta|$ is small, so that $\sin\theta \approx \theta$ and $\cos\theta \approx 1$. Then, if $\omega = \sqrt{(g/a)}$, $\zeta = A\cos\omega t$ and $\eta = B\sin 2\omega t$, the approximate equation of motion is

$$a\ddot{\theta} + g[1 - (A/a)\sin\omega t]\theta = (8Bg/a)\sin\omega t\cos\omega t.$$

Apply the change of scale $\omega t = \tau$: then θ satisfies

$$\theta'' + \left(1 - \frac{A}{a}\sin\tau\right)\theta = \frac{8B}{a}\sin\tau\cos\tau. \tag{i}$$

If $\theta = -(8B/A)\cos\tau$, then

$$\theta'' + \left(1 - \frac{A}{a}\sin\tau\right)\theta = \left(\frac{8B}{A}\right)\cos\tau - \left(1 - \frac{A}{a}\sin\tau\right)\left(\frac{8B}{A}\right)\cos\tau$$

$$= \frac{8B}{a}\sin\tau\cos\tau$$

which implies that $\theta = -(8B/A)\cos\tau$ is a particular solution.

The stability of solutions of eqn (i) is the same as the stability of solutions of the homogeneous equation (see, NODE, Theorem 8.1)

$$\theta'' + \left(1 - \frac{A}{a}\sin\tau\right)\theta = 0.$$

Express the equation in standard Mathieu form by the transformation $\tau = s - \frac{1}{2}\pi$. Therefore

$$\frac{d^2\theta}{ds^2} + (\alpha + \beta\cos s)\theta = 0,$$

where $\alpha = 1$ and $\beta = A/a$. From Figure 9.3 (in NODE) it can be seen that solutions will be unstable.

- **9.18** The equation $\ddot{x} + (\frac{1}{4} - 2\varepsilon b \cos^2 \frac{1}{2}t)x + \varepsilon x^3 = 0$ has the exact solution $x^*(t) = \sqrt{(2b)}\cos\frac{1}{2}t$. Show that the solution is stable by constructing the variational equation.

9.18. Consider the equation

$$\ddot{x} + (\tfrac{1}{4} - 2\varepsilon b\cos^2 \tfrac{1}{2}t)x + \varepsilon x^3 = 0.$$

Let $x^* = \sqrt{(2b)}\cos\frac{1}{2}t$. Then

$$\ddot{x}^* + (\tfrac{1}{4} - 2\varepsilon b\cos^2\tfrac{1}{2}t)x^* + \varepsilon x^{*3} = -\tfrac{1}{4}\sqrt{(2b)}\cos\tfrac{1}{2}t$$

$$+ \left[\tfrac{1}{4} - 2\varepsilon b\cos^2\tfrac{1}{2}t\right]\sqrt{(2b)}\cos\tfrac{1}{2}t$$

$$+ \varepsilon(2b)^{\frac{3}{2}}\cos^3\tfrac{1}{2}t = 0$$

which implies that x^* ia an exact solution.
Let $x = x^* + \xi$. Then the linearized variational equation is given by

$$\ddot{\xi} + (\tfrac{1}{4} - 2\varepsilon b\cos^2\tfrac{1}{2}t)\xi + 3\varepsilon x^{*2}\xi = 0,$$

or

$$\ddot{\xi} + (\tfrac{1}{4} + 2\varepsilon b + 2\varepsilon b\cos t)\xi = 0.$$

In the standard Mathieu format ξ satisfies

$$\ddot{\xi} + (\alpha + \beta\cos t)\xi = 0,$$

where $\alpha = \frac{1}{4} + 2\varepsilon b$ and $\beta = 2\varepsilon b$. Note that $\alpha = \frac{1}{4} + \beta$. In Figure 9.3 (in NODE), the boundary for 4π periodic solutions passes through $\alpha = \frac{1}{4}$, $\beta = 0$. From Section 9.4, the boundaries are approximately given by the lines $\alpha = \frac{1}{4} \pm \frac{1}{2}\beta$. Therefore we expect the solutions to be stable.

- **9.19** Consider the equation $\ddot{x} + (\alpha + \beta \cos t)x = 0$, where $|\beta| \ll 1$ and $\alpha = \frac{1}{4} + \beta c$. In the unstable region near $\alpha = \frac{1}{4}$ (NODE, Section 9.4) this equation has solutions of the form $c_1 e^{\sigma t} q_1(t) + c_2 e^{-\sigma t} q_2(t)$, where σ is real, $\sigma > 0$ and q_1, q_2 have period 4π. Construct the equation for q_1, q_2, and show that $\sigma \approx \pm \beta \sqrt{(\frac{1}{4} - c^2)}$.

9.19. In the equation
$$\ddot{x} + (\alpha + \beta \cos t)x = 0,$$
where $|\beta| \ll 1$ and $\alpha = \frac{1}{4} + \beta c$, let $x = e^{\sigma t} q_1(t)$. Then q_1 satisfies
$$\ddot{q}_1 + 2\sigma \dot{q}_1 + (\sigma^2 + \tfrac{1}{4} + \beta c + \beta \cos t) q_1 = 0.$$

Now assume that, approximately, $q_1 = a_0 \cos \frac{1}{2}t + b_0 \sin \frac{1}{2}t$, that is, q_1 is 4π periodic. Therefore
$$(\sigma b_0 + \sigma^2 a_0 + \beta c a_0 + \tfrac{1}{2}\beta a_0) \cos \tfrac{1}{2}t$$
$$+ (-\sigma a_0 + \sigma^2 b_0 + \beta c b_0 - \tfrac{1}{2}\beta b_0) \sin \tfrac{1}{2}t + \text{(higher harmonics)} = 0$$

The coefficients of the first harmonics are zero if, and only if,
$$(\sigma^2 + \beta c + \tfrac{1}{2}\beta)a_0 + \sigma b_0 = 0,$$
$$-\sigma a_0 + (\sigma^2 + \beta c - \tfrac{1}{2}\beta) b_0 = 0.$$

These equations have non-trivial solutions for a_0 and b_0 if
$$\begin{vmatrix} \sigma^2 + \beta c + \tfrac{1}{2}\beta & \sigma \\ -\sigma & \sigma^2 + \beta c - \tfrac{1}{2}\beta \end{vmatrix} = 0.$$

so that
$$\sigma^4 + (2\beta c + 1)\sigma^2 - \beta^2(\tfrac{1}{4} - c^2) = 0.$$

Given that $|\beta|$ is small it follows that
$$\sigma^2 \approx \beta^2(\tfrac{1}{4} - c^2),$$

or
$$\sigma = \pm \beta \sqrt{(\tfrac{1}{4} - c^2)}.$$

The equation for q_2 is
$$\ddot{q}_2 - 2\sigma \dot{q}_2 + (\sigma^2 + \tfrac{1}{4} + \beta c + \beta \cos t) q_2 = 0.$$

However it leads to the same result for σ.

● **9.20** By using the method of NODE, Section 9.5 show that a solution of the equation
$$\ddot{x} + \varepsilon(x^2 - 1)\dot{x} + x = \Gamma \cos \omega t,$$
where $|\varepsilon| \ll 1$, $\omega = 1 + \varepsilon\omega_1$, of the form $x^* = r_0 \cos(\omega t + \alpha)$ (α constant) is asymptotically stable when $4\omega_1^2 + \frac{3}{16}r_0^4 - r_0^2 + 1 < 0$. (Use the result of Problem 9.19.)

9.20. Consider the forced van der Pol equation

$$\ddot{x} + \varepsilon(x^2 - 1)\dot{x} + x = \Gamma \cos \omega t, \tag{i}$$

where $|\varepsilon| \ll 1$ and $\omega = 1 + \varepsilon\omega_1$. Let $x = x^* + \xi$, where $x^* = r_0 \cos(\omega t + \alpha)$. Then

$$\ddot{x}^* + \ddot{\xi} + \varepsilon[(x^* + \xi)^2 - 1](\dot{x}^* + \dot{\xi}) + x + \xi = \Gamma \cos \omega t.,$$

where

$$\ddot{x}^* + \varepsilon(x^{*2} - 1)\dot{x}^* + x^* = \Gamma \cos \omega t,$$

Therefore the linearized equation for ξ is

$$\ddot{\xi} + \varepsilon(x^{*2} - 1)\dot{\xi} + (1 + 2\varepsilon x^* \dot{x}^*)\xi = 0. \tag{ii}$$

In the coefficients

$$x^{*2} = r_0^2 \cos^2(\omega t + \alpha) = \tfrac{1}{2}r_0^2[1 + \cos 2(\omega t + \alpha)],$$

$$x^* \dot{x}^* = -r_0^2 \omega \cos(\omega t + \alpha)\sin(\omega t + \alpha) = -\tfrac{1}{2}r_0^2 \omega \sin 2(\omega t + \alpha).$$

Therefore (ii) becomes

$$\ddot{\xi} + \varepsilon[(\tfrac{1}{2}r_0^2 - 1) + \tfrac{1}{2}r_0^2 \cos 2(\omega t + \alpha)]\dot{\xi} + [1 - \varepsilon r_0^2 \omega \sin 2(\omega t + \alpha)]\xi = 0. \tag{iii}$$

Now let $\tau = 2(\omega t + \alpha)$ so that (iii) is transformed into

$$\xi'' + \varepsilon\left[\left(\frac{r_0^2 - 2}{4\omega}\right) + \frac{r_0^2}{4\omega}\cos\tau\right]\xi' + \left(\frac{1}{4\omega^2} - \frac{\varepsilon r_0^2}{4\omega}\sin\tau\right)\xi = 0. \tag{iv}$$

Use the perturbation $\omega = 1 + \varepsilon\omega_1$ and put $\tau = \tfrac{1}{2}\pi + s$ so that after expanding in powers of the small parameter ε to order ε, (iv) is approximately

$$\frac{d^2\xi}{ds^2} + \varepsilon[(\tfrac{1}{4}r_0^2 - \tfrac{1}{2}) - \tfrac{1}{4}r_0^2 \sin s]\frac{d\xi}{ds} + [\tfrac{1}{4}(1 - 2\varepsilon\omega_1) - \tfrac{1}{4}\varepsilon r_0^2 \cos s]\xi = 0.$$

Remove the first derivative by the further change of variable

$$\xi = \zeta \exp[-\tfrac{1}{2}\varepsilon \int (\tfrac{1}{4}r_0^2 - \tfrac{1}{2} - \tfrac{1}{4}r_0^2 \sin s)ds]$$

$$= \zeta \exp[-\tfrac{1}{2}\varepsilon][(\tfrac{1}{4}r_0^2 - \tfrac{1}{2})s + \tfrac{1}{4}r_0^2 \cos s].$$

Finally ζ satisfies

$$\frac{d^2\zeta}{ds^2} + [\tfrac{1}{4}(1 - 2\varepsilon\omega_1) - \tfrac{1}{8}\varepsilon r_0^2 \cos s]\xi = 0.$$

This is Mathieu's equation with

$$\alpha = \tfrac{1}{4}(1 - 2\varepsilon\omega_1), \quad \beta = -\tfrac{1}{8}\varepsilon r_0^2.$$

For small ε, α is close to the critical value $\tfrac{1}{4}$. In the notation of (9.53),

$$\sigma^2 = \tfrac{1}{4}\beta^2 - (\alpha - \tfrac{1}{4})^2 = \tfrac{1}{256}\varepsilon^2 r_0^4 - \tfrac{1}{4}\varepsilon^2 \omega_1^2.$$

Since the damping in the final transformation is $-\tfrac{1}{2}\varepsilon(\tfrac{1}{4}r_0^2 - \tfrac{1}{2})$, stability occurs if

$$\sigma^2 < [\tfrac{1}{2}\varepsilon(\tfrac{1}{4}r_0^2 - \tfrac{1}{2})]^2,$$

or

$$\tfrac{1}{256}r_0^4 - \tfrac{1}{4}\omega_1^2 < \tfrac{1}{16}(\tfrac{1}{4}r_0^4 - r_0^2 + 1),$$

or

$$4\omega_1^2 + \tfrac{3}{16}r_0^4 - r_0^2 + 1 < 0.$$

• 9.21 The equation $\ddot{x} + \alpha x + \varepsilon x^3 = \varepsilon \gamma \cos \omega t$ has the exact subharmonic solution $x = (4\gamma)^{1/3} \cos \tfrac{1}{3}\omega t$, when

$$\omega^2 = 9\left(\alpha + \tfrac{3}{4^{\frac{1}{3}}}\varepsilon\gamma^{2/3}\right).$$

If $0 < \varepsilon \ll 1$, show that the solution is stable.

9.21. Let $x = (4\gamma)^{\frac{1}{3}} \cos \frac{1}{3}\omega t$. Then

$$\ddot{x} + \alpha x + \varepsilon x^3 - \varepsilon \gamma \cos \omega t$$
$$= -\tfrac{1}{9}\omega^2 (4\gamma)^{1/3} \cos \tfrac{1}{3}\omega t + \alpha (4\gamma)^{1/3} \cos \tfrac{1}{3}\omega t + 4\varepsilon \gamma \cos^3 \tfrac{1}{3}\omega t - \varepsilon \gamma \cos \omega t$$
$$= [-\tfrac{1}{9}\omega^2 (4\gamma)^{1/3} + \alpha (4\gamma)^{1/3}] \cos \tfrac{1}{3}\omega t + 4\varepsilon \gamma \tfrac{1}{4}(3 \cos \tfrac{1}{3}\omega t + \cos \omega t) - \varepsilon \gamma \cos \omega t$$
$$= [-\tfrac{1}{9}\omega^2 (4\gamma)^{1/3} + \alpha (4\gamma)^{1/3} + \tfrac{1}{3}\varepsilon \gamma] \cos \tfrac{1}{3}\omega t,$$

and this is zero if

$$-\tfrac{1}{9}\omega^2 (4\gamma)^{1/3} + \alpha (4\gamma)^{1/3} + \tfrac{1}{3}\varepsilon \gamma = 0,$$

or

$$\omega^2 = 9\left[\alpha + \frac{3\varepsilon\gamma}{(4\gamma)^{1/3}}\right]. \tag{i}$$

Therefore $x = x^* = (4\gamma)^{1/3} \cos \tfrac{1}{3}\omega t$ is an exact solution subject to condition (i). Let $x = x^* + \xi$. Then the differential equation becomes

$$\ddot{x}^* + \ddot{\xi} + \alpha(x^* + \xi) + \varepsilon(x^* + \xi)^3 = \varepsilon \gamma \cos \omega t.$$

It follows that the linearized equation for ξ is

$$\ddot{\xi} + \alpha \xi + 3x^{*2}\xi = 0,$$

or

$$\ddot{\xi} + [\alpha + 3\varepsilon (4\gamma)^{2/3} \cos^2 \tfrac{1}{3}\omega t]\xi = 0.$$

or

$$\ddot{\xi} + [\alpha + \tfrac{3}{2}\varepsilon (4\gamma)^{2/3} + \tfrac{3}{2}\varepsilon (4\gamma)^{2/3} \cos \tfrac{2}{3}\omega t]\xi = 0.$$

Let $\tau = \tfrac{2}{3}\omega t$. Then ξ satisfies

$$\xi'' + \frac{9}{4\omega^2}\left[\alpha + \tfrac{3}{2}\varepsilon (4\gamma)^{2/3} + \tfrac{3}{2}\varepsilon (4\gamma)^{2/3} \cos \tfrac{2}{3}\tau\right]\xi = 0.$$

Now assuming that $0 < \varepsilon \ll 1$, expand $1/\omega^2$ in powers of ε. To order ε the equation is approximately

$$\xi'' + \left[\left\{\frac{1}{4} + \frac{3(4\gamma)^{2/3}\varepsilon}{16\alpha}\right\} + \frac{3}{8\alpha}(4\gamma)^{2/3}\varepsilon \cos \tau\right]\xi = 0.$$

This equation is in standard Mathieu form

$$\xi'' + (\alpha_1 + \beta_1 \cos \tau)\xi = 0,$$

where

$$\alpha_1 = \frac{1}{4} + \frac{3(4\gamma)^{2/3}\varepsilon}{16\alpha}, \quad \frac{1}{4}ad\beta_1 = \frac{3}{8\alpha}(4\gamma)^{2/3}\varepsilon.$$

It can be checked that $\alpha_1 = \frac{1}{4} + \frac{1}{2}\beta_1$. From Section 9.4, solutions lie on the boundary curve for 4π periodic solutions (see Figure 9.3 in NODE).

- **9.22** Analyse the stability of the equation $\ddot{x} + \varepsilon x \dot{x}^2 + x = \Gamma \cos \omega t$ for small ε: assume $\Gamma = \varepsilon \gamma$. (First find approximate solutions of the form $a \cos \omega t + b \sin \omega t$ by the harmonic balance method of Chapter 4, then perturb the solution by the method of NODE, Section 9.4.)

9.22. Consider the equation

$$\ddot{x} + \varepsilon x \dot{x}^2 + x = \Gamma \cos \omega t.$$

Use harmonic balance with $x = a \cos \omega t + b \sin \omega t$. The first harmonics balance if

$$a(1 - \omega^2) + \tfrac{1}{4}\varepsilon a \omega^2 (a^2 + b^2) = \Gamma,$$

$$b(1 - \omega^2) + \tfrac{1}{4}\varepsilon b \omega^2 (a^2 + b^2) = 0.$$

It follows that

$$b = 0, \quad a(1 - \omega^2) + \tfrac{1}{4}\varepsilon \omega^2 a^3 = \Gamma. \qquad (i)$$

Let the unperturbed solution be given approximately by $x = x^* = a \cos \omega t$, where a is given by (i), and let the perturbation be $x = x^* + \xi$. Then

$$\ddot{x}^* + \ddot{\xi} + \varepsilon(x^* + \xi)(\dot{x}^* + \dot{\xi})^2 + x^* + \xi = \Gamma \cos \omega t.$$

The linearized equation for ξ is, therefore,

$$\ddot{\xi} + \varepsilon(\dot{x}^{*2}\xi + 2x^*\dot{x}^*\dot{\xi}) + \xi = 0,$$

or

$$\ddot{\xi} + 2\varepsilon x^*\dot{x}^*\dot{\xi} + (1 + \varepsilon \dot{x}^{*2})\xi = 0,$$

or

$$\ddot{\xi} - \varepsilon a^2 \omega \sin 2\omega t \, \dot{\xi} + (1 + \tfrac{1}{2}\varepsilon a^2 \omega^2 - \tfrac{1}{2}a^2 \omega^2 \cos 2\omega t)\xi = 0.$$

To remove the $\dot{\xi}$ term, let

$$\xi = \eta\exp\left[\tfrac{1}{2}\varepsilon a^2\omega \int \sin 2\omega t\,dt\right] = \eta\exp\left[-\tfrac{1}{4}\varepsilon a^2\cos 2\omega t\right] = \eta e^{h(t)},$$

say. Since the exponential term is periodic, stability is not affected. Hence

$$\dot{\xi} = e^{h(t)}[\dot{\eta} + \tfrac{1}{2}\varepsilon a^2\omega \sin 2\omega t\,\eta],$$

and

$$\ddot{\xi} = e^{h(t)}[\ddot{\eta} + \varepsilon a^2\omega \sin 2\omega t\,\dot{\eta} + \varepsilon a^2\omega\cos 2\omega t\,\eta + O(\varepsilon^2)].$$

Elimination of ξ leads to

$$\ddot{\eta} + (1 + \tfrac{1}{2}\varepsilon\omega^2 a^2 + \tfrac{1}{2}\varepsilon\omega^2 a^2\cos 2\omega t)\eta = 0$$

to order ε. To obtain the standard Mathieu form, let $\tau = 2\omega t$, so that

$$\eta'' + (\alpha + \beta\cos\tau)\eta = 0,$$

where

$$\alpha = \frac{2 + \varepsilon a^2\omega^2}{8\omega^2},\quad \beta = \frac{1}{8}\varepsilon a^2\omega^2. \tag{ii}$$

Assume that $\Gamma = \varepsilon\gamma$. Then from (i)

$$a(1 - \omega^2) + \tfrac{1}{4}\varepsilon a^3 = \varepsilon\gamma.$$

Therefore

$$\omega^2 = 1 + \left(\tfrac{1}{4}a^2 - \frac{\gamma}{a}\right)\varepsilon + O(\varepsilon^2).$$

Consequently

$$\alpha = \frac{1}{4} + \left(\frac{3a^2}{16} + \frac{\gamma}{4a}\right)\varepsilon + O(\varepsilon^2),\quad \beta = \frac{1}{8}a^2\varepsilon + O(\varepsilon^2).$$

From NODE, Section 9.5, (for small ε) instability occurs in the interval

$$\tfrac{1}{4} - \tfrac{1}{2}\beta < \alpha < \tfrac{1}{4} + \tfrac{1}{2}\beta,$$

that is, if

$$\frac{1}{4} - \frac{1}{16}a^2\varepsilon < \frac{1}{4} + \left(\frac{3a^2}{16} - \frac{\gamma}{4a}\right)\varepsilon < \frac{1}{4} + \frac{1}{16}a^2\varepsilon,$$

or, $-a^3 < \gamma < a^3$.

- **9.23** The equation $\ddot{x} + x + \varepsilon x^3 = \Gamma \cos \omega t$, $(\varepsilon \ll 1)$ has an approximate solution $x^* = a \cos \omega t$ where (eqn (7.10)) $\frac{3}{4}\varepsilon a^3 - (\omega^2 - 1)a - \Gamma = 0$. Show that the first variational equation (Section 9.4) is $\ddot{\xi} + \{1 + 3\varepsilon x^{*2}(t)\}\xi = 0$. Reduce this to Mathieu's equation and find conditions for stability of $x^*(t)$ if $\Gamma = \varepsilon \gamma$.

9.23. By harmonic balance it can be shown that the equation

$$\ddot{x} + x + \varepsilon x^3 = \Gamma \cos \omega t$$

has the approximate solution $x^* = a \cos \omega t$ where

$$\frac{3}{4}\varepsilon a^3 - (\omega^2 - 1)a - \Gamma = 0. \qquad (i)$$

Let $x = x^* + \xi$. Then the linearized equation for ξ is

$$\ddot{\xi} + (1 + 3\varepsilon x^{*2})\xi = 0,$$

or

$$\ddot{\xi} + (1 + \tfrac{3}{2}\varepsilon a^2 + \tfrac{3}{2}\varepsilon a^2 \cos 2\omega t)\xi = 0.$$

Let $\tau = 2\omega t$ so that ξ satisfies the standard Mathieu equation

$$\xi'' + (\alpha + \beta \cos \tau)\xi = 0,$$

where

$$\alpha = \frac{1 + 3\varepsilon a^2}{4\omega^2}, \quad \beta = \frac{3\varepsilon a^2}{8\omega^2}. \qquad (ii)$$

We now expand ω^2 in powers of ε using (i). Therefore

$$\omega^2 = 1 + \left(\frac{3a^2}{4} - \frac{\gamma}{a}\right)\varepsilon + O(\varepsilon^2),$$

so that

$$\alpha = \frac{1}{4}\left[1 + \left(\frac{9}{4}a^2 + \frac{\gamma}{a}\right)\varepsilon + O(\varepsilon^2)\right], \quad \beta = \frac{3}{8}a^2\varepsilon + O(\varepsilon^2).$$

Instability occurs where

$$\tfrac{1}{4} - \tfrac{1}{2}\beta < \alpha < \tfrac{1}{4} + \tfrac{1}{2}\beta,$$

that is, where

$$-3a^3 < \gamma < -\tfrac{3}{2}a^3.$$

From Figure 9.3 (in NODE), stability will occur if γ takes values just outside this interval.

- **9.24** The equation $\ddot{x} + x - \frac{1}{6}x^3 = 0$ has an approximate solution $a\cos\omega t$ where $\omega^2 = 1 - \frac{1}{8}a^2$, $a \ll 1$ (Example 4.10). Use the method of NODE, Section 9.4 to show that the solution is unstable.

9.24. Using harmonic balance, it can be shown that the equation

$$\ddot{x} + x - \tfrac{1}{6}x^3 = 0$$

has the approximate solution $x^* = a\cos\omega t$, where

$$\omega^2 = 1 - \tfrac{1}{8}a^2. \tag{i}$$

Let $x = x^* + \xi$. Then the linearized equation for ξ is

$$\ddot{\xi} + (1 - \tfrac{1}{2}x^{*2})\xi = 0, \ \text{ or } \ \ddot{\xi} + (1 - \tfrac{1}{2}a^2\cos^2\omega t)\xi = 0,$$

or

$$\ddot{\xi} + (1 - \tfrac{1}{4}a^2 - \tfrac{1}{4}a^2\cos 2\omega t)\xi = 0.$$

Let $\tau = 2\omega t$ so that ξ satisfies the standard Mathieu equation

$$\xi'' + (\alpha + \beta\cos\tau)\xi = 0,$$

where

$$\alpha = \frac{4 - a^2}{16\omega^2}, \ \beta = -\frac{a^2}{16\omega^2}.$$

Assume that $0 < a \ll 1$. Then, using (i)

$$\alpha = \tfrac{1}{4} - \tfrac{1}{32}a^2 + O(a^4), \ \beta = -\tfrac{1}{16}a^2 + O(a^4). \tag{ii}$$

To order a^4, it follows from (ii) that $\alpha = \tfrac{1}{4} \pm \tfrac{1}{4}\beta$, which means that a period 4π solution exists. However the other solution is unbounded, which implies that the general solution is unstable.

- **9.25** Show that a fundamental matrix of the differential equation $\dot{x} = Ax$, where

$$A(t) = \begin{bmatrix} \beta \cos^2 t - \sin^2 t & 1 - (1+\beta)\sin t \cos t \\ -1 - (1+\beta)\sin t \cos t & -1 + (1+\beta)\sin^2 t \end{bmatrix}$$

is

$$\Phi(t) = \begin{bmatrix} e^{\beta t} \cos t & e^{-t} \sin t \\ -e^{\beta t} \sin t & e^{-t} \cos t \end{bmatrix}$$

Find the characteristic multipliers of the system. For what value of β will periodic solutions exist?

Find the eigenvalues of $A(t)$ and show that they are independent of t. Show that for $0 < \beta < 1$ the eigenvalues have negative real parts. What does this problem indicate about the relationship between the eigenvalue of a linear system with a variable coefficients and the stability of the zero solution?

9.25. Consider the homogeneous equation

$$\dot{x} = A(t)x, \qquad (i)$$

where

$$A(t) = \begin{bmatrix} \beta \cos^2 t - \sin^2 t & 1 - (1+\beta)\sin t \cos t \\ -1 - (1+\beta)\sin t \cos t & -1 + (1+\beta)\sin^2 t \end{bmatrix}.$$

Let

$$\phi_1(t) = \begin{bmatrix} e^{\beta t} \cos t \\ -e^{\beta t} \sin t \end{bmatrix}.$$

Then

$$A(t)\phi_1(t) = \begin{bmatrix} \beta \cos^2 t - \sin^2 t & 1 - (1+\beta)\sin t \cos t \\ -1 - (1+\beta)\sin t \cos t & -1 + (1+\beta)\sin^2 t \end{bmatrix} \begin{bmatrix} e^{\beta t} \cos t \\ -e^{\beta t} \sin t \end{bmatrix}$$

$$= \begin{bmatrix} e^{\beta t}(\beta \cos t - \sin t) \\ e^{\beta t}(-\beta \sin t - \cos t) \end{bmatrix} = \dot{\phi}_1(t)$$

Similarly if

$$\phi_2(t) = \begin{bmatrix} e^{-t} \sin t \\ e^{-t} \cos t \end{bmatrix},$$

Then

$$A(t)\phi_2(t) = \dot{\phi}_2(t).$$

The solution
$$\Phi(t) = [\ \phi_1(t)\ \ \phi_2(t)\ ] = \begin{bmatrix} e^{\beta t}\cos t & e^{-t}\sin t \\ -e^{\beta t}\sin t & e^{-t}\cos t \end{bmatrix} \quad \text{(ii)}$$

is a fundamental matrix of (i).

The constant matrix **E** is given by

$$\mathbf{E} = \Phi^{-1}(0)\Phi(2\pi) = \begin{bmatrix} 1 & 0 \\ 0 & 1 \end{bmatrix}\begin{bmatrix} e^{2\pi\beta} & 0 \\ 0 & e^{-2\pi} \end{bmatrix} = \begin{bmatrix} e^{2\pi\beta} & 0 \\ 0 & e^{-2\pi} \end{bmatrix}.$$

The characteristic numbers of **E** are obviously $\mu_1 = e^{2\pi\beta}$ and $\mu_2 = e^{-2\pi}$.
From (ii) it can be seen that periodic solutions only exist for $\beta = 0$.
The eigenvalues of $\mathbf{A}(t)$ are given by

$$\mathbf{A}(t) = \begin{vmatrix} \beta\cos^2 t - \sin^2 t - \lambda & 1 - (1+\beta)\sin t \cos t \\ -1 - (1+\beta)\sin t \cos t & -1 + (1+\beta)\sin^2 t - \lambda \end{vmatrix}.$$

The eigenvalues are (it is helpful to use a symbolic algebra program)

$$\lambda_1, \lambda_2 = \tfrac{1}{2}\{-1 + \beta \pm \sqrt{[(\beta+3)(\beta-1)]}\},$$

which are independent of t. Figure 9.5 shows how the real parts of λ_1 and λ_2 vary in terms of β. The eigenvalues coincide at $\beta = -3$ and at $\beta = 1$, and their real parts are the same between these values of β. It might be inferred that stability of solutions would be indicated by the sign of the real part of these eigenvalues. Note that λ_1 has a negative real part for $\beta < 1$, and λ_2 has a negative real part for all β except at $\beta = 1$. However, (ii) indicates that solutions can be unstable for $0 < \beta < 1$. Therefore the signs of the eigenvalues of a linear system with variable coefficients cannot in general indicate stability.

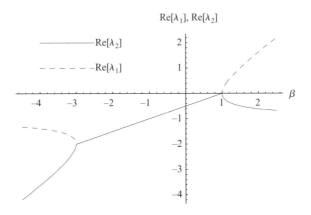

Figure 9.5 Problem 9.25: Re[λ_1] and Re[λ_2] plotted against β.

- **9.26** Find a fundamental matrix for the system $\dot{\mathbf{x}} = \mathbf{A}(t)\mathbf{x}$ where
$$\mathbf{A}(t) = \begin{bmatrix} \sin t & 1 \\ -\cos t + \cos^2 t & -\sin t \end{bmatrix}.$$
Show that the characteristic multipliers of the system are $\mu_1 = e^{2\pi}$ and $\mu_2 = e^{-2\pi}$. By integration confirm that
$$\exp\left(\int_0^{2\pi} \mathrm{tr}\{\mathbf{A}(s)\}ds\right) = \mu_1\mu_2 = 1.$$

9.26. The system
$$\dot{\mathbf{x}} = \mathbf{A}(t)\mathbf{x}, \quad \mathbf{A}(t) = \begin{bmatrix} \sin t & 1 \\ -\cos t + \cos^2 t & -\sin t \end{bmatrix}, \tag{i}$$
where $\mathbf{x} = [x_1, x_2]^T$, is equivalent to
$$\dot{x}_1 = x_1 \sin t + x_2, \quad \dot{x}_2 = (-\cos t + \cos^2 t)x_1 - x_2 \sin t.$$
Elimination of x_2 results in the equation
$$\ddot{x}_1 - x_1 = 0,$$
which has the general solution
$$x_1 = Ae^t + Be^{-t}.$$
It follows that
$$x_2 = \dot{x}_1 - x_1 \sin t = A(1 - \sin t)e^t - B(1 + \sin t)e^{-t}.$$
A fundamental matrix is therefore
$$\Phi(t) = \begin{bmatrix} e^t & e^{-t} \\ (1 - \sin t)e^t & -(1 + \sin t)e^{-t} \end{bmatrix}.$$
Since $\mathbf{A}(t)$ has period 2π, we can define $\mathbf{E}$ as
$$\mathbf{E} = \Phi^{-1}(0)\Phi(2\pi) = \tfrac{1}{2}\begin{bmatrix} 1 & 1 \\ 1 & -1 \end{bmatrix}\begin{bmatrix} e^{2\pi} & e^{-2\pi} \\ e^{2\pi} & -e^{-2\pi} \end{bmatrix} = \begin{bmatrix} e^{2\pi} & 0 \\ 0 & e^{-2\pi} \end{bmatrix}.$$
Obviously the eigenvalues of $\mathbf{E}$ are $\mu_1 = e^{2\pi}$ and $\mu_2 = e^{-2\pi}$.
From (i), $\mathrm{tr}\{\mathbf{A}(s)\} = \sin t - \sin t = 0$. Therefore
$$\exp\left(\int_0^{2\pi} \mathrm{tr}\{\mathbf{A}(s)\}ds\right) = e^0 = 1 = \mu_1\mu_2.$$

10 Liapunov methods for determining stability of the zero solution

• **10.1** Find a a simple V or U function (NODE, Theorems 10.5, 10.11 or 10.13) to establish the stability or instability respectively of the zero solutions of the following equations:

(i) $\dot{x} = -x + y - xy^2,\ \dot{y} = -2x - y - x^2y$;
(ii) $\dot{x} = y^3 + x^2y,\ \dot{y} = x^3 - xy^2$;
(iii) $\dot{x} = 2x + y + xy,\ \dot{y} = x - 2y + x^2 + y^2$;
(iv) $\dot{x} = -x^3 + y^4,\ \dot{y} = -y^3 + y^4$;
(v) $\dot{x} = \sin y,\ \dot{y} = -2x - 3y$;
(vi) $\dot{x} = x + e^{-y-1},\ \dot{y} = x$;
(vii) $\dot{x} = e^x - \cos y,\ \dot{y} = y$;
(viii) $\dot{x} = \sin(y+x),\ \dot{y} = -\sin(y-x)$;
(ix) $\ddot{x} = x^3$;
(x) $\dot{x} = x + 4y,\ \dot{y} = -2x - 5y$;
(xi) $\dot{x} = -x + 6y,\ \dot{y} = 4x + y$.

10.1. In each case the origin is an equilibrium point of the autonomous system.

(i) $\dot{x} = -x + y - xy^2,\ \dot{y} = -2x - y - x^2y$. Try the Liapunov function $V(x, y) = x^2 + y^2$. Then

$$\dot{V} = \frac{\partial V}{\partial x}\dot{x} + \frac{\partial V}{\partial y}\dot{y}$$
$$= 2x(-x + y - xy^2) + 2y(-2x - y - x^2y)$$
$$= -2x^2 + 2xy - x^2y^2 - 2xy - 2y^2 - 2x^2y^2$$
$$= -2(x+y)^2 - 4x^2y^2 \leq 0,$$

for all x, y. Hence the origin is stable by Theorem 10.11.
  We can cross-check this result by linearization. Near the origin

$$\dot{x} \approx -x + y,\quad \dot{y} \approx -2x - y.$$

The eigenvalues are given by

$$\begin{vmatrix} -1-\lambda & 1 \\ -2 & -1-\lambda \end{vmatrix} = 0,$$

that is, $\lambda = -1 \pm 2i$, which confirms the stability.

(ii) $\dot{x} = y^3 + x^2 y$, $\dot{y} = x^3 - xy^2$. Let $U(x, y) = xy$. Consider the conditions of Theorem 13. Then

$$U(0,0) = 0, \quad U(k,k) > 0 \text{ for every } k > 0,$$

$$\dot{U} = \frac{\partial U}{\partial x}\dot{x} + \frac{\partial U}{\partial y}\dot{y} = y(y^3 + x^2 y) + x(x^3 - xy^2) = x^4 + y^4 > 0,$$

for $(x, y) \neq (0, 0)$. Therefore the origin is unstable.

(iii) $\dot{x} = 2x + y + xy$, $\dot{y} = x - 2y + x^2 + y^2$. The linearized equations near the origin are

$$\dot{x} \approx 2x + y, \quad \dot{y} \approx x - 2y.$$

Its eigenvalues are given by

$$\begin{vmatrix} 2-\lambda & 1 \\ 1 & -2-\lambda \end{vmatrix} = \lambda^2 - 5 = 0.$$

Therefore $\lambda = \pm\sqrt{5}$, which implies instability at the origin (a saddle). This can be proved by choosing $U(x, y) = xy$. Then

$$U(0,0) = 0, \quad U(k,k) > 0 \text{ for every } k > 0,$$

$$\dot{U} = \frac{\partial U}{\partial x}\dot{x} + \frac{\partial U}{\partial y}\dot{y} = y(2x + y + xy) + x(x - 2y + x^2 + y^2)$$

$$= x^2 + y^2 + 2x^2 y + x^3$$

The function $\dot{U}$ has a relative minimum at the origin since, if $q(x, y) = x^2 + y^2 + 2x^2 y + x^3$, then

$$q_{xx}(0,0) = q_{yy}(0,0) = 2, \quad q_{xy}(0,0) = 0.$$

By the usual conditions for functions of two variables, $q(x, y)$ has a relative minimum at $(0, 0)$ if

$$q_{xx}(0,0) = 2 > 0, \quad \text{and } \Delta(0,0) = q_{x,x}(0,0)q_{y,y}(0,0) - q_{xy}(0,0)^2 = 4 > 0.$$

Therefore the origin is unstable.

(iv) $\dot{x} = -x^3 + y^4$, $\dot{y} = -y^3 + y^4$. Let $V(x, y) = x^2 + y^2$. Then

$$\dot{V} = \frac{\partial V}{\partial x}\dot{x} + \frac{\partial V}{\partial y}\dot{y} = 2(-x^4 - y^4 + xy^4 + y^5).$$

We can argue that for $|x|$ and $|y|$ sufficiently small the term $-2(x^4 + y^4)$ dominates over the remainder $2y^4(x + y)$. so that $\dot{V} < 0$ in some neighbourhood of the origin which implies that the origin is stable.

To find such a neighbourhood $\mathcal{N}$ can be more complicated. In this case try using polar coordinates $x = r\cos\theta$, $y = r\sin\theta$. Then

$$\dot{V} = -2r^4(\cos^4\theta + \sin^4\theta) + 2r^5\sin^4\theta(\cos\theta + \sin\theta).$$

Since
$$(\cos^2\theta + \sin^2\theta)^2 = 1,$$

then
$$\cos^4\theta + \sin^4\theta = 1 - 2\sin^2\theta\cos^2\theta = 1 - \tfrac{1}{2}\sin^2 2\theta.$$

Therefore
$$\tfrac{1}{2} \leq \cos^4\theta + \sin^4\theta \leq 1.$$

Finally
$$\dot{V} \leq -r^4 + 2r^5\sin^4\theta\cos\theta \leq -r^4 + 2r^5 \leq 0,$$

if $r \leq \tfrac{1}{2}$. Hence we could choose the interior of the the circle radius $\tfrac{1}{2}$ as the neighbourhood $\mathcal{N}$.

(v) $\dot{x} = \sin y$, $\dot{y} = 2x - 3y$. Approximate to the sine function near the origin: $\sin y = y + O(y^3)$ as $x \to 0$. We can test the origin by linearization which is

$$\dot{x} \approx y, \quad \dot{y} = -2x - 3y.$$

Hence in the standard notation $a = 0$, $b = 1$, $c = -2$, $d = -3$. The usual parameters are $p = -3 < 0$, $q = 2 > 0$, $\Delta = p^2 - 4q = 1 > 0$, which imply that the origin is a stable node.
An appropriate Liapunov function is given by NODE, (10.27):

$$V(x, y) = -\{(dx - by)^2 + (cx - ay)^2 + q(x^2 + y^2)\}/(2pq)$$
$$= \frac{1}{12}\{(-3x - y)^2 + 4x^2 + 2(x^2 + y^2)\}$$
$$= \tfrac{1}{4}(5x^2 + 2xy + y^2)$$

Therefore
$$\dot{V} = \frac{\partial V}{\partial x}\dot{x} + \frac{\partial V}{\partial y}\dot{y} = -x^2 - y^2.$$

and the origin is stable.

(vi) $\dot{x} = x + e^{-y} - 1$, $\dot{y} = x$. The linearized equations are

$$\dot{x} \approx x - y, \quad \dot{y} = x.$$

Let $U(x, y) = \alpha x^2 + 2\beta xy + \gamma y^2$. Then

$$\dot{U} = \frac{\partial U}{\partial x}\dot{x} + \frac{\partial U}{\partial y}\dot{y}$$

$$= 2(\alpha x + 2\beta y)(x - y) + 2(\beta x + \gamma y)x$$

$$= 2(\alpha + \beta)x^2 + 2(\beta - \alpha - \gamma)xy - 2\beta y^2$$

$$= x^2 + y^2$$

if

$$\alpha + \beta = \tfrac{1}{2}, \quad \beta - \alpha - \gamma = 0, \quad \beta = -\tfrac{1}{2}.$$

Therefore $\alpha = 1$ and $\gamma = -\tfrac{3}{2}$, and

$$U(x, y) = x^2 - xy - \tfrac{3}{2}y^2 = (x - \tfrac{1}{2}y)^2 - \tfrac{7}{4}y^2.$$

There are points where $U(x, y)$ is positive. Therefore the origin is unstable.

(vii) $\dot{x} = e^x - \cos y$, $\dot{y} = x$. The linearized approximation is

$$\dot{x} \approx x, \quad \dot{y} = x.$$

Let

$$U(x, y) = x^2 + 2xy - y^2.$$

Then $U(1, 0) > 0$ and

$$\dot{U} = (2x + 2y)x + (2x - 2y)x = 4x^2,$$

which is positive definite. Hence the origin is unstable.

(viii) $\dot{x} = \sin(y + x)$, $\dot{y} = -\sin(y - x)$. The linear approximation is

$$\dot{x} \approx x + y, \quad \dot{y} \approx x - y.$$

The eigenvalues of this system are $\lambda_{1,2} = \pm\sqrt{2}$. This is essentially the instability case considered in NODE, Section 10.5. A suitable U function is

$$U(x, y) = \frac{x^2}{\lambda_1} + \frac{y^2}{\lambda_2} = \frac{1}{\sqrt{2}}(x^2 - y^2).$$

Therefore
$$\dot{U} = \frac{\partial U}{\partial x}\dot{x} + \frac{\partial U}{\partial y}\dot{y} = \sqrt{2}(x^2 + y^2),$$
which is positive definite. Hence the origin is unstable.

(ix) $\ddot{x} = x^3$, or $\dot{x} = y$, $\dot{y} = x^3$. Let $U(x, y) = xy$. Then
$$\dot{U} = y^2 + x^4,$$
which is positive definite. Hence the origin is unstable.

(x) $\dot{x} = x + 4y$, $\dot{y} = -2x - 5y$. This is a linear system with coefficients $a = 1$, $b = 4$, $c = -2$ and $d = -5$, and parameters $p = -4 < 0$, $q = 3 > 0$ and $\Delta = 4 > 0$. The origin is therefore a stable node. By NODE, (10.27), a suitable Liapunov function is

$$V(x, y) = -\frac{1}{2pq}[(c^2 + d^2 + q)x^2 - 2(ac + bd)xy + (a^2 + b^2 + q)y^2]$$
$$= \tfrac{1}{6}[8x^2 + 11xy + 5y^2]$$

We can check that
$$\dot{V} = \tfrac{1}{6}[(16x + 11y)(x + 4y) + (11x + 10y)(-2x - 5y)] = -6x^2 - 6y^2,$$
which is negative definite. The origin is stable.

(xi) $\dot{x} = -x + 6y$, $\dot{y} = 4x + y$. This is a linear system with coefficients $a = -1$, $b = 6$, $c = 4$ and $d = 1$, and parameters $p = 0$, $q = -25 > 0$, $\Delta = -100 < 0$. The origin is therefore a saddle. The eigenvalues are $\lambda_{1,2} = \pm 5$. As in NODE, Section 10.8, apply the change of variable $\mathbf{x} = \mathbf{Cu}$, where
$$\mathbf{C} = \begin{bmatrix} -b & -b \\ a - \lambda_1 & d - \lambda_2 \end{bmatrix} = \begin{bmatrix} -6 & -6 \\ -4 & 4 \end{bmatrix}.$$
Then $\mathbf{u}$ satisfies $\dot{\mathbf{u}} = \mathbf{Du}$, where is the $\mathbf{D}$ is the diagonal matrix given by
$$\mathbf{D} = \begin{bmatrix} \lambda_1 & 0 \\ 0 & \lambda_2 \end{bmatrix} = \begin{bmatrix} 5 & 0 \\ 0 & -5 \end{bmatrix}.$$

Finally, we choose $U = u_1^2/\lambda_1 + u_2^2/\lambda_2$ with the result that $\dot{U} = 2u_1^2 + 2u_2^2$. The origin is therefore unstable.

- **10.2** Show that α may be chosen so that $V = x^2 + \alpha y^2$ is a strong Liapunov function for the system
$$\dot{x} = y - \sin^3 x, \quad \dot{y} = -4x - \sin^3 y.$$

10.2. For the system (which has an equilibrium point at $(0,0)$)

$$\dot{x} = y - \sin^3 x, \quad \dot{y} = -4x - \sin^3 y,$$

let $V(x, y) = x^2 + \alpha y^2$. Then

$$\dot{V} = 2x(y - \sin^3 x) + 2\alpha y(-4x - \sin^3 y) = (2 - 8\alpha)xy - 2x\sin^3 x - 2\alpha y \sin^3 y.$$

The xy terms can be eliminated by choosing $\alpha = \frac{1}{4}$. Also, $x\sin^3 x > 0$ and $y\sin^3 y > 0$ in the neighbourhood $|x| < \pi$, $|y| < \pi$. Therefore $V(x,y)$ is a strong Liapunov function, so that the zero solution is uniformly and asymptotically stable.

- **10.3** Find domains of asymptotic stability for the following systems, using $V = x^2 + y^2$:
  (i) $\dot{x} = -\frac{1}{2}x(1 - y^2)$, $\dot{y} = -\frac{1}{2}y(1 - x^2)$;
  (ii) $\dot{x} = y - x(1 - x)$, $\dot{y} = -x$.

10.3. (i) $\dot{x} = -\frac{1}{2}x(1 - y^2)$, $\dot{y} = -\frac{1}{2}y(1 - x^2)$. The system has five equilibrium points, at $(0,0)$ and at all points $(\pm 1, \pm 1)$. The function $V(x,y) = x^2 + y^2$ is positive definite for all x, y. Then

$$\dot{V} = -x^2(1 - y^2) - y^2(1 - x^2),$$

which is negative definite in the square $|x| < 1$, $|y| < 1$. The zero solution is asymptotically stable. The largest domain of asymptotic stability, $\mathcal{N}_\mu$, is the largest circle in the square, namely, $x^2 + y^2 = 1$.

(ii) $\dot{x} = y - x(1 - x)$, $\dot{y} = -x$. The system has one equilibrium point at the origin. The function $V(x,y) = x^2 + y^2$ is positive definite for all x, y. Then

$$\dot{V} = 2x[y - x(1 - x)] + 2y(-x) = -2x^2(1 - x) < 0$$

which is negative definite for $x < 1$. Therefore the zero solution is asymptotically stable in the neighbourhood $\mathcal{N}_\mu: x^2 + y^2 = 1$.

- **10.4** Find a strong Liapunov function at $(0,0)$ for the system
  $\dot{x} = x(y - b)$, $\dot{y} = y(x - a)$
  and confirm that all solutions starting in the domain $(x/a)^2 + (y/b)^2 < 1$ approach the origin.

10.4. $\dot{x} = x(y-b)$, $\dot{y} = y(x-a)$. The system has two equilibrium points at $(0,0)$ and (a,b). For the origin, the inequality in the question suggests that we try the positive definite function

$$V(x,y) = \frac{x^2}{a^2} + \frac{y^2}{b^2}.$$

Then

$$\dot{V} = \frac{2x^2}{a^2}(y-b) + \frac{2y^2}{b^2}(x-a) < 0,$$

if $x < a$ and $y < b$. Hence $\dot{V}$ is negative definite and the zero solution is asymptotically stable. The largest domain of asymptotic stability is the largest ellipse centred at the origin which satisfies $x \leq a$, $y \leq b$, that is,

$$\mathcal{N}_\mu: \quad \frac{x^2}{a^2} + \frac{y^2}{b^2} = 1.$$

- 10.5 Show that the origin of the system
$$\dot{x} = xP(x,y), \quad \dot{y} = yQ(x,y)$$
is asymptotically stable when $P(x,y) < 0$, $Q(x,y) < 0$ in a neighbourhood of the origin.

10.5. $\dot{x} = xP(x,y)$, $\dot{y} = yQ(x,y)$. The origin is an equilibrium point. Choose the positive definite function $V(x,y) = x^2 + y^2$. Then, for $(x,y) \neq (0,0)$,

$$\dot{V} = 2x^2 P(x,y) + 2y^2 Q(x,y) < 0$$

in some neighbourhood of the origin if $P(x,y) < 0$ and $Q(x,y) < 0$ in the same neighbourhood. In this case $\dot{V}$ will be negative definite, which implies that the origin is asymptotically stable.

- 10.6 Show that the zero solution of
$$\dot{x} = y + xy^2, \quad \dot{y} = x + x^2 y$$
is unstable.

10.6. The system
$$\dot{x} = y + x^3, \quad \dot{y} = x - y^3$$
has the zero solution $x = 0$, $y = 0$. Let $U(x,y) = x^2 - y^2$. Then

$$\dot{U} = \frac{\partial U}{\partial x}\dot{x} + \frac{\partial U}{\partial y}\dot{y}$$
$$= 2x(y + x^3) - 2y(x - y^3) = 2x^4 + 2y^4,$$

which is positive definite. Therefore, by NODE, Theorem 10.13, the origin is unstable.

- **10.7** Investigate the stability of the zero solution of
$$\dot{x} = x^2 - y^2, \quad \dot{y} = -2xy$$
by using the function $U(x, y) = \alpha x y^2 + \beta x^3$ for suitable constants α and β.

10.7. The system
$$\dot{x} = x^2 - y^2, \quad \dot{y} = -2xy$$
has a solution $x = y = 0$. Let $U(x, y) = \alpha x y^2 + \beta x^3$. Then

$$\dot{U} = \frac{\partial U}{\partial x}\dot{x} + \frac{\partial U}{\partial y}\dot{y}$$
$$= (\alpha y^2 + 3\beta x^2)(x^2 - y^2) + 2\alpha xy(-2xy)$$
$$= -3(\alpha + \beta)x^2 y^2 + 3\beta x^4 - \alpha y^4$$
$$= -3\alpha x^4 - \alpha y^4,$$

if $\beta = -\alpha$. Choose α to be a negative number, say, $\alpha = -1$. Then $\dot{U}(x, y)$ is positive definite. Theorem 10.13 applies since $U(x, 0) = x^3 > 0$ for every $x > 0$. Hence the origin is unstable.

- **10.8** Show that the origin of the system
$$\dot{x} = -y - x\sqrt{(x^2 + y^2)}, \quad \dot{y} = x - y\sqrt{(x^2 + y^2)}$$
is a centre in the linear approximation, but in fact is a stable spiral. Find a Liapunov function for the zero solution.

10.8. Consider the system
$$\dot{x} = -y - x\sqrt{(x^2 + y^2)}, \quad \dot{y} = x - y\sqrt{(x^2 + y^2)}.$$

The linear approximation near the origin is
$$\dot{x} = -y, \quad \dot{y} = x,$$
which are the equations for a centre in the (x, y) phase plane.

To obtain the exact solution switch to polar coordinates (r, θ). Therefore
$$r\dot{r} = x\dot{x} + y\dot{y} = -(x^2 + y^2)^{3/2} = -r^3, \quad \dot{\theta} = 1,$$
so that $\dot{r} = -r^2$. Hence $r = 1/(t + A)$ and $\theta = t + B$, which means that the phase paths are stable spirals.

To prove this by Liapunov's method, let $V(x, y) = x^2 + y^2$. Then

$$\dot{V} = \frac{\partial V}{\partial x}\dot{x} + \frac{\partial V}{\partial y}\dot{y} = -x^2 - y^2,$$

which is negative definite. The implication is that the origin is asymptotically stable.

- 10.9 Euler's equations for a body spinning freely about a fixed point under no forces are

$$A\dot{\omega}_1 - (B-C)\omega_2\omega_3 = 0, \quad B\dot{\omega}_2 - (C-A)\omega_3\omega_1 = 0, \quad C\dot{\omega}_3 - (A-B)\omega_1\omega_2 = 0,$$

where A, B and C (all different) are the principal moments of inertia, and $(\omega_1, \omega_2, \omega_3)$ is the spin of the body in principal axes fixed in the body. Find all the states of steady spin of the body.

Consider perturbations about the steady state $(\omega_0, 0, 0)$ by putting $\omega_1 = \omega_0 + x_1$, $\omega_2 = x_2$, $\omega_3 = x_3$, and show that the linear approximation is

$$\dot{x}_1 = 0, \quad \dot{x}_2 = \frac{C-A}{B}\omega_0 x_3, \quad \dot{x}_3 = \frac{A-B}{C}\omega_0 x_2.$$

Deduce that this state is unstable if $C < A < B$ or $B < A < C$.

Show that

$$V(x_1, x_2, x_3) = \{B(A-B)x_2^2 + C(A-C)x_3^2\} + \{Bx_2^2 + Cx_3^2 + A(x_1^2 + 2\omega_0 x_1)\}^2$$

is a Liapunov function for the case when A is the largest moment of inertia, so that this state is stable. Suggest a Liapunov function which will establish the stability of the case in which A is the smallest moment of inertia. Are these states asymptotically stable?

Why would you expect V as given above to be a first integral of the Euler equations? Show that each of the terms in braces is such an integral.

10.9. The Euler equations in dynamics for a body spinning about a fixed point under no forces are

$$A\dot{\omega}_1 = (B-C)\omega_2\omega_3,$$
$$B\dot{\omega}_2 = (C-A)\omega_3\omega_1,$$
$$C\dot{\omega}_3 = (A-B)\omega_1\omega_2,$$

where the spin is $(\omega_1, \omega_2, \omega_3)$. Equilibrium occurs where $\dot{\omega}_1 = \dot{\omega}_2 = \dot{\omega}_3 = 0$, that is where any pair of $\omega_1, \omega_2, \omega_3$ are zero (assuming that A, B and C are all different),
Let $\omega_1 = \omega_0 + x_1$, $\omega_2 = x_2$, and $\omega_3 = x_3$. Then the linear approximation is

$$\dot{x}_1 \approx 0, \quad \dot{x}_2 \approx \frac{C-A}{B}\omega_0 x_3, \quad \dot{x}_3 \approx \frac{A-B}{C}\omega_0 x_2.$$

Therefore

$$x_1 = \text{constant}, \quad \ddot{x}_2 = \frac{(C-A)(A-B)\omega_0^2}{BC}x_2.$$

The zero solution is stable if $(C-A)(A-B) < 0$, and unstable if $(C-A)(A-B) > 0$.

Let $A > \max(B, C)$, and consider the Liapunov function (this is an example of a function in three dimensions)

$$V(x_1, x_2, x_3) = \{B(A - B)x_2^2 + C(A - C)x_3^2\} + \{Bx_2^2 + Cx_3^2 + A(x_1^2 + 2\omega_0 x_1)\}^2. \quad \text{(i)}$$

Then

$$\begin{aligned}
\dot{V} &= 4\{Bx_2^2 + Cx_3^2 + A(x_1^2 + 2\omega_0 x_1)\}(x_1 + \omega_0)(B - C)x_2 x_3 \\
&\quad + [2(A - B)x_2 + 4x_2\{Bx_2^2 + Cx_3^2 + A(x_1^2 + 2\omega_0 x_1)\}] \\
&\quad (C - A)x_3(\omega_0 + x_1) + [2(A - C)x_3 + 4x_3\{Bx_2^2 + Cx_3^2 \\
&\quad + A(x_1^2 + 2\omega_0 x_1)\}](A - B)x_2 x_2(\omega_0 + x_1) \\
&= 4\{Bx_2^2 + Cx_3^2 + A(x_1^2 + 2\omega_0 x_1)\}[(B - C) + (C - A) \\
&\quad + (A - B)]x_2 x_3(\omega_0 + x_1) \\
&= 0
\end{aligned}$$

Hence $V(x_1, x_2, x_3)$ is positive definite and $\dot{V}$ is negative semidefinite, so that the equilibrium state $(\omega_0, 0, 0)$ is uniformly stable.

If A is the smallest moment of inertia choose the Liapunov function

$$V(x_1, x_2, x_3) = \{B(B - A)x_2^2 + C(C - A)x_3^2\} + \{Bx_2^2 + Cx_3^2 + A(x_1^2 + 2\omega_0 x_1)\}^2.$$

Since $\dot{V} = 0$, then the level curves of V coincide with the solutions of the Euler equations. One conclusion is that the equilibrium states are not asymptotically stable. The second conclusion is that V must be composed of first integrals of the Euler equations in some manner. Let

$$F(x_1, x_2, x_3) = \{B(A - B)x_2^2 + C(A - C)x_3^2\}.$$

Then

$$\frac{dF(x_1, x_2, x_3)}{dt} = 2\{B(A - B)x_2 \dot{x}_2 + C(A - C)x_3 \dot{x}_3\} = 0,$$

by (i). Let

$$G(x_1, x_2, x_3) = \{Bx_2^2 + Cx_3^2 + A(x_1^2 + 2\omega_0 x_1)\}.$$

Then

$$\frac{dG(x_1, x_2, x_3)}{dt} = 2\{Bx_2 \dot{x}_2 + Cx_3 \dot{x}_3 + A(x_1 \dot{x}_1 + 2\omega_0 \dot{x}_1)\} = 0,$$

by (i). These results prove that $F(x_1, x_2, x_3) = $ constant and $G(x_1, x_2, x_3) = $ constant are first integrals of the Euler equations.

- **10.10** Show that the zero solution of the equation $\ddot{x} + h(x,\dot{x})\dot{x} + x = 0$ is stable if $h(x,y) \geq 0$ in a neighbourhood of the origin.

10.10. Express the equation as

$$\dot{x} = y, \quad \dot{y} = -h(x,y)y - x.$$

Consider the Liapunov function $V(x,y) = x^2 + y^2$. Then

$$\dot{V} = \frac{\partial V}{\partial x}\dot{x} + \frac{\partial V}{\partial y}\dot{y} = 2xy + 2y(-h(x,y) - x) = -2y^2 h(x,y) \leq 0.$$

Therefore $\dot{V}$ is semidefinite, which implies that the zero solution is uniformly stable.

- **10.11** The n-dimension system $\dot{\mathbf{x}} = \mathrm{grad}\, W(\mathbf{x})$ has an isolated equilibrium point at $\mathbf{x} = 0$. Show that the zero solution is asymptotically stable if W has a local minimum at $\mathbf{x} = 0$. Give a condition for instability of zero solution.

10.11. Since the n-dimensional system

$$\dot{\mathbf{x}} = \mathrm{grad}\, W(\mathbf{x})$$

has an isolated equilibrium point at $\mathbf{x} = 0$, $\mathrm{grad}\, W(\mathbf{x}) = 0$ at the origin. Consider the Liapunov function $V = W$. Then V will be positive definite if W has a local maximum at $\mathbf{x} = 0$. The derivative of V is

$$\dot{V} = \mathrm{grad}\, W(\mathbf{x}) \cdot \dot{\mathbf{x}} = \mathrm{grad}\, W(\mathbf{x}) \cdot \mathrm{grad}\, W(\mathbf{x}) > 0,$$

except at $\mathbf{x} = 0$ where $\dot{V}$ is zero. Therefore the zero solution is asymp- totically stable.

Instability will occur if W is negative in at least one point in every deleted neighbourhood of the origin. We can then apply NODE, Theorem 10.13 with $U = W$: as above $\dot{U}$ will be positive definite in some neighbourhood of the origin.

- **10.12** A particle of mass m and position vector $\mathbf{r} = (x, y, z)$ moves in a potential field $W(x, y, z)$, so that its equation of motion is $m\ddot{\mathbf{r}} = -\mathrm{grad}\, W$. By putting $\dot{x} = u$, $\dot{y} = v$, $\dot{z} = w$, express this in terms of first-order derivatives. Suppose that W has a minimum at $\mathbf{r} = 0$. Show that the origin of the system is stable, by using the Liapunov function $V = W + \frac{1}{2}m(u^2 + v^2 + w^2)$. What do the level curves of V represent physically? Is the origin asymptotically stable?
An additional non-conservative force $\mathbf{f}(u, v, w)$ is introduced, so that $m\ddot{\mathbf{r}} = -\mathrm{grad}\, W + \mathbf{f}$. Use the Liapunov function to give a sufficient condition for $\mathbf{f}$ to be of frictional type.

10.12. The equation of motion of the particle is

$$m\ddot{\mathbf{r}} = \text{grad} W.$$

Let $\dot{x} = u$, $\dot{y} = v$, $\dot{z} = w$, and express the equation of motion as the first-order system

$$\dot{x} = u, \quad \dot{y} = v, \quad \dot{z} = w, \quad \dot{u} = -W_x/m, \quad \dot{v} = -W_y/m, \quad \dot{w} = -W_z/m,$$

which has an equilibrium point at $(0,0,0,0,0,0)$. Assume that W has a minimum at $\mathbf{r} = 0$, and consider the Liapunov function

$$V = W + \tfrac{1}{2}m(u^2 + v^2 + w^2), \tag{i}$$

which is positive definite. It follows that

$$\dot{V} = W_x\dot{x} + W_y\dot{y} + W_z\dot{z} + W_u\dot{u} + W_v\dot{v} + W_w\dot{w} = 0. \tag{ii}$$

Therefore V is a weak Liapunov function, and the zero solution is stable. The level curves of V are curves of constant energy, which implies that the zero solution cannot be asymptotically stable.

Suppose that an additional non-conservative force $\mathbf{f}(u, v, w)$ is introduced which is a function of the 'velocity' $(u, v, w) = (\dot{x}, \dot{y}, \dot{z})$ only, so that

$$m\ddot{\mathbf{r}} = \text{grad} W + \mathbf{f}.$$

Consider the same function V given by (i). Then, using (ii),

$$\dot{V} = m\mathbf{u} \cdot \mathbf{f},$$

where $\mathbf{u} = (u, v, w)$. V is a strong Liapunov function if $\mathbf{u} \cdot \mathbf{f} < 0$ in some deleted neighbourhood of the origin. In this case the zero solution will be asymptotically stable.

- **10.13** Use the test for instability to show that if $\dot{x} = X(x, y)$, $\dot{y} = Y(x, y)$ has an equilibrium point at the origin, then the zero solution is unstable if there exist constants α and β such that $\alpha X(x, y) + \beta Y(x, y) > 0$ in a neighbourhood of the origin except at the origin where it is zero.

10.13. Consider the system $\dot{x} = X(x, y)$, $\dot{y} = Y(x, y)$. Let $U(x, y) = \alpha x + \beta y$, which has some positive values in every neighbourhood of the origin for any values of α and β such that $(\alpha, \beta) \neq (0, 0)$. Then

$$\dot{U} = U_x\dot{x} + U_y\dot{y} = \alpha X(x, y) + \beta Y(x, y).$$

If $\alpha X(x, y) + \beta Y(x, y) > 0$ in a deleted neighbourhood of the origin, then by NODE, Theorem 11.13, the zero solution is unstable.

- **10.14** Use the result of Problem 10.13 to show that the origin is unstable for each of the following:
  (i) $\dot{x} = x^2 + y^2,\ \dot{y} = x + y$;
  (ii) $\dot{x} = y \sin y,\ \dot{y} = xy + x^2$;
  (iii) $\dot{x} = y^{2m},\ \dot{y} = x^{2n}$ (m, n positive integers).

10.14. We use the result from Problem 10.13 with $U(x, y) = \alpha x + \beta y$.

(i) $\dot{x} = x^2 + y^2,\ \dot{y} = x + y$. Put $\alpha = 1$ and $\beta = 0$. Then

$$\dot{U} = \alpha X(x, y) + \beta Y(x, y) = x^2 + y^2 > 0,$$

for $(x, y) \neq (0, 0)$. Since $\dot{U}$ is positive definite, the origin is unstable.

(ii) $\dot{x} = y \sin y,\ \dot{y} = xy + x^2$. Then, expanding the sine function,

$$\dot{U} = \alpha y \sin y + \beta(xy + x^2) \approx \alpha y^2 + \beta(xy + x^2)$$

$$= \beta\left(x + \tfrac{1}{2}y\right)^2 + \left(\alpha - \tfrac{1}{4}\beta\right)y^2,$$

which is positive definite if $\beta > 0$ and $\alpha > \tfrac{1}{4}\beta$.

(iii) $\dot{x} = y^{2m},\ \dot{y} = x^{2n}$. Then

$$\dot{U} = \alpha y^{2m} + \beta x^{2n} > 0$$

which is positive definite if $\alpha > 0$ and $\beta > 0$. Therefore the origin is unstable.

- **10.15** For the system $\dot{x} = y,\ \dot{y} = f(x, y)$ where $f(0, 0) = 0$, show that V given by

$$V(x, y) = \tfrac{1}{2}y^2 - \int_0^x f(u, 0)\,du$$

is a weak Liapunov function for the zero solution when

$$\{f(x, y) - f(x, 0)\}y \leq 0, \quad \int_0^x f(u, 0)\,du < 0,$$

in a neighbourhood of the origin.

10.15. The system is

$$\dot{x} = y, \quad \dot{y} = f(x, y),$$

where $f(0, 0) = 0$. Consider the function

$$V(x, y) = \tfrac{1}{2}y^2 - \int_0^x f(u, 0)\,du.$$

$V(x, y)$ is positive definite if

$$\int_0^x f(u, 0) du < 0$$

in some neighbourhood of the origin.
The derivative

$$\dot{V} = V_x \dot{x} + V_y \dot{y} = -f(x, 0)y + yf(x, y).$$

This function will be negative semidefinite if

$$y\{f(x, y) - f(x, 0)\} \leq 0.$$

In this case the origin is uniformly stable.

• **10.16** Use the result of Problem 10.15 to show the stability of the zero solutions of the following:

(i) $\ddot{x} = -x^3 - x^2 \dot{x}$;
(ii) $\ddot{x} = -x^3/(1 - x\dot{x})$;
(iii) $\ddot{x} = -x + x^3 - x^2 \dot{x}$.

10.16. Use the function $V(x, y)$ defined in Problem 10.15.

(i) $\dot{x} = y$, $\dot{y} = -x^3 - x^2 y$. In this case $f(x, y) = -x^3 - x^2 y$. The required conditions are

$$\{f(x, y) - f(x, 0)\}y = \{-x^3 - x^2 y + x^3\}y = -x^2 y^2 < 0,$$

for all $(x, y) \neq (0, 0)$. Also

$$\int_0^x f(u, 0) du = -\int_0^x u^3 du = -\tfrac{1}{4}x^4 < 0.$$

Therefore the origin is stable.

(ii) $\dot{x} = y$, $\dot{y} = -x^3/(1 - xy)$. Assume $|xy| < 1$. In this case $f(x, y) = -x^3/(1 - xy)$. The required conditions are

$$\{f(x, y) - f(x, 0)\}y = \left(-\frac{x^3}{1 - xy} + x^3\right) y = -\frac{x^4 y^2}{1 - xy} \leq 0.$$

Also

$$\int_0^x f(u, 0) du = -\int_0^x u^3 du = -\tfrac{1}{4}x^4 \leq 0.$$

Therefore the origin is stable.

(iii) $\dot{x} = y$, $\dot{y} = -x + x^3 - x^2 y$. In this case $f(x, y) = -x + x^3 - x^2 y$. The required conditions are

$$\{f(x, y) - f(x, 0)\}y = \{-x + x^3 - x^2 y + x - x^3\}y = -x^2 y^2 \leq 0.$$

Also

$$\int_0^x (-u + u^3) du = -\tfrac{1}{2}x^2 + \tfrac{1}{4}x^4 \leq 0$$

for $|x|$ sufficiently small.

- 10.17 Let $\dot{x} = -\alpha x + \beta f(y)$, $\dot{y} = \gamma x - \delta f(y)$, where $f(0) = 0$, $yf(y) > 0$ ($y \neq 0$), and $\alpha \delta > 4\beta\gamma$, where $\alpha, \beta, \gamma, \delta$ are positive. Show that, for suitable values of A and B

$$V = \tfrac{1}{2}Ax^2 + B \int_0^y f(u) du$$

is a strong Liapunov function for the zero solutions.

10.17. The system is

$$\dot{x} = -\alpha x + \beta f(y), \quad \dot{y} = \gamma x - \delta f(y),$$

where $f(0) = 0$ and $yf(y) > 0$ ($y \neq 0$). Consider the function

$$V(x, y) = \tfrac{1}{2}Ax^2 + B \int_0^y f(u) du.$$

Since $yf(y) > 0$ for $y \neq 0$, $V(x, y)$ is positive definite. The derivative

$$\dot{V} = Ax[-\alpha x + \beta f(y)] + Bf(y)[\gamma x - \delta f(y)]$$

$$= -A\alpha x^2 + (A\beta + B\gamma)xf(y) - B\delta f(y)^2$$

$$= -A\alpha \left(x - \frac{A\beta + B\gamma}{2A\alpha} f(y) \right)^2 + \frac{f(y)^2}{A\alpha}[-AB\alpha\delta + \tfrac{1}{4}(A\beta + B\gamma)^2],$$

which is negative definite if $A > 0$ and

$$(A\beta + B\gamma)^2 - 4AB\alpha\delta < 0.$$

The equation

$$(A\beta + B\gamma)^2 - 4AB\alpha\delta = 0,$$

or

$$\gamma^2 B^2 + (2\beta\gamma - \alpha\delta)AB + \beta^2 A^2 = 0,$$

has the solutions
$$\frac{B}{A} = \frac{1}{2\gamma^2}[-(2\beta\gamma - \alpha\delta) \pm \sqrt{\{\alpha\delta(\alpha\delta - 4\beta\gamma)\}}],$$

which are real if $\alpha\delta > 4\beta\gamma$ (as given in the question). Therefore B must satisfy

$$\frac{-(2\beta\gamma - \alpha\delta) - \sqrt{\{\alpha\delta(\alpha\delta - 4\beta\gamma)\}}}{2\gamma^2} < \frac{B}{A} < \frac{-(2\beta\gamma - \alpha\delta) + \sqrt{\{\alpha\delta(\alpha\delta - 4\beta\gamma)\}}}{2\gamma^2}.$$

With the given conditions on A and B, $V(x, y)$ is a strong Liapunov function which means that the origin is asymptotically stable.

- **10.18** A particle moving under a central attractive force $f(r)$ per unit mass has the equations of motion

$$\ddot{r} - r\dot{\theta}^2 = f(r), \quad \frac{d}{dt}(r^2\dot{\theta}) = 0.$$

For a circular orbit, $r = a$, show that $r^2\dot{\theta} = h$, a constant, and $h^2 + a^3 f(a) = 0$. The orbit is subjected to a small radial perturbation $r = a + \rho$, in which h is kept constant. Show that the equation for ρ is

$$\ddot{\rho} - \frac{h^2}{(a+\rho)^3} - f(a + \rho) = 0.$$

Show that

$$V(\rho, \dot{\rho}) = \frac{1}{2}\dot{\rho}^2 + \frac{h^2}{2(a+\rho)^2} - \int_0^\rho f(a+u)du - \frac{h^2}{2a^2}$$

is a Liapunov function for the zero solution of this equation provided that $3h^2 > a^4 f'(a)$, and that the gravitational orbit is stable in this sense.

10.18. A particle moving under a central attractive force has the equations of motion

$$\ddot{r} - r\dot{\theta}^2 = f(r), \quad \frac{d}{dt}(r^2\dot{\theta}) = 0.$$

The second equation implies generally that $r^2\dot{\theta} = h$, a constant. If the orbit is a circle, then

$$a^2\dot{\theta} = h, \text{ and } -a\dot{\theta}^2 = f(a),$$

so that $h^2 + a^3 f(a) = 0$.

Consider the perturbation $r = a + \rho$ so that $\rho = 0$ corresponds to the circular orbit. Then, the equation of motion becomes

$$\ddot{\rho} - \frac{h^2}{(a+\rho)^3} - f(a+\rho) = 0.$$

Let $\sigma = \dot{\rho}$, and consider the function

$$V(\rho, \sigma) = \frac{1}{2}\sigma^2 + \frac{1}{2}h^2\left(\frac{1}{(a+\rho)^2} - \frac{1}{a^2}\right) - \int_0^\rho f(a+u)du$$

$$\approx \frac{1}{2}\sigma^2 + \frac{h^2}{2a^2}\left(1 - \frac{2\rho}{a} + \frac{3\rho^2}{a^2}\right) - f(a)\rho - \frac{1}{2}f'(a)\rho^2 - \frac{h^2}{2a^2}$$

$$= \frac{1}{2}\sigma^2 + \left(\frac{3h^2}{2a^4} - \frac{f'(a)}{2}\right)\rho^2$$

Therefore $V(\rho,\sigma)$ is positive definite if $3h^2 > a^4 f'(a)$. Also

$$\dot{V} = \left[-\frac{h^2}{(a+\rho)^3} - f(a+\rho)\right]\sigma + \sigma\left[\frac{h^2}{(a+\rho)^3} + f(a+\rho)\right] = 0,$$

which implies that $\dot{V}$ is negative semidefinite. It follows that $\rho = 0$ is stable, which, in turn, implies that the circular orbit is stable.

- **10.19** Show that the following Liénard-type equations have zero solutions which are asymptotically stable:

(i) $\ddot{x} + |x|(\dot{x} + x) = 0$;
(ii) $\ddot{x} + (\sin x / x)\dot{x} + x^3 = 0$;
(iii) $\dot{x} = y - x^3$, $\dot{y} = -x^3$.

10.19. These equations are of the Liénard type (see NODE, Section 10.11)

$$\ddot{x} + f(x)\dot{x} + g(x) = 0.$$

In the Liénard plane

$$\dot{x} = y - \int_0^x f(u)du, \quad \dot{y} = -g(x).$$

Then a possible Liapunov function is

$$V(x,y) = G(x) + \tfrac{1}{2}y^2,$$

where

$$G(x) = \int_0^x g(u)du,$$

if $f(x)$ is positive in a deleted neighbourhood of the origin and if $g(x)$ is positive/negative when x is positive/negative. In this case

(i) $\ddot{x} + |x|(\dot{x} + x) = 0$. In this case $f(x) = |x|$ and $g(x) = x|x|$. Therefore the Liénard system is

$$\dot{x} = y - \int_0^x |u|du = y - \tfrac{1}{2}x|x|, \quad \dot{y} = -x|x|.$$

Since $f(x)$ and $g(x)$ satisfy (a) and (b) above, the zero solution is asymptotically stable.

(ii) $\ddot{x} + (\sin x/x)\dot{x} + x^3 = 0$. In this case $f(x) = \sin x/x$ and $g(x) = x^3$. Therefore the system in the Liénard plane is

$$\dot{x} = y - \int_0^x \frac{\sin u}{u}du, \quad \dot{y} = -x^3.$$

Since $f(x) = \sin x/x$ is an even function, and therefore positive in every deleted neighbourhood. Also $g(x)$ is an odd function so that the conditions for asymptotic stability are satisfied.

(iii) $\dot{x} = y - x^3$, $\dot{y} = -x^3$. This is a Liénard system with $f(x) = 3x^2$ and $g(x) = x^3$. The conditions for asymptotic stability are satisfied.

- **10.20** Give a geometrical account of NODE, Theorem 10.13 (an instability test).

10.20. The geometry of the instability condition of Theorem 10.13 will be illustrated for a particular example of a linear system with eigenvalues with positive real parts (see Section 10.8, Case (ii)). Consider the system

$$\dot{\mathbf{x}} = \mathbf{A}\mathbf{x}, \quad \mathbf{A} = \begin{bmatrix} 1 & 1 \\ -1 & 1 \end{bmatrix}, \tag{i}$$

which has the eigenvalues $\lambda_1 = 1 + i$, $\lambda_2 = 1 - i$. A matrix $\mathbf{C}$ which diagonalizes $\mathbf{A}$ is

$$\mathbf{C} = \begin{bmatrix} -1 & -1 \\ -i & i \end{bmatrix}.$$

Since the eigenvalues are complex, then we have to use the transformation $\mathbf{x} = \mathbf{G}\mathbf{u}$, where

$$\mathbf{G} = \mathbf{C}\begin{bmatrix} 1 & i \\ 1 & -i \end{bmatrix} = \begin{bmatrix} -2 & 0 \\ 0 & 2 \end{bmatrix},$$

to obtain the transformed equation

$$\dot{\mathbf{u}} = \begin{bmatrix} 1 & -1 \\ 1 & 1 \end{bmatrix}\mathbf{u}.$$

10 : Liapunov methods for determining stability of the zero solution   467

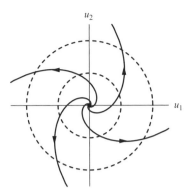

Figure 10.1 Problem 10.21: The dashed circles represent the level curves $U(u_1, u_2) = $ constant: the other curves are phase paths.

We now choose the function $U(\mathbf{u}) = \mathbf{u}^T \mathbf{u} = u_1^2 + u_2^2$ (which is positive except at $(0,0)$), so that

$$\dot{U}(\mathbf{u}) = 2(u_1^2 + u_2^2),$$

which is positive definite. Therefore Theorem 10.13 can be applied. Figure 10.1 shows the unstable spiral which cut the curves $U = $ constant from inside to outside, which is implied also by $\dot{U}$ positive definite.

- **10.21** For the system $\dot{x} = f(x) + \beta y$, $\dot{y} = \gamma x + \delta y$, $(f(0) = 0)$, establish that V given by

$$V(x,y) = (\delta x - \beta y)^2 + 2\delta \int_0^x f(u)du - \beta\gamma x^2$$

is a strong Liapunov function for the zero solution when, in some neighbourhood of the origin,

$$\delta \frac{f(x)}{x} - \beta\gamma > 0, \quad \frac{f(x)}{x} + \delta < 0$$

for $x \neq 0$. (Barbashin 1970.)
  Deduce that for initial conditions in the circle $x^2 + y^2 < 1$, the solutions of the system
$$\dot{x} = -x^3 + x^4 + y, \quad \dot{y} = -x,$$
tend to zero.

10.21. The system
$$\dot{x} = f(x) + \beta y, \quad \dot{y} = \gamma x + \delta y, \quad (f(0) = 0)$$
has an equilibrium point at $(0,0)$. Consider the function
$$V(x,y) = (\delta x - \beta y)^2 + 2\delta \int_0^x f(u)du - \beta\gamma x^2,$$

which we can express also in the form

$$V(x, y) = (\delta x - \beta y)^2 + 2 \int_0^x [\delta f(u) - \beta \gamma u] du.$$

This function is positive definite if

$$\int_0^x [\delta f(u) - \beta \gamma u] du > 0,$$

which is true if $(\delta f(x)/x) - \beta \gamma > 0$.

The derivative of V is given by

$$\dot V = [2\delta(\delta x - \beta y) + 2\delta f(x) - 2\beta \delta x][f(x) + \beta y] - 2\beta[\delta x - \beta y][\gamma x + \delta y]$$
$$= 2[f(x)\delta - x\beta\gamma][f(x) + \delta x],$$

if $f(x) + \delta x < 0$ for $x \neq 0$. Therefore $V(x, y)$ is a Liapunov function for this system, which implies that the zero solution is asymptotically stable.

For the system

$$\dot x = -x^3 + x^4 + y, \quad \dot y = -x,$$
$$f(x) = -x^3 + x^4, \quad \beta = 1, \quad \gamma = -1, \quad \delta = 0,$$

in the notation above. Then

$$\int_0^x [\delta f(u) - \beta \gamma u] du = \tfrac{1}{2} x^2 > 0,$$

for $x \neq 0$. Also

$$f(x) + \delta x = -x^3 + x^4 < 0 \text{ for } |x| < 1.$$

The conditions above are satisfied so that the zero solution is asymptotically stable.

- **10.22** For the system
$$\dot x = f(x) + \beta y, \quad \dot y = g(x) + \delta y, \quad f(0) = g(0) = 0,$$
show that V given by
$$V(x, y) = (\delta x - \beta y)^2 + 2 \int_0^x \{\delta f(u) - \beta g(u)\} du$$
is a strong Liapunov function for the zero solution when, in some neighbourhood of the origin,
$$\{\delta f(x) - \beta g(x)\}x > 0, \quad xf(x) + \delta x^2 < 0$$
for $x \neq 0$. (Barbashin 1970.)

Deduce that the zero solution of the system
$$\dot{x} = -x^3 + 2x^4 + y, \quad \dot{y} = -x^4 - y$$
is asymptotically stable. Show how to find a domain of initial conditions from which the solutions tend to the origin. Sketch phase paths and a domain of asymptotic stability of the origin.

10.22. The system
$$\dot{x} = f(x) + \beta y, \quad \dot{y} = g(x)\delta y, \quad f(0) = g(0) = 0,$$
has an equilibrium point at $(0,0)$. Consider the possible Liapunov function
$$V(x, y) = (\delta x - \beta y)^2 + 2 \int_0^x \{\delta f(u) - \beta g(u)\} du.$$

The function is positive definite if
$$\{\delta f(x) - \beta g(x)\} x > 0. \tag{i}$$

Its derivative is
$$\dot{V} = \{2\delta(\delta x - \beta y) + 2[\delta f(x) - \beta g(x)]\}[f(x) + \beta y] - 2\beta(\delta x - \beta y)[g(x) + \delta y]$$
$$= 2[\delta f(x) - \beta g(x)][\delta x + f(x)]$$

which is negative definite if
$$[\delta f(x) - \beta g(x)][\delta x + f(x)] < 0$$

in a deleted neighbourhood of the origin. Combined with inequality (i) this is equivalent to
$$(\delta x + f(x)) x < 0, \quad (x \neq 0). \tag{ii}$$

For the particular system
$$\dot{x} = -x^3 + 2x^4 + y, \quad \dot{y} = -x^4 - y$$
choose $f(x) = -x^3 + 2x^4$, $g(x) = -x^4$, $\beta = 1$ and $\delta = -1$. The system has equilibrium points at $(0,0)$ and $(1,-1)$. Inequalities (i) and (ii) become
$$\{\delta f(x) - \beta g(x)\} x = x^4 - x^5 > 0,$$

for $0 < |x| < 1$, and
$$(\delta x + f(x))x = -x^2 - x^3 + 2x^5 < 0,$$
in some deleted interval about the origin. Since
$$2x^5 - x^3 - x^2 = x^2(x-1)(2x^2 + 2x + 1)$$
has only zeros at $x = 0$ and $x = 1$ (for real x), we can be more specific and say that $\dot{V}$ is negative definite also in $0 < |x| < 1$. Therefore the zero solution is asymptotically stable.

The Liapunov function is
$$V(x, y) = (x + y)^2 + 2 \int_0^x [-(-u^3 + 2u^4) + u^4] du = (x + y)^2 + 2 \left(\tfrac{1}{4}x^4 - \tfrac{1}{5}x^5 \right).$$

A domain of asymptotic stability will be the interior of the largest level curve $V(x, y) = $ constant which is within $|x| < 1$. Consider the closed curve
$$(x + y)^2 + 2 \left(\tfrac{1}{4}x^4 - \tfrac{1}{5}x^5 \right) = C.$$

The straight line $x = 1$ will cut this curve in only one point if $C = 2(\tfrac{1}{4} - \tfrac{1}{5}) = \tfrac{1}{10}$: at this point $y = -\tfrac{9}{10}$. Figure 10.2 shows the phase diagram and the level curve $V(x, y) = 0.1$. All phase paths which start within this curve will approach the origin asymptotically. A linear approximation indicates that the equilibrium point at $(1, -1)$ is a saddle point.

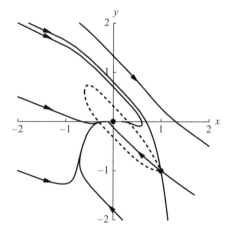

Figure 10.2 Problem 10.22: The closed dashed curve shows the domain of asymptotic stability detected.

- **10.23** Consider van der Pol's equation $\ddot{x} + \varepsilon(x^2 - 1)\dot{x} + x = 0$, for $\varepsilon < 0$, in the Liénard phase plane, NODE, eqn (10.83):

$$\dot{x} = y - \varepsilon(\tfrac{1}{3}x^3 - 1), \quad \dot{y} = -x,$$

Show that, in this plane, $V = \tfrac{1}{2}(x^2 + y^2)$ is a strong Liapunov function for the zero solution, which is therefore asymptotically stable. Show that all solutions starting from initial conditions inside the circle $x^2 + y^2 = 3$ tend to the origin (and hence the limit cycle lies outside this region for every $\varepsilon < 0$). Sketch this domain of asymptotic stability in the ordinary phase plane with $\dot{x} = y$.

10.23. The van der Pol equation

$$\ddot{x} + \varepsilon(x^2 - 1)\dot{x} + x = 0$$

can be expressed in the form

$$\dot{x} = y - \varepsilon\left(\tfrac{1}{3}x^3 - x\right), \quad \dot{y} = -x.$$

Consider the positive definite function $V(x, y) = \tfrac{1}{2}(x^2 + y^2)$. Then

$$\dot{V} = \frac{\partial V}{\partial x}\dot{x} + \frac{\partial V}{\partial y}\dot{y} = x\left[y - \varepsilon\left(\tfrac{1}{3}x^3 - x\right)\right] + y(-x) = \varepsilon\left(x^2 - \tfrac{1}{3}x^4\right) < 0,$$

for $|x| < \sqrt{3}$, ($x \neq 0$) and $\varepsilon < 0$. Therefore $V(x, y)$ is a Liapunov function for this system, and the zero solution is asymptotically stable. The largest topographic curve of $V(x, y)$ which can be inserted into $|x| < \sqrt{3}$ is the circle $x^2 + y^2 = 3$, and this is the domain of asymptotic stability of the origin.

Expressed in the usual phase plane, the van der Pol equation is

$$\dot{x} = y, \quad \dot{y} = -\varepsilon(x^2 - 1)y - x.$$

In this phase plane the domain of asymptotic stability detected becomes the interior of the closed curve

$$x^2 + \left[y + \varepsilon\left(\tfrac{1}{3}x^3 - x\right)\right]^2 = 3.$$

The curve is shown in Figure 10.3 for $\varepsilon = -1$.

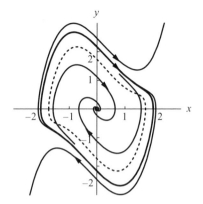

Figure 10.3 Problem 10.23: Phase diagram of $\dot{x} = y$, $\dot{y} = -\varepsilon(x^2 - 1)y - x$ with $\varepsilon = -1$; the closed dashed closed curve shows the boundary of the domain of asymptotic stability.

- **10.24** Show that the system $\dot{x} = -x - xy^2$, $\dot{y} = -y - x^2 y$ is globally asymptotically stable, by guessing a suitable Liapunov function.

10.24. Consider the system

$$\dot{x} = -x - xy^2, \quad \dot{y} = -y - x^2 y.$$

Let $V(x, y) = x^2 + y^2 > 0$ for $(x, y) \neq (0, 0)$. Then

$$\dot{V} = \frac{\partial V}{\partial x}\dot{x} + \frac{\partial V}{\partial y}\dot{y} = 2x(-x - xy^2) + 2y(-y - x^2 y) = -2(x^2 + y^2)^2 < 0,$$

for all (x, y), except at $(0, 0)$. Therefore the zero solution is globally asymptotically stable.

- **10.25** Assuming that the conditions of Problem 10.22 are satisfied, obtain further conditions which ensure that the system is globally asymptotically stable. Show that the system $\dot{x} = y - x^3$, $\dot{y} = -x - y$ is globally asymptotically stable.

10.25. The system is

$$\dot{x} = f(x) + \beta y, \quad \dot{y} = \gamma x + \delta y, \quad (f(0) = 0).$$

The Liapunov function is

$$V(x, y) = (\delta x - \beta y)^2 + 2\delta \int_0^x f(u)du - \beta \gamma x^2,$$

where

$$\delta \frac{f(x)}{x} - \beta\gamma > 0, \quad \frac{f(x)}{x} + \delta < 0. \tag{i}$$

The zero solution is globally asymptotically stable if inequalities (i) are true for all $x \neq 0$.

For the system

$$\dot{x} = y - x^3, \quad \dot{y} = -x - y,$$

$f(x) = -x^3$, $\beta = 1$, $\gamma = -1$ and $\delta = -1$. Then (i) become

$$\delta \frac{f(x)}{x} - \beta\gamma = \frac{x^3}{x} + 1 = x^2 + 1 > 0, \quad \text{for all } x,$$

and

$$\frac{f(x)}{x} + \delta = -\frac{x^3}{x} - 1 = -x^2 - 1 < 0, \quad \text{for all } x.$$

Therefore the zero solution is globally asymptotically stable.

- **10.26** Assuming that the conditions of Problem 10.23 are satisfied, obtain further conditions which ensure that the system is globally asymptotically stable. Show that the system $\dot{x} = -x^3 - x + y$, $\dot{y} = -x^3 - y$ is globally asymptotically stable.

10.26. For the system

$$\dot{x} = f(x) + \beta y, \quad \dot{y} = g(x) + \delta y, \quad (f(0) = g(0) = 0),$$

$$V(x, y) = (\delta x - \beta y)^2 + 2\int_0^x [\delta f(u) - \beta g(u)]du$$

is a Liapunov function if

$$[\delta f(x) - \beta g(x)]x > 0, \quad xf(x) + \delta x^2 < 0. \tag{i}$$

The zero solution is globally asymptotically stable if inequalities (i) are true for all x.

For the system

$$\dot{x} = -x^3 - x + y, \quad \dot{y} = -x^3 - y,$$

$f(x) = -x^3 - x$, $\beta = 1$, $g(x) = -x^3$ and $\delta = -1$. Then (i) become

$$[\delta f(x) - \beta g(x)]x = (x^3 + x + x^3)x = 2x^4 + x^2 > 0, \quad \text{for all } x \neq 0,$$

and

$$xf(x) + \delta x^2 = -x^4 - x^2 - x^2 = -x^4 - 2x^2 < 0, \quad \text{for all } x \neq 0.$$

Therefore the zero solution is globally asymptotically stable.

- **10.27** Give conditions on the functions f and g of the Liénard equation, $\ddot{x} + f(x)\dot{x} + g(x) = 0$ which ensure that the corresponding system $\dot{x} = y - F(x)$, $\dot{y} = -g(x)$ (NODE, Section 10.11) is globally asymptotically stable.
  Show that all solutions of the equation $\ddot{x} + x^2\dot{x} + x^3 = 0$ tend to zero.

10.27. The equation

$$\ddot{x} + f(x)\dot{x} + g(x) = 0,$$

can be expressed as

$$\dot{x} = y - F(x), \quad \dot{y} = -g(x),$$

where

$$F(x) = \int_0^x f(u)du.$$

Let

$$G(x) = \int_0^x g(u)du,$$

and assume that $g(x)$ is positive/negative when x is positive/negative for all x. It follows that $G(x) > 0$ for $x \neq 0$. Therefore the function $V(x, y) = G(x) + \frac{1}{2}y^2$ is positive definite for all x. Also

$$\dot{V}(x, y) = g(x)\dot{x} + y\dot{y} = -g(x)F(x).$$

Let $f(x)$ be positive for all $x \neq 0$. Then $g(x)F(x) < 0$ for all x which implies that $\dot{V}(x, y)$ is negative definite for all x. Hence solutions from all initial positions ultimately approach the origin.

For the equation

$$\ddot{x} + x^2\dot{x} + x^3 = 0,$$

$f(x) = x^2$ and $g(x) = x^3$, so that

$$F(x) = \int_0^x u^2 du = \tfrac{1}{3}x^3, \quad G(x) = \int_0^x u^3 du = \frac{1}{4}x^4.$$

Therefore

$$V(x, y) = G(x) + \tfrac{1}{2}y^2 = \tfrac{1}{4}x^4 + \tfrac{1}{2}y^2$$

is a Liapunov function for all x, from which it follows that the zero solution is globally asymptotically stable.

- 10.28 (Zubov's method.) Suppose that a function $W(x, y)$, negative definite in the whole plane, is chosen as the time derivative $\dot{V}$ of a possible Liapunov function for a system $\dot{x} = X(x, y)$, $\dot{y} = Y(x, y)$, for which the origin is an asymptotically stable equilibrium point. Show that $V(x, y)$ satisfies the linear partial differential equation

$$X\frac{\partial V}{\partial x} + Y\frac{\partial V}{\partial y} = W$$

with $V(0, 0) = 0$.
Show also that for the path $x(t)$, $y(t)$ starting at (x_0, y_0) at time t_0

$$V\{x(t), y(t)\} - V(x_0, y_0) = \int_{t_0}^{t} W\{x(u), y(u)\}du.$$

Deduce that the boundary of the domain of asymptotically stability (the domain of initial conditions from which the solutions go into the origin) is the set of points (x, y) for which $V(x, y)$ is infinite, by considering the behaviour of the integral as $t \to \infty$, first when (x_0, y_0) is inside this domain and then when it is outside. (Therefore the solution $V(x, y)$ of the partial differential equation above could be used to give the boundary of the domain directly. However, solving this equation is equivalent in difficulty to finding the paths: the characteristics are in fact the paths themselves.)

10.28. Suppose that $W(x, y)$ is a negative definite function for the system

$$\dot{x} = X(x, y), \quad \dot{y} = Y(x, y).$$

Suppose also that $W(x, y) = \dot{V}(x, y)$. Then

$$W = \frac{\partial V}{\partial x}\dot{x} + \frac{\partial V}{\partial y}\dot{y} = \frac{\partial V}{\partial x}X + \frac{\partial V}{\partial y}Y,$$

can be interpreted as a partial differential equation for V.
Since $\dot{V}(x, y) = W(x, y)$ on a phase path, we can integrate with respect to t from an initial point (x_0, y_0) to give

$$V\{x(t), y(t)\} - V(x_0, y_0) = \int_{t_0}^{t} W(x(u), y(u))du.$$

The initial point (x_0, y_0) lies within a domain of asymptotic stability if

$$\lim_{t \to \infty} \int_{t_0}^{t} W(x(u), y(u))du = -V(x_0, y_0).$$

All such points for which this limit is true define the domain of asymptotic stability.

- **10.29** For the system
$$\dot{x} = X(x,y) = -\tfrac{1}{2}x(1-x^2)(1-y^2),$$
$$\dot{y} = Y(x,y) = -\tfrac{1}{2}y(1-x^2)(1-y^2)$$
show that the Liapunov function $V = x^2 + y^2$ leads to $\dot{V} = -(x^2+y^2)(1-x^2)(1-y^2)$ and explain why the domain of asymptotic stability (see Problem 10.30) contains at least the unit circle $x^2 + y^2 = 1$.
Alternatively, start with $\dot{V} = -x^2 - y^2 - 2x^2 y^2$, and obtain V from the equation
$$X\frac{\partial V}{\partial x} + Y\frac{\partial V}{\partial y} = \dot{V}, \quad V(0,0) = 0$$
(see Problem 10.30). It is sufficient to verify that $V = -\ln\{(1-x^2)(1-y^2)\}$. Explain why the square $|x| < 1$, $|y| < 1$ is the complete domain of asymptotic stability for the zero solution.

10.29. The system is

$$\dot{x} = X(x,y) = -\tfrac{1}{2}x(1-x^2)(1-y^2), \quad \dot{y} = Y(x,y) = -\tfrac{1}{2}y(1-x^2)(1-y^2).$$

Equilibrium occurs at the origin $(0,0)$, and *all* points on the lines $x = \pm 1$, $y = \pm 1$. Let $V(x,y) = x^2 + y^2$, a positive definite function. Then

$$\dot{V}(x,y) = -x^2(1-x^2)(1-y^2) - y^2(1-x^2)(1-y^2)$$
$$= -(x^2+y^2)(1-x^2)(1-y^2),$$

which is negative definite in $|x| < 1$, $|y| < 1$. The largest level curve $V(x,y)$ which lies on or within this square is the unit circle $x^2+y^2 = 1$ within which is a domain of asymptotic stability of the origin.

Suppose that we approach the problem using Zubov's method. Let $W(x,y) = -x^2(1-y^2) - y^2(1-x^2)$ which is negative definite in $|x| < 1$, $|y| < 1$. From Problem 10.29, $V(x,y)$ satisfies

$$-\tfrac{1}{2}x(1-x^2)(1-y^2)\frac{\partial V}{\partial x} - \tfrac{1}{2}(1-x^2)(1-y^2)\frac{\partial V}{\partial y} = W(x,y) = -x^2 - y^2 - 2x^2 y^2.$$

It can be verified that a solution of this equation is

$$V(x,y) = -\ln[(1-x^2)(1-y^2)],$$

10 : Liapunov methods for determining stability of the zero solution

which is positive definite in the square $|x| < 1$, $|y| < 1$. Therefore this square is the boundary of a domain of asymptotic stability of the origin, which is an improvement on the unit circle in the first part of the problem. This must be the maximum possible domain since $x = \pm 1$ and $y = \pm 1$ are lines of equilibrium points.

- **10.30** Use the series definition of e^{At} to prove the following properties of the exponential function of a matrix:

  (i) $e^{A+B} = e^A e^B$ if $AB = BA$;

  (ii) e^A is non-singular and $(e^A)^{-1} = e^{-A}$;

  (iii) $\dfrac{d}{dt} e^{At} = A e^{At} = e^{At} A$;

  (iv) $(e^{At})^T = e^{A^T t}$.

10.30. The exponential matrix is defined by

$$e^{At} = \sum_{n=0}^{\infty} A^n \frac{t^n}{n!}.$$

(i) By the product rule for power series

$$e^{At} e^{Bt} = \sum_{n=0}^{\infty} A^n \frac{t^n}{n!} \sum_{n=0}^{\infty} B^n \frac{t^n}{n!} = \sum_{n=0}^{\infty} C^n \frac{t^n}{n!},$$

where

$$C = \sum_{k=0}^{n} A^k B^{n-k} = (A+B)^n.$$

Therefore

$$e^A e^B = e^{A+B}.$$

(ii) By (i)

$$e^A e^{-A} = e^{A-A} = e^0 = I,$$

the identity matrix. Therefore e^{-A} is the inverse of e^A.

(iii) Differentiation of the series term by term leads to

$$\frac{d}{dt}(e^{At}) = \frac{d}{dt} \sum_{n=0}^{\infty} A^n \frac{t^n}{n!} = \sum_{n=1}^{\infty} A^n \frac{t^{n-1}}{(n-1)!} = A e^{At}.$$

(iv) Thus

$$e^{A^T t} = \sum_{n=0}^{\infty}(A^T)^n \frac{t^n}{n!} = \sum_{n=0}^{\infty}(A^n)^T \frac{t^n}{n!} = \left(\sum_{n=0}^{\infty} A^n \frac{t^n}{n!}\right)^T = (e^{At})^T,$$

using standard algebraic rules for transposes of matrices.

• **10.31** Let the distinct eigenvalues of the $n \times n$ matrix A be $\lambda_1, \lambda_2, \ldots, \lambda_n$. Show that, whenever $\gamma > \max_{1 \leq i \leq n} \text{Re}(\lambda_i)$, there exists a constant $c > 0$ such that $\| e^{At} \| \leq c e^{\gamma t}$.

10.31. The matrix A has the distinct eigenvalues $\lambda_1, \lambda_2, \ldots, \lambda_n$. It is known that there exists a matrix P such that

$$PAP^{-1} = \begin{bmatrix} \lambda_1 & 0 & \cdots & 0 \\ 0 & \lambda_2 & \cdots & 0 \\ \cdots & \cdots & \cdots & \cdots \\ 0 & 0 \cdots & 0 & \lambda_n \end{bmatrix},$$

a diagonal matrix of the eigenvalues. As a consequence of this

$$(PAP^{-1})^n = PA^n P^{-1}.$$

Therefore

$$e^{PAP^{-1}t} = \sum_{n=0}^{\infty} \frac{1}{n!}(PAP^{-1})^n t^n = P\left(\sum_{n=0}^{\infty} \frac{1}{n!} A^n t^n\right) P^{-1} = Pe^{At}P^{-1}.$$

Then, using the matrix norm defined in NODE, Section 8.7,

$$\|e^{At}\| = \|P^{-1} e^{At} P\| \leq \|P^{-1}\| \|e^{PAP^{-1}t}\| \|P\|$$

$$= \|P^{-1}\| \sqrt{\left[\sum_{i=1}^{n} |e^{\lambda_i t}|^2\right]} \|P\|$$

$$= \|P^{-1}\| \sqrt{\left[\sum_{i=0}^{n} e^{2\text{Re}(\lambda_i t)}\right]} \|P\|$$

$$\leq \|P^{-1}\| \sqrt{\left[\sum_{i=0}^{n} e^{2\gamma t}\right]} \|P\|$$

$$\leq c e^{\gamma t}$$

for some constant c. Note that the norms of P and P^{-1} are constants.

- 10.32 Express the solution of

$$\begin{bmatrix} \dot{x} \\ \dot{y} \end{bmatrix} = \begin{bmatrix} 0 & 1 \\ 1 & 0 \end{bmatrix} \begin{bmatrix} x \\ y \end{bmatrix}, \quad x(0) = 0, \quad \dot{x}(0) = 1$$

in matrix form, and, by calculating the exponential matrix obtain the ordinary form of the solution.

10.32. The equation is

$$\begin{bmatrix} \dot{x} \\ \dot{y} \end{bmatrix} = \begin{bmatrix} 0 & 1 \\ 1 & 0 \end{bmatrix} \begin{bmatrix} x \\ y \end{bmatrix}.$$

The eigenvalues of

$$\mathbf{A} = \begin{bmatrix} 0 & 1 \\ 1 & 0 \end{bmatrix},$$

are $\lambda_1 = 1$ and $\lambda_2 = -1$. Corresponding eigenvectors are

$$\mathbf{r}_1 = \begin{bmatrix} 1 \\ 1 \end{bmatrix}, \quad \mathbf{r}_2 = \begin{bmatrix} 1 \\ -1 \end{bmatrix}.$$

We can therefore choose

$$\mathbf{P} = \begin{bmatrix} 1 & 1 \\ 1 & -1 \end{bmatrix},$$

so that

$$\mathbf{P}^{-1} = -\frac{1}{2}\begin{bmatrix} -1 & -1 \\ -1 & 1 \end{bmatrix}.$$

It can be confirmed that

$$\mathbf{PAP}^{-1} = \begin{bmatrix} \lambda_1 & 0 \\ 0 & \lambda_2 \end{bmatrix} = \begin{bmatrix} 1 & 0 \\ 0 & -1 \end{bmatrix}.$$

In terms of the exponential matrix, the solution can be expressed as

$$\begin{bmatrix} x \\ y \end{bmatrix} = e^{\mathbf{A}t}\begin{bmatrix} x(0) \\ \dot{x}(0) \end{bmatrix} = e^{\mathbf{A}t}\begin{bmatrix} 0 \\ 1 \end{bmatrix}.$$

We can express the solution in the usual form as follows. The exponential matrix becomes

$$e^{At} = P^{-1}e^{PAP^{-1}}P$$

$$= P^{-1}\left\{\exp\left(\begin{bmatrix} 1 & 0 \\ 0 & -1 \end{bmatrix}t\right)\right\}P$$

$$= -\frac{1}{2}\begin{bmatrix} -1 & -1 \\ -1 & 1 \end{bmatrix}\begin{bmatrix} e^t & 0 \\ 0 & e^{-t} \end{bmatrix}\begin{bmatrix} 1 & 1 \\ 1 & -1 \end{bmatrix}$$

$$= -\frac{1}{2}\begin{bmatrix} -e^t - e^{-t} & -e^t + e^{-t} \\ -e^t + e^{-t} & -e^t - e^{-t} \end{bmatrix}.$$

Finally the solution is

$$\begin{bmatrix} x \\ y \end{bmatrix} = -\frac{1}{2}\begin{bmatrix} -e^t - e^{-t} & -e^t + e^{-t} \\ -e^t + e^{-t} & -e^t - e^{-t} \end{bmatrix}\begin{bmatrix} 0 \\ 1 \end{bmatrix} = -\frac{1}{2}\begin{bmatrix} -e^t + e^{-t} \\ -e^t - e^{-t} \end{bmatrix}.$$

- **10.33** Evaluate

$$K = \int_0^\infty e^{A^T t}e^{At}dt, \text{ where } A = \frac{1}{2}\begin{bmatrix} -3 & 1 \\ 1 & -3 \end{bmatrix},$$

and confirm that $A^T K + KA = -I$

10.33. We wish to evaluate

$$K = \int_0^\infty e^{A^T t}e^{At}dt,$$

where

$$A = \frac{1}{2}\begin{bmatrix} -3 & 1 \\ 1 & -3 \end{bmatrix}.$$

The eigenvalues of A are $\lambda_1 = -1$ and $\lambda_2 = -2$ with corresponding eigenvectors

$$r_1 = \begin{bmatrix} 1 \\ 1 \end{bmatrix}, \quad r_2 = \begin{bmatrix} 1 \\ -1 \end{bmatrix}.$$

Now define the diagonalizing matrix P as

$$P = [\, r_1 \quad r_2 \,] = \begin{bmatrix} 1 & 1 \\ 1 & -1 \end{bmatrix}.$$

It follows that

$$e^{At} = P^{-1}e^{PAP^{-1}}P$$

$$= \frac{1}{2}\begin{bmatrix} 1 & 1 \\ 1 & -1 \end{bmatrix}\begin{bmatrix} e^{-t} & 0 \\ 0 & e^{-2t} \end{bmatrix}\begin{bmatrix} 1 & 1 \\ 1 & -1 \end{bmatrix}$$

$$= \frac{1}{2}\begin{bmatrix} e^{-t}+e^{-2t} & e^{-t}-e^{-2t} \\ e^{-t}-e^{-2t} & e^{-t}+e^{-2t} \end{bmatrix}.$$

By Problem 10.30(iv),

$$e^{A^T t} = (e^{At})^T = e^{At}$$

in this case. Therefore

$$K = \int_0^\infty e^{A^T t} e^{At} dt$$

$$= \frac{1}{2}\int_0^\infty \begin{bmatrix} e^{-t}+e^{-2t} & e^{-t}-e^{-2t} \\ e^{-t}-e^{-2t} & e^{-t}+e^{-2t} \end{bmatrix}^2 dt$$

$$= \frac{1}{2}\int_0^\infty \begin{bmatrix} e^{-2t}+e^{-4t} & e^{-2t}-e^{-4t} \\ e^{-2t}-e^{-4t} & e^{-2t}+e^{-4t} \end{bmatrix} dt$$

$$= \frac{1}{8}\begin{bmatrix} 3 & 1 \\ 1 & 3 \end{bmatrix}.$$

Finally

$$A^T K + KA = \frac{1}{8}\begin{bmatrix} -3 & 1 \\ 1 & -3 \end{bmatrix}\begin{bmatrix} 3 & 1 \\ 1 & 3 \end{bmatrix} + \frac{1}{8}\begin{bmatrix} 3 & 1 \\ 1 & 3 \end{bmatrix}\begin{bmatrix} -3 & 1 \\ 1 & -3 \end{bmatrix}$$

$$= \begin{bmatrix} -1 & 0 \\ 0 & -1 \end{bmatrix} = -I.$$

- 10.34 Show that, if B is an $n \times n$ matrix, $A = e^B$ and C is non-singular, then $C^{-1}AC = e^{C^{-1}BC}$.

10.34. If C is a non-singular matrix, then

$$(C^{-1}BC)^n = C^{-1}B^n C.$$

Therefore, using this result,

$$e^{C^{-1}BC} = I + \sum_{n=1}^{\infty} \frac{1}{n!}(C^{-1}BC)^n$$

$$= C^{-1}\left(I + \sum_{n=1}^{\infty} \frac{1}{n!}B^n\right)C$$

$$= C^{-1}e^B C$$

$$= C^{-1}AC,$$

as required.

- 10.35 (i) Let $L = \text{diag}(\lambda_1, \lambda_2, \ldots, \lambda_n)$, where λ_i are distinct and $\lambda_i \neq 0$ for any i. Show that $L = e^D$, where $D = \text{diag}(\ln \lambda_1, \ln \lambda_2, \ldots, \ln \lambda_n)$. Deduce that for non-singular A with distinct eigenvalues, $A = e^B$ for some matrix B.
(ii) Show that, for the system $\dot{x} = P(t)x$, where $P(t)$ has period T and E (eqn (9.15) in NODE) is non-singular with distinct eigenvalues, every fundamental matrix has the form $\Phi(t) = R(t)e^{Mt}$, where $R(t)$ has period T, and M is a constant matrix. (See the result of Problem 10.35.)

10.35. (i) Let
$$L = \text{diag}(\lambda_1, \lambda_2, \ldots, \lambda_n),$$

where the notation on the right denotes the matrix with diagonal elements $\lambda_1, \lambda_2, \ldots, \lambda_n$, and all other elements zero. Let
$$D = (\ln \lambda_1, \ln \lambda_2, \ldots, \ln \lambda_n).$$

Then

$$e^D = I + \sum_{i=1}^{\infty} \frac{1}{i!}D^i = I + \sum_{i=1}^{\infty} \frac{1}{i!}\text{diag}[\ln \lambda_1, \ldots, \ln \lambda_n]^i$$

$$= I + \sum_{i=1}^{\infty} \frac{1}{i!}\text{diag}[(\ln \lambda_1)^i, \ldots, (\ln \lambda_n)^i]$$

$$= \text{diag}[e^{\ln \lambda_1}, \ldots, e^{\ln \lambda_n}]$$

$$= \text{diag}[\lambda_1, \ldots, \lambda_n] = L.$$

Since the eigenvalues of A are distinct, there exists a matrix P such that

$$P^{-1}AP = L,$$

where **L** is the diagonal matrix of eigenvalues of **A**. Then

$$\mathbf{A} = \mathbf{PLP}^{-1} = \mathbf{Pe^D P^{-1}} = \mathbf{I} + \sum_{i=1}^{\infty} \frac{1}{i!}(\mathbf{PD}^i\mathbf{P}^{-1})$$

$$= \mathbf{I} + \sum_{i=1}^{\infty} \frac{1}{i!}(\mathbf{PDP}^{-1})^i$$

$$= e^{\mathbf{PDP}^{-1}}.$$

Therefore there exists a matrix **B** such that $\mathbf{A} = e^{\mathbf{B}}$: in fact $\mathbf{B} = \mathbf{PDP}^{-1}$.
(ii) In the equation $\dot{\mathbf{x}} = \mathbf{P}(t)\mathbf{x}$, $\mathbf{P}(t)$ has minimal period T. Let $\mathbf{\Phi}(t)$ be a fundamental matrix of the system. As in Section 9.2, $\mathbf{\Phi}(t + T)$ is also a fundamental matrix, and

$$\mathbf{\Phi}(t + T) = \mathbf{\Phi}(t)\mathbf{E},$$

where **E** is non-singular. Since **E** is a constant matrix with distinct eigenvalues, there exists a constant matrix $T\mathbf{M}$, say, such that $\mathbf{E} = e^{T\mathbf{M}}$ (see (i) above). Express the fundamental matrix in the form $\mathbf{\Phi}(t) = \mathbf{R}(t)e^{t\mathbf{M}}$. Then

$$\mathbf{R}(t+T) = \mathbf{\Phi}(t+T)e^{-(t+T)\mathbf{M}} = \mathbf{\Phi}(t)e^{T\mathbf{M}}e^{-(t+T)\mathbf{M}} = \mathbf{\Phi}(t)e^{-t\mathbf{M}} = \mathbf{R}(t).$$

Therefore $\mathbf{R}(t)$ has period T.

- **10.36** Using the results from Problem 10.35, show that the transformation $\mathbf{x} = \mathbf{R}(t)\mathbf{y}$ reduces the system $\dot{\mathbf{x}} = \mathbf{P}(t)\mathbf{x}$, where $\mathbf{P}(t)$ has period T, to the form $\dot{\mathbf{y}} = \mathbf{M}\mathbf{y}$, where **M** is a constant matrix.

10.36. Using the notation of Problem 10.35, substitute $\mathbf{x} = \mathbf{R}(t)\mathbf{y}$ into the equation $\dot{\mathbf{x}} = \mathbf{P}(t)\mathbf{x}$, so that

$$\dot{\mathbf{y}} = \mathbf{R}(t)^{-1}[\mathbf{P}(t)\mathbf{R}(t) - \dot{\mathbf{R}}(t)]\mathbf{y}. \tag{i}$$

However, from the previous problem,

$$\mathbf{R}(t) = \mathbf{\Phi}(t)e^{-t\mathbf{M}},$$

so that

$$\dot{\mathbf{R}}(t) = \dot{\mathbf{\Phi}}(t)e^{-t\mathbf{M}} - \mathbf{\Phi}(t)\mathbf{M}e^{-t\mathbf{M}}$$
$$= \mathbf{P}(t)\mathbf{\Phi}(t)e^{-t\mathbf{M}}$$
$$= \mathbf{P}(t)\mathbf{R}(t) - \mathbf{R}(t)e^{t\mathbf{M}}\mathbf{M}e^{-t\mathbf{M}}$$
$$= \mathbf{P}(t)\mathbf{R}(t) - \mathbf{R}(t)\mathbf{M}$$

Elimination of $\dot{\mathbf{R}}(t)$ in (ii) leads to

$$\dot{\mathbf{y}} = \mathbf{R}(t)^{-1}\mathbf{R}(t)\mathbf{M} = \mathbf{M}\mathbf{y}.$$

11 The existence of periodic solutions

- 11.1 Prove that the equilibrium point of
$$\ddot{x} + \frac{x}{1+x^2}\dot{x} + x\ln(1+x^2) = 0, \quad \dot{x} = y$$
is a centre in the (x, y) plane. Compute the phase diagram in the neighbourhood of $(0, 0)$.

11.1. The system
$$\ddot{x} + \frac{x}{1+x^2}\dot{x} + x\ln(1+x^2) = 0, \quad \dot{x} = y$$
has one equilibrium point at $(0, 0)$. Apply NODE, Theorem 11.3 with $f(x) = x/(1+x^2)$ and $g(x) = x\ln(1+x^2)$. It is obvious that $f(x)$ and $g(x)$ are both odd functions, and that $f(x) > 0$ and $g(x) > 0$ for $x > 0$. Then

$$g(x) - \alpha f(x) F(x) = x\ln(1+x^2) - \frac{\alpha x}{1+x^2}\int_0^x \frac{u}{1+u^2}du$$

$$= x\ln(1+x^2) - \frac{\alpha x}{2(1+x^2)}\ln(1+x^2)$$

$$= \frac{x\ln(1+x^2)}{1+x^2}\left[1+x^2 - \frac{1}{2}\alpha x\right]$$

$$= \frac{x\ln(1+x^2)}{1+x^2}\left[\left(x - \frac{1}{4}\alpha\right)^2 + \left(1 - \frac{1}{16}\alpha^2\right)\right]$$

$$> 0$$

for $1 < \alpha < 4$. The conditions of Theorem 11.3 are satisfied which implies that the origin is a centre. Some computed phase paths are shown in Figure 11.1.

- 11.2 A system has exactly one equilibrium point, n limit cycles and no other periodic solutions. Explain why an asymptotically stable limit cycle must be adjacent to unstable limit cycles, but an unstable limit cycle may have stable or unstable cycles adjacent to it.
  Let c_n be the number of possible configurations, with respect to stability, of n nested limit cycles. Show that $c_1 = 2$, $c_2 = 3$, $c_3 = 5$, and that in general $c_n = c_{n-1} + c_{n-2}$.

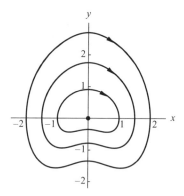

Figure 11.1 Problem 11.1: Phase diagram for $\ddot{x} + [x/(1+x^2)]\dot{x} + x\ln(1+x^2) = 0$.

(This recurrence relation generates the Fibonacci sequence.) Deduce that

$$c_n = \frac{1}{2^{n-1}\sqrt{2}}\{(2+\sqrt{5})(1+\sqrt{5})^{n-1} + (-2+\sqrt{5})(1-\sqrt{5})^{n-1}\}.$$

11.2. Since the system has only one equilibrium point, the n limit cycles must be 'nested'. List the limit cycles as $\mathcal{L}_1, \mathcal{L}_2, \ldots, \mathcal{L}_n$ from the inside. Suppose that the limit cycle $\mathcal{L}_r$ is stable in an asymptotic sense so that all adjacent external and internal phase paths approach it as $t \to \infty$. It follows that paths must diverge from $\mathcal{L}_{r+1}$ and $\mathcal{L}_{r-1}$ which implies that both these limit cycles are unstable. On the other hand if $\mathcal{L}_r$ is unstable then there are three possibilities (a) both internal and external phase paths diverge from $\mathcal{L}_r$, in which case it is possible that both $\mathcal{L}_{r+1}$ and $\mathcal{L}_{r-1}$ are stable, or (b) external paths diverge and internal paths converge to $\mathcal{L}_r$ in which case it is possible that $\mathcal{L}_{r+1}$ is stable, or (c) external paths converge and internal paths diverge to $\mathcal{L}_r$ in which case it is possible that $\mathcal{L}_{r-1}$ is stable.

If c_n is the number of possible configurations, stable(s) or unstable(u), of the first n limit cycles. Thus $c_1 = 2$ since the limit cycle can be s or u. The possible combinations for c_2 are

$$\begin{cases} \mathcal{L}_1 & & \mathcal{L}_2 \\ s & \to & u \\ u & \to & s \\ u & \to & u \end{cases}$$

Therefore $c_2 = 3$. For three limit cycles the possible combinations are

$$\begin{cases} \mathcal{L}_1 & & \mathcal{L}_2 & & \mathcal{L}_3 \\ s & \to & u & \to & s \\ s & \to & u & \to & u \\ u & \to & s & \to & u \\ u & \to & u & \to & s \\ u & \to & u & \to & u. \end{cases}$$

Therefore
$$c_3 = 5 = 2 + 3 = c_1 + c_2.$$

Generally the value of c_n depends on the states of $\mathcal{L}_{n-2}$ and $\mathcal{L}_{n-1}$: if $\mathcal{L}_{n-2}$ is stable then there are two sequences to the state of $\mathcal{L}_n$, and if $\mathcal{L}_{n-2}$ is unstable there are three possible sequences to the state of $\mathcal{L}_n$. Therefore

$$c_n = c_{n-1} + c_{n-2}, \quad (n \geq 3).$$

This difference equation generates the *Fibonacci sequence*.

To solve the difference equation, let $c_n = \lambda^n$. Then

$$\lambda^n - \lambda^{n-1} - \lambda^{n-2} = 0, \text{ or } \lambda^2 - \lambda - 1 = 0.$$

The solutions of this equation are

$$\lambda_1 = \tfrac{1}{2}(1 + \sqrt{5}), \quad \lambda_2 = \tfrac{1}{2}(1 - \sqrt{5}).$$

Therefore
$$c_n = A \frac{1}{2^n}(1 + \sqrt{5})^n + B \frac{1}{2^n}(1 - \sqrt{5})^n.$$

The initial conditions lead to

$$c_1 = 2 = \frac{A}{2}(1 + \sqrt{5}) + \frac{B}{2}(1 - \sqrt{5}),$$

$$c_2 = 3 = \frac{A}{4}(1 + \sqrt{5})^2 + \frac{B}{4}(1 - \sqrt{5})^2 = \frac{A}{2}(3 + \sqrt{5}) + \frac{B}{2}(3 - \sqrt{5}).$$

Therefore
$$A = \frac{1}{10}(5 + 3\sqrt{5}), \quad B = \frac{1}{10}(5 - 3\sqrt{5}),$$

and

$$c_n = \frac{1}{5 \cdot 2^{n+1}}[(5 + 3\sqrt{5})(1 + \sqrt{5})^n + (5 - 3\sqrt{5})(1 - \sqrt{5})^n]$$

$$= \frac{1}{2^{n-1}\sqrt{2}}\{(2 + \sqrt{5})(1 + \sqrt{5})^{n-1} + (-2 + \sqrt{5})(1 - \sqrt{5})^{n-1}\}.$$

- **11.3** By considering the path directions across each of the suggested topographic systems show that in each of the cases given there exists a limit cycle. Locate the region in which a limit cycle might exist as closely as possible. Show that in each case only one limit cycle exists:

(i) $\dot{x} = 2x + 2y - x(2x^2 + y^2), \dot{y} = -2x + y - y(2x^2 + y^2),$

(topographic system $x^2 + y^2 =$ constant);
(ii) $\dot{x} = -x - y + x(x^2 + 2y^2)$, $\dot{y} = x - y + y(x^2 + 2y^2)$,
(topographic system $x^2 + y^2 =$ constant);
(iii) $\dot{x} = x + y - x^3 - 6xy^2$, $\dot{y} = -\frac{1}{2}x + 2y - 8y^3 - x^2 y$,
(topographic system $x^2 + 2y^2 =$ constant); compute the phase diagram, and show the topographic system;
(iv) $\dot{x} = 2x + y - 2x^3 - 3xy^2$, $\dot{y} = -2x + 4y - 4y^3 - 2x^2 y$,
(topographic system $2x^2 + y^2 =$ constant).

11.3. Consult NODE, Example 11.1.

(i) $\dot{x} = X(x, y) = 2x + 2y - x(2x^2 + y^2)$, $\dot{y} = Y(x, y) = -2x + y - y(2x^2 + y^2)$. A normal to the topographic system $x^2 + y^2 =$ constant is $\mathbf{n} = (x, y)$. Then

$$\mathbf{n} \cdot \mathbf{X} = (x, y) \cdot [2x + 2y - x(2x^2 + y^2), -2x + y - y(2x^2 + y^2)]$$
$$= (2x^2 + y^2)[1 - (x^2 + y^2)].$$

Hence $\mathbf{n} \cdot \mathbf{X} > 0$ for $x^2 + y^2 < 1$ and $\mathbf{n} \cdot \mathbf{X} < 0$ for $x^2 + y^2 > 1$. It follows that the circle $x^2 + y^2 = 1$ is a limit cycle of the system.

(ii) $\dot{x} = -x - y + x(x^2 + 2y^2)$, $\dot{y} = x - y + y(x^2 + 2y^2)$. A normal to the topographic system $x^2 + y^2 =$ constant is $\mathbf{n} = (x, y)$. Then

$$\mathbf{n} \cdot \mathbf{X} = (x, y) \cdot [-x - y + x(x^2 + 2y^2), x - y + y(x^2 + 2y^2)]$$
$$= (x^2 + y^2)(x^2 + 2y^2 - 1).$$

Therefore $\mathbf{n} \cdot \mathbf{X} > 0$ for $x^2 + 2y^2 > 1$, and $\mathbf{n} \cdot \mathbf{X} < 0$ for $x^2 + 2y^2 < 1$. The ellipse $x^2 + y^2 = 2$ is bounded by the circles $x^2 + y^2 = 1$ and $x^2 + y^2 = \frac{1}{2}$, so that the limit cycle must lie on or within these circles.

(iii) $\dot{x} = x + y - x^3 - 6xy^2$, $\dot{y} = -\frac{1}{2}x + 2y - 8y^3 - x^2 y$. A normal to the topographic system $x^2 + 2y^2 =$ constant is $\mathbf{n} = (x, 2y)$. Then

$$\mathbf{n} \cdot \mathbf{X} = (x, 2y) \cdot (x + y - x^3 - 6xy^2, -\frac{1}{2}x + 2y - 8y^3 - x^2 y)$$
$$= (x^2 + 4y^2)[1 - (x^2 + 4y^2)].$$

Therefore $\mathbf{n} \cdot \mathbf{X} > 0$ for $x^2 + 4y^2 < 1$, and $\mathbf{n} \cdot \mathbf{X} < 0$ for $x^2 + 4y^2 > 1$. The ellipse $x^2 + 4y^2 = 1$ is bounded by the ellipses $x^2 + 2y^2 = 1$ and $x^2 + 2y^2 = \frac{1}{2}$ from the topographic system. The position of the stable limit cycle in relation to the two bounding topographic curves is shown in the computed phase diagram in Figure 11.2.

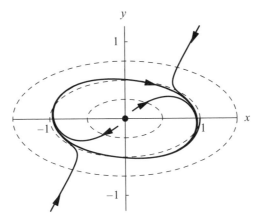

Figure 11.2 Problem 11.1(iii): Phase diagram for $\dot{x} = x+y-x^3-6xy^2$, $\dot{y} = -\frac{1}{2}x+2y-8y^3-x^2y$; the topographic system $x^2+2y^2 =$ constant is shown by dashed curves.

(iv) $\dot{x} = 2x+y-2x^3-3xy^2$, $\dot{y} = -2x+4y-4y^3-2x^2y$. A normal to the topographic system $2x^2+y^2 =$ constant is $\mathbf{n} = (2x, y)$. Then

$$\mathbf{n} \cdot \mathbf{X} = (2x, y) \cdot (2x+y-2x^3-3xy^2, -2x+4y-4y^3-2x^2y)$$
$$= 4(x^2+y^2)[1-(x^2+y^2)].$$

There $\mathbf{n} \cdot \mathbf{X} > 0$ for $x^2+y^2 < 1$, and $\mathbf{n} \cdot \mathbf{X} < 0$ for $x^2+y^2 > 1$. The circle $x^2+y^2 = 1$ is bounded by the ellipses $2x^2+y^2 = 2$ and $2x^2+y^2 = 1$ from the topographic system. The phase path of the periodic solution will lie on or between these ellipses.

- **11.4** Show that the equation $\ddot{x} + \beta(x^2+\dot{x}^2-1)\dot{x}+x^3 = 0$, $(\beta > 0)$, has at least one periodic solution.

11.4. Consider the equation

$$\ddot{x} + \beta(x^2+\dot{x}^2-1)\dot{x}+x^3 = 0, \quad (\beta > 0).$$

Apply NODE, Theorem 11.2 after checking the conditions on $f(x, y) = \beta(x^2+y^2-1)$ and $g(x) = x^3$ as follows:

(i) $f(x, y) > 0$ for $\sqrt{(x^2+y^2)} > 1$;
(ii) $f(0, 0) = -\beta < 0$;
(iii) $g(0) = 0$, $g(x) > 0$ for $x > 0$ and $g(x)$ is odd;
(iv) $G(x) = \int_0^x u^3 du = \frac{1}{4}x^4 \to \infty$ as $x \to \infty$.

The conditions are satisfied which implies that the system has at least one periodic solution.

- **11.5** Show that the origin is a centre for the equations:
  (i) $\ddot{x} - x\dot{x} + x = 0$;
  (ii) $\ddot{x} + x\dot{x} + \sin x = 0$.

11.5. (i) $\ddot{x} - x\dot{x} + x = 0$. In the notation of NODE, Theorem 11.3, $f(x) = -x$ and $g(x) = x$. Check the conditions required:

(a) $f(x) = -x$ is odd and negative for $x > 0$;
(b) $g(x) = x > 0$ for $x > 0$, and $g(x)$ is odd;
(c) For $x > 0$

$$g(x) - \alpha f(x) \int_0^x f(u)du = x - \alpha \int_0^x udu = x\left(1 - \tfrac{1}{2}\alpha x^2\right) > 0,$$

if $\alpha = 2$ (say) and $|x| < 1$.

The conditions are satisfied which means that $(0, 0)$ is a centre in the (x, y) plane where $\dot{x} = y$.
(ii) $\ddot{x} + x\dot{x} + \sin x = 0$. In the notation of Theorem 11.3, $f(x) = x$ and $g(x) = \sin x$. Check the conditions of Theorem 11.3

(a) $f(x) = x$ is odd and positive for $x > 0$;
(b) $g(x) = \sin x > 0$ for $0 < x < \pi$, and $g(x)$ is odd;
(c) For $|x|$ sufficiently small

$$g(x) - \alpha f(x) \int_0^x f(u)du = \sin x - \tfrac{1}{2}\alpha x^3 > 0$$

with $\alpha = 2$ (say).
The conditions are satisfied which means that locally $(0, 0)$ is a centre.

- **11.6** Suppose that $f(x)$ in the equation $\ddot{x} + f(x)\dot{x} + x = 0$ is given by $f(x) = x^n$. Show that the origin is a centre if n is an odd positive integer.

11.6. In the equation

$$\ddot{x} + f(x)\dot{x} + x = 0,$$

the function $f(x) = x^n$, where n is an odd positive integer. In the notation of Theorem 1.3, $g(x) = x$, and the required conditions are satisfied as follows:

(i) $f(x) = x^n$ is odd and positive for $x > 0$;
(ii) $g(x) = x$ is odd;

(iii) The difference
$$g(x) - \alpha f(x) \int_0^x f(u)du = x - \alpha \frac{x^{2n+1}}{n+1} > 0$$
in the interval
$$0 < x < \left(\frac{n+1}{\alpha}\right)^{1/(2n)},$$
where α is any number greater than 1. The optimal interval occurs for $\alpha = 1$.
By the Theorem the origin must be locally a centre.

- **11.7** Show that the equation $\ddot{x} + \beta(x^2 - 1)\dot{x} + \tanh kx = 0$ has exactly one periodic solution when $k > 0$, $\beta > 0$. Decide on its stability.
  The 'restoring force' resembles a step function when k is large. Is the conclusion the same when it is exactly a step function?

11.7. Apply NODE, Theorem 11.4 to the equation
$$\ddot{x} + \beta(x^2 - 1)\dot{x} + \tanh kx = 0.$$
In this case $f(x) = \beta(x^2 - 1)$ and $g(x) = \tanh kx$. Check the conditions required by the theorem:

(i) $F(x) = \int_0^x f(u)du = \beta(\frac{1}{3}x^3 - x)$ is an odd function;
(ii) $F(x) = 0$ only at $x = 0$, $x = \sqrt{3}$ and $x = -\sqrt{3}$;
(iii) $F(x)$ is monotonic increasing for $x > \sqrt{3}$;
(iv) $g(x) = \tanh kx$ is an odd function, and positive for $x > 0$.

The theorem then asserts that the system has a unique periodic solution. As explained in the theorem the limit cycle will be stable.
For the limiting step function
$$g(x) = \begin{cases} 1 & x > 0 \\ -1 & x < 0. \end{cases}$$
In the Liénard plane the differential equation for the phase paths becomes
$$\frac{dy}{dx} = \frac{\text{sgn}(x)}{y - \beta((1/3)x^3 - x)},$$
so that there are no points at which $dy/dx = 0$. Paths can never 'close' across the y axis.

- **11.8** Show that $\ddot{x} + \beta(x^2 - 1)\dot{x} + x^3 = 0$ has exactly one periodic solution.

11.8. In
$$\ddot{x} + \beta(x^2 - 1)\dot{x} + x^3 = 0$$
assume that $\beta > 0$. Apply NODE, Theorem 11.4 with $f(x) = \beta(x^2 - 1)$ and $g(x) = x^3$. Check the conditions as follows:

(i) $F(x) = \int_0^x f(u)du = \beta(\frac{1}{3}x^3 - x)$ is an odd function;
(ii) $F(x) = 0$ only at $x = 0$, $x = \sqrt{3}$ and $x = -\sqrt{3}$;
(iii) $F(x)$ is monotonic increasing for $x > \sqrt{3}$;
(iv) $g(x) = x^3$ is an odd function, and positive for $x > 0$.

The theorem asserts that the system has a unique limit cycle which is stable. In the case $\beta < 0$, reverse time and re-apply the theorem.

- **11.9** Show that $\ddot{x} + (|x| + |\dot{x}| - 1)\dot{x} + x|x| = 0$ has at least one periodic solution.

11.9. Apply NODE, Theorem 11.2 to
$$\ddot{x} + (|x| + |\dot{x}| - 1)\dot{x} + x|x| = 0,$$
with $f(x, y) = (|x| + |y| - 1)$ and $g(x) = x|x|$. The requirements of Theorem 11.2 are as follows:

(i) $f(x, y) > 0$ for $x^2 + y^2 > 1$, since $|x| + |y| = 1$ is a square within this circle;
(ii) $f(0, 0) = -1 < 0$;
(iii) $g(x)$ is an odd function with $g(x) > 0$ for $x > 0$;
(iv) $G(x) = \int_0^x g(u)du = \int_0^x u|u|du = \frac{1}{3}|x|^3 \to \infty$ as $x \to \infty$.

Theorem 11.2 implies that the system has at least one periodic solution.

- **11.10** Show that the origin is a centre for the equation $\ddot{x} + (k\dot{x} + 1)\sin x = 0$.

11.10. In the equation
$$\ddot{x} + (k\dot{x} + 1)\sin x = 0,$$

let $f(x) = k \sin x$ and $g(x) = \sin x$, and apply NODE, Theorem 11.3. The requirements of the theorem as follows:

(i) $f(x) = k \sin x$ is odd, and of one sign in $0 < x < \pi$;
(ii) $g(x) = \sin x$ is also odd and positive in $0 < x < \pi$;
(iii) For $\alpha > 1$,

$$g(x) - \alpha k \sin x \int_0^x k \sin u\, du = \sin x[1 - \alpha k^2(1 - \cos x)] > 0$$

for x sufficiently small. For example, choose $\alpha = 2$: then the function is positive for $\cos x > (k^2 - 1)/k^2$, for which a positive interval can be found for every k.

• **11.11** Using the method of NODE, Section 11.4, show that the amplitude of the limit cycle of
$$\varepsilon \ddot{x} + (|x| - 1)\dot{x} + \varepsilon x = 0, \quad \dot{x} = y, \quad (0 < \varepsilon \ll 1)$$
is approximately $a = 1 + \sqrt{2}$ to order ε. Show also that the solution for $y > 0$ is approximately
$$\varepsilon y = (x - a) - \tfrac{1}{2}x^2 \operatorname{sgn}(x) + \tfrac{1}{2}a^2, \quad (-1 < x < a).$$
Compare this curve with the computed phase path for $\varepsilon = 0.1$.

11.11. In the equation
$$\varepsilon \ddot{x} + (|x| - 1)\dot{x} + \varepsilon x = 0 \tag{i}$$
$\varepsilon > 0$ is a small parameter. This equation is very similar to the van der Pol equation in Section 11.4. The solution below follows the method given in the text with the differences pointed out.

The phase paths are given by
$$\frac{dy}{dx} = -\frac{|x| - 1}{\varepsilon} - \frac{x}{y}.$$

The isocline of zero slope is the curve
$$y = \frac{\varepsilon x}{1 - |x|}.$$

In equation (i) put $t = \varepsilon \tau$, so that
$$x'' + (|x| - 1)x' + \varepsilon^2 x = 0. \tag{ii}$$

Therefore to lowest order
$$x'' + (|x| - 1)x' = 0,$$

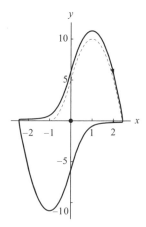

Figure 11.3 Problem 11.11: The computed limit cycle is the solid curve: the approximation given by (iii) is the dashed curve in the case $\varepsilon = 0.1$.

which can be integrated to give

$$x' = x - \tfrac{1}{2}x^2 \text{sgn}(x) + C = x - a - \tfrac{1}{2}x^2 \text{sgn}(x) + \tfrac{1}{2}a^2 = \varepsilon y, \qquad \text{(iii)}$$

if $x = a > 0$ where $y = 0$. This part of the solution must return to the x axis at $x = -1$. Therefore a is given by

$$0 = -1 - a + \tfrac{1}{2} + \tfrac{1}{2}a^2, \text{ or } a^2 - 2a - 1 = 0.$$

It follows that $a = 1 + \sqrt{2}$. The approximate equation for the limit cycle is given by (ii). The computed limit cycle and the approximation are shown in Figure 11.3 for $\varepsilon = 0.1$.

- **11.12** Let F and g be functions satisfying the conditions of NODE, Theorem 11.4. Show that the equation $\ddot{u} + F(\dot{u}) + g(u) = 0$ has a unique periodic solution (put $\dot{u} = z$). Deduce that Rayleigh's equation $\ddot{u} + \beta(\tfrac{1}{3}\dot{u}^3 - \dot{u}) + u = 0$ has a unique limit cycle.

11.12. The functions F and g in the equation

$$\ddot{u} + F(\dot{u}) + g(u) = 0$$

satisfy the conditions of Theorem 11.4. Put $\dot{u} = -z$. Then

$$\dot{z} = -\ddot{u} = F(\dot{u}) + g(u) = F(-z) + g(u) = -F(z) + g(u),$$

since $F(z)$ is an odd function. These are simply the equations in the Liénard plane. Hence Theorem 11.4 applies so that the equation has a stable limit cycle.

In Rayleigh's equation

$$\ddot{u} + \beta\left(\tfrac{1}{3}\dot{u}^3 - \dot{u}\right) + u = 0,$$

$F(\dot{u}) = \beta\left(\tfrac{1}{3}\dot{u}^3 - \dot{u}\right)$ and $g(u) = u$. It follows that conditions (i), (ii) and (iii) of the theorem are satisfied. Stability depends on the sign of β.

- **11.13** Show that the equation $\ddot{x} + \beta(x^2 - \dot{x}^2 - 1)\dot{x} + x = 0$, unlike the van der Pol equation, does not have a relaxation oscillation for large positive β.

11.13. The equation

$$\ddot{x} + \beta(x^2 + \dot{x}^2 - 1)\dot{x} + x = 0,$$

has the exact sinusoidal solution $x = \cos t$, which is not a relaxation oscillation. Damping ensures that the solution is stable and the only periodic solution.

- **11.14** For the van der Pol oscillator $\delta\ddot{x} + (x^2 - 1)\dot{x} + \delta x = 0$ for small positive δ, use the formula for the period, NODE, eqn (11.13), to show that the period of the limit cycle is approximately $(3 - 2\ln 2)\delta^{-1}$. (Hint: the principal contribution arises from that part of the limit cycle given in (ii) in Section 11.4.)

11.14. Consider the van der Pol equation

$$\delta\ddot{x} + (x^2 - 1)\dot{x} + \delta x = 0,$$

where δ is large. It can be seen from Figure 11.16 (in NODE) that the main contribution to the period occurs over the intervals $-2 < x < -1$ and $1 < x < 2$. In the former interval the relation between x and y (see (ii) in Section 11.4) is given by

$$y = \frac{\delta x}{1 - x^2}.$$

If this curve is denoted by $\mathcal{C}$, then by eqn (1.13) (in the book) the elapsed time is

$$\tfrac{1}{2}T = \int_{\mathcal{C}} \frac{\mathrm{d}x}{y} = \frac{1}{\delta}\int_{-2}^{-1}\left(\frac{1}{x} - x\right)\mathrm{d}x = \frac{1}{2\delta}(3 - 2\ln 2).$$

The period is therefore $T = (3 - 2\ln 2)/\delta$.

- **11.15** Use the Poincaré–Bendixson theorem to show that the system
$$\dot{x} = x - y - x(x^2 + 2y^2), \quad \dot{y} = x + y - y(x^2 + 2y^2)$$
had at least one periodic solution in the annulus $1/\sqrt{2} < r < 1$, where $r = \sqrt{(x^2 + y^2)}$.

11.15. We apply the Poincaré–Bendixson theorem to
$$\dot{x} = X(x, y) = x - y - x(x^2 + 2y^2), \quad \dot{y} = Y(x, y) = x + y - y(x^2 + 2y^2).$$

The outward normal to the general circle $x^2 + y^2 = r^2 =$ constant is $\mathbf{n} = (x, y)$. Let $\mathbf{X} = (X, Y)$. Then
$$\mathbf{n} \cdot \mathbf{X} = x[x - y - x(x^2 + 2y^2)] + y[x + y - y(x^2 + 2y^2)]$$
$$= r^2[1 - (x^2 + 2y^2)]$$
$$= r^2(1 - r^2 - y^2)$$

If $r = 1$ then $\mathbf{n} \cdot \mathbf{X} = -y^2 \leq 0$. Alternatively we can write
$$\mathbf{n} \cdot \mathbf{X} = r^2(1 - 2r^2 + x^2).$$

In this case if $r = 1/\sqrt{2}$, then $\mathbf{n} \cdot \mathbf{X} = \frac{1}{2}x^2 \geq 0$. We conclude that phase paths cross $r = 1$ from the outside, whilst on $r = 1/\sqrt{2}$ the phase paths cross from the inside. By the Poincaré–Bendixson theorem there must be at least one closed path between the circles since the system has only one equilibrium point at $(0, 0)$.

12 Bifurcations and manifolds

• 12.1 Find the bifurcation points of the linear system $\dot{\mathbf{x}} = \mathbf{A}(\lambda)\mathbf{x}$ with $\mathbf{x} = [x_1\ x_2]^T$ and $\mathbf{A}(\lambda)$ given by

(i) $\mathbf{A}(\lambda) = \begin{bmatrix} -2 & \frac{1}{4} \\ -1 & \lambda \end{bmatrix}$;

(ii) $\mathbf{A}(\lambda) = \begin{bmatrix} \lambda & \lambda - 1 \\ 1 & \lambda \end{bmatrix}$.

12.1. We require the bifurcation points of the linear system $\dot{\mathbf{x}} = \mathbf{A}(\lambda)\mathbf{x}$ with $\mathbf{x} = [x_1, x_2]^T$.

(i)
$$\mathbf{A}(\lambda) = \begin{bmatrix} -2 & \frac{1}{4} \\ -1 & \lambda \end{bmatrix}.$$

The eigenvalues of $\mathbf{A}$ are given by

$$\begin{vmatrix} -2 - m & \frac{1}{4} \\ -1 & \lambda - m \end{vmatrix} = 0, \text{ or } m^2 - (\lambda - 2)m + \tfrac{1}{4} - 2\lambda = 0,$$

which has the solutions

$$m_1, m_2 = \tfrac{1}{2}[\lambda - 1 \pm \sqrt{\{(\lambda + 3)(\lambda + 1)\}}].$$

The solutions are real for $\lambda \leq -3$ and $\lambda \geq -1$, and complex for $-3 < \lambda < -1$. The graphs of Re(m_1) and Re(m_2) against λ are shown in Figure 12.1. Noting the signs of m_1 and m_2, we can observe that the equilibrium point at the origin is:

- $\lambda < -3$, stable node;
- $-3 < \lambda < -1$, stable spiral;
- $-1 < \lambda < -\frac{1}{3}$, stable node;
- $\lambda > -\frac{1}{3}$, saddle point.

A bifurcation occurs at the parametric value $\lambda = -\frac{1}{3}$ where, as λ increases, the equilibrium point changes from a stable node to an unstable saddle point.

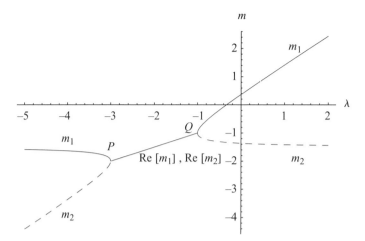

Figure 12.1 Problem 12.1(i): m_1 and m_2 are shown for $\lambda \leq -3$ and for $\lambda \geq -1$: $\mathrm{Re}(m_1)$ and $\mathrm{Re}(m_2)$ are shown between P and Q.

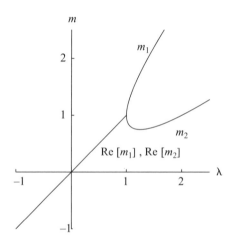

Figure 12.2 Problem 12.1(ii): $\mathrm{Re}(m_1)$ and $\mathrm{Re}(m_2)$ against λ are shown.

(ii)
$$\mathbf{A}(\lambda) = \begin{bmatrix} \lambda & \lambda - 1 \\ 1 & \lambda \end{bmatrix}.$$

The eigenvalues of $\mathbf{A}$ are given by

$$\begin{vmatrix} \lambda - m & \lambda - 1 \\ 1 & \lambda - m \end{vmatrix} = 0 \quad \text{or} \quad m^2 - 2m\lambda + \lambda^2 - \lambda + 1 = 0,$$

which has the solutions

$$m_1, m_2 = \lambda \pm \sqrt{(\lambda - 1)}.$$

The solutions are real for $\lambda \geq 1$ and complex for $\lambda \leq 1$ (see Figure 12.2). The origin is:

- $\lambda < 0$, stable spiral;
- $0 < \lambda < 1$, unstable spiral;
- $1 < \lambda$, unstable node.

There is a bifurcation point at $\lambda = 0$ where the equilibrium point changes from a stable spiral to an unstable spiral.

- 12.2 In a conservative system, the potential is given by $V(x, \lambda) = \frac{1}{3}x^3 + \lambda x^2 + \lambda x$ (cf, NODE, eqn (12.2)). Find the equilibrium points of the system, and show that it has bifurcation points at $\lambda = 0$ and $\lambda = 1$. What type of bifurcations occur for $\lambda < 0$ and $\lambda > 0$?

12.2. For the conservative system with potential

$$V(x, \lambda) = \tfrac{1}{3}x^3 + \lambda x^2 + \lambda x,$$

the corresponding equation is

$$\ddot{x} = -\frac{dV}{dx} = -x^2 - 2\lambda x - \lambda.$$

Equilibrium points occur where

$$x^2 + 2\lambda x + \lambda = 0.$$

Its solutions are

$$x = -\lambda \pm \sqrt{(\lambda^2 - \lambda)},$$

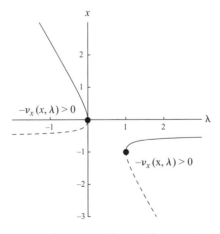

Figure 12.3 Problem 12.2: The curves show locations of the equilibrium points together with the bifurcation points; the dashed curves indicate stable equilibrium.

which are real only if $\lambda \geq 1$, or $\lambda \leq 0$. The system has two equilibrium points if $\lambda > 1$ or $\lambda < 0$, one equilibrium point if $\lambda = 0$ or $\lambda = 1$, and none if $0 < \lambda < 1$. Bifurcations occur at $\lambda = 0$ and at $\lambda = 1$ as indicated in Figure 12.3. Both these are *saddle-node* bifurcations (see Section 12.4). The method of NODE, Section 1.7 indicates that the dashed curves are stable.

- 12.3 Let $\mathcal{V}(x, \lambda, \mu) = \frac{1}{4}x^4 - \frac{1}{2}\lambda x^2 + \mu x$ as in eqn (12.4) (in NODE). Draw projections of the bifurcations given the cusp surface $x^3 - \lambda x + \mu = 0$ on to both the (x, λ) plane and the (x, μ) plane. Sketch the projection of the cusp on to the (μ, λ) plane.

12.3. For the potential $\mathcal{V}(x, \lambda, \mu) = \frac{1}{4}x^4 - \frac{1}{2}x^2 + \mu x$, the corresponding equation is

$$\ddot{x} = \mathcal{V}_x(x, \lambda, \mu) = -x^3 + \lambda x - \mu.$$

The equilibrium points lie on the surface

$$x^3 - \lambda x + \mu = 0 \qquad (i)$$

in (λ, μ, x) space as shown in Figure 12.4.

On the tangents (λ fixed)$d\mu/dx = 0$ with slope on the surface (see Figure 12.4). From (i)

$$\frac{d\mu}{dx} = -3x^2 + \lambda.$$

The projection on to the (x, λ) plane is given by $\lambda = 3x^2$, shown in Figure 12.5. The projection on to the (x, μ) plane is given by eliminating λ between (i) and $d\mu/dx = 0$, namely $\mu = 2x^3$. The graph is also shown in Figure 12.5.

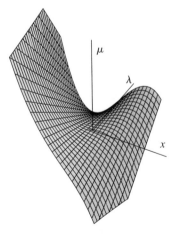

Figure 12.4 Problem 12.3: The equilibrium surface $\mu = x^3 - \lambda x$ is shown.

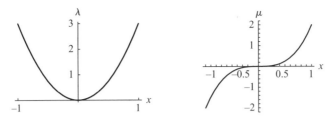

Figure 12.5 Problem 12.3: Projections of the cusp on to the (x, λ) and (x, μ) planes.

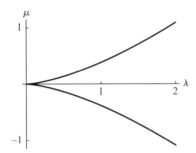

Figure 12.6 Problem 12.3: The cusp in the (μ, λ) plane.

The cusp in (λ, μ) plane is obtained by eliminating x between (i) and $d\mu/dx = 0$, namely,

$$\mu = 2\left(\frac{\lambda}{3}\right)^{3/2}.$$

The cusp is shown in Figure 12.6.

- **12.4** Discuss the stability and bifurcation of the equilibrium points of the parameter-dependent conservative system $\ddot{x} = -V_x(x, \lambda)$, where $V(x, \lambda) = \frac{1}{4}x^4 - \frac{1}{2}\lambda x^2 + \lambda x$.

12.4. With $V(x, \lambda) = \frac{1}{4}x^4 - \frac{1}{2}\lambda x^2 + \lambda x$, the equation for x is

$$\ddot{x} = -x^3 + \lambda x - \lambda.$$

Equilibrium occurs where

$$-x^3 + \lambda x - \lambda = 0. \tag{i}$$

The equilibrium points are shown in Figure 12.7.
The abscissa of the point P in Figure 12.7 is by

$$\frac{d\lambda}{dx} = \frac{d}{d\lambda}\left(\frac{x^3}{x-1}\right) = \frac{x^2(2x-3)}{(x-1)^2} = 0.$$

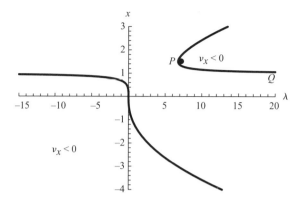

Figure 12.7 Problem 12.4: Equilibrium points in the (λ, x) plane.

Therefore $x = 0$ or $x = \frac{3}{2}$. A bifurcation takes place at the point P which has the coordinates $\left(\frac{27}{4}, \frac{3}{2}\right)$. The system has one equilibrium state for $\lambda < \frac{27}{4}$, and three for $\lambda > \frac{27}{4}$. The regions in which $\mathcal{V}_x$ is negative is also shown in the figure: only equilibrium points on PQ are stable. A saddle-node bifurcation occurs at P.

- **12.5** Discuss bifurcations of the system $\dot{x} = y^2 - \lambda$, $\dot{y} = x + \lambda$.

12.5. $\dot{x} = y^2 - \lambda$, $\dot{y} = x + \lambda$. Equilibrium occurs where

$$x = -\lambda, \quad y^2 = \lambda.$$

Parametrically in (x, y, λ) space this can be represented by the curve $(x, y, \lambda) = (-w^2, w, w^2)$ as shown in Figure 12.8. The system has no equilibrium points for $\lambda < 0$ and two for $\lambda > 0$. Clearly there is a bifurcation point at $\lambda = 0$.

The classification of the equilibrium points for $\lambda > 0$ can be done by linearization. Thus, on the branch $x = -\lambda$, $y = \sqrt{\lambda}$, let $x = -\lambda + x'$ and $y = \sqrt{\lambda} + y'$. Then equations are approximately

$$\dot{x}' = (y' + \sqrt{\lambda})^2 - \lambda \approx 2\sqrt{\lambda} y', \quad \dot{y}' = x' - \lambda + \lambda = x',$$

which indicates a saddle point. For the branch $x = -\lambda$, $y = -\sqrt{\lambda}$,

$$\dot{x}' \approx -2\sqrt{\lambda} y', \quad \dot{y}' = x',$$

which indicates a centre but linearization could fail to predict the type. However, the system is Hamiltonian with function

$$H(x, y) = \tfrac{1}{3}y^3 + \lambda y - \tfrac{1}{2}x^2 - \lambda x.$$

Consequently, any simple equilibrium points are either centres or saddle points.

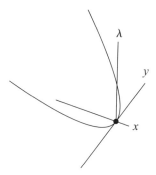

Figure 12.8 Problem 12.5.

- **12.6** Find the bifurcation points of $\dot{x} = y^2 - \lambda$, $\dot{y} = x + \lambda$.

12.6. The system

$$\dot{x} = y^2 - \lambda, \quad \dot{y} = x(x + \lambda)$$

has equilibrium points where

$$y^2 = \lambda, \quad x(x + \lambda) = 0.$$

There are no equilibrium points for $\lambda < 0$, and four for $\lambda > 0$, on the parabolas

$$x = -\lambda, \quad y = \pm\sqrt{\lambda}, \quad \text{and} \quad x = 0, \quad y = \pm\sqrt{\lambda}$$

as shown in Figure 12.9. There is a bifurcation point at the origin. The system is Hamiltonian which means that the equilibrium points will be either saddles or centres.

The classification of the equilibrium points for $\lambda > 0$ can be done by linearization. On $x = 0$, $y = \pm\sqrt{\lambda}$, let $y = y' \pm \sqrt{\lambda}$. Then

$$\dot{x} = (y' \pm \sqrt{\lambda})^2 - \lambda \approx \pm 2y'\sqrt{\lambda}, \quad \dot{y}' = x(x + \lambda) \approx x\lambda.$$

The point $(0, \sqrt{\lambda})$ is a saddle, and $(0, -\sqrt{\lambda})$ is a centre.
On $x = -\lambda$, $y = \pm\sqrt{\lambda}$, let $x = x' - \lambda$, $y = y' \pm \sqrt{\lambda}$. Then the equations become

$$\dot{x}' = (y' \pm \sqrt{\lambda})^2 - \lambda \approx \pm 2y'\sqrt{\lambda}, \quad \dot{y}' = (x' - \lambda)x' \approx -\lambda x'.$$

The point $(-\lambda, 2\sqrt{\lambda})$ is a centre, and $(-\lambda, -2\sqrt{\lambda})$ is a saddle.

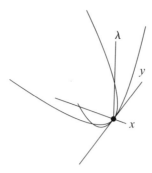

Figure 12.9 Problem 12.6: Showing the parabolas $x = 0$, $y^2 = \lambda$ and $x = -\lambda$, $y^2 = \lambda$.

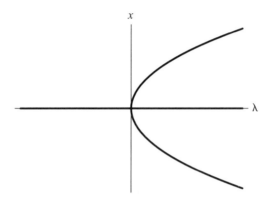

Figure 12.10 Problem 12.7: Equilibrium points of $\dot{x} = y$, $\dot{y} = x(\lambda - x^2)$.

- 12.7 Consider the system $\dot{x} = y$, $\dot{y} = x(\lambda - x^2)$, $-\infty < \lambda < \infty$. Investigate the phase diagrams for $\lambda < 0$, $\lambda = 0$ and $\lambda > 0$. Describe the bifurcation of the system as λ increases through zero.

12.7. The system
$$\dot{x} = y, \quad \dot{y} = x(\lambda - x^2),$$
has equilibrium points:

(i) at $(0, 0)$, $(\sqrt{\lambda}, 0)$, $(-\sqrt{\lambda}, 0)$ if $\lambda > 0$;
(ii) at $(0, 0)$ if $\lambda \leq 0$.

There is a bifurcation point at $\lambda = 0$. The bifurcation is shown in Figure 12.10 in the (λ, x) of the section $y = 0$. This is a pitchfork bifurcation. The equilibrium point at $x = 0$ is a centre for $\lambda < 0$, and a saddle for $\lambda > 0$. The equilibrium points at $x = \pm\sqrt{\lambda}$ ($\lambda > 0$) are centres.

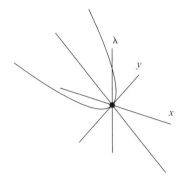

Figure 12.11 Problem 12.8: Pitchfork bifurcation of the system $\dot{x} = y(y^2 - \lambda)$, $\dot{y} = x + \lambda$.

- 12.8 Discuss the bifurcations of $\dot{x} = (y^2 - \lambda)y$, $\dot{y} = x + \lambda$.

12.8. The system
$$\dot{x} = y(y^2 - \lambda), \quad \dot{y} = x + \lambda$$
is in equilibrium if
$$y(y^2 - \lambda) = 0, \quad x + \lambda = 0.$$

In (x, y, λ) equilibrium occurs on the parabola $x = -\lambda$, $y = \pm\sqrt{\lambda}$, and on the line $x = -\lambda$, $y = 0$ as shown in Figure 12.11. The system is Hamiltonian which implies that the equilibrium points are either centres or saddle points.

For $\lambda < 0$, let $x = -\lambda + x'$. Then the approximate equations are
$$\dot{x}' \approx -\lambda y, \quad \dot{y}' = x'.$$

The linear equations predict a centre. For $\lambda > 0$, $x = -\lambda$, $y = 0$ is a saddle, and $x = -\lambda$, $y = \pm\sqrt{\lambda}$ are centres.

- 12.9 Investigate the bifurcation of the system $\dot{x} = x$, $\dot{y} = y^2 - \lambda$ at $\lambda = 0$. Show that, for $\lambda > 0$, the system has an unstable node at $(0, \sqrt{\lambda})$ and a saddle point at $(0, -\sqrt{\lambda})$. Sketch the phase diagrams for $\lambda < 0$, $\lambda = 0$ and $\lambda > 0$.

12.9. The system
$$\dot{x} = x, \quad \dot{y} = y^2 - \lambda$$
has two equilibrium points at $x = 0$, $y = \pm\sqrt{\lambda}$ if $\lambda > 0$, and none if $\lambda < 0$, as shown in Figure 12.12. Let $y = \pm\sqrt{\lambda} + y'$. Then the equations become
$$\dot{x} = x, \quad \dot{y}' = (y' \pm \sqrt{\lambda})^2 - \lambda \approx \pm 2\sqrt{\lambda} y'.$$

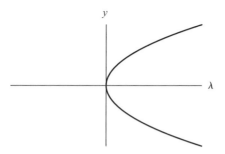

Figure 12.12 Problem 12.9: equilibrium states for $\dot{x} = x$, $\dot{y} = y^2 - \lambda$ in the $x = 0$ plane.

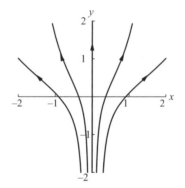

Figure 12.13 Problem 12.9: Phase diagram for $\lambda = -1$.

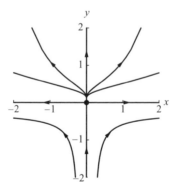

Figure 12.14 Problem 12.9: Phase diagram for $\lambda = 0$.

Hence $(0, \sqrt{\lambda})$ is an unstable node, and $(0, -\sqrt{\lambda})$ is a saddle point. This is an example of a saddle–node bifurcation.

Phase diagrams for $\lambda = -1, 0, 1$ are shown respectively in Figures 12.13, 12.14, 12.15. In Figure 12.13 ($\lambda = -1$) the system has no equilibrium points. The critical case shown in Figure 12.14 in which the origin is a higher-order saddle–node hybrid. These bifurcate into a separate node and saddle point for $\lambda > 0$.

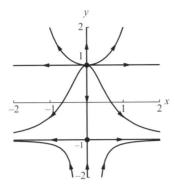

Figure 12.15 Problem 12.9: Phase diagram for $\lambda = 1$.

Note that the differential equation for the phase paths is

$$\frac{dy}{dx} = \frac{(y^2 - \lambda)}{x},$$

which is separable, so that the equations for the phase paths can be found explicitly.

- 12.10 A homoclinic path (NODE, Section 3.6) is a phase path which joins an equilibrium point to itself in an autonomous system. Show that $\dot{x} = y$, $\dot{y} = x - x^2$ has such a path and find its equation. Sketch the phase paths for the perturbed system $\dot{x} = y + \lambda x$, $\dot{y} = x - x^2$, for both $\lambda > 0$ and $\lambda < 0$. (The homoclinic saddle connection is destroyed by the perturbation; the system undergoes what is known as a homoclinic bifurcation (Section 3.6) at $\lambda = 0$.)

12.10. The system

$$\dot{x} = y, \quad \dot{y} = x - x^2$$

has two equilibrium points at $(0,0)$ (a saddle point) and at $(1,0)$ (a centre). The phase paths are given by

$$\frac{dy}{dx} = \frac{x - x^2}{y},$$

which can be integrated to give

$$\tfrac{1}{2}y^2 = \tfrac{1}{2}x^2 - \tfrac{1}{3}x^3 + C.$$

The homoclinic path passes through the saddle at the origin so that $C = 0$ leaving the equation

$$y^2 = x^2 - \tfrac{2}{3}x^3, \quad (x \geq 0).$$

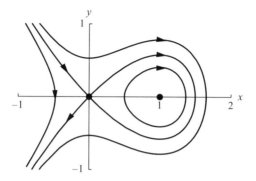

Figure 12.16 Problem 12.10: Phase paths for $\dot{x} = y$, $\dot{y} = x - x^2$.

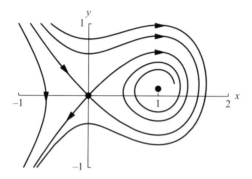

Figure 12.17 Problem 12.10: Phase diagram with $\lambda = -0.1$.

The phase diagram is shown in Figure 12.16. For the perturbed system

$$\dot{x} = y + \lambda x, \quad \dot{y} = x - x^2,$$

$(0, 0)$ and $(1, -\lambda)$ are equilibrium points. The origin remains a saddle point. For the other point, let $x = 1 + \xi$, $y = -\lambda + \eta$. Then the linearized equations are

$$\dot{\xi} = \lambda \xi + \eta, \quad \dot{\eta} = (\xi + 1) - (\xi + 1)^2 \approx -\xi.$$

The classification is as follows:

- $\lambda \leq -2$, stable node;
- $-2 < \lambda < 0$, stable spiral;
- $0 < \lambda < 2$, unstable spiral;
- $2 \leq \lambda$, unstable node.

Phase diagrams for $\lambda = -0.1$ and for $\lambda = 0.1$ are shown in Figures 12.17 and 12.18.

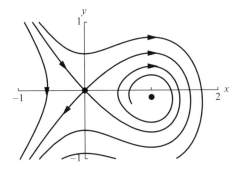

Figure 12.18 Problem 12.10: Phase diagram with $\lambda = 0.1$.

- 12.11 A heteroclinic path (NODE, Section 3.6) is a phase path which joins two different equilibrium points. Find the heteroclinic saddle connection for the system $\dot{x} = xy$, $\dot{y} = 1 - y^2$. Sketch the phase paths of the perturbed system $\dot{x} = xy + \lambda$, $\dot{y} = 1 - y^2$ for both $\lambda > 0$ and $\lambda < 0$.

12.11. The system
$$\dot{x} = xy, \quad \dot{y} = 1 - y^2,$$
has equilibrium points at $(0, 1)$ and at $(0, -1)$. Both equilibrium points are saddle points. Note also that $x = 0$ and $y = \pm 1$ are phase paths. The phase diagram is shown in Figure 12.19. This phase diagram has a heteroclinic path joining the saddle points at $(0, 1)$ and $(0, -1)$.

The perturbed system
$$\dot{x} = y + \lambda x, \quad \dot{y} = x - x^2$$
has equilibrium points at $(-\lambda, 1)$ and $(\lambda, -1)$, both of which are saddle points. The lines $y = \pm 1$ are still phase paths. The phase diagram for $\lambda = 0.2$ is shown in Figure 12.20 in which the saddle connection is broken. The phase diagram for $\lambda = -0.2$ is shown in Figure 12.21. The saddle connection bifurcates in the opposite direction in this case. The three figures indicate a heteroclinic bifurcation.

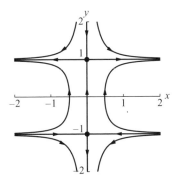

Figure 12.19 Problem 12.11: Phase diagram of $\dot{x} = xy$, $\dot{y} = 1 - y^2$.

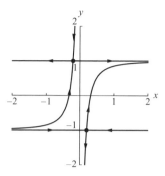

Figure 12.20 Problem 12.11: Phase diagram for $\lambda = 0.2$.

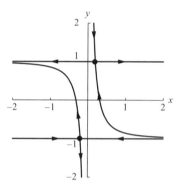

Figure 12.21 Problem 12.11: phase diagram for $\lambda = -0.2$.

- **12.12** Let

$$\dot{x} = -\mu x - y + \frac{x}{1+x^2+y^2}, \quad \dot{y} = x - \mu y + \frac{y}{1+x^2+y^2}.$$

Show that the equations display a Hopf bifurcation as $\mu > 0$ decreases through $\mu = 1$. Find the radius of the periodic path for $0 < \mu < 1$.

12.12. Express the system

$$\dot{x} = -\mu x - y + \frac{x}{1+x^2+y^2}, \quad \dot{y} = x - \mu y + \frac{y}{1+x^2+y^2}$$

in polar coordinates with $x = r\cos\theta$, $y = r\sin\theta$. Then

$$r\dot{r} = x\dot{x} + y\dot{y} = -\mu r^2 + \frac{r^2}{1+r^2} = -\frac{r^2}{1+r^2}[\mu r^2 + (\mu - 1)],$$

$$\dot{\theta} = \frac{1}{r^2}(\dot{y}x - \dot{x}y) = 1.$$

12: Bifurcations and manifolds

For $0 < \mu < 1$, $r = r_0 = \sqrt{[(1-\mu)/\mu]}$ is a limit cycle which is stable since for $r < r_0$, $\dot{r} > 0$, and for $r > r_0$, $\dot{r} < 0$. In other words adjacent paths spiral into the limit cycle. The equilibrium point at the origin is an unstable spiral.

For $\mu > 1$, the system has a stable equilibrium point at the origin and no limit cycle. The system passes through a Hopf bifurcation as μ decreases through $\mu = 1$.

- **12.13** Show that the system
$$\dot{x} = x - \gamma y - x(x^2+y^2), \quad \dot{y} = \gamma x + y - y(x^2+y^2) - \gamma, \quad (\gamma > 0)$$
has a bifurcation point at $\gamma = \frac{1}{2}$, by investigating the numbers of equilibrium points for $\gamma > 0$. Compute the phase diagram for $\gamma = \frac{1}{4}$.

12.13. The system
$$\dot{x} = x - \gamma y - x(x^2+y^2), \quad \dot{y} = \gamma x + y - y(x^2+y^2) - \gamma, \quad (\gamma > 0)$$
has equilibrium points where
$$x - \gamma y - x(x^2+y^2) = 0, \quad \gamma x + y - y(x^2+y^2) = \gamma \qquad (i)$$

Squaring and adding
$$[x - \gamma y - x(x^2+y^2)]^2 + [\gamma x + y - y(x^2+y^2)]^2 = \gamma^2,$$
or
$$r^2(1-r^2)^2 - \gamma^2(1-r^2) = 0,$$
where $r^2 = x^2 + y^2$. Therefore
$$r = 1, \text{ or } r^4 - r^2 + \gamma^2 = 0. \qquad (ii)$$

The cases are as follows:

- $\gamma > \frac{1}{2}$. Equation (ii) has one real solution where $r = 1$, which from (i) implies that $(1,0)$ is an equilibrium point. To classify the point, let $x = 1 + \xi$. Then
$$\dot{\xi} \approx -2\xi - \gamma y, \quad \dot{y} \approx \gamma \xi.$$
Therefore $(1,0)$ is a stable node for $\frac{1}{2} < \gamma < 1$.

- $\gamma = \frac{1}{2}$. In this case (ii) has the solutions $r = 1$ and $r = 1/\sqrt{2}$ which lead to equilibrium points at $(1,0)$ and at $(\frac{1}{2}, \frac{1}{2})$. As in the previous case $(1,0)$ remains a stable node.

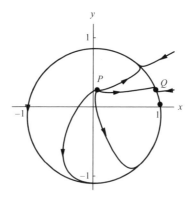

Figure 12.22 Problem 12.13.

Linearization is not helpful for $(\frac{1}{2}, \frac{1}{2})$ since the equations for the perturbations ξ and η become

$$\dot\xi \approx -\eta, \quad \dot\eta \approx 0,$$

which is a degenerate case.

- $0 < \gamma < \frac{1}{2}$. (ii) now has three real solutions $r = 1$, and $r = r_1, r_2 = \frac{1}{2}(1 \pm \sqrt{(1-4\gamma^2)})$. The origin remains a stable node. Classification of the other equilibrium points is complicated.

As the parameter γ decreases through $\gamma = \frac{1}{2}$ an equilibrium point appears at $(\frac{1}{2}, \frac{1}{2})$ which then splits into two equilibrium points. The phase diagram for $\gamma = 0.25$ is shown in Figure 12.22. The computed diagram seems to indicate that the bifurcation creates an unstable node at P, and a saddle point at Q. The equilibrium point at $(1, 0)$ is a stable node for $\gamma < 1$.

- 12.14 Let $\dot{\mathbf{x}} = \mathbf{A}\mathbf{x}$, where $\mathbf{x} = [x \ y \ z]^T$. Find the eigenvalues and eigenvectors of $\mathbf{A}$ in each of the following cases. Describe the stable and unstable manifolds of the origin.

(a) $\mathbf{A} = \begin{bmatrix} 1 & 1 & 2 \\ 1 & 2 & 1 \\ 2 & 1 & 1 \end{bmatrix}$.

(b) $\mathbf{A} = \begin{bmatrix} 3 & 0 & -1 \\ 0 & 1 & 0 \\ 2 & 0 & 0 \end{bmatrix}$.

(c) $\mathbf{A} = \begin{bmatrix} 2 & 0 & 0 \\ 0 & 2 & 2 \\ 0 & 2 & -1 \end{bmatrix}$.

(d) $\mathbf{A} = \begin{bmatrix} 6 & 5 & 5 \\ 5 & 6 & 5 \\ 5 & 5 & 6 \end{bmatrix}$.

12.14. The origin is an equilibrium point for the linear system

$$\dot{\mathbf{x}} = \mathbf{A}\mathbf{x}.$$

(a)

$$\mathbf{A} = \begin{bmatrix} 1 & 1 & 2 \\ 1 & 2 & 1 \\ 2 & 1 & 1 \end{bmatrix}.$$

The eigenvalues of $\mathbf{A}$ are given by

$$\begin{vmatrix} 1-\lambda & 1 & 2 \\ 1 & 2-\lambda & 1 \\ 2 & 1 & 1-\lambda \end{vmatrix} = 0, \text{ or } -(\lambda-4)(\lambda-1)(\lambda+1) = 0.$$

Therefore the eigenvalues are $-1, 1, 4$ with corresponding eigenvectors

$$(-1, 0, 1)^T, \quad (1, -2, 1)^T, \quad (1, 1, 1)^T.$$

Hence the general solution is given by

$$\begin{bmatrix} x \\ y \\ z \end{bmatrix} = \alpha \begin{bmatrix} -1 \\ 0 \\ 1 \end{bmatrix} e^{-t} + \beta \begin{bmatrix} 1 \\ -2 \\ 1 \end{bmatrix} e^{t} + \gamma \begin{bmatrix} 1 \\ 1 \\ 1 \end{bmatrix} e^{4t}.$$

The stable manifold is given parametrically by $\beta = \gamma = 0$, that is

$$\begin{bmatrix} x \\ y \\ z \end{bmatrix} = \alpha \begin{bmatrix} -1 \\ 0 \\ 1 \end{bmatrix} e^{-t},$$

which is the straight line $z = -x$, $y = 0$.
The unstable manifold is given by

$$\begin{bmatrix} x \\ y \\ z \end{bmatrix} = \beta \begin{bmatrix} 1 \\ -2 \\ 1 \end{bmatrix} e^{t} + \gamma \begin{bmatrix} 1 \\ 1 \\ 1 \end{bmatrix} e^{4t},$$

which defines the plane $x - 3z = 0$.
(b)

$$\mathbf{A} = \begin{bmatrix} 3 & 0 & -1 \\ 0 & 1 & 0 \\ 2 & 0 & 0 \end{bmatrix}.$$

The eigenvalues of **A** are given by

$$\begin{vmatrix} 3-\lambda & 0 & -1 \\ 0 & 1-\lambda & 0 \\ 2 & 0 & -\lambda \end{vmatrix} = 0, \text{ or } -(\lambda-2)(\lambda-1)^2 = 0.$$

Therefore the eigenvalues are 1(repeated) and 2 with corresponding eigenvectors

$$(1,0,2)^T, \quad (0,1,0)^T, \quad (1,0,1)^T.$$

Hence the general solution

$$\begin{bmatrix} x \\ y \\ z \end{bmatrix} = \alpha \begin{bmatrix} 1 \\ 0 \\ 2 \end{bmatrix} e^t + \beta \begin{bmatrix} 0 \\ 1 \\ 0 \end{bmatrix} e^t + \gamma \begin{bmatrix} 1 \\ 0 \\ 1 \end{bmatrix} e^{2t}.$$

Since all the eigenvalues are positive there is no stable manifold, and the unstable manifold is the whole space.

(c)

$$\mathbf{A} = \begin{bmatrix} 2 & 0 & 0 \\ 0 & 2 & 2 \\ 0 & 2 & -1 \end{bmatrix}.$$

The eigenvalues of **A** are given by

$$\begin{vmatrix} 2-\lambda & 0 & 0 \\ 0 & 2-\lambda & 2 \\ 0 & 2 & -1-\lambda \end{vmatrix} = 0, \text{ or } -(\lambda-3)(\lambda-2)(\lambda+2) = 0.$$

Therefore the eigenvalues are $-2, 2, 3$ with corresponding eigenvectors

$$(0,-1,2)^T, \quad (1,0,0)^T, \quad (0,2,1)^T.$$

Hence the general solution is

$$\begin{bmatrix} x \\ y \\ z \end{bmatrix} = \alpha \begin{bmatrix} 0 \\ -1 \\ 2 \end{bmatrix} e^{-2t} + \beta \begin{bmatrix} 1 \\ 0 \\ 0 \end{bmatrix} e^{2t} + \gamma \begin{bmatrix} 0 \\ 2 \\ 1 \end{bmatrix} e^{3t}.$$

The stable manifold ($\beta = \gamma = 0, \alpha = 1$, say) is the straight line given parametrically by $x = 0$, $y = -e^{-2t}$, $z = 2e^{-2t}$. The unstable manifold ($\alpha = 0$) is the plane $-y + 2z = 0$.

(d)

$$\mathbf{A} = \begin{bmatrix} 6 & 5 & 5 \\ 5 & 6 & 5 \\ 5 & 5 & 6 \end{bmatrix}.$$

The eigenvalues of **A** are given by

$$\begin{vmatrix} 6-\lambda & 5 & 5 \\ 5 & 6-\lambda & 5 \\ 5 & 5 & 6-\lambda \end{vmatrix} = 0, \text{ or } -(\lambda-16)(\lambda-1)^2 = 0.$$

Therefore the eigenvalues are 1(repeated) and 16 with corresponding eigenvectors

$$(-1,0,1)^T, \quad (-1,1,0)^T, \quad (1,1,1)^T.$$

Hence the general solution is

$$\begin{bmatrix} x \\ y \\ z \end{bmatrix} = \alpha \begin{bmatrix} -1 \\ 0 \\ 1 \end{bmatrix} e^t + \beta \begin{bmatrix} -1 \\ 1 \\ 0 \end{bmatrix} e^t + \gamma \begin{bmatrix} 1 \\ 1 \\ 1 \end{bmatrix} e^{16t}.$$

Since the eigenvalues are all positive there is no stable manifold, whilst the stable manifold is the whole space.

- **12.15** Show that $\dot{\mathbf{x}} = \mathbf{A}\mathbf{x}$ where $\mathbf{x} = [x \; y \; z]^T$ and

$$\mathbf{A} = \begin{bmatrix} -3 & 0 & -2 \\ -4 & -1 & -4 \\ 3 & 1 & 3 \end{bmatrix}$$

has two imaginary eigenvalues. Find the equation of the centre manifold of the origin. Is the remaining manifold stable or unstable?

12.15. In the system $\dot{\mathbf{x}} = \mathbf{A}\mathbf{x}$,

$$\mathbf{A} = \begin{bmatrix} -3 & 0 & -2 \\ -4 & -1 & -4 \\ 3 & 1 & 3 \end{bmatrix}.$$

The eigenvalues of **A** are given by

$$\begin{vmatrix} -3-m & 0 & -2 \\ -4 & -1-m & -4 \\ 3 & 1 & 3-m \end{vmatrix} = -(m+1)(m^2+1) = 0.$$

The eigenvalues are $-1, -i, i$ with corresponding eigenvectors

$$[-1,-1,1]^T, \quad [-3-i,-6-2i,5]^T, \quad [-3+i,-6+2i,5]^T.$$

Therefore the general solution is

$$\begin{bmatrix} x \\ y \\ z \end{bmatrix} = \alpha \begin{bmatrix} -1 \\ -1 \\ 1 \end{bmatrix} e^{-t} + \beta \begin{bmatrix} -3-i \\ -6-2i \\ 5 \end{bmatrix} e^{-it} + \bar{\beta} \begin{bmatrix} -3+i \\ -6+2i \\ 5 \end{bmatrix} e^{it},$$

where α is a real constant and β is a complex constant.

The system has a stable manifold which is the line given parametrically (put $\beta = 0$ and $\alpha = 1$) by $(x, y, z) = (-e^{-t}, -e^{-t}, e^t)$. Since the other eigenvalues are imaginary, the origin has an associated centre manifold defined by $\alpha = 0$, namely by

$$\begin{bmatrix} x \\ y \\ z \end{bmatrix} = \beta \begin{bmatrix} -3-i \\ -6-2i \\ 5 \end{bmatrix} e^{-it} + \bar{\beta} \begin{bmatrix} -3+i \\ -6+2i \\ 5 \end{bmatrix} e^{it},$$

which defines the plane $-2x + y = 0$. Depending on the initial values, as $t \to \infty$, solutions approach a periodic solution of the centre which lies in the plane $-2x + y = 0$.

- **12.16** Show that the centre manifold of

$$\begin{bmatrix} \dot{x} \\ \dot{y} \\ \dot{z} \end{bmatrix} = \begin{bmatrix} -1 & 0 & 1 \\ 0 & 1 & -2 \\ 0 & 1 & -1 \end{bmatrix} \begin{bmatrix} x \\ y \\ z \end{bmatrix},$$

is given by $2x + y - 2z = 0$

12.16. Let

$$\mathbf{A} = \begin{bmatrix} -1 & 0 & 1 \\ 0 & 1 & -2 \\ 0 & 1 & -1 \end{bmatrix}.$$

The eigenvalues of $\mathbf{A}$ are given by

$$\begin{vmatrix} -1-m & 0 & 1 \\ 0 & 1-m & -2 \\ 0 & 1 & -1-m \end{vmatrix} = -(m+1)(m^2+1) = 0.$$

The eigenvalues are $-1, -i, i$ with corresponding eigenvectors

$$[1, 0, 0]^T, \quad [1+i, 2-2i, 2]^T, \quad [1-i, 2+2i, 2]^T.$$

Therefore the general solution is

$$\begin{bmatrix} x \\ y \\ z \end{bmatrix} = \alpha \begin{bmatrix} 1 \\ 0 \\ 0 \end{bmatrix} e^{-t} + \beta \begin{bmatrix} 1+i \\ 2-2i \\ 2 \end{bmatrix} e^{-it} + \bar{\beta} \begin{bmatrix} 1-i \\ 2+2i \\ 2 \end{bmatrix} e^{it},$$

where α is a real constant and β is a complex constant. Solutions for which $\alpha = 0$ lie on the centre manifold, that is,

$$\begin{bmatrix} x \\ y \\ z \end{bmatrix} = \beta \begin{bmatrix} 1+i \\ 2-2i \\ 2 \end{bmatrix} e^{-it} + \bar{\beta} \begin{bmatrix} 1-i \\ 2+2i \\ 2 \end{bmatrix} e^{it},$$

which is the plane $-2x - y + 2z = 0$.

- **12.17** Show that the phase paths of $\dot{x} = y(x+1)$, $\dot{y} = x(1-y^2)$ are given by
$y = \pm\sqrt{[1 - Ae^{-2x}(1+x)^2]}$,
with singular solutions $x = -1$ and $y = \pm 1$. Describe the domains in the (x, y) plane of the stable and unstable manifolds of each of the three equilibrium points of the system.

12.17. The phase paths of

$$\dot{x} = y(x+1), \quad \dot{y} = x(1-y^2),$$

are given by

$$\frac{dy}{dx} = \frac{x(1-y^2)}{y(x+1)}.$$

Separating the variables and integrating,

$$\int \frac{y \, dy}{1-y^2} = \int \frac{x \, dx}{x+1} + B.$$

Therefore

$$-\tfrac{1}{2} \ln|1 - y^2| = -\ln[e^{-x}|x+1|] + B,$$

or

$$y = \pm\sqrt{[1 - Ae^{-2x}(1+x)^2]}.$$

By inspection the equations also have the solutions $x = -1$ and $y = \pm 1$.
The system has equilibrium points at $(0,0)$, $(-1, 1)$ and $(-1, -1)$. The classification of the equilibrium points is as follows (easier to use the differential equations rather than the solutions):
- $(0,0)$: $\dot{x} \approx y$, $\dot{y} \approx x$ – saddle point;

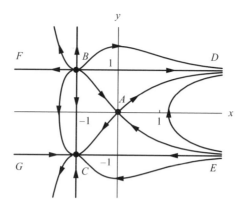

Figure 12.23 Problem 12.17: Phase diagram for $\dot{x} = y(x+1)$, $\dot{y} = x(1-y^2)$.

- $(-1, 1)$: let $x = -1 + x'$, $y = 1 + y'$; then $\dot{x}' \approx x'$, $\dot{y}' \approx 2y'$ – unstable node;
- $(-1, -1)$; let $x' = 1 + x'$, $y = -1 + y'$; then $\dot{x}' \approx -x'$, $\dot{y}' \approx -2y'$– stable node.

For the saddle point at the origin, the stable manifolds are the separatrices EA and BA in Figure 12.23.
For the equilibrium point at $(-1, 1)$, the domain above $GCAD$ is its unstable manifold.
For the equilibrium point at $(-1, -1)$, the domain below $FBAE$ is its stable manifold.

- **12.18** Show that the linear approximation at $(0, 0, 0)$ of
$$\dot{x} = -y + yz + (y-x)(x^2+y^2), \quad \dot{y} = x - xz - (x+y)(x^2+y^2), \quad \dot{z} = -z + (1-2z)(x^2+y^2),$$
has a centre manifold there. Show that $z = x^2 + y^2$ is a solution of this system of equations. To which manifold of the origin is this surface tangential? Show also that, on the surface, x and y satisfy
$$\dot{x} = -y + (2y - x)(x^2 + y^2), \quad \dot{y} = x - (2x + y)(x^2 + y^2).$$
Using polar coordinates determine the stability of solutions on this surface and the stability of the origin.

12.18. The linear approximation near $(0, 0, 0)$ of the system
$$\dot{x} = -y + yz + (y-x)(x^2 + y^2), \quad \dot{y} = x - xz - (x+y)(x^2 + y^2),$$
$$\dot{z} = -z + (1 - 2z)(x^2 + y^2),$$
is
$$\dot{x} \approx -y, \quad \dot{y} \approx x, \quad \dot{z} \approx -z.$$

The eigenvalues of the linear approximation are given by

$$\begin{vmatrix} -\lambda & -1 & 0 \\ 1 & -\lambda & 0 \\ 0 & 0 & -1-\lambda \end{vmatrix} = 0, \text{ or } -(\lambda+1)(\lambda^2+1) = 0.$$

The eigenvalues are $-1, i, -i$ with corresponding eigenvectors

$$(0,0,1)^T, \quad (-i,1,0)^T, \quad (i,1,0).$$

Hence the general solution near the origin is

$$\begin{bmatrix} x \\ y \\ z \end{bmatrix} \approx \alpha \begin{bmatrix} 0 \\ 0 \\ 1 \end{bmatrix} e^{-t} + \beta \begin{bmatrix} -i \\ 1 \\ 0 \end{bmatrix} e^{it} + \gamma \begin{bmatrix} i \\ 1 \\ 0 \end{bmatrix} e^{-it}.$$

The stable manifold is given by $\beta = 0$, $\gamma = 0$, that is the line $x = y = 0$, $z = 1$. Since two eigenvalues are imaginary, the linear approximation has a centre manifold given by $\alpha = 0$ which defines the plane $z = 0$.
If $z = x^2 + y^2$, then

$$\dot{z} = 2x\dot{x} + 2y\dot{y}$$
$$= 2x[-y + yz + (y-x)(x^2+y^2)] + 2y[x - xz - (x+y)(x^2+y^2)]$$
$$= -2(x^2+y^2) = -z + (1-2z)(x^2+y^2)$$

which confirms that this function is a particular solution. The centre manifold of the linear approximation is the tangent plane to this paraboloid at the origin.
On the surface $z = x^2 + y^2$, x and y satisfy

$$\dot{x} = -y + (2y-x)r^2, \quad \dot{y} = x - (2x-y)r^2.$$

where $r^2 = x^2 + y^2$. Introduce polar coordinates through $x = r\cos\theta$, $y = r\sin\theta$. Then

$$r\dot{r} = x\dot{x} + y\dot{y} = x[-y + (2y-x)r^2] + y[x - (2x+y)r^2] = -r^4, \text{ or } \dot{r} = -r^3.$$

Also

$$\dot{\theta} = \frac{x\dot{y} - y\dot{x}}{r^2} = \frac{x[x - (2x-y)r^2] - y[-y + (2y-x)r^2]}{r^2} = 1 - 2r^2.$$

For all $r > 0$, $\dot{r} < 0$ and $\dot{r} \to 0$, which implies that any solution starting on the paraboloid will approach the origin, from which we infer that the origin is globally asymptotically stable since the remaining manifold is stable. This result counters the centre manifold predicted by linearization.

520 Nonlinear ordinary differential equations: problems and solutions

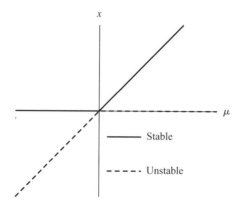

Figure 12.24 Problem 12.19: Stability diagram for $\dot{x} = \mu x - x^2$, $\dot{y} = y(\mu - 2x)$.

- **12.19** Investigate the stability of the equilibrium points of $\dot{x} = \mu x - x^2$, $\dot{y} = y(\mu - 2x)$ in terms of the parameter μ. Draw a stability diagram in the (μ, x) plane for $y = 0$. What type of bifurcation occurs at $\mu = 0$? Obtain the equations of the phase paths, and sketch the phase diagrams in the cases $\mu = -1$, $\mu = 0$ and $\mu = 1$.

12.19. The system
$$\dot{x} = \mu x - x^2, \quad \dot{y} = y(\mu - 2x)$$
has equilibrium points at $(0, 0)$ and $(\mu, 0)$. The linear classification of the points is as follows.

- $(0, 0)$. The linear approximation is
$$\dot{x} \approx \mu x, \quad \dot{y} \approx \mu y,$$
which is a critical node, stable if $\mu < 0$ and unstable if $\mu > 0$.
- $(\mu, 0)$. Let $x = \mu + \xi$. Then
$$\dot{\xi} \approx -\mu \xi, \quad \dot{y} \approx -\mu y,$$
which is also a critical node, stable if $\mu > 0$ and unstable if $\mu < 0$.

The stability diagram of equilibrium points in the (μ, x) plane is shown in Figure 12.24. The origin is an example of a transcritical bifurcation.

The phase paths are given by solutions of
$$\frac{dy}{dx} = \frac{y(\mu - x)}{x(\mu - x)}.$$

Separation of variables leads to
$$\int \frac{dy}{y} = \int \frac{\mu - 2x}{x(\mu - x)} dx,$$

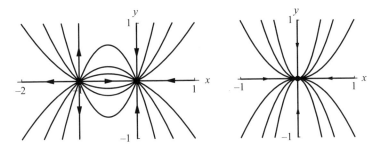

Figure 12.25 Problem 12.19: Phase diagrams for, respectively $\mu = -1$ and $\mu = 0$.

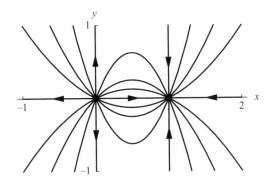

Figure 12.26 Problem 12.19: phase diagram for $\mu = 1$.

or
$$\ln |y| = \ln |x(x - \mu)| + C, \quad \text{or} \quad y = Bx(x - \mu).$$

The straight lines $x = 0$, $x = \mu$ and $y = 0$ are also phase paths. The phase diagrams for $\mu = -1$ and $\mu = 0$ are shown in Figure 12.25. The phase diagram for $\mu = 1$ is shown in Figure 12.26.

- 12.20 Where is the bifurcation point of the parameter-dependent system
$\dot{x} = x^2 + y^2 - \mu, \quad \dot{y} = 2\mu - 5xy$?
Discuss how the system changes as μ increases. For $\mu = 5$, find all linear approximations for all equilibrium points and classify them.

12.20. The system
$$\dot{x} = x^2 + y^2 - \mu, \quad \dot{y} = 2\mu - 5xy,$$
is in equilibrium where
$$x^2 + y^2 - \mu = 0, \quad 2\mu - 5xy = 0.$$
Elimination of y leads to
$$25x^4 - 25\mu x^2 + 4\mu^2 = 0, \quad \text{or} \quad (5x^2 - 4\mu)(5x^2 - \mu) = 0.$$

where $y = 2\mu/(5x)$. If $\mu < 0$, the system has no equilibrium points, if $\mu = 0$ the system has one point, at $(0,0)$, and if $\mu > 0$, the system has four points, at

$$\left(\frac{2\sqrt{\mu}}{\sqrt{5}}, \frac{\sqrt{\mu}}{\sqrt{5}}\right), \quad \left(\frac{-2\sqrt{\mu}}{\sqrt{5}}, \frac{-\sqrt{\mu}}{\sqrt{5}}\right), \quad \left(\frac{\sqrt{\mu}}{\sqrt{5}}, \frac{2\sqrt{\mu}}{\sqrt{5}}\right), \quad \left(\frac{-\sqrt{\mu}}{\sqrt{5}}, \frac{-2\sqrt{\mu}}{\sqrt{5}}\right).$$

The system has a bifurcation point at $\mu = 0$. As μ increases through zero, a single equilibrium point emerges at the origin at $\mu = 0$ which splits into four equilibrium points as μ becomes positive.

If $\mu = 5$, the equilibrium points simplify to the coordinates $(2, 1), (-2, -1), (1, 2), (-1, -2)$. The linear classification is as follows:

- $(2, 1)$. Let $x = 2 + \xi$, $y = 1 + \eta$. Then

$$\dot{\xi} = (2+\xi)^2 + (1+\eta)^2 - 5 \approx 4\xi + 2\eta,$$
$$\dot{\eta} = 10 - 5(2+\xi)(1+\eta) \approx -5\xi - 10\eta.$$

In the usual notation $p = -6 < 0$, $q = -30 < 0$, which implies that $(2, 1)$ is a saddle.

- $(-2, -1)$. Let $x = -2 + \xi$, $y = -1 + \eta$. Then

$$\dot{\xi} = (-2+\xi)^2 + (-1+\eta)^2 - 5 \approx -4\xi - 2\eta,$$
$$\dot{\eta} = 10 - 5(-2+\xi)(-1+\eta) \approx 5\xi + 10\eta.$$

Therefore $p = 6 > 0$, $q = -30 < 0$, which implies that $(-2, -1)$ is a saddle.

- $(1, 2)$. Let $x = 1 + \xi$, $y = 2 + \eta$. Then

$$\dot{\xi} = (1+\xi)^2 + (2+\eta)^2 - 5 \approx 2\xi + 4\eta,$$
$$\dot{\eta} = 10 - 5(1+\xi)(2+\eta) \approx -10\xi - 5\eta.$$

Therefore $p = -3 < 0$, $q = 30 > 0$, $\Delta = -111 < 0$, which implies that $(1, 2)$ is a stable spiral.

- $(-1, -2)$. Let $x = -1 + \xi$, $y = -2 + \eta$. Then

$$\dot{\xi} = (-1+\xi)^2 + (-2+\eta)^2 - 5 \approx -2\xi - 4\eta,$$
$$\dot{\eta} = 10 - 5(-1+\xi)(-2+\eta) \approx 10\xi + 5\eta.$$

Therefore $p = 3 > 0$, $q = 30$, $\Delta = -111$, which implies that $(-1, -2)$ is an unstable spiral.

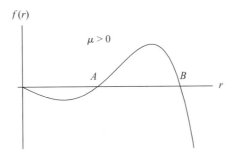

Figure 12.27 Problem 12.21:

- 12.21 Obtain the polar equations for (r,θ) of
$$\dot{x} = y + x[\mu - (x^2 + y^2 - 1)^2], \quad \dot{y} = -x + y[\mu - (x^2 + y^2 - 1)^2],$$
where $|\mu| < 1$. Show that, for $0 < \mu < 1$, the system has two limit cycles, one stable and one unstable, which collide at $\mu = 0$ and disappear for $\mu < 0$. This is an example of a blue sky catastrophe in which a finite amplitude limit cycles simply disappears as a parameter is changed incrementally.

12.21. In the system
$$\dot{x} = y + x[\mu - (x^2 + y^2 - 1)^2], \quad \dot{y} = -x + y[\mu - (x^2 + y^2 - 1)^2],$$
let $x = r\cos\theta$, $y = r\sin\theta$. In terms of r and θ the equations become
$$\dot{r} = x\dot{x} + y\dot{y} = r[\mu - (r^2 - 1)^2] = f(r) \text{ say},$$
$$\dot{\theta} = \frac{\dot{y}x - \dot{x}y}{r^2} = -1.$$
For $\mu > 0$, the general shape of $f(r)$ versus r is shown in Figure 12.27. The zero at $r = 0$ is an equilibrium point, whilst the zeros at A and B define limit cycles of the system. For point A, $r = \sqrt{(1 - \sqrt{\mu})}$ and for B, $r = \sqrt{(1 + \sqrt{\mu})}$. As μ decreases through zero the two limit cycles merge at $r = 1$ and then disappear for $\mu < 0$.

- 12.22 Discuss the bifurcations of the equilibrium points of $\dot{x} = y$, $\dot{y} = -x - 2x^2 - \mu x^3$ for $-\infty < \mu < \infty$. Sketch the bifurcation diagram in the (μ, x) plane. Confirm that there is a bifurcation at $\mu = 1$. What happens at $\mu = 0$?

12.22. Equilibrium of
$$\dot{x} = y, \quad \dot{y} = -x - 2x^2 - \mu x^3,$$

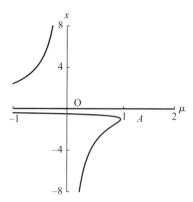

Figure 12.28 Problem 12.22.

occurs where $y = 0$ and $\mu x^3 + 2x^2 + x = 0$. The solutions of the latter equation are

$$x = 0, \quad x = \frac{1}{\mu}[-1 \pm \sqrt{(1-\mu)}], \quad (\mu \neq 0, \mu < 1).$$

If $\mu = 0$, the system has two equilibrium points, at $x = 0$ and $x = -\frac{1}{2}$. If $\mu > 1$, then the system has one equilibrium point, at $x = 0$

As μ decreases through $\mu = 1$, two additional equilibrium points appear at $\mu = 1$ (point A in Figure 12.28). One of these disappears as $\mu \to 0$ (point O), and $x \to -\infty$, but a second point re-appears immediately for large positive x.

- **12.23** Consider the system $\dot{x} = y - x(x^2 + y^2 - \mu)$, $\dot{y} = -x - y(x^2 + y^2 - \mu)$, where μ is a parameter. Express the equations in polar form in terms of (r, θ) show that the origin is a stable spiral for $\mu < 0$, and unstable spiral for $\mu > 0$. What type of bifurcation occurs at $\mu = 0$?

12.23. The system

$$\dot{x} = y - x(x^2 + y^2 - \mu), \quad \dot{y} = -x - y(x^2 + y^2 - \mu)$$

has one equilibrium point, at $(0, 0)$. Let $x = r \cos \theta$ and $y = r \sin \theta$, so that the polar equations are

$$\dot{r} = -r(r^2 - \mu), \quad \dot{\theta} = -1.$$

This system has a stable limit cycle at $r = \sqrt{\mu}$ if $\mu > 0$. If $\mu < 0$ the origin is a global stable spiral. A stable limit cycle appears from the origin as μ increases through zero. This is a Hopf bifurcation.

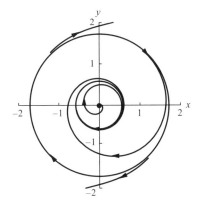

Figure 12.29 Problem 12.24: Phase diagram with $\mu = 2.3$ showing stable and unstable limit cycles.

- **12.24** In polar form a system is given by $\dot{r} = r(r^2 - \mu r + 1)$, $\dot{\theta} = -1$, where μ is a parameter. Discuss the bifurcations which occur as μ increases through $\mu = 2$.

12.24. The polar equations of a system are given by

$$\dot{r} = r(r^2 - \mu r + 1), \quad \dot{\theta} = -1.$$

Equilibrium occurs where $r = 0$. $\dot{r} = 0$ where

$$r = \tfrac{1}{2}[\mu \pm \sqrt{(\mu^2 - 4)}]. \tag{i}$$

If $\mu < -2$, then $r < 0$. If $\mu > 2$ then the system has two limit cycles with radii given by (i). In this case the origin is an unstable spiral, and the inner limit cycle is stable and the outer one unstable as shown in Figure 12.29. As μ increases through $\mu = 2$ a limit cycle appears which immediately bifurcates into two limit cycles.

- **12.25** The equations of a displaced van der Pol oscillator are given by
$\dot{x} = y - a$, $\dot{y} = -x + \delta(1 - x^2)y$,
where $a > 0$ and $\delta > 0$. If the parameter $a = 0$ then the usual equations for the van der Pol oscillator appear. Suppose that a is increased from zero. Show that the system has two equilibrium points one of which is a saddle point at $x \approx -1/(a\delta)$, $y = a$ for small a. Compute phase paths for $\delta = 2$, and $a = 0, 1, 0.2, 0.4$, and observe that the saddle point approaches with the limit cycle of the van der Pol equation. Show that at $a \approx 0.31$ the saddle point collides with the limit cycle, which then disappears.

12.25. The displaced van der Pol equations are

$$\dot{x} = y - a, \quad \dot{y} = -x + \delta(1 - x^2)y, \quad (a > 0, \delta > 0).$$

Equilibrium occurs where

$$y = a, \quad x - \delta(1 - x^2)a = 0.,$$

that is, where

$$x = \{x_1, x_2\} = \frac{1}{2\delta a}[-1 \pm \sqrt{(1 + 4\delta^2 a^2)}].$$

Therefore there are two equilibrium points at (x_1, a) and (x_2, a). The linear classifications at these points are as follows.

- (x_1, a). Let $x = x_1 + \xi$ and $y = a + \eta$. Then $\dot{\xi} = \eta$, and

$$\dot{\eta} = -(x_1 + \xi) + \delta[1 - (x_1 + \xi)^2](a + \eta)$$
$$\approx (-1 - 2x_1 a\delta)\xi + \delta(1 - x_1^2)\eta$$

In the usual notation

$$p = \delta(1 - x_1^2) = \frac{x_1}{a} = \frac{1}{2\delta a^2}[-1 + \sqrt{(1 + 4\delta^2 a^2)}] > 0,$$

$$q = 1 + 2a\delta x_1 = \sqrt{(1 + 4\delta^2 a^2)} > 0,$$

which implies that (x_1, a) is an unstable node or spiral.
- (x_2, a). Let $x = x_2 + \xi$ and $y = a + \eta$. Then $\dot{\xi} = \eta$, and

$$\dot{\eta} = -(x_2 + \xi) + \delta[1 - (x_2 + \xi)^2](a + \eta)$$
$$\approx (-1 - 2x_2 a\delta)\xi + \delta(1 - x_2^2)\eta$$

In this case

$$p = \delta(1 - x_2^2) = \frac{x_2}{a} = \frac{1}{2\delta a^2}[-1 - \sqrt{(1 + 4\delta^2 a^2)}] < 0,$$

$$q = 1 + 2a\delta x_1 = -\sqrt{(1 + 4\delta^2 a^2)}, 0,$$

which implies that (x_2, a) is a saddle.

As a increases from zero, the saddle point B approaches from infinity and collides with the limit cycle as shown in the sequence of Figures 12.30, 12.31, 12.32. At $a = 0$ the limit cycle is van der Pol cycle, which becomes distorted by the approaching saddle, and eventually disappears at $a \approx 0.31$.

12 : Bifurcations and manifolds   527

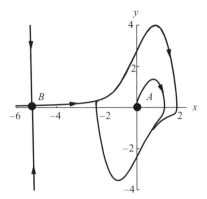

Figure 12.30 Problem 12.25: For $\delta = 2$, $a = 0.1$, the saddle is at B and the unstable spiral at A.

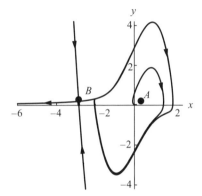

Figure 12.31 Problem 12.25: For $\delta = 2$, $a = 0.2$, the saddle is at B and the unstable spiral at A.

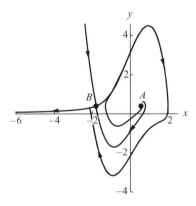

Figure 12.32 Problem 12.25: For $\delta = 2$, $a = 0.4$, the saddle is at B and the unstable spiral at A; the limit cycle has now disappeared.

- **12.26** Find the stable and unstable manifolds of the equilibrium points of
$$\dot{x} = x^2 + \mu, \quad \dot{y} = -y, \quad \dot{z} = z,$$
for $\mu < 0$. What type of bifurcation occurs at $\mu = 0$?

12.26. The equilibrium points of the system
$$\dot{x} = x^2 + \mu, \quad \dot{y} = -y, \quad \dot{z} = z$$
occur at $(\pm\sqrt{(-\mu)}, 0, 0)$. The solutions are
$$x = \sqrt{(-\mu)} \tanh[-\sqrt{(-\mu)}t + A], \quad y = Be^{-t}, \quad z = Ce^{t}.$$
For x with $A = 0$, as $t \to \infty$, $x \to -\sqrt{(-\mu)}$, and as $t \to -\infty$, $x \to \sqrt{(-\mu)}$.

- For $(\sqrt{(-\mu)}, 0, 0)$, the stable manifold is the straight line $x = \sqrt{(-\mu)}, z = 0$. The unstable manifold is the half-plane $y = 0$, $x > -\sqrt{(-\mu)}$.
- For $(-\sqrt{(-\mu)}, 0, 0)$, the unstable manifold is the straight line $x = \sqrt{(-\mu)}, y = 0$. The stable manifold is the half-plane $y = 0$, $x < \sqrt{(-\mu)}$.

- **12.27** Consider the system $\dot{x} = \mu x - y - x^3$, $\dot{y} = x + \mu y - y^3$. By putting $z = \mu - x^2$, show that any equilibrium points away from the origin are given by the solutions of $z^4 - \mu z^3 + \mu z + 1 = 0$. Plot the graph of μ against z and show that there is only one equilibrium point at the origin if $\mu < 2\sqrt{2}$, approximately, and nine equilibrium points if $\mu > 2\sqrt{2}$.
Investigate the linear approximation for the equilibrium point at the origin and show that the system has a Hopf bifurcation there at $\mu = 0$. Compute the phase diagram for $\mu = 1.5$.

12.27. The equilibrium points of
$$\dot{x} = \mu x - y - x^3, \quad \dot{y} = x + \mu y - y^3,$$
satisfy
$$\mu x - y - x^3 = 0, \quad x + \mu y - y^3 = 0.$$
Eliminate y so that
$$x + \mu(\mu x - x^3) - (\mu x - x^3)^3 = 0, \quad \text{or} \quad x[1 + \mu(\mu - x^2) - x^2(\mu - x^2)^3] = 0.$$
One solution is $x = 0$. For the others express the remaining equation in the form
$$1 + \mu(\mu - x^2) + (\mu - x^2)^4 - \mu(\mu - x^2)^3 = 0.$$

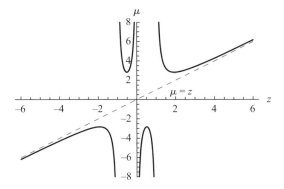

Figure 12.33 Problem 12.27: Graph of $\mu = (z^4 + 1)/(z(z^2 - 1))$; only solutions above the dashed line $\mu = z$ are of interest.

Let $z = \mu - x^2$, so that z satisfies

$$z^4 - \mu z^3 + \mu z + 1 = 0.$$

Therefore

$$\mu = \frac{z^4 + 1}{z(z^2 - 1)} = f(z),$$

say. Real solutions for x can only occur if $\mu > z$. The graph of μ against z is shown in Figure 12.33. Stationary values of $\mu = f(z)$ occur where $f'(z) = 0$, namely where

$$z^6 - 3z^4 - 3z^2 + 1 = (1 + z^2)(1 - 4z^2 + z^4) = 0.$$

Real solutions can only occur if $z^4 - 4z^2 + 1 = 0$, that is where

$$z^2 = 2 \pm \sqrt{3}.$$

To satisfy $\mu > z$ we must choose $z = \sqrt{[2 + \sqrt{3}]}$ and $z = -\sqrt{[2 - \sqrt{3}]}$. Both values of z give the same value for μ, namely $\mu = 2\sqrt{2}$. Therefore there is one equilibrium point if $\mu < 2\sqrt{2}$, 5 equilibrium points if $\mu = 2\sqrt{2}$, and 9 if $\mu > 2\sqrt{2}$.

Near the origin

$$\dot{x} \approx \mu x - y, \quad \dot{y} \approx x + \mu y.$$

In the usual notation

$$p = 2\mu, \quad q = \mu^2 + 1 > 0, \quad \Delta = p^2 - 4q = -4 < 0.$$

As μ increases through zero a stable spiral becomes an unstable spiral. Switch to polar coordinates. Then

$$r\dot{r} = x\dot{x} + y\dot{y} = \mu r^2 - (x^4 + y^4).$$

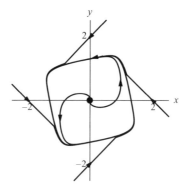

Figure 12.34 Problem 12.27: Limit cycle for $\dot{x} = \mu x - y - x^3$, $\dot{y} = x + \mu y - y^3$ with $\mu = 1.5$.

With $x = r\cos\theta$ and $y = r\sin\theta$,

$$x^4 + y^4 - \mu r^2 = r^4(\cos^4\theta + \sin^4\theta) - \mu r^2 \geq \tfrac{1}{2}r^4 - \mu r^2$$
$$= \tfrac{1}{2}r^2(r^2 - 2\mu) > 0$$

for $r^2 > 2\mu$. It follows that $\dot{r} < 0$ for r sufficiently large which means that the radial paths are decreasing if μ is positive. Therefore there must be at least one stable periodic solution generated at the origin at $\mu = 0$. For $\mu < 0$, $\dot{r} > 0$ on all paths. Hence this is an example of a Hopf bifurcation. A phase diagram for the system with $\mu = 1.5$ is shown in Figure 12.34. The limit cycle has been created by a Hopf bifurcation at $\mu = 0$.

- 12.28 Show that the system $\dot{x} = x^2 + y + z + 1$, $\dot{y} = z - xy$, $\dot{z} = x - 1$ has one equilibrium point at $(1, -1, -1)$. Determine the linear approximation $\dot{\mathbf{x}}' = \mathbf{A}\mathbf{x}'$ to the system at this point. Find the eigenvalues and eigenvectors of $\mathbf{A}$, and the equations of the stable and unstable manifolds E^s and E^c of the linear approximation.

12.28. The system
$$\dot{x} = x^2 + y + z + 1, \quad \dot{y} = z - xy, \quad \dot{z} = x - 1,$$
is in equilibrium where
$$x^2 + y + z + 1 = 0, \quad z = xy, \quad x = 1.$$

The only solution is $x = 1$, $y = -1$, $z = -1$. Let $x = 1 + x'$, $y = -1 + y'$, $z = -1 + z'$. Then the linearized matrix equation is $\dot{\mathbf{x}}' = \mathbf{A}\mathbf{x}'$, where

$$\mathbf{A} = \begin{bmatrix} 2 & 1 & 1 \\ 1 & -1 & 1 \\ 1 & 0 & 0 \end{bmatrix}.$$

The eigenvalues are given by

$$\begin{vmatrix} 2-\lambda & 1 & 1 \\ 1 & -1-\lambda & 1 \\ 1 & 0 & -\lambda \end{vmatrix} = 0, \text{ or } -(\lambda+1)(\lambda^2 - 2\lambda - 2) = 0.$$

Therefore the eigenvalues are $\lambda_1 = -1$, $\lambda_2 = 1 + \sqrt{3}$, $\lambda_3 = 1 - \sqrt{3}$, and the corresponding eigenvectors are

$$\mathbf{r}_1 = [-1, 2, 1]^T, \quad \mathbf{r}_2 = [1 - \sqrt{3}, 1, 1]^T, \quad \mathbf{r}_3 = [1 + \sqrt{3}, 1, 1]^T.$$

The general solution is

$$\begin{bmatrix} x' \\ y' \\ z' \end{bmatrix} = \alpha \begin{bmatrix} -1 \\ 2 \\ 1 \end{bmatrix} e^{-t} + \beta \begin{bmatrix} 1-\sqrt{3} \\ 1 \\ 1 \end{bmatrix} e^{(1-\sqrt{3})t} + \gamma \begin{bmatrix} 1+\sqrt{3} \\ 1 \\ 1 \end{bmatrix} e^{(1+\sqrt{3})t}.$$

The stable manifold is given by $\gamma = 0$, which is given parametrically by the equations

$$x' = -\alpha + (1 - \sqrt{3})\beta,$$
$$y' = 2\alpha + \beta,$$
$$z' = \alpha + \beta.$$

In terms of x, y, z the stable manifold is the plane

$$(x - 1) + (2 - \sqrt{3})(y + 1) + (2\sqrt{3} - 3)(z + 1) = 0.$$

The unstable manifold is defined by $\alpha = \beta = 0$, which defines the straight line

$$x' = x - 1 = (1 + \sqrt{3})s, \quad y' = y + 1 = s, \quad z' = z + 1 = s.$$

- **12.29** Consider the equation $\dot{z} = \lambda z - |z|^2 z$, where $z = x + iy$ is a complex variable, and $\lambda = \alpha + i\beta$ is a complex constant. Classify the equilibrium point at the origin, and show that the system has a Hopf bifurcation as α increases through zero for $\beta \neq 0$. How does the system behave if $\beta = 0$?

12.29. In the equation $\dot{z} = \lambda z - |z|^2 z$, let $z = x + iy$ and $\lambda = \alpha + i\beta$. The real and imaginary equations are

$$\dot{x} = \alpha x - \beta y - (x^2 + y^2)x,$$
$$\dot{y} = \beta x + \alpha y - (x^2 + y^2)y.$$

The system has one equilibrium point, at $(0,0)$. Near the origin

$$\dot{x} \approx \alpha x - \beta y, \quad \dot{y} \approx \beta x + \alpha y.$$

In the usual notation, the classification parameters are

$$p = 2\alpha, \quad q = \alpha^2 + \beta^2 > 0, \quad \Delta = p^2 - 4q = -4\beta^2 < 0.$$

As α increases through zero the origin changes from a stable spiral to an unstable spiral. By Theorem 12.1, a Hopf bifurcation occurs at $\mu = 0$.
In polar coordinates (r, θ),

$$\dot{r} = r(\alpha - r^2).$$

If $\alpha > 0$, the limit cycle is the circle of radius $\sqrt{\alpha}$.
If $\beta = 0$, the equations become

$$\dot{x} = (\alpha - r^2)x, \quad \dot{y} = (\alpha - r^2)y.$$

Equilibrium occurs at $(0,0)$ only if $\alpha \leq 0$, and at $(0,0)$ and all points on the circle $r = \sqrt{\alpha}$ if $\alpha > 0$. The phase paths are given by

$$\frac{dy}{dx} = \frac{y}{x},$$

which has the general solution $y = Cx$. The phase paths are radial lines through the origin.

13 Poincaré sequences, homoclinic bifurcation, and chaos

• **13.1** Obtain the solutions for the usual polar coordinates r and θ in terms of t, for the system
$$\dot{x} = x + y - x(x^2 + y^2), \quad \dot{y} = -x + y - y(x^2 + y^2).$$
Let Σ be the section $\theta = 0$, $r > 0$. Find the difference equation for the Poincaré sequence in this section.

13.1. In the equations
$$\dot{x} = x + y - x(x^2 + y^2), \quad \dot{y} = -x + y - y(x^2 + y^2),$$

let $x = r\cos\theta$, $y = r\sin\theta$. Then

$$\dot{r} = r(1 - r^2), \quad \dot{\theta} = \frac{x\dot{y} - \dot{x}y}{r^2} = -1.$$

Integration of the equations leads to

$$r = \frac{r_0}{\sqrt{[r_0^2 + (1 - r_0^2)e^{-2t}]}}, \quad \theta = -t + \theta_0,$$

where $r(0) = r_0$, $\theta(0) = \theta_0$. The system has an equilibrium point at the origin, which is an unstable spiral. The system also has a stable limit cycle given by circle $r = 1$.

The polar equations of the paths are given by

$$r = \frac{r_0}{\sqrt{[r_0^2 + (1 - r_0^2)e^{2(\theta - \theta_0)}]}}.$$

In the section Σ, $\theta_0 = 0$, and successive returns occur at $\theta = -2\pi, -4\pi, \ldots$. Denoting these radii by r_n, we have

$$r_n = \frac{r_0}{\sqrt{[r_0^2 + (1 - r_0^2)e^{-4\pi n}]}}.$$

As expected, as $n \to \infty$, $r_n \to 1$ irrespective of the initial value r_0.

- **13.2** Find the map of 2π first returns on the section $\Sigma : t = 0$ for $\ddot{x} + 2\dot{x} + 2x = 2\sin t$ in the usual phase plane. Find also the coordinates of the fixed point of the map and discuss its stability. Where is the fixed point of the map if the section is $t = \frac{1}{2}\pi$?

13.2. The linear equation
$$\ddot{x} + 2\dot{x} + 2x = 2\sin t,$$
has the general solution
$$x = (A\cos t + B\sin t)e^{-t} - \tfrac{4}{5}\cos t + \tfrac{2}{5}\sin t. \tag{i}$$

It follows that
$$\dot{x} = [-(A-B)\cos t + (A+B)\sin t]e^{-t} + \tfrac{2}{5}\cos t + \tfrac{4}{5}\sin t. \tag{ii}$$

Equations (i) and (ii) give represent points parametrically in the phase plane $x, \dot{x} = y$. At $t = 0$,
$$x(0) = x_0 = A - \tfrac{4}{5}, \quad y(0) = y_0 = -A + B + \tfrac{2}{5}.$$

The first return is given by
$$x(2\pi) = x_1 = Ae^{-2\pi} - \tfrac{4}{5}, \quad y(2\pi) = y_1 = (-A+B)e^{-2\pi} + \tfrac{2}{5}.$$

Elimination of A and B leads to
$$x_1 = (x_0 + \tfrac{4}{5})e^{-2\pi} - \tfrac{4}{5}, \quad y_1 = (y_0 + A - B)e^{-2\pi} + \tfrac{2}{5}.$$

The fixed point of the map occurs where $x_1 = x_0$ and $y_1 = y_0$ which results in $A = B = 0$ leading to the fixed point $(-\tfrac{4}{5}, \tfrac{2}{5})$.

If the section is $t = \tfrac{1}{2}\pi$, then from (i) and (ii) again
$$x(\tfrac{1}{2}\pi) = u_0 = Be^{-(1/2)\pi} + \tfrac{2}{5}, \quad y(\tfrac{1}{2}\pi) = v_0 = (A+B)e^{-(1/2)\pi} + \tfrac{4}{5}.$$

The first return is given by
$$x(\tfrac{5}{2}\pi) = u_1 = Be^{-(5/2)\pi} + \tfrac{2}{5}, \quad y(\tfrac{5}{2}\pi) = v_1 = (A+B)e^{-(5/2)\pi} + \tfrac{4}{5}.$$

Then $u_0 = u_1$ and $v_0 = v_1$ if $A = B = 0$ which leads to the fixed point $(\tfrac{2}{5}, \tfrac{4}{5})$.

Alternatively for this linear equation the fixed points can be found by simply eliminating the exponential terms in the solution by putting $A = B = 0$. All fixed points for any section can then be read off.

- 13.3 Let x_1 satisfy $\ddot{x}_1 + \frac{1}{4}\omega^2 x_1 = \Gamma \cos \omega t$. Obtain the solutions for x_1 and $x_2 = \dot{x}_1$ given that $x_1(0) = x_{10}$ and $x_2(0) = x_{20}$. Let Σ be the section $t = 0$ and find the first returns of period $2\pi/\omega$. Show that the mapping is
$$P_\Sigma(x_{10}, x_{20}) = \left(-x_{10} - \frac{8\Gamma}{3\omega^2}, -x_{20}\right), \text{ and that } P_\Sigma^2(x_{10}, x_{20}) = (x_{10}, x_{20}).$$
Deduce that the system exhibits period doubling for all initial values except one. Find the coordinates of this fixed point.

13.3. The equation
$$\ddot{x}_1 + \tfrac{1}{4}\omega^2 x_1 = \Gamma \cos \omega t,$$
has the general solution
$$x_1 = A \cos \tfrac{1}{2}\omega t + B \sin \tfrac{1}{2}\omega t - \frac{4\Gamma}{3\omega^2} \cos \omega t. \tag{i}$$

It follows that
$$x_2 = \dot{x}_1 = -\tfrac{1}{2}\omega A \sin \tfrac{1}{2}\omega t + \tfrac{1}{2}\omega B \cos \omega t + \frac{4\Gamma}{3\omega} \sin \omega t. \tag{ii}$$

From the given initial conditions
$$A = x_{10} + \frac{4\Gamma}{3\omega^2}, \quad B = \frac{2x_{20}}{\omega},$$

so that
$$x_1 = \left(x_{10} + \frac{4\Gamma}{3\omega^2}\right) \cos \tfrac{1}{2}\omega t + \frac{2x_{20}}{\omega} \sin \tfrac{1}{2}\omega t - \frac{4\Gamma}{3\omega^2} \cos \omega t,$$

$$x_2 = -\frac{1}{2}\omega \left(x_{10} + \frac{4\Gamma}{3\omega^2}\right) \sin \tfrac{1}{2}\omega t + x_{20} \cos \tfrac{1}{2}\omega t + \frac{4\Gamma}{3\omega} \sin \omega t.$$

At $t = 2\pi/\omega$,
$$x_1(2\pi/\omega) = x_{11} = -x_{10} - \frac{8\Gamma}{3\omega^2}, \quad x_2(2\pi/\omega) = x_{21} = -x_{20}.$$

Therefore
$$P_\Sigma(x_{10}, x_{20}) = (x_{11}, x_{21}) = \left(-x_{10} - \frac{8\Gamma}{3\omega^2}, -x_{20}\right).$$

Since (i) and (ii) are both of period $4\pi/\omega$, it follows that the mapping P_Σ shows period doubling, that is,
$$P_\Sigma^2(x_{10}, x_{2,0}) = P_\Sigma(x_{11}, x_{21}) = (x_{10}, x_{20}).$$

The exception occurs if

$$x_{10} = -x_{10} - \frac{8\Gamma}{3\omega^2}, \text{ and } x_{20} = -x_{20},$$

that is, if

$$x_{10} = -\frac{4\Gamma}{3\omega^2}, \quad x_{20} = 0,$$

which is the fixed point of this mapping.

- 13.4 (a) Let $\dot{x} = y$, $\dot{y} = -3y - 2x + 10\cos t$ and assume the initial conditions $x(0) = 4$, $y(0) = -1$. Consider the associated three-dimensional system with $\dot{z} = 1$. Assuming that $z(0) = 0$, plot the solution in the (x, y, z) space and indicate the 2π periodic returns which occur at $t = 0$, $t = 2\pi$, $t = 4\pi$, ...
  (b) Sketch some typical period-1 Poincaré maps in the (x, y, z) space for $\dot{x} = \lambda x$, $\dot{y} = \lambda y$, $\dot{z} = 1$ for each of the cases $\lambda < 0$, $\lambda = 0$, $\lambda > 0$. Discuss the nature of any fixed points in each case. Assume that $x(0) = x_0$, $y(0) = y_0$, $z(0) = 0$, and show that
$$x_{n+1} = e^\lambda x_n, \quad y_{n+1} = e^\lambda y_n, \quad n = 0, 1, 2, \ldots.$$

13.4. The system

$$\dot{x} = y, \quad \dot{y} = -3y - 2x + 10\cos t, \quad \dot{z} = 1,$$

is used as an example of the three-dimensional representation of the first returns. A particular solution is shown in Figure 13.1 with initial conditions $x(0) = 4$, $y(0) = -1$ and $T = 2\pi$. The solution is

$$x = 2e^{-t} + e^{-2t} + 3\sin t + \cos t.$$

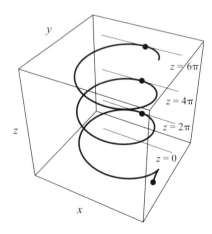

Figure 13.1 Problem 13.4: The curve shows a phase path for the system $\dot{x} = y$, $\dot{y} = -3y - 2x + 10\cos t$, $\dot{z} = 1$, with initial conditions $x(0) = 4$, $y(0) = -1$, $z(0) = 0$.

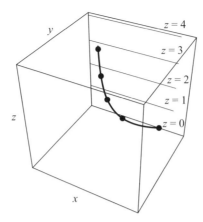

Figure 13.2 Problem 13.4: Returns for the solution $x = 3e^{-t}$, $y = 3e^{-t}$, $z = t$.

The dots in the figure are the points of intersection of the path with the planes $z = 0$, $2\pi, 4\pi, 6\pi, \ldots$. The fixed point lies on the line $x = 1$, $y = 3$.

The system

$$\dot{x} = \lambda x, \quad \dot{y} = \lambda y, \quad \dot{z} = 1$$

has the general solution

$$x = Ae^{\lambda t}, \quad y = Be^{\lambda t}, \quad z = t + C.$$

The fixed point lies on the line $x = y = 0$ in the x, y, z space. If $\lambda < 0$, the fixed point $x = y = 0$ is stable since all returns approach it as $t \to \infty$. Returns for the solution with initial values $x(0) = 3$, $y(0) = 3$, $z(0) = 0$ are shown in Figure 13.2.

If $\lambda = 0$, then the general solution is $x = A$, $y = B$, $z = t + C$. Every point is a fixed point. Paths in the x, y, z space are all straight lines parallel to the z axis.

If $\lambda > 0$, the fixed points still lie on the line $x = y = 0$ but in this case the fixed point is unstable.

If $x(n) = x_n$ and $y(n) = y_n$, then

$$x_{n+1} = Ae^{\lambda(n+1)} = e^{\lambda} x_n, \quad y_{n+1} = Be^{\lambda(n+1)} = e^{\lambda} y_n.$$

- **13.5** Two rings can slide on two fixed horizontal wires which lie in the same vertical plane with separation a. The two rings are connected by a spring of unstretched length l and stiffness μ. The upper ring is forced to move with displacement $\phi(t)$ from a fixed point O as shown below or in Figure 13.41 (in NODE). The resistance on the lower ring which has mass m is assumed to be $mk \times$ speed. Let y be the relative displacement between the rings. Show that the equation of motion of the lower ring is given by

$$\ddot{y} + k\dot{y} - \frac{\mu}{ma}(l-a)y + \frac{\mu l}{2ma^3} y^3 = -\ddot{\phi} - k\dot{\phi}.$$

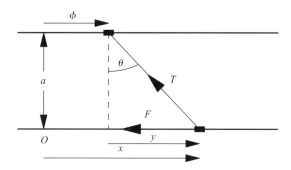

Figure 13.3 Problem 13.5: Forced spring-loaded pendulum between fixed horizontal slides.

13.5. Figure 13.3 shows the constraints on the lower bob. Let T be the tension in the spring, and F the frictional force with directions as shown in the figure. If x is measured from the fixed origin O, then the horizontal equation of motion for the bob is

$$-T \sin\theta - F = m\ddot{x}, \qquad (i)$$

where

$$x = \phi + y, \quad \sin\theta = \frac{y}{\sqrt{(y^2 + a^2)}}.$$

Assuming Hooke's law, the tension in the spring is given by

$$T = \mu[\sqrt{(y^2 + a^2)} - l],$$

whilst the frictional force has magnitude $mk|\dot{x}|$, and opposes the direction of motion. Elimination of T, F, θ and x in (i) leads to

$$-\mu y + \frac{l\mu y}{\sqrt{(y^2 + a^2)}} - mk(\dot{\phi} + \dot{y}) = m(\ddot{\phi} + \ddot{y}), \qquad (ii)$$

which is the exact equation for the motion of the bob.
Apply the binomial approximation

$$(a^2 + y^2)^{-1/2} \approx \frac{1}{a}\left(1 - \frac{y^2}{2a^2}\right)$$

to (ii) assuming that $|y|$ is small. The result is the Duffing equation

$$\ddot{y} + k\dot{y} - \frac{\mu}{ma}(l-a)y + \frac{\mu l}{2ma^3}y^3 = -\ddot{\phi} - k\dot{\phi}.$$

The standard equation follows if we put

$$-\ddot{\phi} - k\dot{\phi} = \Gamma \cos\omega t.$$

- 13.6 Search for period doubling in the undamped Duffing equation $\ddot{x} - x + x^3 = \Gamma \cos \omega t$ using the form $x = c + a_1 \cos \omega t + a_2 \cos \tfrac{1}{2}\omega t$, where c, a_1 and a_2 are constants. If frequencies $\tfrac{3}{2}\omega$ and above are neglected, show that the shift and amplitudes satisfy

$$c[-1 + c^2 + \tfrac{3}{2}(a_1^2 + a_2^2)] + \tfrac{3}{4}a_2^2 a_1 = 0,$$

$$a_1(-\omega^2 - 1 + 3c^2 + \tfrac{3}{4}a_1^2 + \tfrac{3}{2}a_2^2) + \tfrac{3}{2}a_2^2 c = \Gamma,$$

$$a_2(-\tfrac{1}{4}\omega^2 - 1 + 3c^2 + \tfrac{3}{2}a_1^2 + 3ca_1 + \tfrac{3}{4}a_2^2) = 0.$$

Deduce that for harmonic solutions ($a_2 = 0$), c and a_1 are given by solutions of

(i) $c = 0$, $a_1(-\omega^2 - 1 + \tfrac{3}{4}a_1^2) = \Gamma$,

or

(ii) $c^2 = 1 - \tfrac{3}{2}a_1^2$, $a_1(-\omega^2 + 2 - \tfrac{15}{4}a_1^2) = \Gamma$.

Sketch the amplitude $|a_1|$/amplitude $|\Gamma|$ curves corresponding to NODE, Figure 13.13 for $\omega = 1.2$.

13.6. In the equation
$$\ddot{x} - x + x^3 = \Gamma \cos \omega t,$$
let $x = c + a_1 \cos \omega t + a_2 \cos \tfrac{1}{2}\omega t$. Then

$$\dot{x} = -a_1 \omega \sin \omega t - \tfrac{1}{2}a_2 \omega \sin \tfrac{1}{2}\omega t, \quad \ddot{x} = -a_1 \omega^2 \cos \omega t - \tfrac{1}{4}a_2 \omega^2 \cos \tfrac{1}{2}\omega t.$$

We also require the following expansion (computer algebra was used here)

$$x^3 = (c + a_1 \cos \omega t + a_2 \cos \tfrac{1}{2}\omega t)^3$$
$$= \tfrac{1}{4}[3a_1 a_2^2 + 6c(a_1^2 + a_2^2) + 4c^3] + \tfrac{1}{4}[a_1^3 + 2a_1 a_2^2 + 2a_2^2 c + 4a_1 c^2]\cos \omega t$$
$$+ \tfrac{3}{4}[2a_1^2 a_2 + a_2^3 + 4a_1 a_2 c + 4a_2 c^2]\cos \tfrac{1}{2}\omega t + \text{higher harmonics}.$$

The constant term and the coefficients of $\cos \omega t$ and $\cos \tfrac{1}{2}\omega t$ are zero if

$$-c + c^3 + \tfrac{3}{2}c(a_1^2 + a_2^2) + \tfrac{3}{4}a_1 a_2^2 = 0, \qquad \text{(i)}$$

$$-a_1 \omega^2 - a_1 + 3c^2 a_1 + \tfrac{3}{4}a_1^3 + \tfrac{3}{2}a_1 a_2^2 + \tfrac{3}{2}a_2^2 c = \Gamma, \qquad \text{(ii)}$$

$$a_2[-\tfrac{1}{4}\omega^2 - 1 + 3c^2 + \tfrac{3}{2}a_1^2 + 3a_1 c + \tfrac{3}{4}a_2^2] = 0. \qquad \text{(iii)}$$

From (iii), one solution is $a_2 = 0$, in which case (i) and (ii) become

$$c[-1 + c^2 + \tfrac{3}{2}a_1^2] = 0,$$

$$a_1[-\omega^2 - 1 + 3c^2 + \tfrac{3}{4}a_1^2] = \Gamma.$$

Hence

(a) $c = 0$, $a_1[-\omega^2 - 1 + \frac{3}{4}a_1^2] = \Gamma$,

or

(b) $c^2 = 1 - \frac{3}{2}a_1^2$, $a_1[-\omega^2 + 2 - \frac{15}{4}a_1^2] = \Gamma$. This is the case in which no subharmonic is present which was investigated in NODE, Chapter 7.

The other solution in (iii) is

$$-\tfrac{1}{4}\omega^2 - 1 + 3c^2 + \tfrac{3}{2}a_1^2 + 3a_1c + \tfrac{3}{4}a_2^2 = 0. \tag{iv}$$

It is possible to eliminate a_2^2 between (i), (ii) and (iii) to obtain two equations relating Γ and a_1 implicitly. However from (iv),

$$a_2^2 = \tfrac{1}{3}[\omega^2 + 4 - 12c^2 - 6a_1^2 - 12a_1c], \tag{v}$$

and a subharmonic will emerge where the right-hand side is zero. Again there will two cases as in (a) and (b) above.

(c) $c = 0$, so that from (v) $a_1^2 = \tfrac{1}{6}(\omega^2 + 4)$, (must have $a_1^2 < \tfrac{1}{6}(\omega^2 + 4)$ for a_2 to be real),

or

(d) $c^2 = 1 - \tfrac{3}{2}a_1^2$, so that from (v)

$$\tfrac{1}{4}\omega^2 + 1 - 3 + \tfrac{9}{2}a_1^2 - \tfrac{3}{2}a_1^2 - 3a_1\sqrt{(1 - \tfrac{3}{2}a_1^2)} = 0.$$

(again the right-hand side must be positive for real a_2). Rearranging and squaring, we have

$$(\tfrac{1}{4}\omega^2 - 2 + 3a_1^2)^2 = 9a_1^2(1 - \tfrac{3}{2}a_1^2),$$

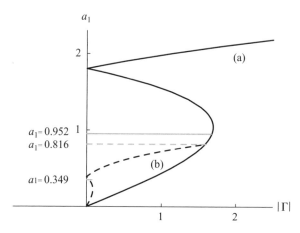

Figure 13.4 Problem 13.6: Amplitude–amplitude curves for $c = 0$ (the solid curve) and for $c^2 = 1 - \tfrac{3}{2}a_1^2$ (the dashed curve) for $\omega = 1.2$.

or
$$\tfrac{45}{2}a_1^4 + (\tfrac{3}{2}\omega^2 - 27) + (\tfrac{1}{4}\omega^2 - 2)^2 = 0, \qquad \text{(vi)}$$
from which it is possible to find a_1.

In our example $\omega = 1.2$. From (c), $a_2 \geq 0$ where $|a_1| \leq 0.9522$ on curve (a) in Figure 13.4. From (d), the solutions of (vi) are $|a_1| = 0.9911$ and $|a_1| = 0.3488$. Real solutions for a_2 lie between these values on curve (b) in Figure 13.4.

- 13.7 Design a computer program to plot $2\pi/\omega$ first returns for the system $\dot{x} = X(x, y, t)$, $\dot{y} = Y(x, y, t)$ where $X(x, y, t)$ and $Y(x, y, t)$ are $2\pi/\omega$-periodic functions of t. Apply the program to the system
$$X(x, y, t) = y, \ Y(x, y, t) = -ky + x - x^3 + \Gamma \cos \omega t,$$
for $k = 0.3$, $\omega = 1.2$ and Γ taking a selection of values between 0 and 0.8. Let the initial section be $t = 0$.

13.7. The first returns have been computed using a *Mathematica* program. Consider the Duffing oscillator
$$\dot{x} = y, \quad \dot{y} = -ky + x - x^3 + \Gamma \cos \omega t.$$

- Parameter values $k = 0.3$, $\omega = 1$, $\Gamma = 0.2$ The first returns starting from $x(0) = 0.9$, $y(0) = 0.8$ are shown in Figure 13.5. The returns approach a fixed point at P which indicates a stable periodic solution.
- Parameter values $k = 0.3$, $\omega = 1$, $\Gamma = 0.28$. The returns starting from $x(0) = 0.5$, $y(0) = 0.4$ but delayed by 15 steps are shown in Figure 13.6. The returns oscillate between two points indicating period doubling.
- Parameters $k = 0.3$, $\omega = 1$, $\Gamma = 0.4$. The returns start from $x(0) = 0.5$, $y(0) = 0.4$ but are delayed by 10 steps to eliminate transience. The returns are shown in Figure 13.7 and indicate a strange attractor.

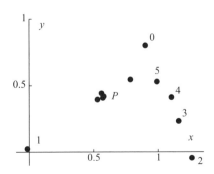

Figure 13.5 Problem 13.7(i): Poincaré section of period $2\pi/\omega$ with $k = 0.3$, $\omega = 1$, $\Gamma = 0.2$ and initial values $x(0) = 0.9$, $y(0) = 0.8$; the successive returns are labelled '0','1','2', ... and approach the fixed point at P.

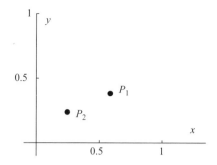

Figure 13.6 Problem 13.7(i): Poincaré section of period $2\pi/\omega$ with $k = 0.3$, $\omega = 1$, $\Gamma = 0.28$ and initial values $x(0) = 0.5$, $y(0) = 0.4$; only returns after 15 steps are shown which reveals period doubling between the points P_1 and P_2.

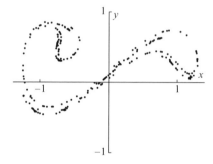

Figure 13.7 Problem 13.1(i): Poincaré section of period $2\pi/\omega$ with $k = 0.3$, $\omega = 1$, $\Gamma = 0.4$ and initial values $x(0) = 0.5$, $y(0) = 0.4$: the section contains 300 returns indicating a strange attractor.

- **13.8** Find the equations of the stable and unstable manifolds in the (x, y)-plane of
$$\ddot{x} + \dot{x} - 2x = 10 \cos t, \quad \dot{x} = y$$
for Poincaré maps of period 2π and initial time $t = 0$.

13.8. The general solution of
$$\ddot{x} + \dot{x} - 2x = 10 \cos t$$
is
$$x = A e^{-2t} + B e^{t} - 3 \cos t + \sin t.$$

The fixed point of the system for sections $t = 0$, period 2π is at $(-3, 1)$ in the (x, y) plane. The stable manifold consists of the set of all points for which $B = 0$ and $t = 2n\pi$, $(n = 0, 1, 2, 3, \ldots)$, that is,
$$x = A e^{-4n\pi} - 3, \quad y = -2A e^{-4n\pi} + 1.$$

Elimination of A gives the stable manifold as the line $2x + y + 5 = 0$.

The unstable manifold is given by the set of points for which $B = 0$, that is

$$x = Be^{2n\pi} - 3, \quad y = Be^{2n\pi} + 1.$$

Elimination of B gives the unstable manifold as the line $x - y + 4 = 0$.

- 13.9 Apply Melnikov's method to
$$\ddot{x} + \varepsilon\kappa\dot{x} + x^3 = \varepsilon\gamma(1 - x^2)\cos\omega t, \quad \kappa > 0, \ \varepsilon > 0, \ \gamma > 0,$$
and show that homoclinic bifurcation occurs if, for $\omega^2 \ll 2$,
$$|\gamma| \geq \frac{2\sqrt{2}\kappa}{\pi\omega(2 - \omega^2)}\cosh(\tfrac{1}{2}\omega\pi).$$

13.9. The perturbed system is

$$\ddot{x} + \varepsilon\kappa\dot{x} - x + x^3 = \varepsilon\gamma(1 - x^2)\cos\omega t.$$

For $\varepsilon = 0$, the equation $\ddot{x} - x + x^3 = 0$ has the homoclinic solutions $x_0 = \pm\sqrt{2}\,\mathrm{sech}\,t$. We consider the solution for which $x > 0$. By NODE, (13.53), the Melnikov function is

$$M(t_0) = \int_{-\infty}^{\infty} \dot{x}_0(t - t_0)[-\kappa\dot{x}_0(t - t_0) + \gamma\{1 - x_0^2(t - t_0)\}]\cos\omega t\,dt$$

$$= \sqrt{2}\gamma\sin\omega t_0 \int_{-\infty}^{\infty} \mathrm{sech}\,t\tanh t\sin\omega t\,dt - 2\kappa\int_{-\infty}^{\infty}\mathrm{sech}^2 t\tanh^2 t\,dt$$

$$- 2\sqrt{2}\gamma\sin\omega t_0\int_{-\infty}^{\infty}\mathrm{sech}^3 t\tanh t\sin\omega t\,dt$$

$$= \sqrt{2}\gamma\omega\sin\omega t_0\int_{-\infty}^{\infty}\mathrm{sech}\,t\cos\omega t\,dt - 2\kappa\int_{-\infty}^{\infty}\mathrm{sech}^2 t\tanh^2 t\,dt$$

$$- \frac{2}{3}\sqrt{2}\gamma\omega\sin\omega t_0\int_{-\infty}^{\infty}\mathrm{sech}^3 t\cos\omega t\,dt.$$

where we have integrated by parts, and eliminated integrals of odd functions. Now use the known definite integrals

$$\int_{-\infty}^{\infty}\mathrm{sech}^2 t\tanh^2 t\,dt = \tfrac{2}{3};$$

$$\int_{-\infty}^{\infty}\mathrm{sech}\,t\cos\omega t\,dt = \pi\,\mathrm{sech}\,(\tfrac{1}{2}\omega\pi);$$

$$\int_{-\infty}^{\infty}\mathrm{sech}^3 t\cos\omega t\,dt = \tfrac{1}{2}\pi(1 + \omega^2)\mathrm{sech}\,(\tfrac{1}{2}\omega\pi).$$

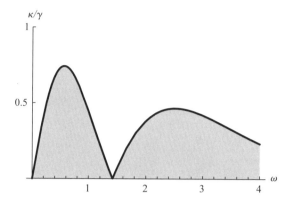

Figure 13.8 Problem 13.9: The shaded region indicates possible homoclinic bifurcation.

Finally

$$M(t_0) = \tfrac{1}{3}[\sqrt{2}\gamma\omega\pi(2-\omega^2)\mathrm{sech}(\tfrac{1}{2}\omega\pi)\sin\omega t_0 - 4\kappa]. \qquad (i)$$

The Melnikov function for $x = -\sqrt{2}\mathrm{sech}\, t$ is given by (i) with γ replaced by $-\gamma$. Homoclinic bifurcation occurs where the Melnikov function vanishes, namely where

$$\sqrt{2}\gamma\omega\pi(2-\omega^2)\mathrm{sech}(\tfrac{1}{2}\omega\pi)\sin\omega t_0 = 4\kappa,$$

or

$$\sin\omega t_0 = \frac{2\sqrt{2}\kappa}{\pi\omega\gamma(2-\omega^2)}\cosh(\tfrac{1}{2}\omega\pi), \quad (\omega^2 \neq 2).$$

It follows that homoclinic bifurcation can only occur if

$$\gamma \geq \frac{2\sqrt{2}\kappa}{\pi\omega|2-\omega^2|}\cosh(\tfrac{1}{2}\omega\pi), \quad (\omega^2 \neq 2).$$

The graph of κ/γ against ω is shown in Figure 13.8: the shaded regions indicate possible homoclinic bifurcation.

- **13.10** The Duffing oscillator with equation $\ddot{x} + \varepsilon\kappa\dot{x} - x + x^3 = \varepsilon f(t)$, is driven by an even T-periodic function $f(t)$ with mean value zero. Assuming that $f(t)$ can be represented by the Fourier series

$$\sum_{n=1}^{\infty} a_n \cos n\omega t, \quad \omega = \frac{2\pi}{T},$$

find the Melnikov function for the oscillator.

Let
$$f(t) = \begin{cases} \gamma & -\frac{1}{2} < t < \frac{1}{2} \\ -\gamma & \frac{1}{2} < t < \frac{3}{2} \end{cases},$$

where $f(t)$ is a function of period 2. Show that the Melnikov function vanishes if

$$\frac{\kappa}{\gamma} = -\frac{3\pi}{2\sqrt{2}} \sum_{r=1}^{\infty} (-1)^r \operatorname{sech}[\tfrac{1}{2}\pi^2(2r-1)] \sin[(2r-1)\pi t_0].$$

Plot the Fourier series as a function of t_0 for $0 \leq t_0 \leq 2$, and estimate the value of κ/γ at which homoclinic tangency occurs.

13.10. In the equation

$$\ddot{x} + \varepsilon\kappa\dot{x} - x + x^3 = \varepsilon f(t),$$

$f(t)$ is an even T-periodic function with zero mean. For $\varepsilon = 0$, the homoclinic solutions $x_0 = \pm\sqrt{2}\operatorname{sech} t$. We consider the solution for which $x > 0$. The forcing term is

$$f(t) = \sum_{n=1}^{\infty} a_n \cos n\omega t, \quad \omega = 2\pi/T.$$

Elimination of odd integrands, leads to the Melnikov function in the form

$$M(t_0) = \int_{-\infty}^{\infty} \dot{x}_0[f(s+t_0) - \kappa\dot{x}_0]ds$$

$$= \sqrt{2}\sum_{n=1}^{\infty} a_n \int_{-\infty}^{\infty} \operatorname{sech} s \tanh s \sin n\omega s\, ds - 2\kappa \int_{-\infty}^{\infty} \operatorname{sech}^2 s \tanh^2 s\, ds$$

$$= \sqrt{2}\pi \sum_{n=1}^{\infty} a_n n\omega \operatorname{sech}(\tfrac{1}{2}\pi n\omega) \sin n\omega t_0 - \tfrac{4}{3}\kappa.$$

Homoclinic bifurcation occurs if the equation

$$\frac{2\sqrt{2}\kappa}{3\pi} = \sum_{n=1}^{\infty} a_n n\omega \operatorname{sech}(\tfrac{1}{2}\pi n\omega) \sin n\omega t_0 \tag{i}$$

can be solved for t_0.
Consider the forcing function

$$f(t) = \begin{cases} \gamma & -\frac{1}{2} < t < \frac{1}{2} \\ -\gamma & \frac{1}{2} < t < \frac{3}{2} \end{cases},$$

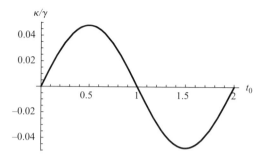

Figure 13.9 Problem 13.10:

where $T = 2$ so that $\omega = \pi$. The Fourier coefficients of $f(t)$ are

$$a_1 = \frac{\gamma}{\pi}, \quad a_2 = 0, \quad a_3 = -\frac{\gamma}{3\pi}, \quad a_4 = 0, \quad a_5 = \frac{\gamma}{5\pi}, \ldots, \qquad (ii)$$

so that

$$f(t) = -\frac{\gamma}{\pi} \sum_{r=1}^{\infty} \frac{(-1)^r}{2r-1} \cos(2r-1)\pi t.$$

Substitution of the Fourier coefficients given by (ii) into (i) gives the condition

$$\frac{\kappa}{\gamma} = q(t_0) = -\frac{3\pi}{2\sqrt{2}} \sum_{r=1}^{\infty} (-1)^r \operatorname{sech}[\tfrac{1}{2}\pi^2(2r-1)] \sin(2r-1)\pi t_0 \qquad (iii)$$

for the onset of homoclinic bifurcation. The graph of κ/γ against t_0 is shown in Figure 13.9. The first term dominates in the series for $q(t_0)$ which accounts for the curve being very close to a sine curve. Hence

$$\frac{\kappa}{\gamma} \approx q(0.5) \approx 0.048.$$

- **13.11** Melnikov's method can be applied also to autonomous systems. The manifolds become the separatrices of a saddle. Let
$$\ddot{x} + \varepsilon \kappa \dot{x} - \varepsilon x^2 \dot{x} + x^3 = 0.$$
Show that the homoclinic path exists to order $O(\varepsilon^2)$ if $\kappa = \tfrac{4}{5}\alpha$. [The following integrals are required:
$$\int_{-\infty}^{\infty} \operatorname{sech}^4 s\, ds = \frac{4}{3}; \quad \int_{-\infty}^{\infty} \operatorname{sech}^6 s\, ds = \frac{16}{15}.$$

13.11. Consider the autonomous equation

$$\ddot{x} + \varepsilon \kappa \dot{x} - \varepsilon \alpha x^2 \dot{x} - x + x^3 = 0.$$

The Melnikov function is given by

$$M(t_0) = \int_{-\infty}^{\infty} \dot{x}_0(t)(\alpha x_0^2 \dot{x}_0 - \kappa \dot{x}_0)dt,$$

where, for $\varepsilon = 0$, $x_0 = \sqrt{2} \operatorname{sech} t$. Therefore

$$M(t_0) = 2\int_{-\infty}^{\infty} \operatorname{sech}^2 t \tanh^2 t (2\alpha \operatorname{sech}^2 t - \kappa) dt$$

$$= 4\alpha \int_{-\infty}^{\infty} \operatorname{sech}^4 t \tanh^2 t\, dt - 2\kappa \int_{-\infty}^{\infty} \operatorname{sech}^2 t \tanh^2 t\, dt$$

$$= 4\alpha \int_{-\infty}^{\infty} \operatorname{sech}^4 t\, dt - 4\alpha \int_{-\infty}^{\infty} \operatorname{sech}^6 t\, dt - \frac{4}{3}\kappa$$

$$= \frac{16\alpha}{3} - \frac{64\alpha}{15} - \frac{4\kappa}{3} = \frac{16\alpha}{15} - \frac{4\kappa}{3}$$

Hence $M(t_0) = 0$ where $\kappa = 4\alpha/5$.

- **13.12** Show that $x = 3^{1/4}\sqrt{(\operatorname{sech} 2t)}$ is a homoclinic solution of $\ddot{x} + \varepsilon(\kappa - \alpha x^2)\dot{x} - x + x^5 = 0$ when $\varepsilon = 0$. Use Melnikov's method to show that homoclinic bifurcation occurs when $\kappa = 4\sqrt{3\alpha}/(3\pi)$.

13.12. The given equation is

$$\ddot{x} + \varepsilon(\kappa - \alpha x^2)\dot{x} - x + x^5 = 0.$$

Let $x_0 = 3^{1/4}\sqrt{(\operatorname{sech} 2t)}$. Then

$$\ddot{x}_0 - x_0 + x_0^5 = 3^{1/4}[-2\sqrt{(\operatorname{sech} 2t)} + 3\operatorname{sech}^{5/2} 2t \sinh^2 2t] - 3^{1/4}\sqrt{(\operatorname{sech} 2t)} + 3^{5/4}(\operatorname{sech} 2t)^{5/2}$$
$$= 0,$$

which confirms that $x_0(t)$ is a homoclinic solution of the unperturbed system. The Melnikov function is given by

$$M(t_0) = \int_{-\infty}^{\infty} \dot{x}_0(\alpha x_0^2 \dot{x}_0 - \kappa \dot{x}_0) dt.$$

Hence

$$M(t_0) = \sqrt{3} \int_{-\infty}^{\infty} \text{sech}^3 2t \sinh^2 2t [\alpha\sqrt{3}\text{sech } 2t - \kappa] dt$$

$$= 3\alpha \int_{-\infty}^{\infty} \text{sech}^4 2t \sinh^2 2t \, dt - \kappa\sqrt{3} \int_{-\infty}^{\infty} \text{sech}^3 2t \sinh^2 2t \, dt$$

$$= 3\alpha \int_{-\infty}^{\infty} [\text{sech}^2 2t - \text{sech}^4 2t] dt - \kappa\sqrt{3} \int_{-\infty}^{\infty} [\text{sech } 2t - \text{sech}^3 2t] dt$$

$$= 3\alpha(1 - \tfrac{2}{3}) - \kappa\sqrt{3}(\tfrac{1}{2}\pi - \tfrac{1}{4}\pi)$$

$$= \alpha - \tfrac{1}{4}\kappa\pi\sqrt{3}.$$

Therefore a homoclinic bifurcation first occurs where $\kappa \approx 4\sqrt{3}\alpha/(3\pi)$.

- **13.13** Apply Melnikov's method to the perturbed system $\ddot{x} + \varepsilon\kappa\dot{x} - x + x^3 = \varepsilon\gamma x \cos\omega t$, which has an equilibrium point at $x = 0$ for all t. Show that the manifolds of the origin intersect if

$$\gamma \geq \frac{4\kappa}{3\omega^2\pi}\sinh(\tfrac{1}{2}\omega\pi).$$

$$\left[\text{Hint: } \int_{-\infty}^{\infty} \text{sech}^2 u \cos\omega u \, du = \frac{\pi\omega}{\sinh(\tfrac{1}{2}\omega\pi)}.\right]$$

13.13. In the Duffing type oscillator

$$\ddot{x} + \varepsilon\kappa\dot{x} - x + x^3 = \varepsilon\gamma x \cos\omega t, \quad \dot{x} = y$$

the forcing term depends also on x. In the solution of the problem it is assumed that $\varepsilon > 0$, $\kappa > 0$ and $\gamma > 0$. The unperturbed solution is the familiar $x_0 = \sqrt{2}\text{sech } t$. The Melnikov function is given by

$$M(t_0) = \int_{-\infty}^{\infty} y_0(t - t_0) h(x_0(t - t_0), y_0(t - t_0), t) dt$$

$$= \int_{-\infty}^{\infty} y_0(s) h(x_0(s), y_0(s), s + t_0) ds,$$

after a change of variable, where $h(x, y, t) = -\kappa y + \gamma x \cos\omega t$. Therefore

$$M(t_0) = -\kappa \int_{-\infty}^{\infty} y_0^2(s) ds + \gamma \int_{-\infty}^{\infty} y_0(s) x_0(s) \cos(\omega s + t_0) ds$$

$$= -\kappa \int_{-\infty}^{\infty} y_0^2(s) ds - \gamma \int_{-\infty}^{\infty} y_0(s) x_0(s) \sin\omega s \sin\omega t_0 ds,$$

since x_0 is an even function. Substitution for x_0 gives

$$M(t_0) = -2\kappa \int_{-\infty}^{\infty} \operatorname{sech}^2 s \tanh^2 s\, ds + 2\gamma \sin \omega t_0 \int_{-\infty}^{\infty} \operatorname{sech}^2 s \tanh s \sin \omega s\, ds$$

$$= -2\kappa \int_{-\infty}^{\infty} [\operatorname{sech}^2 s - \operatorname{sech}^4 s]\, ds - \gamma \sin \omega t_0 \int_{-\infty}^{\infty} \frac{d}{ds}(\operatorname{sech}^2 s) \sin \omega s\, ds$$

$$= -2\kappa(2 - \tfrac{4}{3}) + \gamma \omega \sin \omega t_0 \int_{-\infty}^{\infty} \operatorname{sech}^2 s \cos \omega s\, ds$$

$$= -\frac{4}{3}\kappa + \frac{\gamma \omega^2 \pi}{\sinh(\tfrac{1}{2}\omega\pi)}.$$

The stable and unstable manifolds of the origin intersect if $M(t_0) \geq 0$, that is, if

$$\gamma \geq \frac{4\kappa}{3\omega^2 \pi} \sinh(\tfrac{1}{2}\omega\pi).$$

• 13.14 Show that the logistic difference equation $u_{n+1} = \lambda u_n(1 - u_n)$ has the general solution $u_n = \sin^2(2^n C\pi)$ if $\lambda = 4$, where C is an arbitrary constant (without loss C can be restricted to $0 \leq C \leq 1$). Show that the solution is 2^q-periodic (q any positive integer) if $C = 1/(2^q - 1)$. The presence of all these periodic doubling solutions indicates chaos. (See the article by Brown and Chua (1996) for further exact solutions of nonlinear difference equations relevant to this and succeeding problems.)

13.14. In the difference equation

$$u_{n+1} = \lambda u_n(1 - u_n),$$

let $u_n = \sin^2(2^n C\pi)$. Then

$$\lambda u_n(1 - u_n) = \lambda \sin^2(2^n C\pi)(1 - \sin^2(2^n C\pi)) = \lambda \sin^2(2^n C\pi) \cos^2(2^n C\pi)$$
$$= \tfrac{1}{4}\lambda \sin^2(2^{n+1} C\pi) = u_{n+1}$$

if $\lambda = 4$. Hence $u_n = \sin^2(2^n C\pi)$ is an exact solution. It is sufficient that $0 \leq C \leq 1$. A period q solution exists if C satisfies

$$u_n = u_{n+q}, \quad \text{or} \quad \sin^2(2^n C\pi) = \sin^2(2^{n+q} C\pi).$$

where q is a positive integer. Therefore C must satisfy

$$\cos(2^{n+1} C\pi) - \cos(2^{n+1+q} C\pi) = 0,$$

or

$$\sin[2^n C\pi(2^q + 1)] \sin[2^n C\pi(2^q - 1)] = 0.$$

Hence a period q solution exists if $C = 1/(2^q - 1)$. Hence these solutions exist for *all* q which implies that the solution is chaotic.

- **13.15** Show that the difference equation $u_{n+1} = 2u_n^2 - 1$ has the exact solution $u_n = \cos(2^n C\pi)$ where C is any constant satisfying $0 \leq C \leq 1$. For what values of C do q-periodic solutions exist?

13.15. In the difference equation

$$u_{n+1} = 2u_n^2 - 1,$$

let $u_n = \cos(2^n C\pi)$. Then

$$2u_n^2 - 1 = 2\cos^2(2^n C\pi) - 1 = \cos(2^{n+1} C\pi) = u_{n+1}.$$

Hence $u_n = \cos(2^n C\pi)$ is an exact solution.
 A period q solution exists if C satisfies

$$u_n = u_{n-q}, \quad \text{or} \quad \cos(2^n C\pi) = \cos(2^{n+q} C\pi),$$

or

$$\sin[C\pi(2^{n-1+q} + 2^{n-1})] \sin[C\pi(2^{n-1+q} - 2^{n-1})] = 0.$$

Since C is independent of n, period q solutions exist if

$$C = \frac{1}{2^q + 1}, \quad \text{or} \quad C = \frac{1}{2^q - 1}, \quad (q \geq 1).$$

- **13.16** Using a trigonometric identity for $\cos 3t$, find a first-order difference equation satisfied by $u_n = \cos(3^n C\pi)$.

13.16. Problems 13.14, 13.15 and this problem follow from trigonometric identities for multiple angles. In this case we require a difference equation which has the solution $u_n = \cos(3^n C\pi)$. Consider the identity

$$\cos 3u = 4\cos^3 u - 3\cos u.$$

If we put $u = 3^n C\pi$, it follows that

$$u_{n+1} = 4u_n^3 - 3u_n.$$

- 13.17 A large number of phase diagrams have been computed and analysed for the two-parameter Duffing equation
$$\ddot{x} + k\dot{x} + x^3 = \Gamma \cos t, \quad \dot{x} = y$$
revealing a complex pattern of periodic, subharmonic and chaotic oscillations (see Ueda (1980) for an extensive catalogue of outputs, and also Problem 7.32). Using a suitable computer package plot phase diagram and time solutions in each of the following cases for the initial data given, and discuss the type of solutions in each generated:

(a) $k = 0.08$, $\Gamma = 0.2$; $x(0) = -0.205$, $y(0) = 0.0171$; $x(0) = 1.050$, $y(0) = 0.780$.
(b) $k = 0.2$, $\Gamma = 5.5$; $x(0) = 2.958$, $y(0) = 2.958$; $x(0) = 2.029$, $y(0) = -0.632$.
(c) $k = 0.2$, $\Gamma = 10$; $x(0) = 3.064$, $y(0) = 4.936$.
(d) $k = 0.1$, $\Gamma = 12$; $x(0) = 0.892$, $y(0) = -1.292$.
(e) $k = 0.1$, $\Gamma = 12$; $x(0) = 3$, $y(0) = -1.2$.

13.17. The equation
$$\ddot{x} + k\dot{x} + x^3 = \Gamma \cos t$$
has been analysed numerically in some detail by Ueda (1980). Here we display cases (a), (d) and (e) for various initial values.

- (a) $k = 0.08$, $\Gamma = 0.2$, with two sets of initial values $x_0 = -0.205$, $y_0 = 0.0171$ and $x_0 = 1.050$, $y_0 = 0.780$. These initial values generate approximately two co-existing stable 2π-periodic solutions which are shown in Figure 13.10. The time solutions of the two periods are shown in Figure 13.11.
- (d) $k = 0.1$, $\Gamma = 12$, with initial values $x_0 = 0.892$, $y_0 = -1.292$. These initial values generate a 2π periodic solution shown in Figure 13.12. The time solution is shown in Figure 13.13.
- (e) $k = 0.1$, $\Gamma = 12$, with initial values $x_0 = 3$, $y_0 = 1.2$. These initial values generate a chaotic response shown in Figure 13.14. The parameter values are the same as those in case (d) so that this chaotic solution co-exists with the periodic solution shown in (d). The time solution is shown in Figure 13.15.

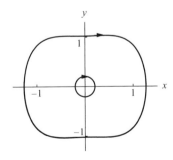

Figure 13.10 Problem 13.17(a): Ueda's equation with $k = 0.08$, $\Gamma = 0.2$, and the two sets of initial values $x_0 = -0.205$, $y_0 = 0.0171$ and $x_0 = 1.050$, $y_0 = 0.780$.

552 Nonlinear ordinary differential equations: problems and solutions

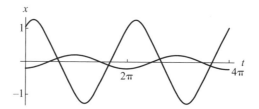

Figure 13.11 Problem 13.17: Periodic time solutions with $k = 0.08$, $\Gamma = 0.2$.

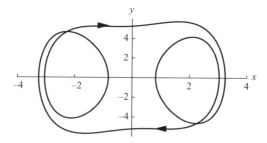

Figure 13.12 Problem 13.17: Ueda's equation with $k = 0.1$, $\Gamma = 12$, and the initial values $x_0 = 0.892$, $y_0 = -1.292$.

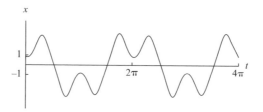

Figure 13.13 Problem 13.17: Periodic time solution with $k = 0.1$, $\Gamma = 12$.

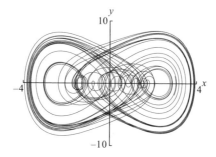

Figure 13.14 Problem 13.17: Ueda's equation with $k = 0.1$, $\Gamma = 12$, and the initial values $x_0 = 3$, $y_0 = 1.2$.

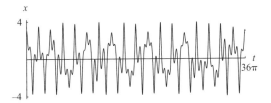

Figure 13.15 Problem 13.17: Periodic time solutions with $k = 0.1$, $\Gamma = 12$.

- 13.18 Consider the Hamiltonian system
$$\dot{p}_i = -\frac{\partial H}{\partial q_i}, \quad \dot{q}_i = \frac{\partial H}{\partial p_i}, \quad (i = 1, 2)$$
where $H = \tfrac{1}{2}\omega_1(p_1^2 + q_1^2) + \tfrac{1}{2}\omega_2(p_2^2 + q_2^2)$. Show that q_1, q_2 satisfy the uncoupled system
$$\ddot{q}_i + \omega_i^2 q_i = 0, \quad (i = 1, 2).$$
Explain why the ellipsoids
$$\tfrac{1}{2}\omega_1(p_1^2 + q_1^2) + \tfrac{1}{2}\omega_2(p_2^2 + q_2^2) = \text{constant}$$
are invariant manifolds in the four-dimensional space (p_1, p_2, q_1, q_2). What condition on ω_1/ω_2 guarantees that all solutions are periodic? Consider the phase path which satisfies $p_1 = 0$, $q_1 = 0$, $p_2 = 1$, $q_2 = 0$. Describe the Poincaré section $p_1 = 0$ of the phase path projected on to the (q_1, p_2, q_2) subspace.

13.18. Consider the mechanical system
$$\dot{p}_i = -\frac{\partial H}{\partial q_i}, \quad \dot{q}_i = \frac{\partial H}{\partial p_i}, \quad (i = 1, 2),$$
where $H = \tfrac{1}{2}\omega_1(p_1^2 + q_1^2) + \tfrac{1}{2}\omega_2(p_2^2 + q_2^2)$. The equations of motion are
$$\dot{p}_1 = -\omega_1 q_1, \quad \dot{q}_1 = \omega_1 p_1, \quad \dot{p}_2 = -\omega_2 q_2, \quad \dot{q}_2 = \omega_2 p_2,$$
or, equivalently,
$$\ddot{q}_1 + \omega_1^2 q_1 = 0, \quad \ddot{q}_2 + \omega_2^2 q_2 = 0.$$
The solutions are
$$q_1 = A_1 \cos \omega_1 t + B_1 \sin \omega_1 t, \quad q_2 = A_2 \cos \omega_2 t + B_2 \sin \omega_2 t.$$
These solutions are periodic if ω_1/ω_2 is a rational number. If this ratio is not rational then q_1 and q_2 are uncoupled with periods $2\pi/\omega_1$ and $2\pi/\omega_2$.

Phase paths in p_1, q_1, p_2, q_2 space are given by

$$\frac{dp_1}{dq_1} = -\frac{q_1}{p_1} \Rightarrow p_1^2 + q_1^2 = C_1, \tag{i}$$

$$\frac{dp_2}{dq_2} = -\frac{q_2}{p_2} \Rightarrow p_2^2 + q_2^2 = C_2. \tag{ii}$$

Also

$$H = \tfrac{1}{2}\omega_1(p_1^2 + q_1^2) + \tfrac{1}{2}\omega_2(p_2^2 + q_2^2) = \text{constant}.$$

Therefore any path which starts on this surface will stay on it, which means that the ellipsoids $H = $ constant are invariant manifolds.

Consider the path for which the initial conditions $p_1 = 0$, $q_1 = 0$, $p_2 = 1$, $q_2 = 0$. In (i) and (ii) $C_1 = 0$ and $C_2 = 1$. Therefore the projection of the manifold on to the (q_1, p_2, q_2) subspace are the straight lines $p_2 = q_2 = 0$, $p_2 = \pm 1$.

- **13.19** Consider the system

$$\dot{x} = -ryz, \quad \dot{y} = rxz, \quad \dot{z} = -z + \cos t - \sin t,$$

where $r = \sqrt{(x^2 + y^2)}$. Show that, projected on to the (x, y) plane, the phase paths have the same phase diagram as a plane centre. Show also that the general solution is given by

$$x = x_0 \cos \omega(t) - y_0 \sin \omega(t), \quad y = y_0 \cos \omega(t) + x_0 \sin \omega(t), \quad z = z_0 e^{-t} + \sin t.$$

where $\omega(t) = r_0[1 - \cos t + z_0(1 - e^{-t})]$, and $x_0 = x(0)$, $y_0 = y(0)$ $z_0 = z(0)$, and $r_0 = \sqrt{(x_0^2 + y_0^2)}$. Confirm that, as $t \to \infty$, all solutions become periodic.

13.19. Consider the forced system

$$\dot{x} = -ryz, \quad \dot{y} = rxz, \quad \dot{z} = -z + \cos t - \sin t, \quad r = \sqrt{(x^2 + y^2)}.$$

From the first two equations,

$$\frac{dy}{dx} = -\frac{x}{y} \Rightarrow x^2 + y^2 = c^2,$$

say (assume $c \geq 0$). Hence, projected on to the x, y plane, the phase paths are the same as those of simple harmonic motion.

Integration of the equation for z leads to the solution

$$z = z_0 e^{-t} + \sin t,$$

where $z(0) = z_0$. Hence

$$\dot{x} = -c(z_0 e^{-t} + \sin t)\sqrt{(c^2 - x^2)}.$$

This separable equation has the general solution

$$x = c\sin(z_0 c e^{-t} + c\cos t + B).$$

Also

$$y = c\cos(z_0 c e^{-t} + c\cos t + B).$$

From the initial conditions

$$x_0 = r_0\sin(z_0 r_0 + r_0 + B), \quad y_0 = r_0\cos(z_0 r_0 + r_0 + B).$$

Therefore

$$x = x_0\cos\omega(t) - y_0\sin\omega(t), \quad y = y_0\cos\omega(t) + x_0\sin\omega(t),$$

where

$$\omega(t) = r_0[1 - \cos t + z_0(1 - e^{-t})].$$

As $t \to \infty$,

$$z \to \sin t, \quad x \to c\sin(c\cos t + B), \quad y \to c\cos(c\cos t + B),$$

which are all periodic with period 2π in t.

- 13.20 A common characteristic feature of chaotic oscillators is sensitive dependence on initial conditions, in which bounded solutions which start very close together utimately diverge. Such solutions locally diverge exponentially. Investigate time-solutions of Duffing's equation
$$\ddot{x} + k\dot{x} - x + x^3 = \Gamma\cos\omega t$$
for $k = 0.3$, $\Gamma = 0.5$, $\omega = 1.2$, which is in the chaotic parameter region (see Figure 13.15 in NODE), for the initial values (a) $x(0) = 0.9$, $y(0) = 0.42$; (b) $y(0) = 0.42$ but with a very small increase in $x(0)$ to say 0.90000001. Divergence between the solutions occurs at about 40 cycles. (Care must be exercised in computing solutions in chaotic domains where sensitive dependence on initial values and computation errors can be comparable in effect.)

13.20. Consider the Duffing equation

$$\ddot{x} + k\dot{x} - x + x^3 = \Gamma\cos\omega t,$$

subject to slightly differing initial conditions. Figure 13.16 shows the numerical solution of the equation with the parameters $k=0.3$, $\Gamma=0.5$ and $\omega=1.2$ for the initial conditions (a) $x(0)=0.90$, $y(0)=0.42$, denoted by x_1; (b) $x(0)=0.90000001$, $y(0)=0.42$, denoted by x_2. The difference between the numerical solutions is shown in the third graph. After about 28 cycles the solutions start to diverge.

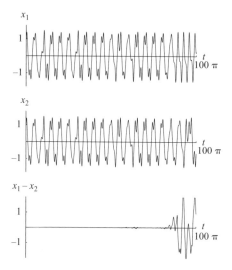

Figure 13.16 Problem 13.20:

- 13.21 The Lorenz equations are given by (see Problem 8.26 and Section 13.2 in NODE)
$$\dot{x} = a(y - x), \quad \dot{y} = bx - y - xz, \quad \dot{z} = xy - cz,$$
Compute solutions of these equations in (x, y, z) phase space. Chaotic solutions appear near parameter values $a = 10$, $b = 27$, $c = 2.65$: a possible initial state is $x(0) = -11.720$, $y(0) = -17.249$, $z(0) = 22.870$.

13.21. The Lorenz equations are given by

$$\dot{x} = a(y - x), \quad \dot{y} = bx - y - xz, \quad \dot{z} = xy - cz.$$

For the parameters $a = 10$, $b = 27$, $c = 2.65$, a single phase path is shown in Figure 13.17. Over long runs the solution continues to display chaotic behaviour.

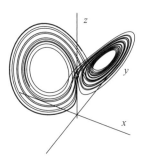

Figure 13.17 Problem 13.21: A single phase path for Lorenz equation $\dot{x} = a(y - x)$, $\dot{y} = bx - y - xz$, $\dot{z} = xy - cz$ with $a = 10$, $b = 27$, $c = 2.65$.

• 13.22 Show that the system $\dot{x} = -y + \Gamma \sin t$, $\dot{y} = -x + 2x^3 + \Gamma \cos t$, ($\Gamma > 0$), has a limit cycle $x = 0$, $y = \Gamma \sin t$. Find also the time-solutions for x and y of the paths which are homoclinic to this limit cycle. Sketch the phase paths of the limit cycle and the homoclinic paths for $\Gamma = 1$.

13.22. In the system

$$\dot{x} = -y + \Gamma \sin t, \quad \dot{y} = -x + 2x^3 + \Gamma \cos t,$$

it can be seen that $x = 0$, $y = \Gamma \sin t$ is a solution. This is obviously a periodic solution of the system.

Let $y = \Gamma \sin t + z$. Then

$$\dot{x} = z, \quad \dot{z} = x - 2x^3.$$

Hence, eliminating z,

$$\ddot{x} = x - 2x^3.$$

We can verify that this equation has the solution $x = \operatorname{sech} t$, which has the required property that $x \to 0$ as $t \to \pm\infty$. Therefore

$$y = \Gamma \sin t - z = \Gamma \sin t - \dot{x} = \Gamma \sin t + \operatorname{sech}^2 t \sinh t.$$

The homoclinic path is

$$x = \operatorname{sech} t, \quad y = \Gamma \sin t + \operatorname{sech}^2 t \sinh t.$$

Similarly there is a complementary path

$$x = -\operatorname{sech} t, \quad y = \Gamma \sin t - \operatorname{sech}^2 t \sinh t.$$

The homoclinic path in the half-plane $x \geq 0$ is shown in Figure 13.18. The periodic solution lies on the y axis between $y = \Gamma$ and $y = -\Gamma$. The homoclinic path for this forced system starts on the periodic solution and ends there.

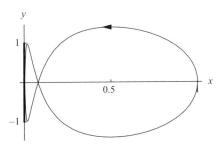

Figure 13.18 Problem 13.22: Homoclinic path $x = \operatorname{sech} t$, $y = \Gamma \sin t + \operatorname{sech}^2 t \sinh t$ with $\Gamma = 1$.

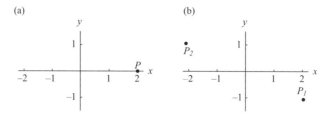

Figure 13.19 Problem 13.23: (a) Fixed point of $x = \cos t$, for $t_0 = 0$ and period $T = 2\pi$; (b) period doubling for $x = 3\cos\frac{1}{2}t$ with $t_0 = \frac{1}{2}\pi$ and period $T = 2\pi$.

- **13.23** For each of the following functions and solutions plot the Poincaré sequence in the $x, y = \dot{x}$ plane, starting with the given initial time t_0 and given period T.
  - (a) $x = 2\cos t$; $t_0 = 0$, $T = 2\pi$.
  - (b) $x = 3\cos t$; $t_0 = \frac{1}{2}\pi$, $T = 2\pi$.
  - (c) $x = \sin t + \sin \pi t$; $t_0 = \frac{1}{2}\pi$, $T = 2\pi$.
  - (d) The periodic solution of $\ddot{x} - (x^2 + \dot{x}^2)\dot{x} + x = \cos t$, where $t_0 = 0$, $T = 2\pi$.

13.23. (a) For $x = 2\cos t$, $y = \dot{x} = -2\sin t$. Therefore if $t_0 = 0$, then

$$x(2n\pi) = 2\cos 2n\pi = 2, \quad y(2n\pi) = -2\sin 2n\pi = 0.$$

In this section the function has a fixed point at $(2, 0)$ shown as P in Figure 13.19(a).

(b) For $x = 3\cos\frac{1}{2}t$, $y = \dot{x} = -\frac{3}{2}\sin\frac{1}{2}t$. If $t_0 = \frac{1}{2}\pi$ and $T = 2\pi$, then

$$x(2n + \tfrac{1}{2}\pi) = 3\cos[(n + \tfrac{1}{4})\pi] = \begin{cases} 3/\sqrt{2} & n \text{ even} \\ -3/\sqrt{2} & n \text{ odd} \end{cases},$$

$$y(2n + \tfrac{1}{2}\pi) = -\tfrac{3}{2}\sin[(n + \tfrac{1}{4})\pi] = \begin{cases} -3/(2\sqrt{2}) & n \text{ even} \\ 3/(2\sqrt{2}) & \text{even} \end{cases}$$

In this section the solution oscillates between the two fixed points at $[3/\sqrt{2}, -3/(2\sqrt{2})]$ and $[-3/\sqrt{2}, 3/(2\sqrt{2})]$ shown as P_1 and P_2 in Figure 13.19(b). In this section the function exhibits period doubling.

(c) For $x = \sin t + \sin \pi t$, $y = \cos t + \pi \cos t$. The function $\sin t + \sin \pi t$ is not periodic. With $t_0 = \frac{1}{2}\pi$ and $T = 2\pi$,

$$x_0 = x(\tfrac{1}{2}\pi) = 1 + \sin\tfrac{1}{2}\pi^2, \quad y_0 = y(\tfrac{1}{2}\pi) = \pi\cos\tfrac{1}{2}\pi^2,$$

$$x_n = x[(2n + \tfrac{1}{2})\pi] = 1 + \sin[(2n + \tfrac{1}{2})\pi^2],$$

$$y_n = y[(2n + \tfrac{1}{2})\pi] = \pi\cos[(2n + \tfrac{1}{2})\pi^2].$$

All these points lie on the ellipse

$$(x-1)^2 + \frac{y^2}{\pi^2} = 1,$$

but there are no repetitions of points.

(d) In the equation

$$\ddot{x} - (x^2 + \dot{x}^2)\dot{x} + x = \cos t,$$

Let $x = A \cos t$. Then

$$\ddot{x} - (x^2 + \dot{x}^2)\dot{x} + x - \cos t = -A \cos t - A^2(-A \cos t) + A \cos t - \cos t$$
$$= (A^3 - 1)\cos t = 0,$$

if $A = 1$. With $t_0 = 0$ and $T = 2\pi$, the periodic solution has a fixed point at $(1, 0)$.

- **13.24** Show that $\ddot{x} + k(1 - x^2 - \dot{x}^2)^2 \dot{x} - x = -2 \cos t$ has a limit cycle whose solution is $x_0 = \cos t$. By looking at perturbations $x = x_0 + x'$ where $|x'|$ is small show that the limit cycle has Poincaré fixed points which are saddles.

13.24. In the forced equation

$$\ddot{x} + k(1 - x^2 - \dot{x}^2)^2 \dot{x} - x = -2\cos t,$$

let $x = A \cos t$. Then

$$\ddot{x} + k(1 - x^2 - \dot{x}^2)^2 \dot{x} - x + 2\cos t = -A\cos t - k(1 - A^2)^2 A \sin t - A\cos t + 2\sin t$$
$$= 2(1 - A)\cos t - k(1 - A^2)A \sin t = 0$$

if $A = 1$. Therefore $x = x_1 = \cos t$ is a limit cycle. For any t_0 and period 2π, the limit cycle has the fixed point $(\cos t_0, -\sin t_0)$.

Let $x = x_1 + x'$. The linearized equation for x' is

$$\ddot{x}' - x' = 0,$$

which has the general solution

$$x' = Be^t + Ce^{-t}.$$

Since this solution has stable and unstable manifolds the fixed points are saddles.

- **13.25** Consider the system $\dot{x} = y$, $\dot{y} = (e^{-2x} - e^{-x}) + \varepsilon \cos t$. For $\varepsilon = 0$, show that the equations of its phase paths is given by $y^2 = 2e^{-x} - e^{-2x} + C$. Show that the system has closed paths about the origin if $-1 < C < 0$ with a bounding separatrix given by $y^2 = 2e^{-x} - e^{-2x}$. What happens to paths for $C > 0$? Sketch the phase diagram. Suppose that the system is moving along the separatrix path, and at some instant the forcing is introduced. Describe what you expect the behaviour of the system to be after the introduction of the forcing. Compute a Poincaré sequence and a time-solution for $\varepsilon = 0.5$ and for the initial conditions, $x(0) = -\ln 2$, $y(0) = 0$.

13.25. The differential equation for the phase paths of

$$\dot{x} = y, \quad \dot{y} = (e^{-2x} - e^{-x})$$

is

$$\frac{dy}{dx} = \frac{e^{-2x} - e^{-x}}{y},$$

which has the general solution

$$y^2 = -(e^{-2x} - 2e^{-x}) + C.$$

This autonomous system has one equilibrium point at $(0, 0)$.

We can determine where paths cut the x axis by putting $y = 0$, in which case

$$e^{-2x} - 2e^{-x} - C = 0, \text{ or } e^{-x} = 1 \pm \sqrt{(1 + C)}. \tag{i}$$

Two real solutions for x occur if $-1 < C < 0$. Since the paths are reflected in the x axis, this implies that closed paths enclose the origin which is a centre. If $C = 0$, then $x = -\ln 2$ is one solution but for the other $x \to \infty$ as $C \to 0-$. The bounding path of the centre is

$$y^2 = -e^{-2x} + e^{-x}.$$

which is the dashed path in the phase diagram shown in Figure 13.20. If $C > 0$, then paths approach $y = \sqrt{(2C)}$ as $x \to \infty$, and approach $y = -\sqrt{(2C)}$ as $x \to -\infty$.

If forcing is introduced at the point $(-\ln 2, 0)$ (on the separatrix), we might expect the solution to oscillate between the stable centre and the unbounded paths but with x progressively increasing. Since the width of the centre decreases with x we might also expect the solution to become unbounded in x. The particular path which starts at $(-\ln 2, 0)$ is shown in Figure 13.21, and confirms the prediction.

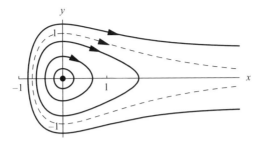

Figure 13.20 Problem 13.25: Phase diagram for $\dot{x} = y$, $\dot{y} = e^{-2x} - e^{-x}$: the dashed path separates the centre from unbounded paths.

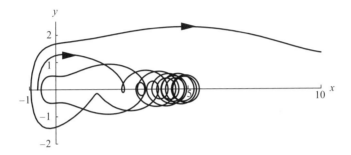

Figure 13.21 Problem 13.25:

- 13.26 Apply the change of variable $z = u + a + b$ to the Lorenz system
$$\dot{x} = a(y - x), \quad \dot{y} = bx - y - xz, \quad \dot{z} = xy - cz,$$
where $a, b, c > 0$. If $s = \sqrt{(x^2 + y^2 + z^2)}$, show that
$$\tfrac{1}{2} s \frac{ds}{dt} = -ax^2 - y^2 - c[u + \tfrac{1}{2}(a+b)]^2 + \tfrac{1}{4}c(a+b)^2.$$
What is the sign of ds/dt on the ellipsoid $ax^2 + y^2 + c[u + \tfrac{1}{2}(a+b)]^2 = \rho$ (*), where $\rho > \tfrac{1}{4}c(a+b)^2$?

Show that all equilibrium points are unstable in the case $a = 4$, $b = 34$, $c = \tfrac{1}{2}$. If this condition is satisfied, what can you say about the attracting set inside the ellipsoid (*) if ρ is sufficiently large?

13.26. Apply the change of variable $z = u + a + b$ to the Lorenz system

$$\dot{x} = a(y - x), \quad \dot{y} = bx - y - xz, \quad \dot{z} = xy - cz, \quad (a, b, c > 0). \tag{i}$$

Then

$$\dot{x} = a(y - x), \quad \dot{y} = -y - x(u + a), \quad \dot{u} = xy - c(u + a + b).$$

Let $s = \sqrt{(x^2 + y^2 + z^2)}$. Then

$$s\frac{ds}{dt} = x\dot{x} + y\dot{y} + z\dot{z}$$
$$= xa(y - x) - y^2 - xy(u + a) + uxy - u(u + a + b)c$$
$$= -ax^2 - y^2 - c[u + \tfrac{1}{2}(a + b)]^2 + \tfrac{1}{4}c(a + b)^2$$

On the ellipsoid

$$ax^2 + y^2 + c[u + \tfrac{1}{2}(a + b)]^2 = \rho, \qquad \text{(ii)}$$

$$s\frac{ds}{dt} = -\rho + \tfrac{1}{4}c(a + b)^2 < 0$$

if $\rho > \tfrac{1}{4}c(a + b)^2 = \rho_1$, say.

From (i), the Lorenz equations have equilibrium points in the (x, y, u) space given by

$$x = y, \quad bx - y - xz = 0, \quad xy - cz = 0.$$

- $b \leq 1$. The system has one equilibrium point at $(0, 0, -a - b)$.
- $b > 1$. Equilibrium occurs at the points

$$(0, 0, -a - b), \quad (\sqrt{c}\sqrt{(b - 1)}, \sqrt{c}\sqrt{(b - 1)}, -a - 1),$$

$$(-\sqrt{c}\sqrt{(b - 1)}, -\sqrt{c}\sqrt{(b - 1)}, -a - 1).$$

The equilibrium point $(0, 0, -a - b)$ lies on the ellipsoid

$$ax^2 + y^2 + c[u + \tfrac{1}{2}(a + b)]^2 = \tfrac{1}{4}c(a + b)^2 = \rho_1,$$

that is, this equilibrium point lie on the critical ellipsoid. Hence this point always lies within the ellipsoid defined by (ii) with $\rho > \rho_1$.

The point $(\sqrt{c}\sqrt{(b - 1)}, \sqrt{c}\sqrt{(b - 1)}, -a - 1)$ lies on the ellipsoid

$$ax^2 + y^2 + c[u + \tfrac{1}{2}(a + b)]^2 = \tfrac{1}{4}c(a^2 + 2b^2 + 2ab).$$

Since $\tfrac{1}{4}c(a^2 + 2b^2 + 2ab) > \tfrac{1}{4}c(a + b)^2 = \rho_1$, this equilibrium point lies outside the ellipsoid with $\rho = \rho_1$. The point $(-\sqrt{c}\sqrt{(b - 1)}, -\sqrt{c}\sqrt{(b - 1)}, -a - 1)$ lies on the same ellipsoid.

The linearized equations associated with the equilibrium points are as follows.

- $(0, 0, -a - b)$. If $u = -a - b + u'$, the linearized equations are

$$\dot{x} = -ax + ay, \quad \dot{y} = (a + 2b)x - y, \quad \dot{u}' = -cu.$$

The eigenvalues are given by

$$\begin{vmatrix} -a-\lambda & a & 0 \\ a+2b & -1-\lambda & 0 \\ 0 & 0 & -c-\lambda \end{vmatrix} = -(\lambda-c)[\lambda^2 + (a+1)\lambda - a^2 - 2ab] = 0.$$

Since one solution for λ is positive this equilibrium point is unstable for all $b > 0$.
- $(\sqrt{c}\sqrt{(b-1)}, \sqrt{c}\sqrt{(b-1)}, -a-1)$, $(b > 1)$. Let $x = \sqrt{c}\sqrt{(b-1)}+x'$, $y = \sqrt{c}\sqrt{(b-1)}+y'$ and $u = -a-1+u'$. The linearized equations are

$$\dot{x}' = -ax' + ay', \quad \dot{y}' = (b+1)x' - y' - \sqrt{c}\sqrt{(b-1)}u',.$$
$$\dot{u}' = \sqrt{c}\sqrt{(b-1)}x' + \sqrt{c}\sqrt{(b-1)}y' - cu'.$$

The eigenvalues are given by

$$\begin{vmatrix} -a-\lambda & a & 0 \\ b+1 & -1-\lambda & -\sqrt{c}\sqrt{(b-1)} \\ \sqrt{c}\sqrt{(b-1)} & \sqrt{c}\sqrt{(b-1)} & -c-\lambda \end{vmatrix} = 0,$$

or

$$\lambda^3 + (a+c+1)\lambda^2 + c(a+b)\lambda + 2ac(b-1) = 0.$$

If $a = 4$, $b = 34$, $c = \frac{1}{2}$, then λ satisfies

$$2\lambda^3 + 11\lambda^2 + 38\lambda + 264 = (\lambda+6)(2\lambda^2 - \lambda + 44) = 0.$$

This equation has the solution $\lambda = -6$, and two complex solutions which have positive real part. Therefore this equilibrium point is unstable.
- $(-\sqrt{c}\sqrt{(b-1)}, -\sqrt{c}\sqrt{(b-1)}, -a-1)$, $(b > 1)$. It can be shown that this equilibrium point has the same eigenvalues as the previous case.

For the given values of a, b and c there are three equilibrium points all of which are unstable. Also there is an ellipsoid (ii) which encloses these equilibrium points, and such that all phase paths pass from the outside to the inside to the outside. Hence subsequently any path which crosses the ellipsoid either approaches a limit cycle or wanders indefinitely inside the ellipsoid.

- 13.27 A plane autonomous system is governed by the equation $\dot{x} = X(x, y)$, $\dot{y} = Y(x, y)$. Consider a set of solutions $x(t, x_0, y_0)$, $y(t, x_0, y_0)$ which start at time $t = t_0$ at (x_0, y_0), where (x_0, y_0) is any point in a region $\mathcal{D}(t_0)$ bounded by a smooth simple closed curve $\mathcal{C}$. At time t, $\mathcal{D}(t_0)$ becomes $\mathcal{D}(t)$. The area of $\mathcal{D}(t)$ is

$$A(t) = \iint_{\mathcal{D}(t)} dxdy = \iint_{\mathcal{D}(t_0)} dx_0 dy_0$$

when expressed in terms of the original region. In this integral, the Jacobian $J(t) = \det(\boldsymbol{\Phi}(t))$, where

$$\boldsymbol{\Phi}(t) = \begin{bmatrix} \dfrac{\partial x}{\partial x_0} & \dfrac{\partial x}{\partial y_0} \\ \dfrac{\partial y}{\partial y_0} & \dfrac{\partial y}{\partial y_0} \end{bmatrix}.$$

Show that $\boldsymbol{\Phi}(t)$ satisfies the linear equation $\dot{\boldsymbol{\Phi}}(t) = B(t)\boldsymbol{\Phi}(t)$, (note that $\boldsymbol{\Phi}(t)$ is a fundamental matrix of this equation) where

$$B(t) = \begin{bmatrix} \dfrac{\partial X}{\partial x} & \dfrac{\partial X}{\partial y} \\ \dfrac{\partial Y}{\partial x} & \dfrac{\partial Y}{\partial y} \end{bmatrix}.$$

Using Theorem 9.4 (on a property of the Wronskian), show that

$$J(t) = J(t_0) \exp\left[\int_{t_0}^{t} \left(\frac{\partial X}{\partial x} + \frac{\partial Y}{\partial y}\right) ds\right].$$

If the system is Hamiltonian deduce that $J(t) = J(t_0)$. What can you say about the area of $\mathcal{D}(t)$? ($A(t)$ is an example of an **integral invariant** and the result is known as **Liouville's theorem**.)

For an autonomous system in n variables $\dot{x} = X(x)$, what would you expect the corresponding condition for a **volume-preserving** phase diagram to be?

13.27. The plane autonomous system is $\dot{x} = X(x, y)$, $\dot{y} = Y(x, y)$. Consider the set of solutions $x(t, x_0, y_0)$, $y(t, x_0, y_0)$ which start at time $t = t_0$ at (x_0, y_0), where (x_0, y_0) is any point in a region $\mathcal{D}(t_0)$ bounded by a smooth simple $\mathcal{C}$. At time t, $\mathcal{D}(t_0)$ becomes $\mathcal{D}(t)$ as shown in Figure 13.22. Let $A(t)$ be the area of $\mathcal{D}(t)$ so that

$$A(t) = \iint_{\mathcal{D}(t)} dxdy.$$

The region $\mathcal{D}(t)$ is obtained from $\mathcal{D}(t_0)$ by the change of variable $x = x(t, x_0, y_0)$, $y = y(t, x_0, y_0)$. In terms of the original region

$$A(t) = \iint_{\mathcal{D}(t)} dxdy = \iint_{\mathcal{D}(t_0)} J(t) dxdy,$$

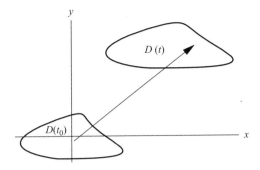

Figure 13.22 Problem 13.27:

where $J(t)$ is the Jacobian

$$J(t) = \det(\Phi(t)),$$

where

$$\Phi(t) = \begin{bmatrix} \dfrac{\partial x}{\partial x_0} & \dfrac{\partial x}{\partial y_0} \\ \dfrac{\partial y}{\partial x_0} & \dfrac{\partial y}{\partial y_0} \end{bmatrix}.$$

The derivative

$$\dot{\Phi}(t) = \begin{bmatrix} \dfrac{\partial \dot{x}}{\partial x_0} & \dfrac{\partial \dot{x}}{\partial y_0} \\ \dfrac{\partial \dot{y}}{\partial x_0} & \dfrac{\partial \dot{y}}{\partial y_0} \end{bmatrix} = \begin{bmatrix} \dfrac{\partial X}{\partial x_0} & \dfrac{\partial X}{\partial y_0} \\ \dfrac{\partial Y}{\partial x_0} & \dfrac{\partial Y}{\partial y_0} \end{bmatrix}$$

$$= \begin{bmatrix} \dfrac{\partial X}{\partial x}\dfrac{\partial x}{\partial x_0} + \dfrac{\partial X}{\partial y}\dfrac{\partial y}{\partial x_0} & \dfrac{\partial X}{\partial x}\dfrac{\partial x}{\partial y_0} + \dfrac{\partial X}{\partial y}\dfrac{\partial y}{\partial y_0} \\ \dfrac{\partial Y}{\partial x}\dfrac{\partial x}{\partial x_0} + \dfrac{\partial Y}{\partial y}\dfrac{\partial y}{\partial x_0} & \dfrac{\partial Y}{\partial x}\dfrac{\partial x}{\partial y_0} + \dfrac{\partial Y}{\partial y}\dfrac{\partial y}{\partial y_0} \end{bmatrix}$$

$$= \begin{bmatrix} \dfrac{\partial X}{\partial x} & \dfrac{\partial X}{\partial y} \\ \dfrac{\partial Y}{\partial x} & \dfrac{\partial Y}{\partial y} \end{bmatrix} \begin{bmatrix} \dfrac{\partial x}{\partial x_0} & \dfrac{\partial y}{\partial x_0} \\ \dfrac{\partial x}{\partial y_0} & \dfrac{\partial y}{\partial y_0} \end{bmatrix} = \mathbf{B}(t)\Phi(t)$$

where

$$\mathbf{B}(t) = \begin{bmatrix} \dfrac{\partial X}{\partial x} & \dfrac{\partial X}{\partial y} \\ \dfrac{\partial Y}{\partial x} & \dfrac{\partial Y}{\partial y} \end{bmatrix}.$$

By NODE, Theorem 9.4,

$$\Phi(t) = \Phi(t_0)\exp\left[\int_{t_0}^{t} \mathrm{tr}[\mathbf{B}(s)]ds\right] = \Phi(t_0)\exp\left[\int_{t_0}^{t} \left\{\dfrac{\partial X}{\partial x} + \dfrac{\partial Y}{\partial y}\right\}ds\right].$$

Finally

$$J(t) = \det[\Phi(t)] = \det[\Phi(t_0)] \exp\left[\int_{t_0}^t \left\{\frac{\partial X}{\partial x} + \frac{\partial Y}{\partial y}\right\} ds\right]$$

$$= J(t_0) \exp\left[\int_{t_0}^t \left\{\frac{\partial X}{\partial x} + \frac{\partial Y}{\partial y}\right\} ds\right]. \tag{i}$$

If the system is Hamiltonian, then

$$\frac{\partial X}{\partial x} + \frac{\partial Y}{\partial y} = 0,$$

so that (i) becomes $J(t) = J(t_0) = 1$. Therefore $A(t) = A(t_0)$, which means that area is preserved for all t.

An n dimensional autonomous system is volume-preserving if $\text{tr}(\mathbf{B}) = 0$.

- **13.28** For the more general version of Liouville's theorem (see Problem 13.27) applied to the case $n = 3$ with $\dot{x} = X(x, y, z)$, $\dot{y} = Y(x, y, z)$, $\dot{z} = Z(x, y, z)$, the volume of a region $\mathcal{D}(t)$ which follows the phase paths is given by

$$W(t) = \iiint_{\mathcal{D}(t)} dx\,dy\,dz = \iiint_{\mathcal{D}(t_0)} J(t) dx_0\,dy_0\,dz_0,$$

where the Jacobian $J(t) = \det[\Phi(t)]$. As in the previous problem

$$J(t) = J(t_0) \exp\left[\int_{t_0}^t \left(\frac{\partial X}{\partial x} + \frac{\partial Y}{\partial y} + \frac{\partial Z}{\partial z}\right) ds\right].$$

Show that $dJ(t)/dt \to 0$ as $t \to \infty$ for the Lorenz system

$$\dot{x} = a(y - x), \quad \dot{y} = bx - y - xz, \quad \dot{z} = xy - cz,$$

where $a, b, c > 0$. What can be said about the volume of any region following phase paths of the Lorenz attractor as time progresses?

13.28. For the system $\dot{x} = X(x, y, z)$, $\dot{y} = Y(x, y, z)$, $\dot{z} = Z(x, y, z)$, the volume of a region $\mathcal{D}(t)$ which follows the phase paths is given by

$$W(t) = \iiint_{\mathcal{D}(t)} dx\,dy\,dz = \iiint_{\mathcal{D}(t_0)} J(t) dx_0\,dy_0\,dz_0,$$

where

$$J(t) = \det[\Phi(t)], \quad \Phi(t) = \begin{bmatrix} \frac{\partial x}{\partial x_0} & \frac{\partial x}{\partial y_0} & \frac{\partial x}{\partial z_0} \\ \frac{\partial y}{\partial x_0} & \frac{\partial y}{\partial y_0} & \frac{\partial y}{\partial z_0} \\ \frac{\partial z}{\partial x_0} & \frac{\partial z}{\partial y_0} & \frac{\partial z}{\partial z_0} \end{bmatrix}.$$

It can be shown using a method which parallels that given in the previous problem that

$$J(t) = J(t_0) \exp\left[\int_{t_0}^{t} \left(\frac{\partial X}{\partial x} + \frac{\partial Y}{\partial y} + \frac{\partial Z}{\partial z}\right) ds\right].$$

A system will be *volume-preserving* if

$$\frac{\partial X}{\partial x} + \frac{\partial Y}{\partial y} + \frac{\partial Z}{\partial z} = 0.$$

For the Lorenz system

$$\dot{x} = a(y - x), \quad \dot{y} = bx - y - xz, \quad \dot{z} = xy - cz,$$

$$\frac{\partial X}{\partial x} + \frac{\partial Y}{\partial y} + \frac{\partial Z}{\partial z} = \frac{\partial [a(y-x)]}{\partial x} + \frac{\partial [bx - y - xz]}{\partial y} + \frac{\partial [xy - cz]}{\partial z}$$

$$= -a - 1 - c < 0.$$

Therefore

$$J(t) = J(t_0) \exp\left[\int_{t_0}^{t} \left(\frac{\partial X}{\partial x} + \frac{\partial Y}{\partial y} + \frac{\partial Z}{\partial z}\right) ds\right] = J(t_0) e^{-(a+c+1)t} \to 0$$

as $t \to \infty$.

- **13.29** Show that $\ddot{x}(1 + \dot{x}) - x\dot{x} - x = -2\gamma(\dot{x} + 1) \cos t$, $(\gamma > 0)$ has the exact solution $x = Ae^t + Be^{-t} + \gamma \cos t$. What can you say about the stability of the limit cycle? Find the Poincaré sequences of the stable and unstable manifolds associated with $t = 0$ and period 2π. Write down their equations and sketch the limit cycle, its fixed Poincaré point and the stable and unstable manifolds.

13.29. Let $x = Ae^t + Be^{-t} + \gamma \cos t$. Then

$$\ddot{x}(1 + \dot{x}) - x\dot{x} - x + 2\gamma(\dot{x} + 1) \cos t = (1 + \dot{x})(\ddot{x} - x + 2\gamma \cos t) = 0.$$

Therefore $x = Ae^t + Be^{-t} + \gamma \cos t$ is an exact solution. The limit cycle is unstable.

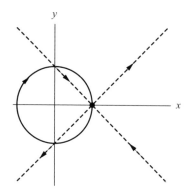

Figure 13.23 Problem 13.29: The fixed point is at $(\gamma, 0)$: the dashed lines indicate the stable and unstable manifolds.

Let $x(0) = x_0$, $y(0) = y_0$. Then, with $\dot{x} = y$,

$$x_0 = A + B + \gamma, \quad y_0 = A - B,$$

so that

$$x = \tfrac{1}{2}(x_0 + y_0 - \gamma)e^t + \tfrac{1}{2}(x_0 - y_0 - \gamma)e^{-t} + \gamma \cos t.$$

The Poincaré section with an initial time $t = 0$ and period 2π. Then

$$x_n = \tfrac{1}{2}(x_0 + y_0 - \gamma)e^{2\pi n} + \tfrac{1}{2}(x_0 - y_0 - \gamma)e^{-2\pi n} + \gamma,$$

$$y_n = \tfrac{1}{2}(x_0 + y_0 - \gamma)e^{2\pi n} - \tfrac{1}{2}(x_0 - y_0 - \gamma)e^{-2\pi n},$$

for $n = 0, 1, 2, \ldots$. The stable and unstable manifolds are given respectively by

$$x_0 + y_0 = \gamma, \quad x - y = \gamma.$$

These manifolds intersect at the fixed point of the periodic solution, namely, $(\gamma, 0)$ as shown in Figure 13.23.

- 13.30 Search for 2π-periodic solutions of $\ddot{x} + k\dot{x} - x + (x^2 + \dot{x}^2)x = \Gamma \cos t$ using $x = c + a \cos t + b \sin t$, and retaining only first harmonics. Show that c, y satisfy
$$c(c^2 - 1 + 2r^2) = 0, \quad (r^2 + 3c^2 - 2)^2 + k^2 r^2 = \Gamma^2,$$
and that the formula is exact for the limit cycle about the origin. Plot a response amplitude (r) against the forcing amplitude (Γ) figure as in Figure 13.13 (in NODE) for $k = 0.25$.

13.30. In the equation

$$\ddot{x} + k\dot{x} - x + (x^2 + \dot{x}^2)x = \Gamma \cos t, \tag{i}$$

let $x = c + a\cos t + b\sin t$. Then

$$\ddot{x} + k\dot{x} - x + (x^2 + \dot{x}^2)x - \Gamma\cos t = -a\cos t - b\sin t - ak\sin t + bk\cos t - ca\cos t - b\sin t$$
$$+ [(c + a\cos t + b\sin t)^2 + (-a\sin t + b\cos t)^2]$$
$$\times (c + a\cos t + b\sin t) - \Gamma\cos t$$
$$= c(c^2 - c + 2r^2) + (-2a + bk + 3c^2 a + ar^2 - \Gamma)\cos t$$
$$+ (-2b - ak + 3bc^2 + br^2)\sin t + \text{(higher harmonics)}$$

where $r^2 = x^2 + y^2$. This is an approximate solution as far as the first harmonics if

$$c(c^2 - 1 + 2r^2) = 0, \qquad \text{(ii)}$$

$$-2a + bk + 3ac^2 + ar^2 - \Gamma = 0, \qquad \text{(iii)}$$

$$-2b - ak + 3bc^2 + br^2 = 0. \qquad \text{(iv)}$$

From (iii) and (iv) it follows that

$$(r^2 + 3c^2 - 2)^2 + k^2 r^2 = \Gamma^2. \qquad \text{(v)}$$

From (ii), one solution is

$$c = 0, \quad (r^2 - 2)^2 + k^2 r^2 = \Gamma^2, \qquad \text{(vi)}$$

and the other solution is

$$c^2 = 1 - 2r^2, \quad (1 - 5r^2)^2 + k^2 r^2 = \Gamma^2. \qquad \text{(vii)}$$

The graphs of r against Γ are shown in Figure 13.24.

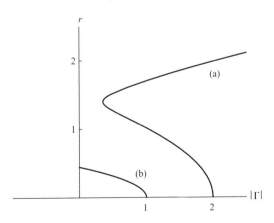

Figure 13.24 Problem 13.30: Curve (a) represents $(r^2-2)^2+k^2r^2=\Gamma^2$ and curve (b) represents $(1-5r^2)^2+k^2r^2=\Gamma^2$.

- 13.31 A nonlinear oscillator has the equation $\ddot{x} + \varepsilon(\dot{x}^2 - x^2 + \frac{1}{2}x^4)\dot{x} - x + x^3 = 0$, $0 < \varepsilon \ll 1$. Show that the system has one saddle and two unstable spiral equilibrium points. Confirm that the saddle point has two associated homoclinic paths given by $x = \pm\sqrt{2}\operatorname{sech} t$. If $u = \dot{x}^2 - x^2 + \frac{1}{2}x^4$, show that u satisfies the equation $\dot{u} + 2\varepsilon\dot{x}^2 u = 0$. What can you say about the stability of the homoclinic paths from the sign of $\dot{u}$? Plot a phase diagram showing the homoclinic and neighbouring paths.

The system is subject to small forcing $\varepsilon\gamma\cos\omega t$ on the right-hand side of the differential equation. Explain, in general terms, how you expect the system to behave if it is started initially from $x(0) = 0$, $\dot{x}(0) = 0$. Plot the phase diagram over a long period interval, say $t \sim 150$ for $\varepsilon = 0.25$, $\omega = 1$, $\gamma = 0.2$.

13.31. The equilibrium points of

$$\ddot{x} + \varepsilon(\dot{x}^2 - x^2 + \tfrac{1}{2}x^4)\dot{x} - x + x^3 = 0$$

occur at $x = -1, 0, 1$, $y = 0$. The linearized approximations near the equilibrium points are as follows.

- $(0, 0)$. The linearized equation are

$$\dot{x} = y, \quad \dot{y} = x.$$

Therefore the origin is a saddle point.
- $(-1, 0)$. Let $x = -1 + x'$. Then

$$\dot{x}' = y, \quad \dot{y} \approx \tfrac{1}{2}y + x' - 3x' = -2x' + \tfrac{1}{2}\varepsilon y.$$

Hence $(-1, 0)$ is an unstable spiral.
- $(1, 0)$. Let $x = 1 + x'$. Then

$$\dot{x}' = y, \quad \dot{y} \approx -2x' + \tfrac{1}{2}\varepsilon y.$$

Therefore $(-1, 0)$ is also an unstable spiral.

Let $x = \sqrt{2}\operatorname{sech} t$. Then

$$\ddot{x} + \varepsilon(\dot{x}^2 - x^2 + \tfrac{1}{2}x^4)\dot{x} - x + x^3 = \sqrt{2}[\operatorname{sech} t - 2\operatorname{sech}^3 t]$$
$$+ \varepsilon[-2\operatorname{sech}^2 t + 2\operatorname{sech}^4 t + 2\operatorname{sech}^2 t \tanh^2 t] - \sqrt{2}\operatorname{sech} t + 2\sqrt{2}\operatorname{sech}^3 t = 0$$

Therefore $x = \sqrt{2}\operatorname{sech} t$ is an exact solution. Similarly it can be shown that $x = -\sqrt{2}\operatorname{sech} t$ is also an exact solution.

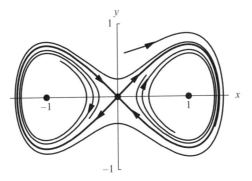

Figure 13.25 Problem 13.31: Phase diagram with $\varepsilon = 0.25$.

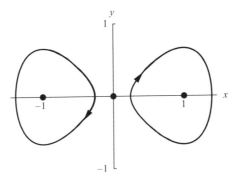

Figure 13.26 Problem 13.31: Periodic solutions for the forced system with $\varepsilon = 0.25$, $\omega = 1$, $\gamma = 0.2$.

Let $u = \dot{x}^2 - x^2 + \frac{1}{2}x^4$. Then

$$\dot{u} = 2\dot{x}\ddot{x} - 2x\dot{x} + 2x^3\dot{x} = 2\dot{x}(\ddot{x} - x + x^3)$$
$$= -2\varepsilon\dot{x}^2(\dot{x}^2 - x^2 + \frac{1}{2}x^4)$$
$$= -2\varepsilon\dot{x}^2 u$$

Hence $\dot{u} < 0$ for $u > 0$, and $\dot{u} > 0$ for $u < 0$. Therefore, since $u = 0$ on the homoclinic path, the sign of $\dot{u}$ implies that any initial perturbation will cause the phase path to approach the homoclinic path $u = 0$. This implies that the homoclinic path is stable as shown in Figure 13.25. In the forced system

$$\ddot{x} + \varepsilon(\dot{x}^2 - x^2 + \frac{1}{2}x^4)\dot{x} - x + x^3 = \epsilon\gamma\cos\omega t.$$

the introduction of forcing causes the homoclinic paths to bifurcate into two stable periodic solutions shown in Figure 13.26.

- **13.32** Show that, for $\alpha > 3$, the logistic difference equation $u_{n+1} = \alpha u_n(1 - u_n)$ has a period 2 solution which alternates between the two values
$$\frac{1}{2\alpha}[1 + \alpha - \sqrt{(\alpha^2 - 2\alpha - 3)}] \text{ and } \frac{1}{2\alpha}[1 + \alpha + \sqrt{(\alpha^2 - 2\alpha - 3)}]$$
Show that it is stable for $3 < \alpha < 1 + \sqrt{6}$.

13.32. The logistic difference equation is
$$u_{n+1} = \alpha u_n(1 - u_n) = f(u_n),$$

say. Fixed points of the equation are given by $u = f(u)$, that is, $u = \alpha u(1 - u)$. There are two such points at $u = 0$ and $u = (\alpha - 1)/\alpha$, $(\alpha > 1)$.

Period doubling occurs where $u = f(f(u))$, or
$$u = \alpha^2 u(1 - u)[1 - \alpha u(1 - u)],$$
$$u[\alpha u - (\alpha - 1)][\alpha^2 u^2 - \alpha(\alpha + 1) + 1 + \alpha] = 0.$$

Period doubling will occur if
$$\alpha^2 u^2 - \alpha(\alpha + 1)u + 1 + \alpha = 0. \tag{i}$$

Therefore
$$u = \frac{1}{2\alpha}\left[\alpha + 1 \pm \sqrt{(\alpha^2 - 2\alpha - 3)}\right] = \frac{1}{2\alpha}[\alpha + 1 \pm \sqrt{(\alpha + 1)(\alpha - 3)}],$$

The solution alternates between these two values of u. However, there can only be real solutions for u if $\alpha \geq 3$.

Stability fails where
$$\frac{d}{du}[f(f(u))] = -4\alpha^3 u^3 + 6\alpha^3 u^2 - 2\alpha^2(\alpha + 1)u + \alpha^2 = -1, \tag{ii}$$

where (i) is also satisfied, that is at the period doubling values. For comparison these equations are
$$\alpha^2 u^2 - \alpha(\alpha + 1)u + (1 + \alpha) = 0, \tag{iii}$$
$$4\alpha^3 u^3 - 6\alpha^3 u^2 + 2\alpha^2(\alpha + 1)u - \alpha^2 - 1 = 0. \tag{iv}$$
Eliminate u^3 between (iii) and (iv) by multiplying (iii) by 4α to give
$$u^2 - \frac{\alpha + 1}{\alpha}u + \frac{\alpha^2 + 1}{2\alpha^2(\alpha - 2)} = 0. \tag{v}$$

For comparison (iii) can be expressed as

$$u^2 - \frac{\alpha+1}{\alpha}u + \frac{\alpha+1}{\alpha^2} = 0. \tag{vi}$$

Equations (v) and (vi) have the same solutions if

$$\frac{\alpha+1}{\alpha^2} = \frac{\alpha^2+1}{2\alpha^2(\alpha-2)} = 0,$$

or

$$\alpha^2 - 2\alpha - 5 = 0.$$

The critical solution is $\alpha = 1 + \sqrt{6}$. Period doubling is stable for $3 < \alpha < 1 + \sqrt{6}$.

- **13.33** The **Shimizu–Morioka equations** are given by the two-parameter system
$\dot{x} = y$, $\dot{y} = x(1-z) - ay$, $\dot{z} = -bz + x^2$.
Show that there are three equilibrium points for $b > 0$, and one for $b \leq 0$. Show that the origin is a saddle point for all a and $b \neq 0$. Obtain the linear approximation for the other equilibrium points assuming $b = 1$. Find the eigenvalues of the linear approximation at $a = 1.2$, $a = 1$ and at $a = 0.844$. What occurs at $a = 1$? For $a = 1.2$ and $a = 0.844$ compute the unstable manifolds of the origin by using initial values close to the origin in the direction of its eigenvector, and plot their projections on to the (x, z) plane (see Figure 13.43 in NODE). Confirm that two homoclinic paths occur for $a \approx 0.844$. What happens to the stability of the equilibrium points away from the origin as a decreases through 1? What type of bifurcation occurs at $a = 1$? Justify any conjecture by plotting phase diagrams for $0.844 < a < 1$.

13.33. The Shimizu–Morioka equations are

$$\dot{x} = y, \quad \dot{y} = x(1-z) - ay, \quad \dot{z} = -bz + x^2.$$

Equilibrium occurs where

$$y = 0, \quad x(1-z) - ay = 0, \quad -bz + x^2 = 0.$$

- $b \leq 0$. System has one equilibrium point at $(0, 0, 0)$.
- $b > 0$. Equilibrium at $(0, 0, 0)$ and $(\pm\sqrt{b}, 0, 1)$.

The linearized classification is as follows.

- Equilibrium point $(0, 0, 0)$. The linearized equations are

$$\dot{x} = y, \quad \dot{y} = x - ay, \quad \dot{z} = -bz.$$

The eigenvalues of the coefficients are given by

$$\begin{vmatrix} -\lambda & 1 & 0 \\ 1 & -a-\lambda & 0 \\ 0 & 0 & -b-\lambda \end{vmatrix} = 0, \text{ or } -(b+k)(-1+ak+k^2) = 0.$$

Therefore the eigenvalues are $-b$, $\frac{1}{2}[-a - \sqrt{(a^2+4)}]$, $\frac{1}{2}[-a + \sqrt{(a^2+4)}]$, which are all real. If $b > 0$, two eigenvalues are negative and one positive, and if $b < 0$, two eigenvalues are positive and one negative. In both cases the origin is a three-dimensional saddle.

- For $b = 1$, one equilibrium point is $(1, 0, 1)$. Let $x = 1 + x'$, $z = 1 + z'$. Then

$$\dot{x}' = y, \quad \dot{y} = -(1+x')z' - ay \approx -ay - z',$$

$$\dot{z}' = -(1+z') + (1+x')^2 \approx 2x' - z'.$$

The eigenvalues are given by

$$\begin{vmatrix} -\lambda & 1 & 0 \\ 0 & -a-\lambda & -1 \\ 2 & 0 & -1-\lambda \end{vmatrix} = 0, \text{ or } -\lambda^3 - (a+1)\lambda^2 - a\lambda - 2 = 0.$$

- For $b = 1$, the other equilibrium point is $(-1, 0, 1)$. Let $x = -1 + x'$, $z = 1 + z'$. Then

$$\dot{x}' = y, \quad \dot{y} \approx -ay + z', \quad \dot{z}' \approx -2x' - z'.$$

The eigenvalues are also given by

$$-\lambda^3 - (a+1)\lambda^2 - a\lambda - 2 = 0.$$

We need only consider the case $b = 1$. The eigenvalues for the three cases $a = 1.2$, $a = 1$, $a = 0.844$ are shown in the table.

a	eigenvalues at $(0, 0, 0)$	eigenvalues at $(1, 0, 1)$
1.200	$1.766, -1, 0.566$	$-2.084, -0.058 \pm 0.978i$
1.000	$-1, \frac{1}{2}(-1 \pm \sqrt{5})$	$-2, \pm i$
0.844	$-1.507, -1, 0.663$	$-1.940, 0.048 \pm 1.014$

For $a = 1.2$, the equilibrium points at $(1, 0, 1)$ and $(-1, 0, 1)$ are stable spiral/nodes.

The unstable manifolds of the origin for the case $a = 1.2$ are shown in Figure 13.27 projected on to the x, z plane. The stable spiral feature of the equilibrium points at $(1, 0, 1)$ and $(-1, 0, 1)$ are clearly visible. The value $a = 0.844$ is the critical case for the appearance of homoclinic paths of the origin as shown in Figure 13.28.

For $a = 1$, the eigenvalues of the equilibrium points $(\pm, 0, 1)$ are $-2, \pm i$ which indicates a transition between stable equilibrium points to unstable points as a decreases through 1.

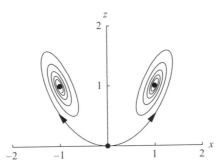

Figure 13.27 Problem 13.33: Unstable manifolds of the origin for $a = 1.2$, $b = 1$ projected on to the x, z plane.

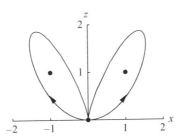

Figure 13.28 Problem 13.33: Unstable manifolds of the origin for $a = 0.844$, $b = 1$ projected on to the x, z plane.

- 13.34 Compute some Poincaré sections given by the plane $\Sigma : z =$ constant of the Rössler system
$$\dot{x} = -y - z, \quad \dot{y} = x + ay, \quad \dot{z} = bx - cz + xz, \quad (a, b, c > 0)$$
where $a = 0.4$, $b = 0.3$ and c takes various values. The choice of the constant for z in Σ is important: if it is too large then the section might not intersect phase paths at all. Remember that the Poincaré sequence arises from intersections which occur as the phase paths cut Σ in the same sense. The period-2 solution (Figure 13.12(b) in NODE), with Poincaré section $z = 2$ should appear as two dots as shown in Figure 13.44(a) (in NODE) after transient behaviour has died down. Figures 13.44(a),(b) (in NODE) show a section of chaotic behaviour at $c = 4.449$ at $z = 4$.

13.34. The Rössler system is given by
$$\dot{x} = -y - z, \quad \dot{y} = x + ay, \quad \dot{z} = bx - cz + xz, \quad (a, b, c > 0).$$

Figure 13.29 shows the section through $z = 4$ for system with $a = 0.4$, $b = 0.3$, $c = 4.449$, which is evidence of a strange attractor. Figure 13.30 shows period doubling for $a = 0.4$, $b = 0.3$, $c = 2$ in the section $z = 1.6$. The curve shows the actual period time solution. It is possible to get period-4 returns, for example, in the section $z = 1$.

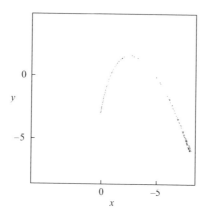

Figure 13.29 Problem 13.34:

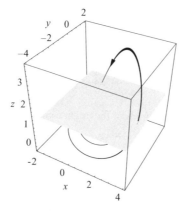

Figure 13.30 Problem 13.34: periodic solution which occurs for $a = 0.4$, $b = 0.3$, $c = 2$, and the section $z = 1.6$.

- **13.35** For the Duffing oscillator $\ddot{x} + k\dot{x} - x + x^3 = \Gamma \cos \omega t$ it was shown in NODE, Section 13.3, that the displacement c and the response amplitude r were related to other parameters by

$$c^2 = 1 - \tfrac{3}{2}r^2, \quad r^2[(2 - \omega^2 - \tfrac{15}{4}r^2)^2 + k^2\omega^2] = \Gamma^2$$

for Type II oscillations (eqn (13.25)). By investigating the roots of $d(\Gamma^2)/dr^2 = 0$, show that a fold develops in this equation for $\omega < \tfrac{1}{2}[4 + 3k^2 - k\sqrt{(24 + 9k^2)}]$. Hence there are three response amplitudes for these forcing frequencies. Design a computer program to plot the amplitude (Γ)/amplitude (r) curves; C_1 and C_2 as in Fig. 13.13. Figure 13.45 (in NODE) shows the two folds in C_1 and C_2 for $k = 0.3$ and $\omega = 0.9$.

13.35. For the Duffing oscillator

$$\ddot{x} + k\dot{x} - x + x^3 = \Gamma \cos \omega t$$

the displacement c and amplitude r are related by

$$c^2 = 1 - \tfrac{3}{2}r^2, \quad r^2[(2-\omega^2-\tfrac{15}{4}r^2)^2 + k^2\omega^2] = \Gamma^2$$

for Type II oscillations (see Section 13.3). Differentiating the second equation

$$\frac{d(\Gamma^2)}{d(r^2)} = (2-\omega^2-\tfrac{15}{4}r^2)^2 + k^2\omega^2 - \tfrac{15}{2}r^2(2-\omega^2-\tfrac{15}{4}r^2).$$

Folds develop where $d(\Gamma^2)/d(r^2) = 0$. Let $\rho = r^2$. Then ρ satisfies

$$(2-\omega^2-\tfrac{15}{4}\rho)^2 + k^2\omega^2 - \tfrac{15}{2}\rho(2-\omega^2-\tfrac{15}{4}\rho) = 0,$$

or

$$\tfrac{675}{16}\rho^2 - 15(2-\omega^2)\rho + (2-\omega^2)^2 + k^2\omega^2 = 0.$$

Therefore

$$\rho = \tfrac{8}{45}\{(2-\omega^2) \pm \tfrac{1}{2}\sqrt{[(2-\omega^2)^2 - 3k^2\omega^2]}\}.$$

This equation will have solutions if ω and k take values which make ρ real and positive. The general restriction $\omega^2 < 2$ (assume that $\omega > 0$) applies. Additionally we require

$$(2-\omega^2)^2 \geq 3k^2\omega^2 \quad \text{or} \quad \omega^4 - (4+3k^2)\omega^2 + 4 \geq 0,$$

which is equivalent to

$$\omega^2 < \omega_1^2 = \tfrac{1}{2}[(4+3k^2) - k\sqrt{(24+9k^2)}], \tag{i}$$

or

$$\omega^2 > \omega_2^2 = \tfrac{1}{2}[(4+3k^2) + k\sqrt{(24+9k^2)}]. \tag{ii}$$

However, only (i) is consistent with $\omega^2 < 2$ so that (i) is the condition for ρ to be real and positive (see Figure 13.31). The Γ, r graphs are shown in Figure 13.32: the Type II case is considered here.

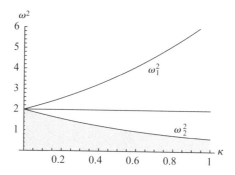

Figure 13.31 Problem 13.35: Graph shows $\omega^2 = \omega_1^2$, $\omega^2 = \omega_2^2$ and $\omega^2 = 2$, all plotted against k: ρ is real and positive in the shaded region.

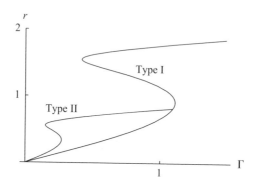

Figure 13.32 Problem 13.35:

- **13.36** It was shown in NODE, Section 13.5 for the Duffing equation $\ddot{x} + k\dot{x} - x + x^3 = \Gamma \cos\omega t$ that the perturbation $a' = [a', b', c', d']^T$ from the translation $c_0 = \sqrt{[1 - \frac{3}{2}(a_0^2 + b_0^2)]}$ and the amplitudes a_0 and b_0 of the harmonic approximation $x = c_0 + a_0 \cos\omega t + b_0 \sin\omega t$ satisfies $\dot{a}' = Aa'$ where

$$A = \begin{bmatrix} R(P - \frac{3}{2}ka_0^2 + 3a_0b_0\omega) & -R(Q - \frac{3}{2}ka_0b_0 + 3b_0^2\omega) & 6Rc_0(-a_0k + 2b_0\omega) & 0 \\ R(Q - 3a_0^2\omega - \frac{3}{2}ka_0b_0) & R(P - 3a_0b_0\omega - \frac{3}{2}b_0^2k) & -12Ra_0c_0k & 0 \\ 0 & 0 & 0 & 1 \\ -3a_0c_0 & -3b_0c_0 & -(2 - 3r_0^2) & -k \end{bmatrix},$$

where $R = 1/(k^2 + 4\omega^2)$, $P = -k(2 + \omega^2 - \frac{15}{4}r_0^2)$, $Q = \omega(4 - 2\omega^2 - k^2 - \frac{15}{4}r_0^2)$, (see eqn (13.37) in NODE). The constants a_0 and b_0 are obtained by solving eqns (13.21) and (13.22). Devise a computer program to find the eigenvalues of the matrix A for $k = 0.3$ and $\omega = 1.2$ as in the main text. By tuning the forcing amplitude Γ, find, approximately, the value of Γ for which one of the eigenvalues changes sign so that the linear system

$\dot{\mathbf{a}}' = \mathbf{A}\mathbf{a}'$ becomes unstable. Investigate numerically how this critical value of Γ varies with the parameters k and ω.

13.36. In the Duffing equation

$$\ddot{x} + k\dot{x} - x + x^3 = \Gamma \cos \omega t,$$

let $a = a_0 + a'(t)$, $b = b_0 + b'(t)$, $c = c_0 + c'(t)$, $d = d'(t)$. As in the text, it follows that $\dot{\mathbf{a}}' = \mathbf{A}\mathbf{a}'$ where

$$\mathbf{A} = \begin{bmatrix} R(P - \tfrac{3}{2}ka_0^2 + 3a_0b_0\omega) & -R(Q - \tfrac{3}{2}ka_0b_0 + 3b_0^2\omega) & 6Rc_0(-a_0k + 2b_0\omega) & 0 \\ R(Q - 3a_0^2\omega - \tfrac{3}{2}ka_0b_0) & R(P - 3a_0b_0\omega - \tfrac{3}{2}b_0^2k) & -12Ra_0c_0k & 0 \\ 0 & 0 & 0 & 1 \\ -3a_0c_0 & -3b_0c_0 & -(2 - 3r_0^2) & -k \end{bmatrix},$$

where

$$R = \frac{1}{k^2 + 4\omega^2}, \quad P = -k(2 + \omega^2 - \tfrac{15}{4}r_0^2), \quad Q = \omega(4 - 2\omega^2 - k^2 - \tfrac{15}{4}r_0^2).$$

The amplitudes a_0 and b_0, and c_0 satisfy (13.20), (13.21) and (13.22), namely

$$c_0^2 = 1 - \tfrac{3}{2}r_0^2, \tag{i}$$

$$a_0(2 - \omega^2 - \tfrac{15}{4}r_0^2) + k\omega b_0 = \Gamma, \tag{ii}$$

$$b_0(2 - \omega^2 - \tfrac{15}{4}r_0^2) - k\omega a_0 = 0. \tag{iii}$$

The procedure is that eqns (i), (ii) and (iii) are solved numerically for a_0, b_0 and c_0 for given values of the parameters k, ω and Γ. Then the eigenvalues of $\mathbf{A}$ are computed which will then indicate whether the solutions of $\dot{\mathbf{a}}' = \mathbf{A}\mathbf{a}'$ are stable or unstable. A table of eigenvalues for $k = 0.3$, $\omega = 1.2$ and $\Gamma = 0.2, 0.25, 0.3, 0.35$ is shown below which can be compared with the computed value of $\Gamma = 0.27$ (see NODE, Section 13.3).

Γ	eigenvalues of $\mathbf{A}$
0.20	$-0.202 \pm 1.276i$, $-0.117 \pm 0.350i$
0.25	$-0.249 \pm 1.149i$, -0.054 ± -0.263
0.30	$-0.171 \pm 0.930i$, -0.224, 0.052
0.35	$-0.408 \pm 1.192i$, $0.072 \pm 0.568i$

For $\Gamma = 0.2, 0.25$, the first harmonic $x = c_0 + a_0 \cos \omega t + b_0 \sin \omega t$ is stable. Instability arises at approximately $\Gamma = 0.3$.

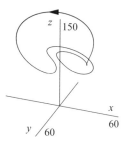

Figure 13.33 Problem 13.37: Periodic solution of the Lorenz system with $a = 10$, $b = 100.5$, $c = 8/3$.

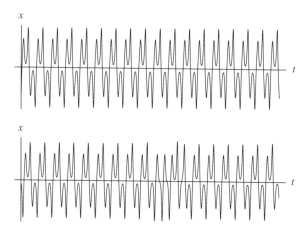

Figure 13.34 Problem 13.37: Time solutions for $a = 10$, $c = 8/3$ with $b = 166$ in the upper figure and $b = 166.1$ in the lower figure.

- **13.37** Compute solutions for the Lorenz system
$$\dot{x} = a(y - x), \quad \dot{y} = bx - y - xz, \quad \dot{z} = xy - cz,$$
for the parameter section $a = 10$, $c = 8/3$ and various values of b: this is the section frequently chosen to illustrate oscillatory features of the Lorenz attractor. In particular try $b = 100.5$ and show numerically that there is a periodic attractor as shown in Figure 13.46(a) (in NODE). Why will this limit cycle be one of a pair?
Shows also that at $b = 166$, the system has a periodic solution as shown in Figure 13.46(b)(in NODE), but at 166.1 (Figure 13.46(c) in NODE) the periodic solution is regular for long periods but is then subject to irregular bursts at irregular intervals before resuming its oscillation again. This type of chaos is known as **intermittency**. (For discussion of intermittent chaos and references see Nayfeh and Balachandran (1995); for a detailed discussion of the Lorenz system see Sparrow (1982)).

13.37. The Lorenz system is
$$\dot{x} = a(y - x), \quad \dot{y} = bx - y - xz, \quad \dot{z} = xy - cz.$$

A computed periodic solution is shown in Figure 13.33 with the parameters $a = 10$, $b = 100.5$, $c = 8/3$. Time solutions showing intermittency for a small change in the parameter b are displayed in Figure 13.34.

- **13.38** The damped pendulum with periodic forcing of the pivot leads to the equation (Bogoliubov and Mitropolski 1961)

$$\ddot{x} + \sin x = \varepsilon(\gamma \sin t \sin x - \kappa \dot{x}),$$

where $0 < \varepsilon \ll 1$. Apply Melnikov's method and show that heteroclinic bifurcation occurs if $\gamma \geq 4\kappa \sinh \tfrac{1}{2}\pi$. [You will need the integral

$$\int_{-\infty}^{\infty} \sin s \, \mathrm{sech}^2 s \tanh s \, ds = \frac{\pi}{2 \sinh(\tfrac{1}{2} a\pi)}.]$$

13.38. The damped pendulum with periodic forcing of the pivot leads to the equation

$$\ddot{x} + \sin x = \varepsilon(\gamma \sin t \sin x - \kappa \dot{x}),$$

where it is assumed that $0 < \varepsilon \ll 1$. This system has equilibrium points at $x = n\pi$, ($n = 0, \pm 1, \pm 2, \ldots$). Of these points, those for which $n = 0, \pm 2, \pm 4, \ldots$ are saddle points, and those for which $n = \pm 1, \pm 3, \ldots$ are centres.

The heteroclinic paths for the unperturbed system with $\varepsilon = 0$ are given

$$x_0 = 2 \tan^{-1}(\sinh t).$$

The Melnikov function (see NODE, Section 13.7) is given by

$$M(t_0) = \int_{-\infty}^{\infty} y_0(t - t_0) h[x_0(t - t_0), y_0(t - t_0), t] dt,$$

where $\dot{x} = y$ and $h(x, y, t) = \gamma \sin t \sin x - \kappa y$. Therefore

$$M(t_0) = 2 \int_{-\infty}^{\infty} \mathrm{sech}\,(t - t_0)\{\gamma \sin t \sin[2 \tan^{-1}(\sinh(t - t_0))] - 2\kappa \,\mathrm{sech}\,(t - t_0)\} dt$$

$$= 4 \int_{-\infty}^{\infty} \mathrm{sech}\,(t - t_0)[\gamma \sin t \,\mathrm{sech}\,(t - t_0) \tanh(t - t_0) - \kappa \,\mathrm{sech}\,(t - t_0)] dt$$

$$= 4\gamma \int_{-\infty}^{\infty} \sin(s+t_0)\operatorname{sech}^2 s \tanh s\, ds - 4\kappa \int_{-\infty}^{\infty} \operatorname{sech}^2 s\, ds$$

$$= 4\gamma \cos t_0 \int_{-\infty}^{\infty} \sin s \operatorname{sech}^2 s \tanh s\, ds - 4\kappa \int_{-\infty}^{\infty} \operatorname{sech}^2 s\, ds$$

$$= \frac{2\gamma \cos t_0}{\sinh \tfrac{1}{2}\pi} - 8\kappa,$$

since

$$\int_{-\infty}^{\infty} \operatorname{sech}^2 s\, ds = 2, \quad \int_{-\infty}^{\infty} \sin s \operatorname{sech}^2 s \tanh s\, ds = \frac{\pi}{2\sinh(\tfrac{1}{2}\pi)}.$$

A heteroclinic bifurcation occurs if $M(t_0) = 0$, that is, if

$$\gamma \cos t_0 = 4\kappa \sinh(\tfrac{1}{2}\pi).$$

A solution for t_0 can only exist if

$$4\kappa \sinh(\tfrac{1}{2}\pi) \leq \gamma,$$

assuming that the parameters are positive.

- **13.39** An alternative method of visualizing the structure of solutions of difference equations and differential equations is to plot **return maps** of u_{n-1} versus u_n. For example, a sequence of solutions of the logistic difference equation $u_{n+1} = \alpha u_n(1 - u_n)$ the ordinate would be u_{n-1} and the abscissa u_n. The return map should be plotted after any initial transient returns have died out. If $\alpha = 2.8$ (see Section 13.4), how will the long-term return amp appear? Find the return map for $\alpha = 3.4$ also. An exact (chaotic) solution of the logistic equation is $u_n = \sin^2(2^n)$ (see NODE, Problem 13.14). Plot the points (u_n, u_{n-1}) for $n = 1, 2, \ldots, 200$, say. What structure is revealed?
Using a computer program generate a time-series (numerical solution) for the Duffing equation

$$\ddot{x} + k\dot{x} - x + x^3 = \Gamma \cos \omega t$$

for $k = 0.3$, $\omega = 1.2$ and selected values of Γ, say $\Gamma = 0.2, 0.28, 0.29, 0.37, 0.5$ (see Figures 13.14, 13.15, 13.16 in NODE). Plot transient-free return maps for the interpolated pairs $[x(2\pi n/\omega), x(2\pi(n-1)/\omega)]$. For the chaotic case $\Gamma = 0.5$, take the time series over an interval $0 \leq t \leq 5000$, say. These return diagrams show that structure is recognizable in chaotic outputs: the returns are not uniformly distributed.

13.39. In this problem return maps are constructed. For the difference equation

$$u_{n+1} = \alpha u_n(1 - u_n),$$

a sequence of solutions are plotted on the (u_{n-1}, u_n) plane.

Return maps in the (u_n, u_{n-1}) plane are shown in Figure 13.35 for the cases $\alpha = 2.8$ and $\alpha = 3.4$. The sequence starting with $u_0 = 0.5$ is shown in the first diagram in Figure 13.35: the sequence approaches the fixed point $(1.8/2.8, 1.8/2.8)$. In the second figure computed for $\alpha = 3.4$, only the ultimate period doubling between the points $(0.452, 0.842)$ and $(0.842, 0.452)$ are marked.

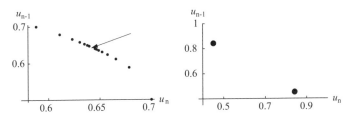

Figure 13.35 Problem 13.39: The return maps for $u_{n+1} = \alpha u_n (1 - u_n)$ with $\alpha = 2.8$ and $\alpha = 3.4$ both starting from $u_0 = 0.5$: the arrow points to the limit of the sequence at $(1.8/2.8, 1.8/2.8)$ for the period 1 solution, The two dots show period doubling after transient effects have been eliminated.

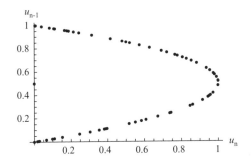

Figure 13.36 Problem 13.39: Return map for the exact solution $u_n = \sin^2(2^n)$ of the logistic equation.

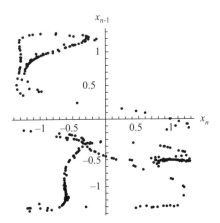

Figure 13.37 Problem 13.39: Return map for the Duffing equation with axes x_n and x_{n-1} with parameter values $k = 0.3$, $\omega = 1.2$, $\Gamma 0.5$.

The logistic equation has the exact solution $u_n = \sin^2(2^n)$ (see Problem 13.14). The chaotic return map for this solution is shown in Figure 13.36. Since $u_n = 4u_{n-1}(1-u_{n-1})$, all the points on the return map lie on the parabola $x = 4y(1-y)$ in continuous variables.

The Duffing equation is

$$\ddot{x} + k\dot{x} - x + x^3 = \Gamma \cos\omega t, \quad \dot{x} = y.$$

We shall only look at the case $k = 0.3$, $\omega = 1.2$, $\Gamma = 0.5$, and in particular the return map. This is obtained by computing the solution numerically, and then listing the discrete values $x_n = x(2n\pi/\omega)$ for $n = n_0, n_0 + 1, \ldots$, where n_0 is some suitable value which reduces transience. The return map is shown in Figure 13.36 for about 1000 returns. It can be seen that there is structure in the chaos: the returns are not simply randomly distributed over a region.

References

This is the list of references given in *Nonlinear Ordinary Differential Equations*.

Abarnarnel DI, Rabinovich MI and Sushchik MM (1993) *Introduction to nonlinear dynamics for physicists*. World Scientific, Singapore.
Abramowitz M and Stegun IA (1965) *Handbook of mathematical functions*. Dover, London.
Acheson D (1997) *From calculus to chaos*. Oxford University Press.
Addison PS (1997) *Fractals and chaos: an illustrated course*. Institute of Physics, Bristol.
Ames WF (1968) *Nonlinear equations in transport processes*. Academic Press, New York.
Andronov AA and Chaikin CE (1949) *Theory of oscillations*. Princeton University Press.
Andronov AA, Leontovich EA, Gordon II and Maier AG (1973a) *Qualitative theory of second-order dynamic systems*. Wiley, New York.
Andronov AA, Leontovich EA, Gordon II and Maier AG (1973b) *Theory of bifurcations of dynamic systems in a plane*. Halstead Press, New York.
Arnold VI (1983) *Geometrical methods in the theory of ordinary differential equations*. Springer-Verlag, Berlin.
Arrowsmith DK and Place CM (1990). *An introduction to dynamical systems*. Cambridge University Press.
Ayres F (1962) *Matrices*. Schaum, New York.
Baker GL and Blackburn JA (2005) *The Pendulum*. Oxford University Press.
Barbashin EA (1970) *Introduction to the theory of stability*. Wolters-Noordhoff, The Netherlands.
Bogoliubov NN and Mitropolsky YA (1961) *Asymptotic methods in the theory of oscillations*. Hindustan Publishing Company, Delhi.
Boyce WE and DiPrima RC (1996) *Elementary differential equations and boundary-value problems*. Wiley, New York.
Brown R and Chua LO (1996) Clarifying chaos examples and counterexamples. *Int. J. Bifurcation Chaos*, 6, 219–249.
Brown R and Chua LO (1998) Clarifying chaos II: Bernoulli chaos, zero Lyapunov exponents and strange attractors. *Int. J. Bifurcation Chaos*. 8, 1–32.
Carr J (1981) *Applications of center manifold theory*. Springer-Verlag, New York.
Cesari L (1971) *Asymptotic behaviour and stability problems in ordinary differential equations*. Springer, Berlin.
Coddington EA and Levinson L (1955) *Theory of ordinary differential equations*. McGraw-Hill, New York.
Cohen AM (1973) *Numerical analysis*. McGraw-Hill, London.
Coppel WA (1978) *Dichotomies in stability thoery*. Springer, Berlin.
Copson ET (1965) *Asymptotic expansions*. Cambridge University Press.

Crocco L (1972) Coordinate perturbations and multiple scales in gas dynamics. *Phil. Trans. Roy. Soc.* **A272**, 275–301.
Diacu F and Holmes P (1996) *Celestial encounters*. Princeton University Press.
Drazin PG (1992) *Nonlinear systems*. Cambridge University Press.
Ermentrout B (2002) *Simulating, analyzing, and animating dynamical systems: a guide to XPPAUT*. SIAM Publications, Philadelphia.
Ferrar WL (1950) *Higher algebra*. Clarendon Press, Oxford.
Ferrar WL (1951) *Finite matrices*. Clarendon Press, Oxford.
Gradshteyn IS and Ryzhik (1994) *Table of integrals, series, and products*. Academic Press, London.
Grimshaw R (1990) *Nonlinear ordinary differential equations*. Blackwell Scientific Publications, Oxford.
Guckenheimer J and Holmes P (1983) *Nonlinear oscillations, dynamical systems, and bifurcations of vector fields*. Springer-Verlag, New York.
Hale J (1969) *Ordinary differential equations*. Wiley-Interscience, London.
Hale J and Kocak H (1991) *Dynamics and bifurcations*. Springer-Verlag, New York.
Hayashi C (1964) *Nonlinear oscillations in physical systems*. McGraw-Hill, New York.
Hilborn RC (1994) *Chaos and nonlinear dynamics*. Oxford University Press.
Hill R (1964) *Principles of dynamics*. Pergamon Press, Oxford.
Hinch EJ (1991) *Perturbation methods*. Cambridge University Press.
Holmes P (1979) A nonlinear oscillator with a strange attractor. *Phil. Trans. Roy. Soc.* **A292**, 419–448.
Jackson EA (1991) *Perspectives in nonlinear dynamics, Vols 1 and 2*. Cambridge University Press.
Jones DS (1963) *Electrical and mechanical oscillations*. Routledge and Kegan Paul, London.
Jordan DW and Smith P (2002) *Mathematical techniques*, 3rd edn. Oxford University Press.
Kevorkian J and Cole JD (1996) *Multiple scale and singular perturbation methods*. Springer, New York.
Krylov N and Bogoliubov N (1949) *Introduction to nonlinear mechanics*. Princeton University Press.
La Salle and Lefshetz S (1961) *Stability of Liapunov's direct method*. Academic Press, New York.
Leipholz H (1970) *Stability theory*. Academic Press, New York.
Logan JD (1994) *Nonlinear partial differential equations*. Wiley-Interscience, New York.
Lorenz EN (1963) Deterministic nonperiodic flow. *J. Atmospheric Sci.* **20**, 130–141.
Magnus K (1965) *Vibrations*. Blackie, London.
Mattheij RMM and Molenaar J (1996) *Ordinary differential equations in theory and practice*. Wiley, Chichester.
McLachlan NW (1956) *Ordinary differential equations in engineering and physical sciences*. Clarendon Press, Oxford.
Minorsky N (1962) *Nonlinear oscillations*. Van Nostrand, New York.
Moon FC (1987) *Chaotic vibrations*. Wiley, New York.
Nayfeh AH (1973) *Perturbation methods*. Wiley, New York.
Nayfeh AH and Balachandran B (1995) *Applied nonlinear dynamics*. Wiley, New York.
Nayfeh AH and Mook DT (1979) *Nonlinear oscillations*. Wiley, New York.

Nemytskii VV and Stepanov VV (1960) *Qualitative theory of differential equations*. Princeton University Press.
Nicolis G (1995) *Introduction to nonlinear science*. Cambridge University Press.
O'Malley RE (1974) *Introduction to singular perturbations*. Academic Press, New York.
Osborne AD (1998) *Complex variables and their applications*. Addison-Wesley Longman.
Pavlidis T (1973) *Biological oscillators: their mathematical analysis*. Academic Press, New York.
Pielou EC (1969) *An introduction to mathematical ecology*. Cambridge University Press.
Poston T and Stewart I (1978) *Catastrophe theory and its applications*. Pitman, London.
Rade L and Westergren B (1995) *Mathematics handbook for science and engineering*. Studentlitteratur, Lund, Sweden.
Rasband SN (1990) *Chaotic dynamics of nonlinear systems*. Wiley, New York.
Reissig R, Sansone G and Conti R (1974) *Nonlinear differential equations of higher order*. Noordhoff, Leiden.
Rosen R (ed) (1973) *Foundations of mathematical systems, Volume III, Supercellular systems*. Academic Press, New York.
Sanchez DA (1968) *Ordinary differential equations and stability theory*. Freeman, San Francisco.
Simmonds JG (1986) *A first look at perturbation theory*. Krieger publishing, Florida.
Small RD (1989) Population growth in a closed system. In *Mathemtical modelling; classroom notes in applied mathematics*, edited by MS Klamkin. SIAM Publications, Philadelphia.
Sparrow C (1982) *The Lorenz equations: bifurcations, chaos, and strange attractors*. Springer-Verlag, New York.
Stoker JJ (1950) *Nonlinear vibrations*. Interscience, New York.
Strogatz SH (1994) *Nonlinear dynamics and chaos*. Perseus, Massachusetts.
Struble RA (1962) *Nonlinear differential equations*. McGraw-Hill, New York.
Thompson JMT and Stewart HB (1986) *Nonlinear dynamics and chaos*. Wiley, Chichester.
Ueda Y (1980) Steady motions exhibited by Duffing's equation: a picture book of regular and chaotic motions. In *New approaches to nonlinear problems in dynamics*, edited by PJ Holmes. SIAM Publications, Philadelphia.
Urabe M (1967) *Nonlinear autonomous oscillations*. Academic Press, New York.
Van Dyke, M (1964) *Perturbation methods in fluid mechanics*. Academic Press, New York.
Verhulst F (1996) *Nonlinear differential equations and dynamical systems*, 2nd edn. Springer, Berlin.
Virgin LN (2000) *Introduction to experimental nonlinear dynamics*. Cambridge University Press.
Watson GN (1966) *A treatise on theory of Bessel functions*. Cambridge University Press.
Whittaker ET and Watson GN (1962) *A course of modern analysis*. Cambridge University Press.
Wiggins S (1990) *Introduction to applied nonlinear dynamical systems and chaos*. Springer-Verlag, New York.
Wiggins S (1992) *Chaotic transport in dynamical systems*. Springer-Verlag, New York.
Willems S (1992) *Stability theory of dynamical systems*. Nelson, London.
Wilson HK (1971) *Ordinary differential equations*. Addison-Wesley, Reading, MA.
Wolfram S (1996) *The Mathematica book*. Cambridge University Press.